IMMUNOLOGY OF FUNGAL DISEASES

IMMUNOLOGY SERIES

Editor-in-Chief
NOEL R. ROSE
*Professor and Chairman
Department of Immunology and
  Infectious Diseases
The Johns Hopkins University
School of Hygiene and Public Health
Baltimore, Maryland*

European Editor
ZDENEK TRNKA
*Basel Institute for
  Immunology
Basel, Switzerland*

1. Mechanisms in Allergy: Reagin-Mediated Hypersensitivity
   Edited by Lawrence Goodfriend, Alec Sehon and Robert P. Orange
2. Immunopathology: Methods and Techniques
   Edited by Theodore P. Zacharia and Sidney S. Breese, Jr.
3. Immunity and Cancer in Man: An Introduction
   Edited by Arnold E. Reif
4. *Bordetella pertussis:* Immunological and Other Biological Activities
   J.J. Munoz and R.K. Bergman
5. The Lymphocyte: Structure and Function (in two parts)
   Edited by John J. Marchalonis
6. Immunology of Receptors
   Edited by B. Cinader
7. Immediate Hypersensitivity: Modern Concepts and Development
   Edited by Michael K. Bach
8. Theoretical Immunology
   Edited by George I. Bell, Alan S. Perelson, and George H. Pimbley, Jr.
9. Immunodiagnosis of Cancer (in two parts)
   Edited by Ronald B. Herberman and K. Robert McIntire
10. Immunologically Mediated Renal Diseases: Criteria for Diagnosis and Treatment
    Edited by Robert T. McCluskey and Giuseppe A. Andres
11. Clinical Immunotherapy
    Edited by Albert F. LoBuglio
12. Mechanisms of Immunity to Virus-Induced Tumors
    Edited by John W. Blasecki
13. Manual of Macrophage Methodology: Collection, Characterization, and Function
    Edited by Herbert B. Herscowitz, Howard T. Holden, Joseph A. Bellanti, and Abdul Ghaffar
14. Suppressor Cells in Human Disease
    Edited by James S. Goodwin
15. Immunological Aspects of Aging
    Edited by Diego Segre and Lester Smith
16. Cellular and Molecular Mechanisms of Immunologic Tolerance
    Edited by Tomáš Hraba and Milan Hašek

17. Immune Regulation: Evolution and Biological Significance
    Edited by Laurens N. Ruben and M. Eric Gershwin
18. Tumor Immunity in Prognosis: The Role of Mononuclear Cell Infiltration
    Edited by Stephen Haskill
19. Immunopharmacology and the Regulation of Leukocyte Function
    Edited by David R. Webb
20. Pathogenesis and Immunology of Treponemal Infection
    Edited by Ronald F. Schell and Daniel M. Musher
21. Macrophage-Mediated Antibody-Dependent Cellular Cytotoxicity
    Edited by Hillel S. Koren
22. Molecular Immunology: A Textbook
    Edited by M. Zouhair Atassi, Carel J. van Oss, and Darryl R. Absolom
23. Monoclonal Antibodies and Cancer
    Edited by George L. Wright, Jr.
24. Stress, Immunity, and Aging
    Edited by Edwin L. Cooper
25. Immune Modulation Agents and Their Mechanisms
    Edited by Richard L. Fenichel and Michael A. Chirigos
26. Mononuclear Phagocyte Biology
    Edited by Alvin Volkman
27. The Lactoperoxidase System: Chemistry and Biological Significance
    Edited by Kenneth M. Pruitt and Jorma O. Tenovuo
28. Introduction to Medical Immunology
    Edited by Gabriel Virella, Jean-Michel Goust, H. Hugh Fudenberg, and Christian C. Patrick
29. Handbook of Food Allergies
    Edited by James C. Breneman
30. Human Hybridomas: Diagnostic and Therapeutic Applications
    Edited by Anthony J. Strelkauskas
31. Aging and the Immune Response: Cellular and Humoral Aspects
    Edited by Edmond A. Goidl
32. Complications of Organ Transplantation
    Edited by Luis H. Toledo-Pereyra
33. Monoclonal Antibody Production Techniques and Applications
    Edited by Lawrence B. Schook
34. Fundamentals of Receptor Molecular Biology
    Donald F. H. Wallach
35. Recombinant Lymphokines and Their Receptors
    Edited by Steven Gillis
36. Immunology of the Male Reproductive Organs
    Edited by Pierre Luigi Bigazzi

37. The Lymphocyte: Structure and Function
    Edited by John J. Marchalonis
38. Differentiation Antigens in Lymphohemopoietic Tissues
    Edited by Masayuki Miyasaka and Zdenek Trnka
39. Cancer Diagnosis In Vitro Using Monoclonal Antibodies
    Edited by Herbert Z. Kupchik
40. Biological Response Modifiers and Cancer Therapy
    Edited by J. W. Chiao
41. Cellular Oncogene Activation
    Edited by George Klein
42. Interferon and Nonviral Pathogens
    Edited by Gerald I. Byrne and Jenifer Turco
43. Human Immunogenetics: Basic Principles and Clinical Relevance
    Edited by Stephen D. Litwin, with David W. Scott, Lorraine Flaherty, Ralph A. Reisfeld, and Donald M. Marcus
44. AIDS: Pathogenesis and Treatment
    Edited by Jay A. Levy
45. Cell Surface Antigen Thy-1: Immunology, Neurology, and Therapeutic Applications
    Edited by Arnold E. Reif and Michael Schlesinger
46. Immune Mechanisms in Cutaneous Disease
    Edited by David A. Norris
47. Immunology of Fungal Diseases
    Edited by Edouard Kurstak
48. Adoptive Cellular Immunotherapy of Cancer
    Edited by Henry C. Stevenson

Additional Volumes in Preparation

IMMUNOLOGY OF FUNGAL DISEASES

Editor

EDOUARD KURSTAK

Associate Editors

Gabriel Marquis
Pierre Auger
Louis de Repentigny
Serge Montplaisir

*Faculty of Medicine
University of Montreal
Montreal, Quebec, Canada*

MARCEL DEKKER, INC.   New York and Basel

Library of Congress Cataloging-in-Publication Data

Immunology of fungal diseases / edited by E. Kurstak ; associate editors, G. Marquis ... [et al.] .
    p. cm. -- (Immunology series ; 47)
  Includes index.
  ISBN 0-8247-8064-7
  1. Fungi, Pathogenic. 2. Fungal antigens. 3. Mycoses--Immunological aspects. I. Kurstak, Edouard. II. Marquis Gabriel. III. Series: Immunology series ; v. 47.
    [DNLM: 1. Mycoses--immunology. WC 450 I325]
QR245.I563 1989
616.9'69079--dc20
DNLM/DLC
for Library of Congress   89-7693
                                                                               CIP

This book is printed on acid-free paper.

Copyright © 1989 by **MARCEL DEKKER, INC.** All Rights Reserved

Neither this book nor any part may be reproduced or transmitted in any form or by any means, electronic or mechanical, including photocopying, microfilming, and recording, or by any information storage and retrieval system, without permission in writing from the publisher.

MARCEL DEKKER, INC.
270 Madison Avenue, New York, New York 10016

Current printing (last digit):
10 9 8 7 6 5 4 3 2 1

PRINTED IN THE UNITED STATES OF AMERICA

Preface

Although invasive fungi are antigenic and induce immunologic reactions in the human host during infection, it seemed to us that current textbooks of medical mycology failed to offer adequate coverage of this important area. Our goal in publishing *Immunology of Fungal Diseases* was to develop a reference manual encompassing all the various facts of immunologic reactions and host resistance, as well as the chemistry of fungal antigens and their detection technology mainly using monoclonal antibodies and enzyme immunoassays.

In the 19 chapters in this volume, laboratory scientists and diagnosis staff and medical students interested in fungal diseases will find up-to-date information on the structure and chemistry of fungal antigens and on selected aspects of host defense mechanisms, immune regulation, and influence of the host's genetic background on responses to these pathogens. Special attention was paid to *Candida, Cryptococcus, Aspergillus, Coccidioides, Paracoccidioides*, and *Sporothrix* species.

Contributors to this volume are well-known experts in their chosen field. Each has prepared a thoughtful and well-documented treatment of the subject. Personal interpretations and conclusions of the authors, as well as the numerous illustrations and unpublished material, provide a large body of information which brings into sharp focus current findings and new directions in research in the fungal diseases.

It is our hope that this volume will provide a useful tool for all concerned with fungal diseases, particularly in hospitals, clinics, medical schools, and research institutions.

We wish to express our sincere gratitude and thanks to the contributors for the effort and care with which they prepared their chapters and to the staff of Marcel Dekker, Inc. for their part in the production of this volume.

Edouard Kurstak
Gabriel Marquis
Pierre Auger
Louis de Repentigny
Serge Montplaisir

Contributors

Robert B. Ashman, Ph.D. Senior Research Officer, Clinical Immunology Research Unit, Princess Margaret Hospital, Perth, Western Australia

Pierre Auger, M.D. Professor, Department of Microbiology and Immunology, University of Montreal, Montreal, Quebec, Canada

Nicole Benhamou Department of Phytology, University Laval, Sainte-Foy, Quebec, Canada

Dick M. Boorsma, Ph.D. Department of Dermatology, Academic Hospital of the Free University, Amsterdam, The Netherlands

Elmer Brummer, Ph.D. Research Associate, Division of Infectious Diseases, Institute for Medical Research, San Jose, California; Lecturer of Medicine, Department of Medicine, Stanford University School of Medicine, Palo Alto, California

Richard A. Calderone, Ph.D. Professor, Department of Microbiology, Georgetown University, Washington, D.C.

Rebecca A. Cox, Ph.D. Director, Department of Research Immunology, San Antonio State Chest Hospital, San Antonio, Texas

Peter de Haan, Ph.D. Department of Dermatology, Immunocytochemistry Unit, Academic Hospital of the Free University, Amsterdam, The Netherlands

Judith E. Domer, Ph.D. Professor, Department of Microbiology and Immunology, Tulane University School of Medicine, New Orleans, Louisiana

Yoshimura Fukazawa, M.D., D. Med. Sci. Professor, Department of Microbiology, Yamanashi Medical College, Tamaho-cho, Yamanashi, Japan

Ronald E. Garner Department of Microbiology and Immunology, Tulane University School of Medicine, New Orleans, Louisiana

Simon Garzon, Ph.D. Electron Microscopist, Department of Immunology, Faculty of Medicine, University of Montreal, Montreal, Quebec, Canada

Veronica M. Hearn, M.Sc., Ph.D. Senior Microbiologist, Mycological Reference Laboratory, Central Public Health Laboratory, London, England

Jeffrey M. Jones, M.D., Ph.D. Professor of Medicine and Assistant Chief of Infectious Diseases, Department of Medicine, University of Wisconsin School of Medicine and William S. Middleton Memorial VA Hospital, Madison, Wisconsin

Thomas R. Kozel, Ph.D. Professor and Chairman, Department of Microbiology, University of Nevada School of Medicine, Reno, Nevada

Edouard Kurstak, Ph.D., Dr. Sc. Director—CVRG, Department of Microbiology and Immunology, Faculty of Medicine, University of Montreal, Montreal, Quebec, Canada

Lisa Linehan, Ph.D. Postdoctoral Fellow, Department of Microbiology, University of Vermont Medical Center, Burlington, Vermont

Donald W. Mackenzie, B.Sc., Ph.D. Director, Mycological Reference Laboratory, Central Public Health Laboratory, London, England

Gabriel Marquis, M.D., Ph.D. Professor, Department of Microbiology and Immunology, Faculty of Medicine, University of Montreal, Montreal, Quebec, Canada

Serge Montplaisir, Ph.D. Department of Microbiology and Immunology, Faculty of Medicine, University of Montreal, Montreal, Quebec, Canada

Giulia Morace Institute of Microbiology, Faculty of Medicine and Surgery, "Agostino Gemelli," Universita Cattolica del Sacro Cuore, Rome, Italy

Juneann W. Murphy, Ph.D. George Lynn Cross Research Professor of Microbiology, Department of Botany—Microbiology, University of Oklahoma, Norman, Oklahoma

Contributors

John M. Papadimitriou, M.D., Ph.D. Professor and Head, Department of Pathology, University of Western Australia, Perth, Western Australia

Luciano Polonelli Professor, Institute of Microbiology, Faculty of Medicine and Surgery, University of Parma, Parma, Italy

Louis de Repentigny, Ph.D. Department of Microbiology and Immunology, Faculty of Medicine, University of Montreal and Ste-Justine Hospital, Montreal, Quebec, Canada

Felipe San-Blas, Lic. Biol., Ph.D. Associate Investigator, Center of Microbiology and Cellular Biology, Instituto Venezolano de Investigaciones Cientificas, Caracas, Venezuela

Gioconda San-Blas, Ph.D. Investigator, Center of Microbiology and Cellular Biology, Instituto Venezolano de Investigaciones Cientificas, Caracas, Venezuela

Luiz R. Travassos, M.D., Ph.D. Professor and Chairman, Mycology Division, Department of Microbiology, Immunology and Parasitology, Escola Paulista de Medicina, São Paulo, Brazil

Elisabeth M. H. van der Raay-Helmer Department of Dermatology, Immunocytochemistry Unit, Academic Hospital of the Free University, Amsterdam, The Netherlands

Alain J-M. Vernes Research Manager, Department of Microbiology, bioMérieux, Marcy l'Etoile, France

Alayn R. Waldorf, Ph.D. Assistant Professor, Department of Biomedical Health Sciences, University of California, Berkeley, Berkeley, California

Jim R. Wikler, M.D. Department of Dermatology, Immunocytochemistry Unit, Academic Hospital of the Free University, Amsterdam, The Netherlands

Contents

Preface   *iii*
Contributors   *v*

Part I Fungal Antigens

1. Antigenic Structure of *Candida albicans*: Electronmicroscopic Localization of Polysaccharide and Immunodeterminants in the Cell Wall   3
   Simon Garzon, Gabriel Marquis, Edouard Kurstak, Serge Montplaisir, and Nicole Benhamou

2. Antigenic Structure of *Candida albicans*: Immunochemical Basis of Serologic Specificity of the Mannans in Yeasts   37
   Yoshimura Fukazawa

3. Antigenic Structure of *Cryptococcus neoformans*: Capsular Polysaccharides   63
   Thomas R. Kozel

4. Antigenic Structure of *Aspergillus* Species   87
   Veronica M. Hearn and Donald W. Mackenzie

5. Antigens of Dermatophytes and Their Characterization Using Monoclonal Antibodies   113
   Peter de Haan, Jim R. Wikler, Elisabeth M. H. van der Raay-Helmer, and Dick M. Boorsma

6. Antigenic Structure of *Coccidioides immitis* — 133
   Rebecca A. Cox

7. Antigenic Structure of *Paracoccidioides brasiliensis* — 171
   Gioconda San-Blas and Felipe San-Blas

8. Antigenic Structures of *Sporothrix schenckii* — 193
   Luiz R. Travassos

Part II Host Response to Invasive Fungi: Mechanisms of Resistance

9. The Role of Complement in Host Resistance to Systemic Fungal Infection — 225
   Richard A. Calderone and Lisa Linehan

10. Pulmonary Defense Mechanisms Against Opportunistic Fungal Pathogens — 243
    Alayn R. Waldorf

11. The Role of Neutrophils and Macrophages in Host Resistance to Systemic Fungal Infections — 273
    Elmer Brummer

Part III Host Response to Invasive Fungi: Immunoregulatory Aspects

12. Immunomodulation in Response to *Candida* — 293
    Judith E. Domer and Ronald E. Garner

13. Immunoregulation in Cryptococcosis — 319
    Juneann W. Murphy

14. Genetic Regulation of Pathogenesis and Host Responses in Fungal Infection — 347
    Robert B. Ashman and John M. Papadimitriou

Part IV Host Resistance to Invasive Fungi: Humoral Response and Laboratory Aspects

15. Humoral Immune Response to *Candida albicans* — 375
    Jeffrey M. Jones

16.	**Humoral Response in Aspergillosis** *Alain J-M. Vernes*	**401**
17.	**Serological Procedures to Detect Dermatophyte Antigens** *Luciano Polonelli and Giulia Morace*	**419**
18.	**Fungal Skin Test Antigens** *Pierre Auger*	**459**
19.	**Circulating Antigens in Systemic Fungal Diseases** *Louis de Repentigny*	**471**

Index　　*485*

IMMUNOLOGY OF FUNGAL DISEASES

Part I
FUNGAL ANTIGENS

1
Antigenic Structure of *Candida albicans*
Electron microscopic Localization of Polysaccharide
and Immunodeterminants in the Cell Wall

SIMON GARZON, GABRIEL MARQUIS, SERGE MONTPLAISIR, and
EDOUARD KURSTAK
University of Montreal, Montreal, Quebec, Canada

NICOLE BENHAMOU
University Laval, Ste-Foy, Quebec, Canada

INTRODUCTION

Studies by many workers on a wide range of yeasts have shown that the cell wall is composed mainly of mannan, glucan, chitin, and proteins, but with considerable difference among the species. The composition of the cell wall of *Candida albicans* has not been studied in as much detail as has *Saccharomyces cerevisiae*. In *C. albicans* the cell wall appears to have a more complex microarchitecture, and the synthesis of wall components is dynamically influenced by growth conditions and the cell metabolic state. Increased cross-linking of wall polymers is seen with prolonged starvation, while growth under nitrogen limitation is associated with a halffold decrease in protein content and a more porous structure (McMurrough and Rose, 1967; Cassone et al., 1978). Reiss (1986) has calculated that mannans represent 15.2-22.9% of mural dry weight, i.e., about 40% of total wall polysaccharide. β-1$\rightarrow$3-D-glucans and β-1$\rightarrow$6-D-glucans are the other major cell wall polysaccharides, accounting for 47-60% of mural dry weight (Chattaway et al., 1968; Sullivan et al., 1983). To sum up various studies, proteins would account for 6-25%, lipids for 1-7%, and chitin for 0.6-9% of the mural dry weight of blastospore, respectively (Kessler and Nickerson, 1959; Chattaway et al., 1968; Reiss et al., 1974; Yamaguchi, 1974b; Elorza et al., 1983; Sullivan et al., 1983). By freeze-etching microscopy, it has been observed that the fracture plane either follows the surface of the cell wall or penetrates the cell wall perpendicularly (Moor and Mühlethaler, 1963; Hubbard et al., 1985). The fact that a fracture of the cell wall into subunits never occurred evinced a highly cohesive and cross-linked structure, where components are interweaved to a considerable degree. Mannans are the major cell wall antigen and readily induce antibody formation, while the glucans are nonimmunogenic (Kemp and Solotorovsky, 1964; Tsuchiya et al., 1974; Meister et al., 1977). Phosphate groups are present in the form of O-phosphomannan, and recent evidence suggests that 1-O-α-phosphorylated β-1→2-linked mannooligosaccharide residues attached to mannan side chains through phosphodiester linkages contribute to the serologic reactivity of *C. albicans* mannan (Shibata et al., 1985). Transmission electron microscopy shows differentially staining layers in the cell wall of *C. albicans* (Djaczenko and Cassone, 1971; Persi and Burnham, 1981; Marquis et al., 1982). These differences in affinity for electron stains, and their modifications, following extraction with alkali and acids and treatment with hydrolytic enzymes, suggest the existence of areas enriched with different components. Hence, the inner

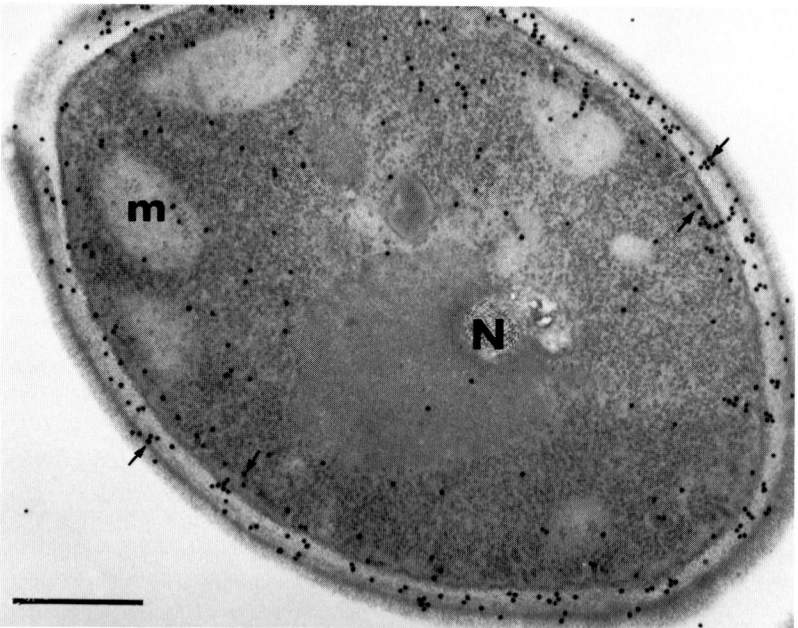

Figure 1 Localization of β-D-galactosyl/N-acetyl-galactosamine residues on *C. albicans* blastospore with the toxin of *Ricinis communis* complexed to gold granules (Au_{15}). The labeling is seen in the cell wall and peripheral cytoplasm near the plasmalemma (arrows), the brushlike layer 1 and bud scar being excepted. Bar = 0.5 μm.

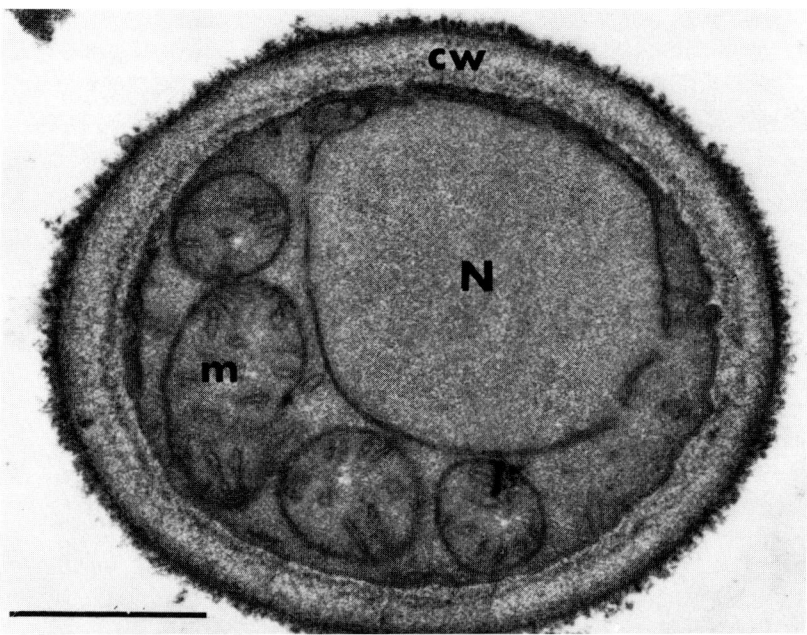

Figure 2 Micrograph of a *C. albicans* blastospore fixed in sequence with a mixture of acrolein, tris aziridinyl phosphine oxide, and glutaraldehyde followed by potassium permanganate. Note a laminated cell wall (cw), several mitochondria (m), large nucleus (N), and nuclear pore. The cytoplasm has a fine granular appearance. Bar = 0.5 μm.

wall layers, which contain the bulk of wall glucan and chitin, maintain the rigidity of the cell wall (Chattaway et al., 1976; Cassone et al., 1978; Torres-Bauzá and Riggsby, 1980; Evron and Drewe, 1984). Finally, the existence of β-galactosyl residues in the cell wall of *C. albicans* is suggested by agglutination reactions with lectins (Herrmann and Uhlenbruck, 1972). At the ultrastructural level, the cell wall is labeled by the *Ricinus communis* toxin complexed to gold granules, which is known to recognize β-galactosyl/N-acetyl-galactosamine residues (Fig. 1). However, more definite information on these residues awaits further development in preparative methods for the separation of cell wall polymers.

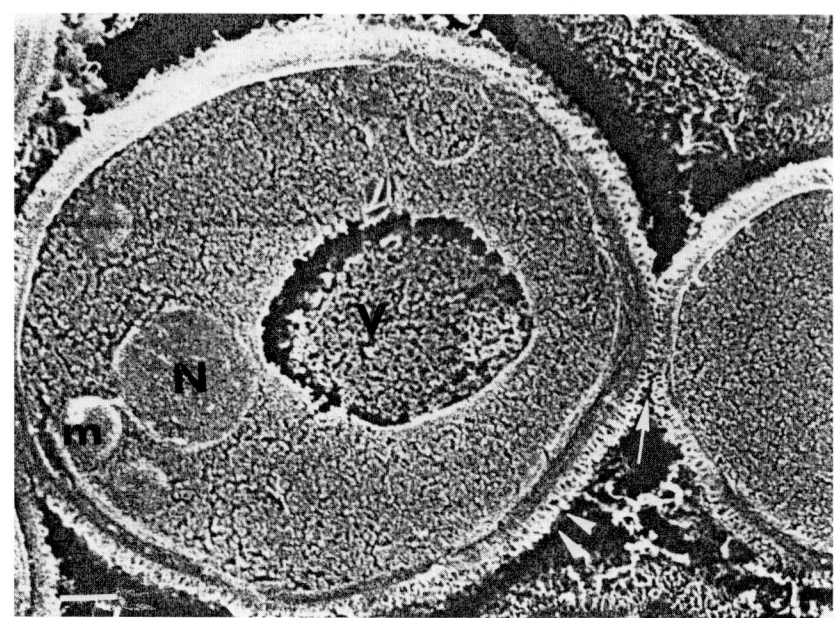

(a)

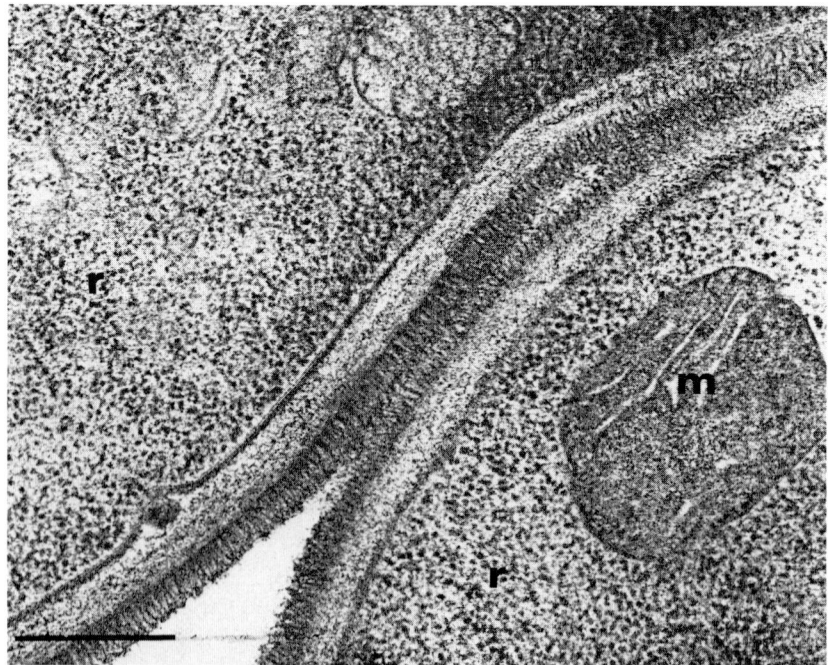

(b)

DYNAMIC VARIATIONS IN THE ORGANIZATION OF THE CELL WALL

Although ultrastructural studies of yeast have revealed the laminated structure of the cell wall of *C. albicans*, the number of layers and their morphology are a matter of controversy. Indeed, descriptions of the cell wall architecture not only depend on the fixation and embedding procedures, but also vary with the culture medium, the age of the culture, and the strains employed. Differences are also observed in the ultrastructure of the cell wall of blastospores during the budding process, through septum formation and with mycelial transformation of the dimorphic *C. albicans*.

CELL WALL ARCHITECTURE OF THE BLASTOSPORE

Conventional chemical fixation with glutaraldehyde or osmium tetroxide proved to be unsatisfactory for the preservation of the general morphology of *C. albicans* (Borgers and Nollin, 1974; Scherwitz et al., 1978). The cell wall was nevertheless resolved in four distinct layers, comprising two outermost and one innermost electron-dense layers distributed on each side of an intermediate electron lucent layer. In some strains of *C. albicans* serotype A sensitive to flucytosine, an additional layer which is removed by trypsin digestion is seen inside the intermediate electron transparent region (Montplaisir et al., 1976). The intermediate electron lucent region, which is thought to be mainly composed of chitin, glucan, and proteins, can be further resolved in individual layers of various density and morphology by modifying the fixation procedures. The addition of tris-(1-azaridinyl)-phosphine oxide to aldehydes and particularly to acrolein (Djaczenko and Cassone, 1971) (TAPO) or of tannic acid (Persi and Burnham, 1981) improved the differentiation of the intermediate electron lucent region into two or three distinct layers. Tannic acid acts as a osmium tetroxide mordant and this mordancing effect was shown to be optimal at pH 5 and to require postfixation with osmium tetroxide and poststaining with lead citrate. Permanganate fixa-

---

Figure 3 (a) Blastospore and germinating cells of *C. albicans* by cryo-SEM. Surfaces of unfractured cells are very rough, being covered with fibrils (arrowheads). Interconnections of the fibrils are seen between the cells (arrows). N, nucleus; V, vacuole. (b) Cell wall of *C. albicans* as seen after freeze-substitution fixation for TEM. The outer surfaces of cells are covered by very delicate fibrillar structure. The fibrillar layer is limited by a boundary line to the middle layers. Abundant ribosomes (r) and mitochondria (m) are seen. (Courtesy of Dr. M. Tokunaga; Bar = 0.5 μm.)

tion, in particular the use of lithium permanganate which was shown to preserve cell membranes better than potassium permanganate (Scherwitz et al., 1978), also improved cell wall differentiation. A combination of aldehydes and TAPO applied in sequence with potassium permanganate (Marquis et al., 1982) revealed a laminated cell wall with seven layers (Fig. 2).

Cryofractures-scanning microscopy and freeze-substitution followed by chemical fixation in transmission electron microscopy confirmed the nature of the brushlike external fibrillary layer (Fig. 3). The cell wall exhibited only three layers of different electron density. The outer layer, 150 nm thick, was composed of electron-dense fibrillar structures. The outermost part of this fibrillar zone was more diffused with fibrillar networks and fine granules. In the basal part of the fibrillar layer a granular fine boundary line separated the fibrillar layer from the middle layer. The middle layer measured approximately 150 nm

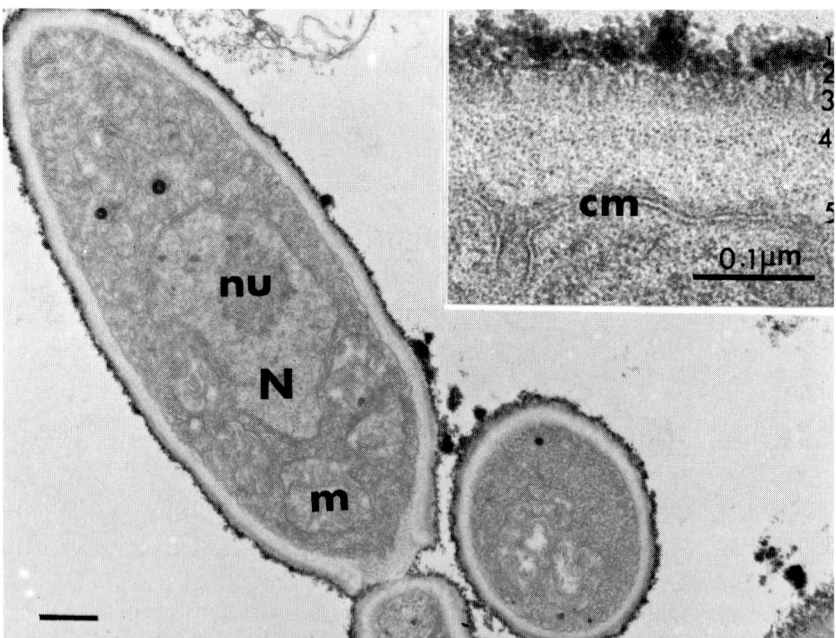

Figure 4 Micrograph of *C. albicans* blastospore fixed with a mixture of nitrogen mustard N-oxide and uranyl acetate, followed by osmium tetroxide (Bar = 0.5 μm). Inset: The convolutions of the cytoplasmic membrane (cm) and several layers of the cell wall are visible. N, nucleus; nu, nucleolus; m, mitochondria.

and was less electron dense. The innermost layer was relatively thin and electron dense (Tokunaga et al., 1986).

More recently, we found that the combined use of nitrogen mustard N-oxide with uranyl acetate, followed by osmium tetroxide, provide optimal resolution of cell wall layers (Fig. 4). The following is a description of individual cell wall layer (see Fig. 5).

1. An outer, brushlike fibrillogranular layer, 60-80 nm thick, which shows some discontinuities and is sometimes diminished or absent in buds and very young blastospores. These fibrils are oriented perpendicular to the cell surface.

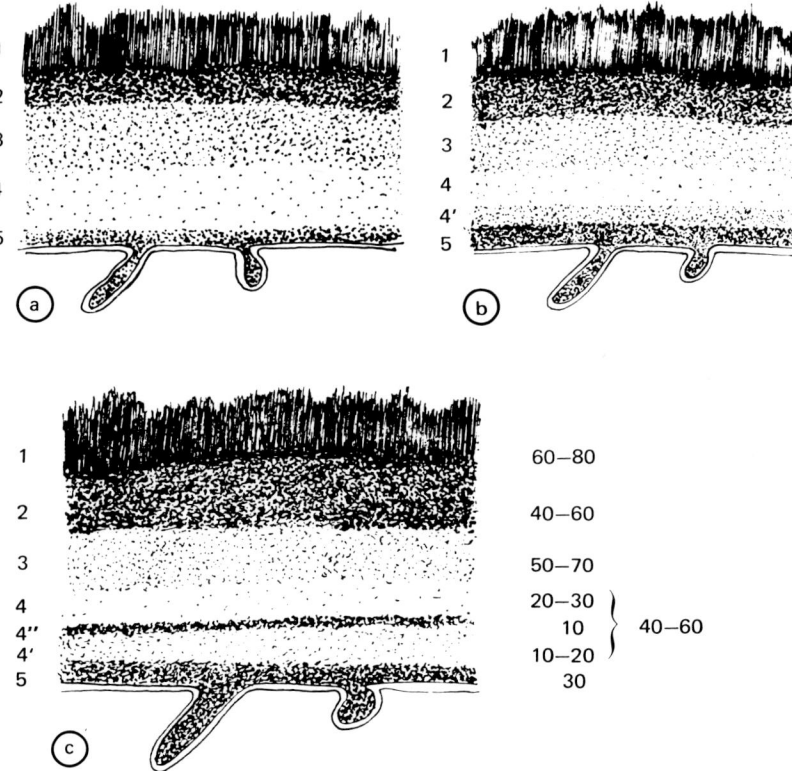

Figure 5 Schematic representation of the cell wall of *C. albicans* based on the interpretation of ultrastructural results obtained by various investigators: a) Djaczenko and Cassone (1974); b) Persi and Burnham (1981); c)Marquis et al. (1982). The different layers of the cell wall expressed in nanometers are diagrammed.

2. An amorphous, highly dense layer, 40–60 nm thick.
3. A filamentous layer, 50–70 nm thick, in which the dense, thick filaments (30–40 Å) are oriented parallel to the cytoplasmic membrane.
4. A 40–60 nm electron lucent layer with interspersed medium density material of various shape and dimension. This layer may be differentiated by tannic acid in two additional layers: a filamentous layer, 4', and a granular dense layer, 4".
5. An homogeneous 30-nm-thick dense layer, immediately adjacent to the external leaflet of the periplasma, following membrane invaginations.

Cassone et al. (1979) have reported variations in the layering of the cell wall in blastospores examined after prolonged starvation compared with exponential phase organisms. In starved cells, both the periplasmic and the outermost electron-dense layers, where mannoproteins are concentrated, gradually disappeared followed by an increase in thickness of the cell wall. This process was reversible if favorable culture conditions were restored. Yamaguchi (1974b) also reported a halffold decrease in the thickness of the outer fibrillogranular layer when *C. albicans* cells were grown with suboptimal amounts of biotin.

ULTRASTRUCTURAL CHANGES IN THE CELL WALL IN ASSOCIATION WITH BUDDING OR GERMINATION

Septum Ultrastructure in the Blastospore

Septation in *C. albicans* is initiated by an electron transparent primary septum, which is then thickened on both sides to form secondary septum. At the time of bud emergence, chitin is deposited from the inner wall layers of mother cells in the bud scar ring and underneath the primary septum (Cabib and Bowers, 1974; Tronchin et al., 1981; Hilenski et al., 1986). The cell wall layer adjacent to the cytoplasmic membrane is also present underneath the primary septum (Fig. 8a). After extraction with chloroform and methanol, the bud scar appears to be formed by densely packed microfibrils of 120 Å in diameter, similar to chitin microfibrils (Poulain et al., 1985b). Whether or not this microfibrillar architecture exists as such or is the result of a preparation artefact remains an unanswered question. Only the outer fibrillogranular layer of the yeast cell wall is seen to uniformly cover the bud scar region (Hilenski et al., 1986). This covering layer of mannoproteins may also be totally interrupted at the bud ring (Fig. 6). These mannoproteins would have a low number of acidic amino acid residues or phosphodiester linkages, accounting for the fact that anionic sites are not found over bud scar regions while they are uniformly distributed on the remaining yeast cell surface (Horisberger and Clerc, 1988). Evidence for the existence of a septal pore between mother and daughter cell has been published

Antigenic Structure of *C. albicans*

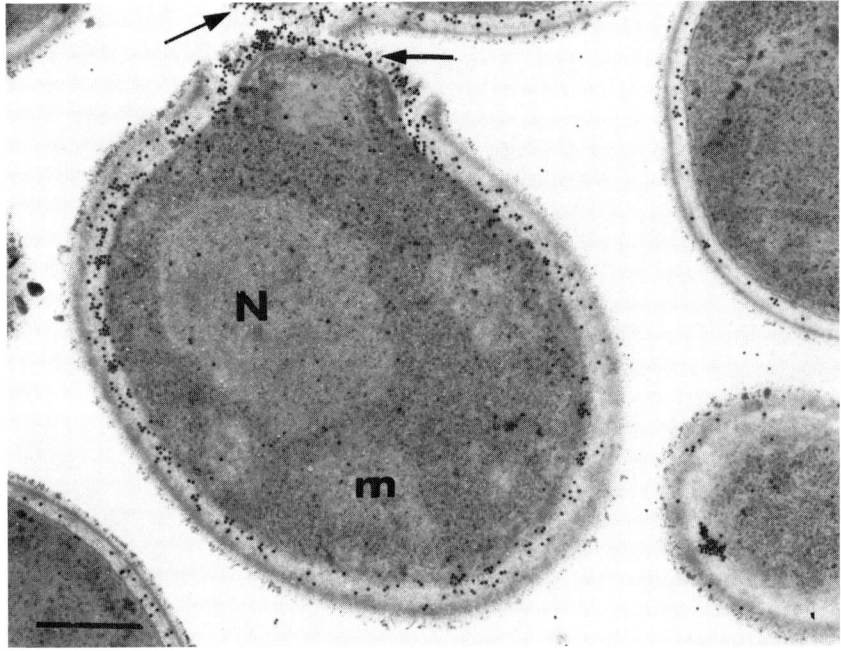

Figure 6 Localization of oligomers of chitin with wheat germ agglutinin-Au$_{15}$ on *Cl albicans* blastospore. Note that the fibrillogranular outer layer of the cell wall is not visible over newly separated bud and birth scars. Heaviest labeling is seen in secondary septum (arrow) and in intermediate cell wall layers. Mannan layers 1 and 2 are unmarked. N, nucleus; m, mitochondrion. Bar = 0.5 μm.

by Tokunaga et al. (1969) and by Odds (1984). Only secondary septum material is incorporated in the birth scar (Shannon and Rothman, 1971).

Germ Tube

At the time of germination, the electron lucent layer 4 develops as the major component of the germ tube cell wall, while the outer layers undergo progressive attrition. The fibrillogranular layer 1 and the amorphous layer 2 were seen to thin out. Layer 3 did not grow and remained at the point of origin of the germ tube. When definitely formed, the germ tube elongated by apical growth. Concomitantly, an increased synthesis of amorphous and fibrillar material occurred so that five distinct layers were reconstituted (Djaczenko and Cassone, 1971; Cassone et al., 1973; Scherwitz et al., 1978). After further development, a par-

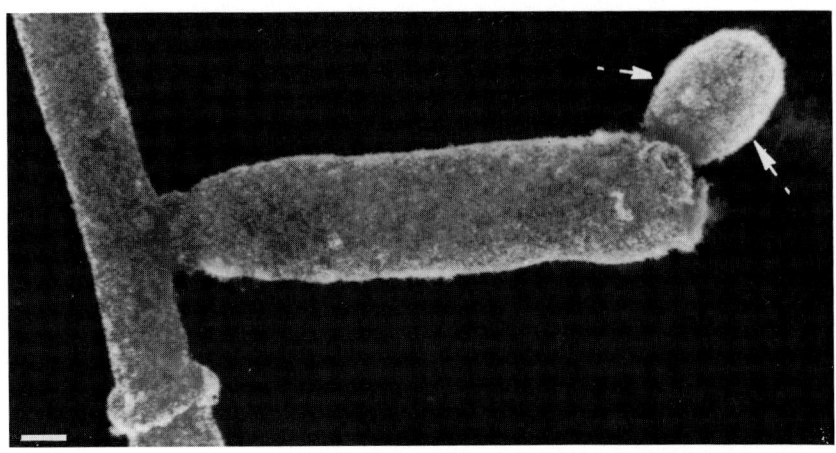

Figure 7 Localization of anionic sites with chitosan-Au$_{44}$ on *C. albicans* hypha. Anionic sites are detected over the whole wall surface with a higher concentration at the apex of the hypha (arrows). (Courtesy of Dr. M. Hirosberger; Bar = 0.5 μm.)

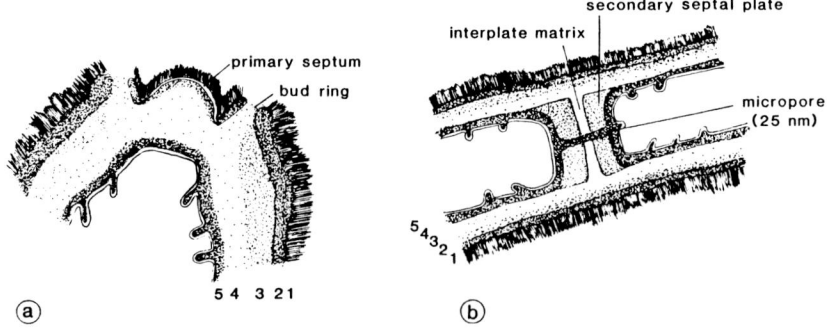

Figure 8 (a) Schematic representation of the bud scar region based on the interpretation of ultrastructural results of Hilenski et al. (1986) and Garzon et al. (unpublished). Cell wall layer 2 and 3 are interrupted by the chitinous bud ring. Except in newly separated bud scars, the brush-like layer 1 covers the primary septum. (b) Schematic representation of the hyphal septum based on the interpretation of ultrastructural results of Gooday and Gow (1983) and Hilenski et al. (1986). The outer wall components do not penetrate the septum. The two secondary septal plates, rich in chitin, have been deposited from the innermost wall layers. The interplate matrix is of unknown composition.

tition wall is formed to separate mother cell and hypha. A fuzzy layer external to the outer region of the cell wall has been demonstrated by poststaining of germ tube with potassium permanganate (Hubbard et al., 1985).

A recent report of Horisberg and Clerc (1988) indicates that a remarkably high density of anionic sites are present on all hyphal surfaces with a higher concentration at the apex (Fig 7). However, since the chemical analysis showed that the amount of O-phosphomannans was lower for the mycelium form of *C. albicans* compared to the yeast form (Shibata et al., 1986), the contribution of phosphorylated mannosyl residues to this net negative surface charge would be of relatively minor importance.

Septum Ultrastructure in the Mycelium

The septal wall may be produced by the accumulation of structural components and not by an inward extension of the cell wall. The mannan-rich surface layers 1 and 2 do not penetrate the septum. The two septal plates, which contain chitin microfibrils (Gow and Gooday, 1983), have developed from layer 4 (Fig. 8b). The interplate matrix, the preferential site of cleavage of the septum, is of unknown composition (Scherwitz et al., 1978; Gow et al., 1980). This narrow electron-transparent band does not reach the outermost cell wall. A membrane-lined septal pore of about 25 nm, which permits cytosolic continuity, runs through the middle septum (Yamaguchi et al., 1974a; Gooday and Gow, 1983; Odds, 1984). The thickness of the septum generally exceeds the cell wall thickness of the blastospore. The fine structure of the septum is found to resemble that of the germ tube rather than that of pseudomycelium.

PRESENCE OF A CELL WALL COAT

Different authors have described the presence of a mucous cell wall coat secreted by both blastospores and hyphae of *C. albicans*. The amount of mucous material produced depends on the growth medium, is proportional to the concentration of sugar (inductive effect of galactose), and increases with the age of the culture. This material appeared to extend into strands which linked the fungal cells together or to other surfaces, i.e., to eukaryotic cells (Fig. 9a) or plastic surfaces (Pugh and Cawson, 1978; McCourtie and Douglas, 1981; Maisch and Calderone, 1981; Marrie and Costerton, 1981; Tronchin et al., 1984; Tokunaga et al., 1986). The combined use of concanavalin A-peroxidase and the periodic acid-thiosemicarbazide-silver proteinate (PATAg) staining procedure demonstrated that this mucous coat is loosely attached to the cell wall (Fig. 9b). This mucous coat is reactive with ruthenium red-osmium tetroxide and concanavalin A, and contains polysaccharide, protein, and secreted enzymes, including acid phosphatase and phospholipase (Pugh and Cawson, 1978; Tronchin et al., 1981). The

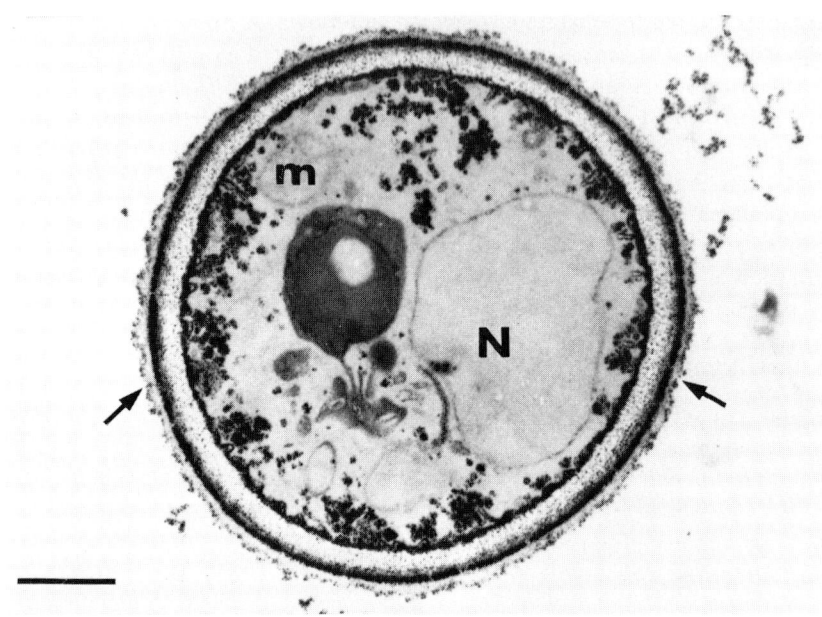

(a)

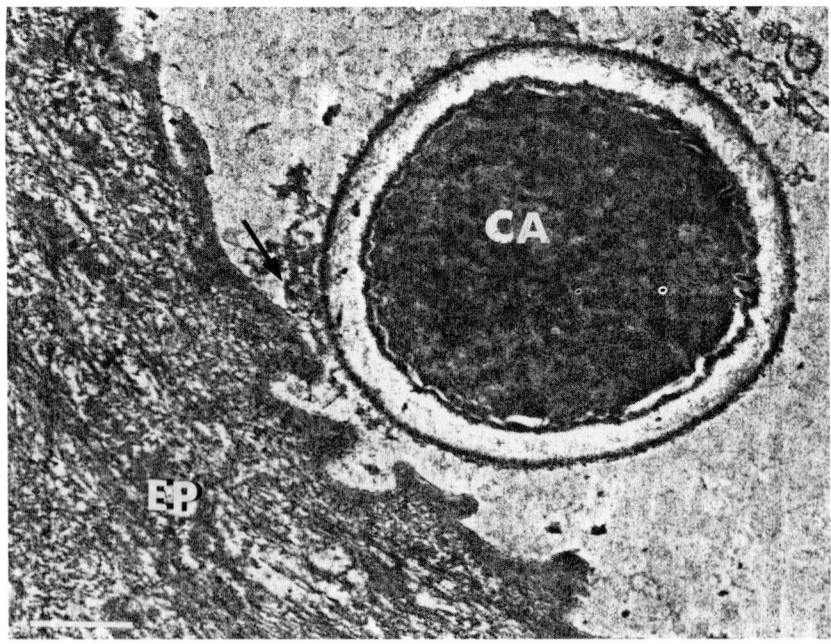

(b)

secretion of this mucous material appears to be related to the development of a fibrillogranular outer cell wall layer, which must play an important role in phagocytosis and adherence to host cells, two important factors in virulence. The increased proliferation of this outer wall layer during the adherence phenomenon and its reorganization is evident, compared to nonadhering forms, by histoultrastructural methods: ruthenium red-osmium tetroxide or polycationic ferritin (McCourtie and Douglas, 1981); concanavalin A-peroxidase or ferritin and PATAg staining (Tronchin et al., 1981, 1984); or by the freeze-substitution transmission electron microscopy technique (Tokunaga et al., 1986).

ULTRASTRUCTURAL LOCALIZATION OF CELL WALL COMPONENTS BY HISTOCHEMICAL AND IMMUNOCYTOCHEMICAL METHODS

Polyclonal human and animal sera were originally used to highlight antigenic determinants of the cell wall of *C. albicans* at the ultrastructural level. In spite of the fact that penetration of reagents was sometimes far less than optimal, these studies revealed that mannans were distributed among two of the peripheral layers and that the layer adjacent to the cytoplasmic membrane was also an important antigenic area (Venezia and Lachapelle, 1973; Meister et al., 1977; Vogt et al., 1980; Poulain et al., 1981). With the use of colloidal gold labeling, it could be shown that anti-mannan and anti-non-mannan reactivities were present in such sera and that these epitopes were uniformly distributed in small clusters on the cell surface (Gerber et al., 1973; Horisberger et al., 1975). Other studies with the use of concanavalin A, which is known to recognize branched *alpha*-linked D-glycopyranosyl polysaccharides (all the cell wall glucans of *C. albicans* are *beta*-linked), confirmed that mannan was located in two continuously reactive layers at the periphery of the cell wall (Cassone et al., 1978; Tronchin et al., 1979). By scanning electron microscopy, concanavalin A-reactive mannoproteins are distributed on the whole yeast cell surface of *C. albicans*, including the bud scars (Fig. 10a). Wheat germ agglutinin labeled with colloid gold or with peroxidase and ferritin was also employed to examine the

---

Figure 9 (a) The periodic acid-thiosemicarbazide-silver proteinate method has been applied to contrast the cell wall of a *C. albicans* blastospore previously incubated with concanavalin A and horseradish peroxidase. The mucous cell coat which is clearly external to the outer cell wall region is delimited by the con A-HRP staining (arrows). N, nucleus; m, mitochondria. (Courtesy of Dr. G. Tronchin; Bar = 0.5 μm.) (b) Yeast-epithelial cell interactions. Adherence of *C. albicans* (CA) to an oral mucosal cell (EP) via ruthenium red positive material of yeast cell origin (arrow). (Courtesy of Dr. J. W. Costerton; Bar = 0.5 μm.)

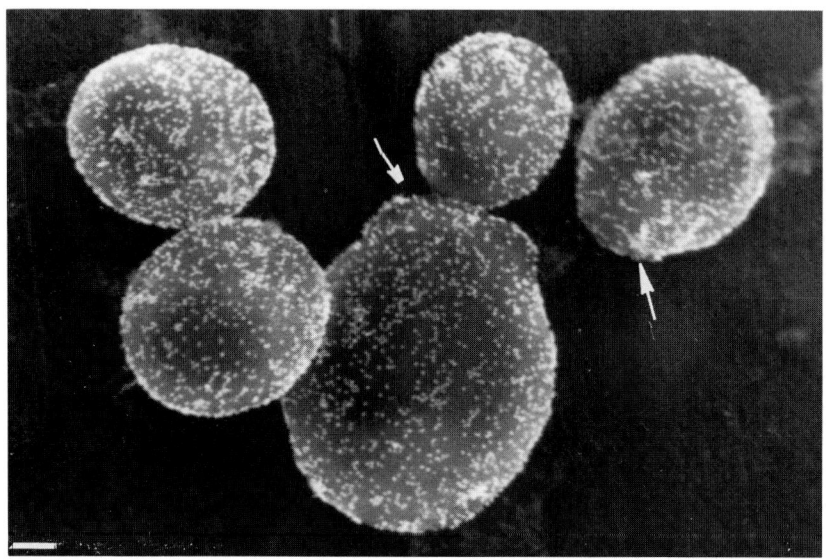

(a)

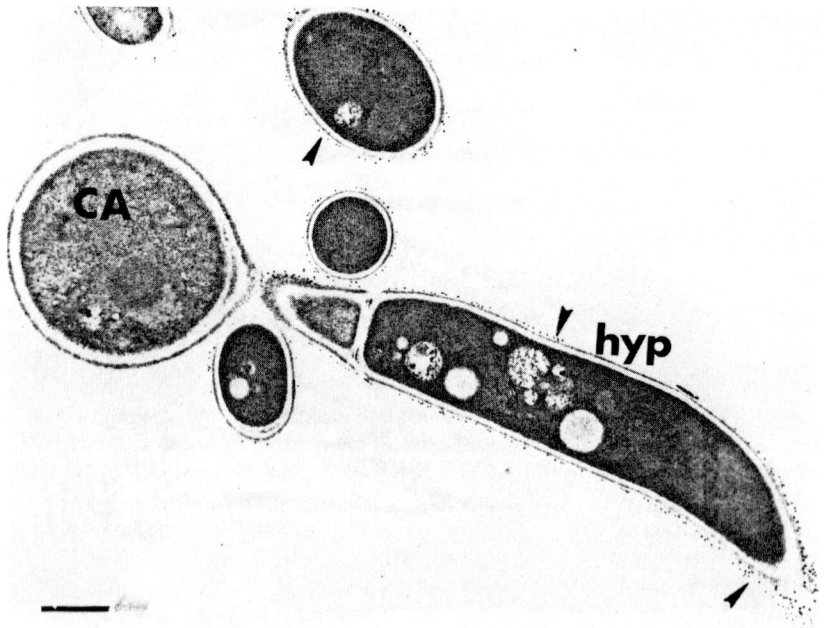

(b)

distribution of chitin within the cell wall. Chitin was predominantly found in bud scars, in septa, and in smaller amounts in the intermediate wall region, mostly layer 4 (Horisberger and Vonlanthen, 1977; Tronchin et al., 1981; Hilenski et al., 1986). More recently, various investigators have used monoclonal antibodies to study antigenic differences among *Candida* species or to follow the expression of cell surface determinants during the growth cycle or germination in *C. albicans* (Brawner and Cutler, 1984, 1986a, 1986b; Chaffin et al., 1988a; Sundstrom et al., 1988). These studies showed that the expression of surface antigens is a dynamic process which is dependent on the strain and on environmental factors (Fig. 7a). Another finding was that all the epitopes recognized by dalton mannoproteins (Chaffin et al., 1988b; Sundstrom et al., 1988), providing confirmation of the heterogeneity of mannan as shown when bulk mannans were separated by anionic exchange chromatorgaphy (Okubo on environmental factors (Fig. 10b). Another finding was that all the epitopes recognized by these monoclonal antibodies were expressed on 180,000-200,000 dalton mannoproteins (Chaffin et al., 1988b; Sundstrom et al., 1988), providing confirmation of the heterogeneity of mannan as shown when bulk mannans were separated by anionic exchange chromatography (Okubo et al., 1979). Poulain et al. (1988) also reported their use of a rat IgM monoclonal antibody directed at a polysaccharide antigen expressed on several cytosolic glycoproteins to follow, at the ultrastructural level, the secretion of the corresponding antigen through the entire cell wall of *C. albicans*.

Another group of cell wall immunodeterminants are the thermostable antigens which are involved in direct agglutination reactions. Factor sera, i.e., monospecific antisera to these antigens, were originally described by Tsuchiya et al. (1959), and their use for the serological identification of medically important yeasts has been validated (Shinoda et al., 1981). We have applied these monospecific antisera (Candida Check, Iatron Laboratories) and a protein A-gold complex to examine the ultrastructural distribution of these surface immunodeterminants in *C. albicans* serotype A (strains 4454M and B311). Since the serologic specificity of yeasts resides in fine differences in the chemical structure of the mannans, the antigens recognized by these monospecific antisera are likely to be

---

Figure 10 (a) Localization of mannans with concanavalin A-Au$_{56}$ on *C. albicans* blastospore by SEM. The mannans are distributed evenly on the whole cell surface including the bud scar (arrow). Con A has specificity for sugar termini with 3,4,6-arabino-α-D-glycopyranosyl structure (Goldstein et al., 1965). (Courtesy of Dr. M. Horisberger; Bar = 0.5 μm.) (b) Variable expression of carbohydrate antigen recognized by murine monoclonal antibody H9 during morphogenesis of *C. albicans*. The determinant recognized by H9 (goat antimouse antibody-Au$_{20}$ staining) was initially lost from the mother cell (CA) and is preferentially expressed on the 4 h germling (courtesy of Dr. J. E. Cutler; Bar = 0.5 μm).

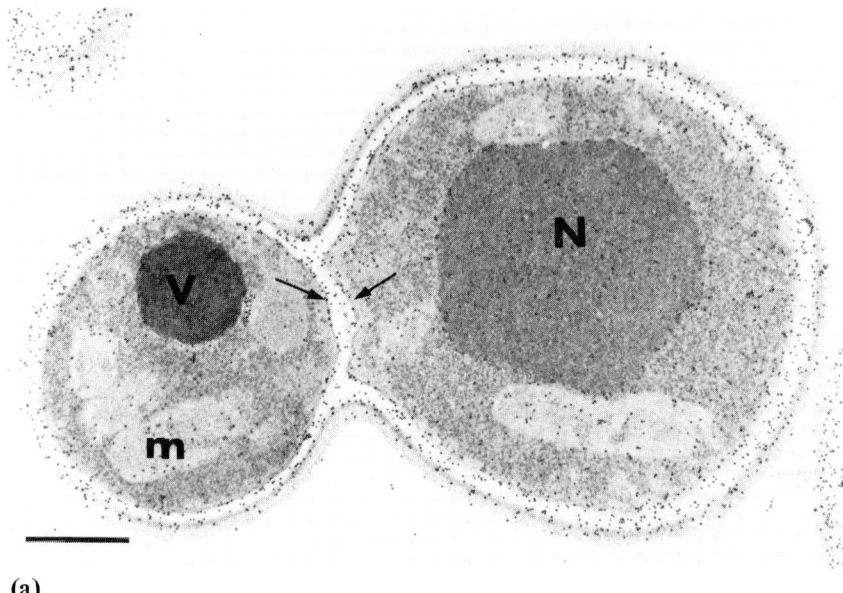

(a)

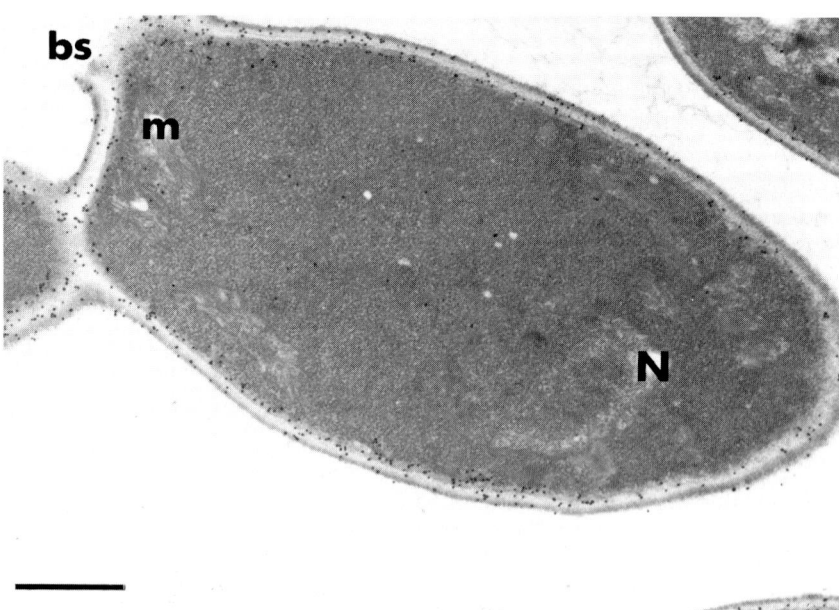

(b)

carbohydrate structures or, perhaps, some peptide portion of the mannoprotein. Factor 4 monospecific antibody continuously labels an intermediate wall layer (Fig. 12) while an additional wall layer adjacent to the cytoplasmic membrane is neatly highlighted after staining with factor 1 and a protein A-gold complex (Fig. 11a). This inner wall layer is also reactive with concanavalin A complexed to gold granules (Fig. 11b), providing another supportive piece of evidence to the concept that it represents an important antigenic area (Poulain et al., 1981). The expression of surface immunodeterminants recognized by monospecific antisera (Iatron Laboratories) is extremely constant for all yeast cells examined by transmission electron microscopy or flow cytometry analysis (Chaffin et al., 1988b). In contrast to the variable expression of determinants recognized by monoclonal antibodies in a given cell population (Brawner and Cutler, 1986a, 1986b; Chaffin et al., 1988a).

The periodic acid-thiosemicarbazide-silver proteinate (PATAg) method of Thiéry (1967) has been applied by Poulain et al. (1978, 1981) to study the organization of cell wall polymers in *C. albicans*. Polysaccharides which contain free vic-glycol groups after treatment with periodic acid are selectively highlighted by the PATAg reaction. Accordingly, the PATAg-reactive layers in the cell wall of *C. albicans* are thought to be composed of mannans and β-1-6-glucans, while chitin and β-1-3-glucans would predominate in the nonreactive layers. The staining observed has been shown to depend on such factors as growth conditions, age of culture, fixation procedures, and the plane of section visualized. By interpretation of all their results, Poulain et al. (1978, 1985a) recognized eight distinct layers. However, since some of these layers are only present in aged cells or under certain growth conditions, no more than five are usually resolved in a single blastospore (Fig. 13).

The dye ruthenium red has also been used by several investigators to define the cell wall layers of *C. albicans* (Pugh and Cawson, 1978; McCourtie and Douglas, 1981; Marrie and Costerton, 1981; Marquis et al., 1982). The ruthenium cation is hexavalent and is assumed to highlight regions of the wall which contain a large number of polyanionic groups, i.e., amino acids with low pK values or

---

Figure 11 (a) Localization of surface immunodeterminant involved in agglutination reaction with Iatron monospecific antiserum 1 and protein A-gold (Au$_8$). Cell wall layers 2 and 5 are uniformly labeled. Note that the labeling of layer 5 is present underneath the primary septum (arrows). N, nucleus; V, vacuole; m, mitochondrion. (b) Localization of mannans with concanavalin A-Au$_{15}$ on *C. albicans* blastospore by TEM. Cell wall layers 2 and 5 are uniformly labeled. Note that labeling of layer 5 is present underneath the primary septum. Layer 2 and its labeling are interrupted in the bud scar region (bs). Bar = 0.5 μm.

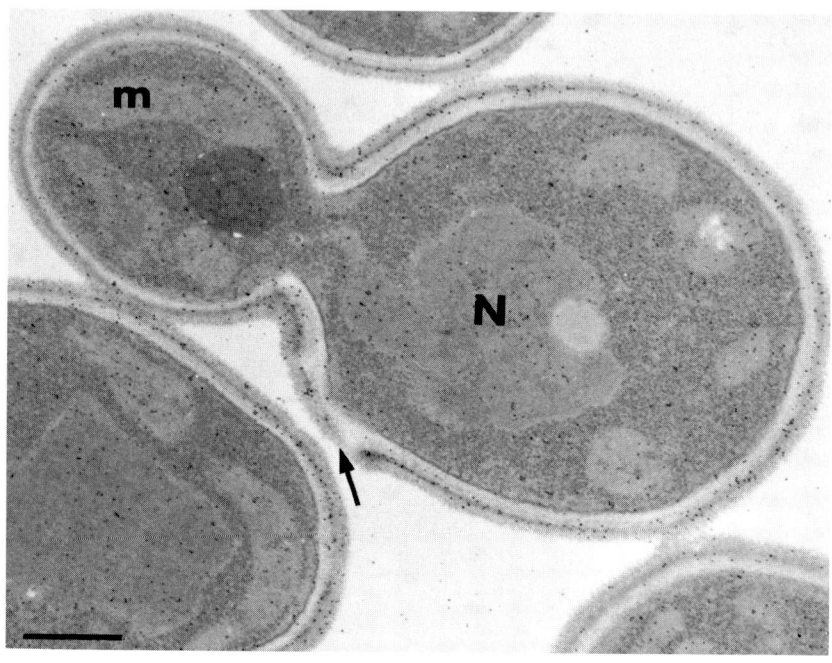

Figure 12 Localization of surface immunodeterminant involved in agglutination reaction with Iatron monospecific antiserum 4 and protein A-gold (Au_8). Cell wall layer 2, in the distribution area of mannan, is uniformly labeled. Note that this layer and the labeling is interrupted in the bud scar region (arrow). Bar = 0.5 μm.

Figure 13 (a) Periodic acid-thiosemicarbazide-silver proteinate reaction on *C. albicans* blastospore. Note the laminar appearance of stained cell wall, with boundary line layer 2 being sharply demarcated from brushlike layer 1 by its heavy PATAg reactivity. The electron lucent primary septum between mother and daughter cell is well delineated. N, nucleus; V, vacuole; bs, bud scar; m, mitochondrion. (b) Micrograph of a *C. albicans* blastospore (CA) within a polymorphonuclear leukocyte (PMN) phagolysosome. Cell wall layer 4 is demonstrated by silver proteinate deposits. Layers 1 and 2 are not individualized having presumably been altered by the digestion process. Periodic acid-thiosemicarbazide-silver proteinate reaction. Bar = 0.5 μm.

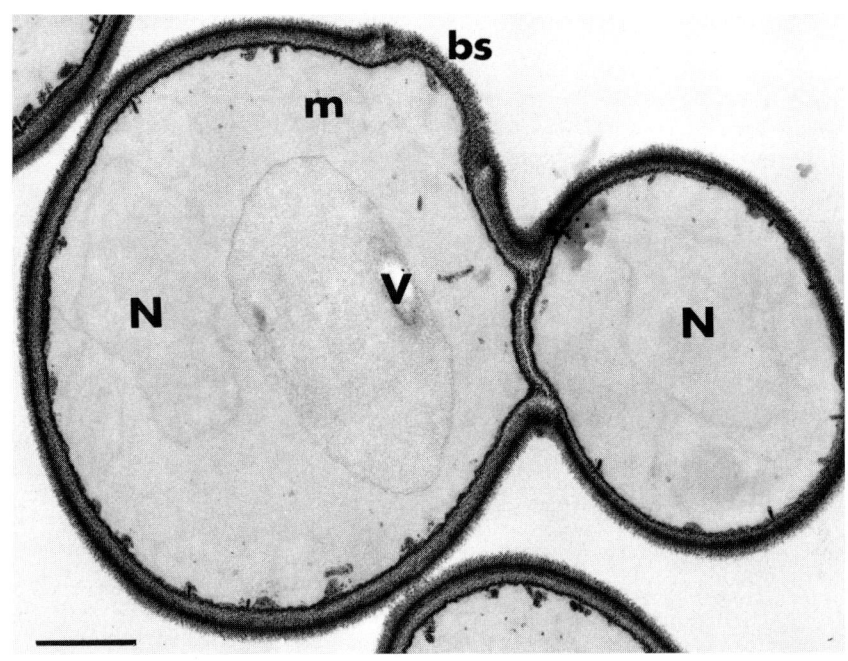

(a)

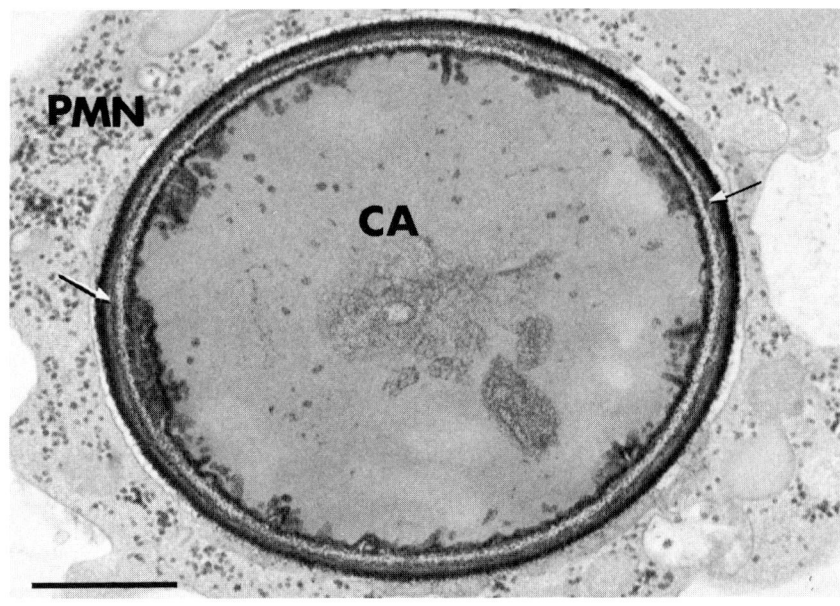

(b)

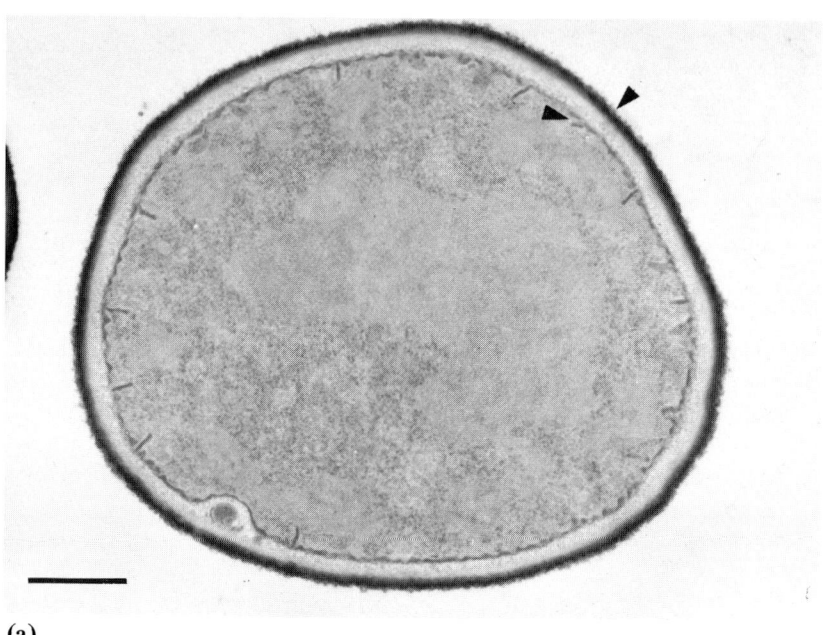

(a)

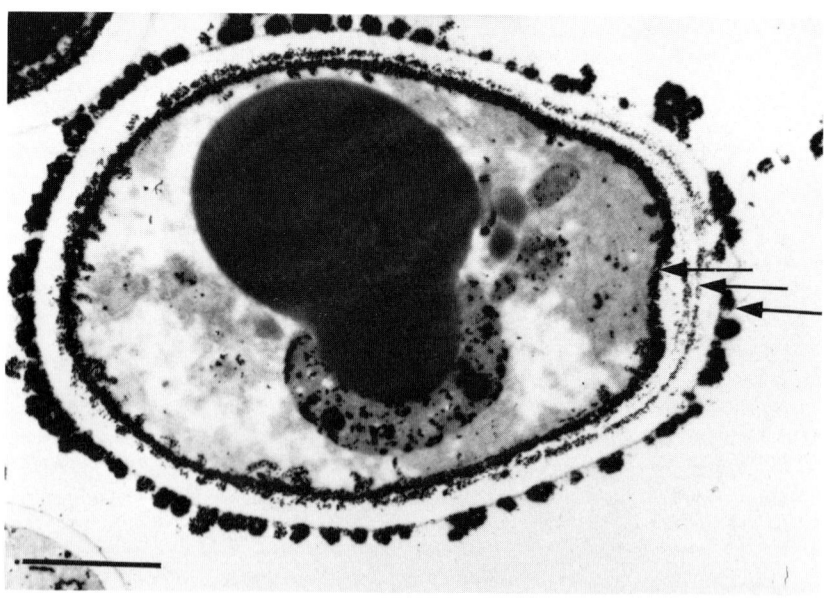

(b)

phosphate residues in the distribution area of mannans. Ruthenium red is completely unreactive with neutral polymers (Mehta, 1925; Luft, 1971). The outer fibrillofloccular layer of the cell wall whose synthesis is stimulated by growth on galactose as a carbon source is heavily contrasted (McCourtie and Douglas, 1981). A wall layer adjacent to the cytoplasmic membrane is also neatly highlighted (Fig. 14a). Finally, under other growth and fixation conditions, the dye has been shown to demonstrate the cell wall coat (Pugh and Cawson, 1978).

CELL WALL ENZYMES

Proteins are part of the cell wall architecture. Some of these proteins have important functions relating to modifications of the cell wall during growth and hyphal transformation, selective transport of nutrients, and virulence. More than 40 enzymes have been identified in *C. albicans* (Odds, 1980), but in only a few instances has the intracellular localization of these enzymes been determined. Reiss (1973) has reviewed the various cytochemical techniques applied to assess the distribution of enzymes in fungi. A common difficulty with isolated wall preparations is to decide whether enzymes are actually integrated into the cell wall or if their presence is transitory. Acid phosphatase is considered to be covalently linked to the cell wall because treatment with a thiol-reducing agent is required to release it (Chattaway and Shenolikar, 1974). Acid proteinases, trehalase, phospholipase A, and lysophospholipase are other known enzymes which are associated with the cell envelope (Arnold and McLellan, 1975; Pugh and Cawson, 1975; Rüchel et al., 1985). Indirect evidences of the presence of exo- and endoglucanases, chitin synthetase, and β-fructofuranosidase (invertase) should also be taken in consideration (Arnold, 1972; Smith and Ballou, 1974; Ram et al., 1984; Molina et al., 1987).

Figure 14 (a) Micrograph of a *C. albicans* blastospore fixed with glutaraldehyde and osmium tetroxide using 1,4-piperazinediethane-sulfonic acid as a buffer. Ruthenium red was added to the vehicle at a final concentration of 0.05%. The outermost and innermost cell wall layers are starkly highlighted (arrows). Stained innermost wall zone intimately follows the invaginations of the cytoplasmic membrane. Bar = 0.5 μm. (b) Cytochemical localization of acid phosphatase in *C. albicans* blastospore. The enzymic activity is found throughout the cell envelope (arrows) and within the cytoplasm, with two major sites of reactivity consisting of the plasma membane area and the cell wall superficial layers. (Courtesy of Dr. G. Tronchin; Bar = 0.5 μm)

Acid Phosphatase

Exposition of *C. albicans* to dithiothreitol is followed by the release of saccharidic or peptidic components including acid phosphatase from blastospores or mycelia. The amount of released acid phosphatase varies between blastospores and mycelia and with the age of the culture, thus indicating changes in the composition of the cell wall (Chattaway and Shenolikar, 1974). The enzyme was liberated as a soluble mannan complex after treatment with dithiothreitol and glucanase (Chattaway et al., 1976) and subsequently localized on cryostat ultrathin sections of *C. albicans* by the Gomori cytochemical technique (Tronchin et al., 1980). Enzymatic activity was observed over various cytoplasmic structures: vacuoles, Golgi-like structures, endoplasmic reticulum, or lysosomelike structures. It was also found at the surface of the cell membrane and at the periphery of the cell wall (Fig. 14b) in the area of distribution of mannans. This enzymatic activity decreased during bud formation and increased gradually after formation of the septum. The fact that some avirulent strains of *C. albicans* were devoid of reactivity for acid phosphatase in their cell wall allowed Tronchin et al. (1980) to correlate this enzymatic activity with virulence and antigenic variations at the cell surface.

Acid Proteinases

Early attempts to grow *C. albicans* on bovine serum albumin as the main substrate revealed that most strains of *C. albicans* released an acid proteinase (Staib, 1969). The detection of high titers of specific antibodies to acid proteinase in the serum of patients suffering from candidosis and the immunohistochemical localization of this enzyme in the immediate vicinity of fungal elements invading host tissues strongly suggest an important role of acid proteinase in the pathogenesis of *Candida* infections (Macdonald and Odds, 1980). Acid proteinases are secreted by the majority of strains of *C. albicans, C. tropicalis*, and *C. parapsilosis* and show strain specificity although close relationships exist between enzymes from strains of the same specie (Rüchel, 1984). Two other proteinases have also been found associated with the cell envelope of *C. albicans* blastospores. One was identified as an angiotensin-1 liberating proteinase and the second as a chymotrypsinlike proteinase (Rüchel et al., 1985). The latter has the ability to activate coagulation factor X at pH 6.8. Ultrastructural observations in disseminated candidiasis have shown that extracellular fungi were often coated with fibrin or associated with capillary thrombi, emphasizing the synthesis of a procoagulant factor (Myerowitz, 1978).

Phospholipase A and Lysophospholipase

Phospholipase activity was first detected by growing the fungus on media containing egg yolk, lecithin, blood serum, or sheep red cells (Costa et al., 1968). Pugh and Cawson (1975) localized by electron microscopic cytochemistry a phospholipase A and a lysophospholipase activity in developing buds, while in older cultures phospholipase appeared at the periphery of nondividing cells and is secreted into the medium. These enzymes could play a role in the controlled growth and remodeling of the cell membrane which accompanied bud formation, as well as in the invasion of host tissues.

β-D-Glucanases

Yeasts produce β-D-glucanases, which are responsible for the plasticity of the cell wall during growth. Two distinct β-D-glucanases were isolated from cell-free extracts, culture fluids, and cell wall autolysates of *C. albicans* (Notario, 1982; Ram et al., 1984; Molina et al., 1987). The exo-1→3-β-glucanase which hydrolyzes the polymer at chain ends, thus releasing glucose is the most abundant glucanase. This enzyme is located in the periplasmic space, secreted by protoplasts, and is released into the culture medium by growing cells. The endoglucanase hydrolyzes β-1→3-glucan by an autolytic mechanism, releasing laminaritetraose as an initial product. To avoid self-destruction, the yeast must regulate the activity of these enzymes, particularly when they are no longer required. Available evidence indicate that interaction with cell wall components (lipids, polysaccharides, and proteins) to form an inactive complex is a privileged mechanism to control the activity of these β-D-glucanases. These enzymes are bound and their activities thereby controlled.

Other Enzymes

Although *C. albicans* is able to assimilate sucrose, no attempt has been made to detect the presence of β-fructofuranosidase (invertase), which normally hydrolyzes sucrose to glucose and fructose. β-Fructofuranosidase seems to be widely distributed in yeasts. This enzyme is present in large quantities in the cell wall of *Saccharomyces cerevisiae* and exists both as an intracellular protein and as an extracellular glycoprotein (Arnold, 1972; Smith and Ballou, 1974). *C. albicans* has also been shown to secrete leucine aminopeptidase and esterase (Kim et al., 1962; Rudek, 1978). However, it is not known if these enzymes are associated with the cell wall or only transported through the entire cell wall from a cytoplasmic location.

Two other enzymes, acid phosphatase and N-acetylglucosaminidase, have been localized in the outer coat of *C. albicans* (Pugh and Cawson, 1978). In this regard, it is interesting to note that electron micrographs clearly showed that the adherence of *C. albicans* to buccal or vaginal epithelial cells involved threadlike structures extending from the blastospore surface to the epithelial surface. Since the adherence of *C. albicans* to epithelial cells is inhibited by trypsin, and by sulfhydryl reagents, but not by α-mannosidase (Lee and King, 1983), it is possible that enzymes located in the outer coat play a role in this biological phenomenon.

PRESENCE OF RECEPTORS FOR FIBRINOGEN AND FOR COMPLEMENT COMPONENTS WITHIN THE CELL WALL

Bouali et al. (1987) have presented evidence that *C. albicans* pseudohyphae and germ tubes express surface receptors for human fibrinogen. The receptor appears to be a mannoprotein which is also excreted in the culture filtrate. Only fragment D degradation product of fibrinogen competes with the parent molecule

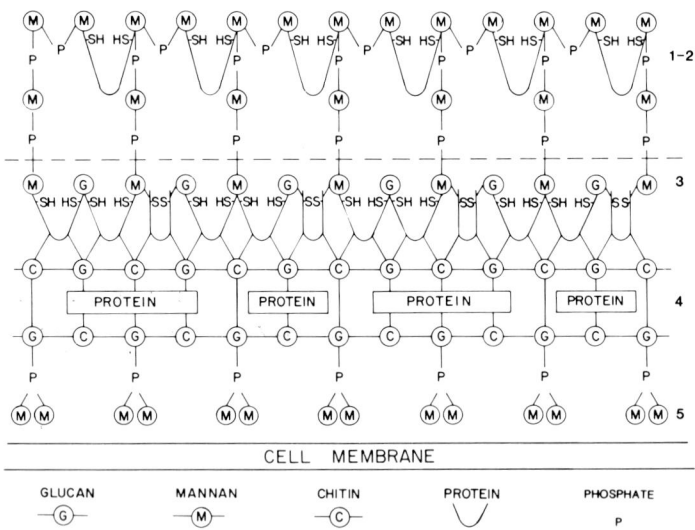

Figure 15 Model for the cell wall architecture of *Candida albicans*. The inner region of the wall is identified by the adjacent cell membrane. The different layers of the cell wall are numbered to correspond with ultrastructural results schematically depicted in Figure 5. —P—, phosphodiester links.

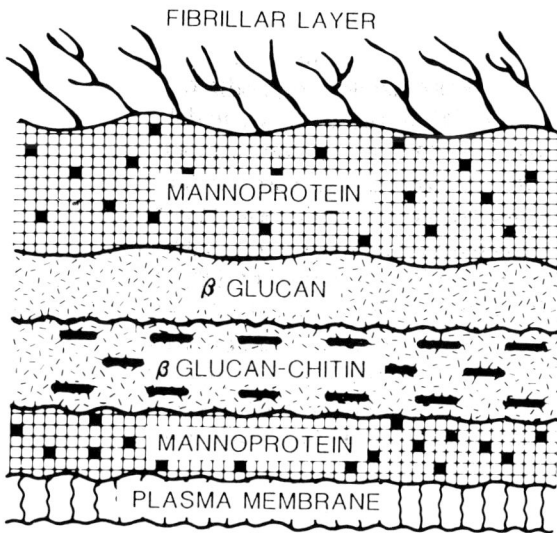

Figure 16 Model for the cell wall architecture of *Candida albicans*. Components such as mannoprotein are found distributed throughout the entire wall and therefore the layered arrangement shows zones of enrichment. (Modified from Shepherd (1987); by kind permission of the author and the publishers.)

for binding to this receptor. Heidenrich and Dierich (1985) and Edwards et al. (1986) observed adherence of complement-coated sheep erythrocytes to germ tubes and pseudohyphae of *C. albicans*. Of the C3-degradation products, iC3b and C3d-coated erythrocytes bound most strongly to *C. albicans*. Gilmore et al. (1988) showed that *Candida* iC3b receptors share homology with the α-chain, but not with the β-chain, of neutrophil receptors for iC3b. Calderone et al. (1988) have produced a monoclonal antibody which blocks the rosetting of C3d-coated erythrocytes when preincubated with pseudohyphae. Using affinity chromatography and blotting, the C3d receptor of *C. albicans* was characterized as a 60,000 dalton glycoprotein. The ultrastructural localization of these receptors has not yet been done.

DISPOSITION OF WALL COMPONENTS

Figure 15 presents a hypothetical model for the molecular organization of the cell wall of *C. albicans* based on that proposed by Kidby and Davies (1970) for

S. cerevisiae after modifications by Koch and Rodemacher (1980) and by us. The presence of a dotted line schematically accounts for the extraction of mannan–protein complex by hydrolysis of labile phosphodiester bonds (Kessler and Nickerson, 1959; Reiss et al., 1974). In addition, this model proposes the existence of phosphodiester linkages between innermost mannans and β-1→6-glucans. Figure 16 presents another model recently proposed by Shepherd (1987). Major differences between the two models are (1) the presence of an individualized β-glucan layer underneath the outer mannan layer versus an interspersed glucomannoprotein layer, and (2) the role of structural protein as a wall cement, especially whether or not proteins are embedded in the β-glucan–chitin area. Except for the bud ring, chitin probably does not exist as an individualized layer but is physically entrapped and covalently linked to β-1→3 glucans. Linear strands of these polymers or this copolymer spontaneously associate in a hydrated triple helix to form densely woven, thin fibrils, which are responsible for the rigidity and shape of the cell wall. This hydrated triple helix is interrupted by the existence of branching points, i.e., β-1→6 linkages, which give a low order of crystallinity to these mixed glucans or glucan–chitin complex. Elorza et al. (1987) recently suggested that chitin would be responsible for wall ontogeny, i.e., that the formation of this polymer would allow the orderly deposition and spatial arrangement of the other polysaccharides. The bud ring is a specialized chitinous structure (Cabib and Bowers, 1971) which functions as a highly rigid channel to protect the nucleus during its migration into the daughter cell during the budding process.

Lipids and proteins are the elusive components of the cell wall in our present state of knowledge. There is in fact very little information available on structural proteins, except for landmark studies on *S. cerevisiae* which emphasized the protein content of mannan being linked to the polysaccharide via diacetylchitobiose glycosamine bonds to asparagine, and by alkali-labile hydroxyamino acid linkages to reducing mannosyl termini (Ballou, 1976). Furthermore, the existence of linkages between glucan and chitin, between glucan and mannan, or between glucan and protein is inferred rather than formally proven.

In spite of these limitations, it is clear that histochemical and immunocytochemical studies at the ultrastructural level have contributed to a better definition of these models. For example, the fact that ruthenium red, concanavalin A, Iatron monospecific antiserum 1, and polyclonal antisera were found to contrast cell wall layer 5 is important supportive evidence for an innermost mannan layer. Cell wall layer 5, which was often seen as an area of deposition of newly synthetized cell wall polymers, should therefore be viewed as an organized mannan layer. Since the epitope recognized by Iatron monospecific antiserum 4 is found exclusively in mannan layer 2 and not in layer 5 (Fig. 12), it is likely that the molecular organization of mannans vary depending on their location throughout the cell wall microarchitecture. Electron microscopic staining with monoclonal antibodies allowed a direct demonstration of the antigenic variability in *C. albicans* (Brawner and Cutler, 1986a, 1986b), i.e., the cell wall is the site of continual changes by which the fungus controls its relationship with its environment. Radial channels, usually 50-100 nm wide, were observed in the cell wall of *C. albicans* with special electron microscopic techniques, such as cryoultramicrotomy and postembedding staining with immunoferritin on styrenemethacrylate sections (Tokunaga and Tokunaga, 1980; Vogt et al., 1980; Takamiya et al., 1985). The pattern of labeling suggested that mannans would be transported from the cytoplasm to the outer cell wall layers via these channels. Finally, the increased availability of purified enzymes of known specificity is likely to permit direct localization of one type of linkage, i.e., β-1→3-D-glucose and β-1→6-D-glucose (gentibiose), directly on thin sections. Horisberger and Clerc (1988) recently reported on the use of anti-gentibiosyl antiserum complexed to gold granules to assess the surface distribution of β-1→6-glucans in *C. albicans* after attempted removal of outer mannan layers by sequential treatment with exo-α-mannan hydrolase (Jones and Ballou, 1969; Reiss et al., 1981) and jack bean α-mannosidase. The presence of β-1→6-glucan was detected to a significant extent only on some blastospores, but since the proportion of total sugar liberated by enzymes was only 5.95%, peeling off of the mannan layers was certainly quite incomplete. Nevertheless, this pioneer work points in the right direction.

ACKNOWLEDGMENTS

The authors are particularly indebted to Drs. J. William Costerton, Jim E. Cutler, Marc Horisberger, Guy Tronchin, and Michiko Tokunaga, who contributed original micrographs to this chapter. We also acknowledge the excellent technical assistance of Henri Strykowski in the preparation of specimens, and of Elizabeth Pérez Nault and Gaston Lambert in the preparation of figures. This work was supported in part by the Fonds de la Recherche en Santé du Québec and by the Medical Research Council of Canada.

REFERENCES

Arnold, W. N. (1972). Localisation of acid phosphatase and β-fructofuranosidase within yeast envelopes. *J. Bacteriol.* 112:1346-1352.

Arnold, W. N., and McLellan, M. N. (1975). Trehalose and glycogen levels during the initial stages of growth of *Candida albicans*. *Physiol. Chem. Phys.* 7:369-380.

Ballou, C. (1976). Structure and biosynthesis of the mannan component of the yeast cell envelope. *Adv. Microbiol. Physiol.* 14:93-158.

Borgers, M., and De Nollin, S. (1974). The preservation of subcellular organelles of *Candida albicans* with conventional fixatives. *J. Cell. Biol. 62*:574–581.

Bouali, A., Robert, R., Tronchin, G., and Senet, J. M. (1987). Characterization of binding of human fibrinogen to the surface of germ-tubes and mycelium of *Candida albicans*. *J. Gen. Microbiol. 133*:545–551.

Brawner, D. L., and Cutler, J. E. (1984). Variability in expression of a cell surface determinant on *Candida albicans* as evidenced by an agglutinating monoclonal antibody. *Infect. Immun. 43*:966–972.

Brawner, D. L., and Cutler, J. E. (1986a). Ultrastructural and biochemical studies of two dynamically expressed cell surface determinants on *Candida albicans*. *Infect. Immun. 51*:327–336.

Brawner, D. L., and Cutler, J. E. (1986b). Variability in expression of cell surface antigens of *Candida* during morphogenesis. *Infect. Immun. 51*:337–343.

Cabib, E., and Bowers, B. (1971). Chitin and yeast binding. *J. Biol. Chem. 246*: 152–159.

Cabib, E., and Bowers, B. (1971). Chitin and yeast budding. *J. Biol. Chem. 246*: yeast. *J. Bacteriol. 124*:1586–1593.

Calderone, R. A., Linehan, L., Wadsworth, E., and Sandberg, A. (1988). Identification of C3d receptors on *Candida albicans*. *Infect. Immun. 56*:1981–1986.

Cassone, A., Simonetti, N., and Strippoli, V. (1973). Ultrastructural changes in the wall during germ-tube formation from blastospores of *Candida albicans*. *J. Gen. Microbiol. 77*:417–426.

Cassone, A., Mattia, E., and Boldrini, L. (1978). Agglutination of blastospores of *Candida albicans* by concanavalin A and its relationship with the distribution of mannan polymers and the ultrastructure of the cell wall. *J. Gen. Microbiol. 105*:263–273.

Cassone, A., Kerridge, D., and Gale, E. F. (1979). Ultrastructural changes in the cell wall of *Candida albicans* following cessation of growth and their possible relationship to the development of polyene resistance. *J. Gen. Microbiol. 110*:339–349.

Chaffin, W. L., Skudlarek, J., and Morrow, K. J. (1988a). Variable expression of a surface determinant during proliferation of *Candida albicans*. *Infect. Immun. 56*:302–309.

Chaffin, W. L., Ringler, L., and Larsen, H. S. (1988b). Interactions of monospecific antisera with cell surface determinants of *Candida albicans*. *Infect. Immun. 56*:3294–3296.

Chattaway, F. W., Holmes, M. R., and Barlow, A. J. E. (1968). Cell wall composition of the mycelial and blastospore forms of *Candida albicans*. *J. Gen. Microbiol. 51*:367–376.

Chattaway, F. W., and Shenolikar, S. (1974). The release of acid phosphatase and polysacchride- and protein-containing components from the surface of the dimorphic forms of *Candida albicans*. *J. Gen. Microbiol. 83*:423–425.

Chattaway, F. W., Shenolikar, S., O'Reilly, J., and Barlow, A. J. E. (1976). Changes in the cell surface of the dimorphic forms of *Candida albicans* by treatment with hydrolytic enzymes. *J. Gen. Microbiol. 95*:335–347.

Costa, A. L., Costa, C., Misefari, A., and Amato, A. (1968). On the enzymatic activity of certain fungi: VII Phosphatidase activity on media containing sheep's blood of pathogenic strains of *Candida albicans*. *Atti Societe Peloritana de Scienze Fisiche Matematiche, Naturali 14*:93-101.

Djaczenko, W., and Cassone, A. (1971). Visualization of new ultrastructural components in the cell wall of *Candida albicans* with fixatives containing TAPO. *J. Cell. Biol. 52*:186-190.

Edwards, J. E., Jr., Gaither, T. A., O'Shea, J. J., Rotrosen, D., Lawley, T. L., Wright, S. A., Frank, M. M., and Green, I. (1986). Expression of specific binding sites on *Candida* with functional and antigenic characteristics of human complement receptors. *J. Immunol. 137*:3577-3583.

Elorza, M. V., Rico, H., Gozalbo, D., and Sentandreu, R. (1983). Cell wall composition and protoplast regeneration in *Candida albicans*. *Antonie van Leeuwenhock 49*:457-469.

Elorza, M. V., Murgui, A., Rico, H., Miragall, F., and Sentandreu, R. (1987). Formation of a new cell wall by protoplasts of *Candida albicans*: Effect of papulacandin B, tunicamycin and nikkomycin, *J. Gen. Microbiol. 133*: 2315-2325.

Evron, R., and Drewe, J. A. (1984). Demonstration of the polysaccharides in the cell wall of *Candida albicans* blastospores using silver methenamine staining and a sequence of extraction procedures. *Mycopathologia 84*:141-149.

Gerber, H., Horisberger, M., and Bauer, H. (1973). Immunosorbent for the isolation of specific antibodies against mannan: Localization of antigens in yeast cell walls. *Infect. Immun. 7*:487-492.

Gilmore, B. J., Retsinas, E. M., Lorenz, J. S., and Hostetter, M. K. (1988). An iC3b receptor on *Candida albicans*: Structure, function, and correlates for pathogenicity. *J. Infect. Dis. 157*:38-46.

Goldstein, I. J., Hollerman, C. E., and Merrick, J. M. (1965). Protein-carbohydrate interactions. I. The interaction of polysaccharides with concanavalin A. *Biochim. Biophys. Acta 97*:68-76.

Gooday, G. W., and Gow, N. A. R. (1983). A model of the hyphal septum of *Candida albicans*. *Exp. Mycol. 7*:370-373.

Gow, N. A. R., Gooday, G. W., Newsam, R. J., and Gull, K. (1980). Ultrastructure of the septum in *Candida albicans*. *Curr. Microbiol. 4*:357-359.

Gow, N. A. R., and Gooday, G. W. (1983). Ultrastructure of chitin in hyphae of *Candida albicans* and other dimorphic and mycelial fungi. *Protoplasma 115*:52-58.

Heidenreich, F., and Dierich, M. P. (1985). *Candida albicans* and *Candida stellatoidea*, in contrast to other *Candida* species, bind iC3b and C3d but not C3b. *Infect. Immun. 50*:598-600.

Herrmann, W. P., and Uhlenbruck, G. (1972). Demonstration of blood group substance H and galatosyl groups on *Candida albicans*. *Z. Naturforsch, 27B*: 1284-1285.

Hilenski, L. L., Naider, F., and Becker, J. M. (1986). Polyoxin D inhibits colloidal gold-wheat germ agglutinin labelling of chitin in dimorphic forms of *Candida albicans*. *J. Gen. Microbiol. 132*:1441-1451.

Horisberger, M., Rosset, J., and Bauer, H. (1975). Colloidal gold granules as markers for cell surface receptors in the scanning electron microscope. *Experientia 31*:1147–1149.

Horisberger, M., and Vonlanthen, M. (1977). Location of mannan and chitin on thin sections of budding yeasts with gold markers. *Arch. Microbiol. 115*: 1–7.

Horisberger, M., and Clerc, M. F. (1988). Ultrastructural localization of anionic sites on the surface of yeast, hyphal and germ-tube forming cells of *Candida albicans*. *Europ. J. Cell. Biol. 46*:444–452.

Hubbard, M. J., Sullivan, P. A., and Shepherd, M. G. (1985). Morphological studies of N-acetylglucosamine induced germ tube formation by *Candida albicans*. *Can. J. Microbiol. 31*:696–701.

Kemp, G., and Solotorovsky, M. (1964). Localization of antigens in mechanically disrupted cells of certain species of the genera *Candida* and *Torulopsis*. *J. Immunol. 93*:305–314.

Kessler, G., and Nickerson, W. J. (1959). Glucomannan-protein complexes from cell walls of yeasts. *J. Biol. Chem. 234*:2281–2285.

Kidby, D. I., and Davies, R. (1970). Invertase and disulphide bridges in the yeast cell wall. *J. Gen. Microbiol. 61*:327–333.

Kim, Y. P., Adachi, K., and Chow, D. (1962). Leucine aminopeptidase in *Candida albicans*. *J. Invest. Dermatol. 38*:115–116.

Koch, Y., and Rademacher, K. H. (1980). Chemical and enzymatic changes in the cell walls of *Candida albicans* and *Saccharomyces cerevisiae* by scanning electron microscopy. *Can. J. Microbiol. 26*:965–970.

Lee, J. C., and King, R. D. (1983). Characterization of *Candida albicans* adherence to human vaginal epithelial cells in vitro. *Infect. Immun. 41*:1024–1030.

Luft, J. H. (1971). Ruthenium red and violet. I. Chemistry, purification, methods of use for electron microscopy and mechanism of action. *Anat. Rec. 171*: 347–368.

Macdonald, F., and Odds, F. C. (1980). Inducible proteinase of *Candida albicans* in diagnostic serology and in the pathogenesis of systemic candidosis. *J. Med. Microbiol. 13*:423–435.

Maisch, A. P., and Calderone, R. A. (1981). Role of surface mannan in the adherence of *Candida albicans* to fibrin-platelet clots formed in vitro. *Infect. Immun. 32*:92–97.

Marquis, G., Montplaisir, S., Garzon, S., Strykowski, H., and Auger, P. (1982). Fungitoxicity of muramidase. Ultrastructural damage to *Candida albicans*. *Lab. Invest. 46*:627–636.

Marrie, T. J., and Costerton, J. W. (1981). The ultrastructure of *Candida albicans* infections. *Can. J. Microbiol. 27*:1156–1164.

McCourtie, J., and Douglas, L. J. (1981). Relationship between cell surface composition of *Candida albicans* and adherence to acrylic after growth on different carbon sources. *Infect. Immun. 32*:1234–1241.

McMurrough, I., and Rose, A. H. (1967). Effect of growth rate and substrate limitation on the composition and structure of the cell wall of *Saccharomyces cerevisiae. Biochem. J. 105*:189-203.

Mehta, M. M. (1925). Biochemical and histological studies on lignification. Part II. Histological studies on the polysaccharides and aromatic constituents of the cell wall. *Biochem. J. 19*:979-997.

Meister, H., Heymer, B., Schäfer, H., and Haferkamp, O. (1977). Role of *Candida albicans* in granulomatous tissue reactions. I. *In vitro* degradation of *C. albicans* and immunospecificity of split products. *J. Infect. Dis. 135*: 224-234.

Molina, M., Cenamor, R., and Nombela, C. (1987). Exo-1,3-glucanase activity in *Candida albicans*: Effect of the yeast-to-mycelium transformation. *J. Gen. Microbiol. 133*:609-617.

Montplaisir, S., Naberra, B., and Drouhet, E. (1976). Susceptibility and resistance of *Candida* to 5-fluorocytosine in relation to the cell wall ultrastructure. *Antimicrob. Agents Chemother. 9*:1028-1032.

Moor, H., and Mühlethaler, K. (1963). Fine structure in frozen-etched yeast cells. *J. Cell Biol. 17*:609-627.

Myerowitz, R. L. (1978). Ultrastructural observations in disseminated candidosis. *Arch. Pathol. Lab. Med. 102*:506-511.

Notorio, V. (1982). β-Glucanases from *Candida albicans*: Purification, characterization and the nature of their attachment to cell wall components. *J. Gen. Microbiol. 128*:747-759.

Okubo, Y., Honma, Y., and Suzuki, S. (1979). Relationship between phosphate content and serological activities of the mannans of *Candida albicans* strains NIH A-207, NIH B-792, and J-1012. *J. Bacteriol. 137*:677-680.

Odds, F. C. (1979). *Candida and Candidiosis*. University Park Press, Baltimore.

Odds, F. C. (1984). Demonstration of a septal pore in budding *Candida albicans* yeast cells. *Sabouraudia 22*:505-507.

Persi, M. A., and Burnham, J. C. (1981). Use of tannic acid as fixative-mordant to improve the ultrastructural appearance of *Candida albicans* blastospores. *Sabouraudia 19*:1-8.

Poulain, D., Tronchin, G., Dubremetz, J. F., and Biguet, J. (1978). Ultrastructure of the cell wall of *Candida albicans* blastospores: Study of its constitutive layers by the use of a cytochemical technique revealing polysaccharides. *Ann. Microbiol. 129A*:141-153.

Poulain, D., Tronchin, G., Jouvert, S., Herbaut, J., and Biguet, J. (1981). Architecture pariétale des blastospores de *Candida albicans*: Localisation de composants chimiques et antigéniques. *Ann. Microbiol. 132A*:219-238.

Poulain, D., Hopwood, V., and Vernes, A. (1985a). Antigenic variability of *Candida albicans. CRC Crit. Rev. Microbiol. 12*:223-270.

Poulain, D., Tronchin, G., and Vernes, A. (1985b). Application d'une méthode d'extraction des lipides aux parois des blastospores de *Candida albicans. Mycopathologia 92*:141-147.

Poulain, D., Stecker, G., Dubremetz, J.-F., Fortier, B., Rousseau, R., and Van Cutsem, J. (1988). Cytological, chemica and serological immunodetection of a *Candida albicans* antigen. In *Fungal Antigens; isolation, Purification, and Detection*, E. Drouhet, G. T. Cole, L. de Repentigny, J. P. Latgé, and B. Dupont (Eds.). Plenum, New York.

Pugh, D., and Cawson, R. A. (1975). The cytochemical localization of phospholipase A and lysophospholipase in *Candida albicans*. *Sabouraudia 13*:110–115.

Pugh, D., and Cawson, R. A. (1978). The surface layer of *Candida albicans*. *Microbios. 23*:19–23.

Ram, S. P., Romana, L. K., Shepherd, M. G., and Sullivan, P. A. (1984). Exo-(1→3)-β-glucanase, autolysin and trehalase activities during yeast growth and germ-tube formation in *Candida albicans*. *J. Gen. Microbiol. 130*:1227–1236.

Reiss, E., Stone, S. H., and Hasenclever, H. F. (1974). Serological and cellular immune activity of peptidoglucomannan fractions of *Candida albicans* cell walls. *Infect. Immun. 9*:881–890.

Reiss, E. (1986). *Molecular Immunology of Mycotic and Actinomycotic Infections*. Elsevier, New York.

Reiss, J. (1973). Enzyme cytochemistry of fungi. In *Progress in Histochemistry and Cytochemistry*, vol. 5. Gustav Fischer, Stuttgart, pp. 1–40.

Rüchel, R. (1983). On the renin-like activity of *Candida* proteinases and activation of blood coagulation in vitro. *Zbl. Bakt. Hyg. 255A*:368–379.

Rüchel, R. (1984). A variety of *Candida* proteinases and their possible targets of proteolytic attack in the host. *Zbl. Bakt. Hyg. 257A*:266–274.

Rüchel, R., Böning, B., and Jahn, E. (1985). Identification and partial characterization of two proteinases from the cell envelope of *Candida albicans* blastospores. *Zbl. Bakt. Hyg. 260A*:523–538.

Rudek, W. (1978). Esterase activity in *Candida* species. *J. Clin. Microbiol. 8*:756–759.

Scherwitz, C., Martin, R., and Ueberberg, H. (1978). Ultrastructural investigations of the formation of *Candida albicans* germ tubes and septa. *Sabouraudia 16*:115–124.

Shannon, J. L., and Rothman, A. H. (1971). Transverse septum formation in budding cells of the yeastlike fungus *Candida albicans*. *J. Bacteriol. 106*:1026–1028.

Shibata, N., Ichikawa, T., Tojo, M., Takahashi, M., Ito, N., Okubo, Y., and Suzuki, S. (1985). Immunochemical study on the mannans of *Candida albicans* NIH A-207, NIH B-792, and J-1012 strains prepared by fractional precipitation with cetyltrimethylammonium bromide. *Arch. Biochem. Biophys. 243*:338–348.

Shibata, N., Kobayashi, H., Tojo, M., and Suzuki, S. (1986). Characterization of phosphomannan-protein complexes isolated from viable cells of yeast and mycelium forms of *Candida albicans* NIH B-792 strain by the action of zymolyase-100T. *Arch. Biochem. Biophys. 251*:697–708.

Shinoda, T., Kaufman, L., and Padhye, A. A. (1981). Comparative evaluation of the Iatron serological Candida check kit for identification of medically important Candida species. *J. Clin. Microbiol. 13*:513-518.
Smith, W. L., and Ballou, C. E. (1974). Immunochemical characterization of the mannan component of the external invertase (β-fructofuranosidase) of *Saccharomyces cerevisiae*. *Biochemistry 13*:355-361.
Staib, F. (1969). Proteolysis and pathogenicity of *Candida albicans* strains. *Mycopathologia 37*:345-348.
Sullivan, P. A., Yin, C. Y., Molloy, C., Templeton, M. D., and Shepherd, M. G. (1983). An analysis of the metabolism and cell wall composition of *Candida albicans* during germ-tube formation. *Can. J. Microbiol. 29*:1514-1525.
Sundstrom, P. M., Tam, M. R., Nichols, E. J., and Kenny, G. E. (1988). Antigenic differences in the surface mannoproteins of *Candida albicans* as revealed by monoclonal antibodies. *Infect. Immun. 56*:601-606.
Suzuki, M., and Fukazawa, Y. (1982). Immunochemical characterization of *Candida albicans* cell wall antigens: Specific determinant of *Candida albicans* serotype A mannan. *Microbiol. Immun. 26*:387-402.
Takamiya, H., Vogt, A., Batsford, S., Kuttin, E., and Müller, J. (1985). Further studies on the immunoelectronmicroscopic localization of polysaccharide antigens on ultra-thin sections of *Candida albicans*. *Mykosen 28*:17-32.
Thiéry, J. P. (1967). Mise en évidence des polysaccharides sur coupes fines en microscopie électronique. *J. Microscopie 6*:987-1018.
Tokunaga, J., and Tokunaga, M. (1980). Ultrastructural changes in various cell forms of *Candida albicans*. In *Human and Animal Mycology*, Proceedings of VII Congress of ISHAM, E. S. Kuttin and G. L. Baum (Eds.). Excerpta Medica, pp. 134-137.
Tokunaga, J., Tokunaga, M., and Egashira, T. (1969). Electron microscopical studies on growing fungal cells. II. Cell wall formation in budding and germinating blastospores of genus *Candida*. *Jpn. J. Bacteriol. 24*:676-682.
Tokunaga, M., Kusamichi, M., and Koike, H. (1986). Ultrastructure of outermost layer of cell wall in *Candida albicans* observed by rapid-freezing technique. *J. Electron Microsc. 35*:237-246.
Torres-Bauzá, L., and Riggsby, W. S. (1980). Protoplasts from yeast and mycelial forms of *Candida albicans*. *J. Gen. Microbiol. 119*:341-349.
Tronchin, G., Poulain, D., and Biguet, J. (1979). Etudes cytochimiques et ultrastructurales de la paroi de *Candida albicans*. I. Localisation des mannanes par utilisation de concanavaline A sur coupes fines. *Arch. Microbiol. 123*: 245-249.
Tronchin, G., Poulain, D., and Biguet, J. (1980). Localisation ultrastructurale de l'activité phosphatasique chez *Candida albicans*. *Biol. Cell. 38*:147-152.
Tronchin, G., Poulain, D., Herbaut, J., and Biguet, J. (1981a). Cytochemical and ultrastructural studies of *Candida albicans*. II. Evidence for a cell wall coat using concanavalin A. *J. Ultrastruct. Res. 75*:50-59.
Tronchin, G., Poulain, D., Herbaut, J., and Biguet, J. (1981b). Localization of chitin in the cell wall of *Candida albicans* by means of wheat germ agglutinin. Fluorescence and ultrastructural studies. *Eur. J. Cell. Biol. 26*:121-128.

Tronchin, G., Poulain, D., and Vernes, A. (1984). Cytochemical and ultrastructural studies of *Candida albicans*. III. Evidence for modifications of the cell wall coat during adherence to human buccal epithelial cells. *Arch. Microbiol. 139*:221-224.

Tsuchiya, T., Fukazawa, Y., Taguchi, M., Nakase, M., Shinoda, T., (1974). Serologic aspects of yeast classification. *Mycopathologia 53*:77-91.

Venezia, R. A., and Lachapelle, R. C. (1973). The use of ferritin-conjugated antibodies in the study of cell wall components of *Candida albicans. Can. J. Microbiol. 19*:1445-1448.

Vogt, A., Batsford, S., Kuttin, E. S., Müller, J., and Takamiya, H. (1980). Immunoelectronmicroscopic localization of polysaccharide antigen in *Candida albicans*. In *Human and Animal Mycology*, Proceedings of VII Congress of ISHAM, E. S. Kuttin and G. L. Baum (Eds.). Excerpta Medica, pp. 134-137.

Yamaguchi, H., Kanda, Y., and Osumi, M. (1974a). Dimorphism in *Candida albicans*. II. Comparison of fine structure of yeast-like and filamentous phase growth. *J. Gen. Appl. Microbiol. 20*:101-110.

Yamaguchi, H. (1974b). Effect of biotin insufficiency on composition and ultrastructure of cell wall of *Candida albicans* in relation to its mycelial morphogenesis. *J. Gen. Appl. Microbiol. 20*:217-228.

2
Antigenic Structure of *Candida albicans*
Immunochemical Basis of the Serologic Specificity of the Mannans in Yeasts

YOSHIMURA FUKAZAWA
Yamanashi Medical College, Tamaho-cho, Yamanashi, Japan

INTRODUCTION

Antigenic determinants (epitopes) are those structural components of antigen molecules that are responsible for determining the specificity of antigen-antibody interactions. Haptens are chemical substances of low molecular weight that do not per se elicit the formation of antibodies but whose chemical configuration is such that it can interact with antibodies elicited by a large molecule that possesses a structural unit similar to or identical with haptens. Haptens can thus be divided into two molecular groups: one reacts with antibodies to form immune aggregates (complete hapten); the other specifically binds with antibodies, but does not form immune aggregates (incomplete hapten).

Whole yeast cells have been demonstrated to be antigenic in humans and animals. Predominant yeast antigens include the extractable mannan or the mannan-protein complex, which are derived from the outermost regions of the cell wall. It has been postulated that the antigenic specificity of the yeast species of *Saccharomyces* and of *Candida* resides in the mannans on the cell surface. Hasenclever and Mitchell (1964) were the first to demonstrate that mannans isolated from whole cells of *S. cerevisiae* and *C. albicans* were able to inhibit the agglutination reactions between parent whole cells and corresponding antisera. The agglutination of *C. albicans* by concanavalin A and the blockade of agglutination by methyl-α-mannoside provide evidence for the presence of mannan on the cell surface (Cassone et al., 1978). It has been recently shown that incomplete haptens of the side chains of oligosaccharides are responsible for the specificity of mannans.

The chemical structure of some yeast mannans have already been demonstrated by several investigators. However, it has not yet been completely elucidated even for one species. This is principally a result of numerous controversies regarding the antigenic formula, as well as the chemical structures responsible for the specificity of the yeast.

In as much as pioneering works on the immunochemical structures of mannans were conducted with the *Saccharomyces* species, the literature in this chapter will initially be concerned with the *Saccharomyces* species. Structures of mannans of the *Candida* species proposed by several other investigators will then be briefly reviewed.

SEROTYPING OF YEASTS

Serological studies for differentiation and the identification of fungi were commenced shortly after the discovery of antigen-antibody reactions in microbiology, especially on brewing and pathogenic yeasts.

However, serology appears to have gained little acceptance in mycology compared to bacteriology, probably due to the difficulty of preparing antisera. These investigations, however, have introduced systematic approaches to fungal serology resulting in antigenic schemes for classification.

Antigenic Structures of Medically Important Yeasts

Martin (1942) described the antigenic structures of four *Candida* species by the use of three antigenic factors (X, Y, and Z). Their relative amounts were shown by agglutination and complement fixation reactions. Tsuchiya et al. (1955) originally proposed the antigenic structures of seven *Candida* species, dividing them into thermostable and thermolabile antigens by using slide agglutination with adsorbed antisera. Antigenic analyses of *Candida* species were also made by Seeliger (1958), Pospisil (1959), Biguet et al. (1962), Müller and Hirsch (1967), and Sweet and Kaufman (1970). Tsuchiya et al. (1974) further proposed the antigenic structures of varieties of species of the genera *Saccharomyces* as well as many other perfect and imperfect yeast species. Recently, the antigenic formula of five serotypes of *Cryptococcus neoformans* was proposed by Ikeda et al. (1982).

Serotypes of *C. albicans*

C. albicans was initially divided into two serotypes, A and B, by Hasenclever and Mitchell (1961), indicating that strains of *C. albicans* serotype A are identical to those of *C. tropicalis*, while serotype B closely resembled *C. stellatoidea*. Adsorption of anti-type A antiserum with type B yeast leaves type A-specific antibodies that can be utilized as a serotype A-typing reagent. In the adsorption

Table 1 Antigenic Formulas of *Candida albicans* and Related Species

Strain[a]	Antigenic factors
C. albicans M1012 (serotype A)	1, 4, 5, 6
C. albicans M1445 (serotype B)	1, 4, 5, 13b
C. tropicalis M1017	1, 4, 5, 6
C. guilliermondii M1023	1, 4, 9
C. krusei M1005	1, 5[b], 11
C. parapsilosis M1015	1, 5[b], 13b
C. pseudotropicalis M1004	1, 8
T. glabrata M4002	1, 4, 6, 34

[a]Culture collection of Meiji College of Pharmacy, Tanashi, Tokyo 188, Japan.
[b]Occasionally negative, depending on the lot of factor sera.

of anti-type B antiserum with type A yeast, all antibodies capable of agglutinating *C. albicans* type B were eliminated (Hasenclever and Mitchell, 1961). In a later development, Tsuchiya et al. (1974) and Fukazawa et al. (1976) demonstrated that *C. albicans* serotype A and *C. tropicalis* possess antigenic factor 6, whereas serotype B has factor 13b instead of 6, and that monospecific antibody to factor 6 is important for the identification of *C. albicans* serotype A as well as *C. tropicalis*. The antigenic formula of medically important *Candida* species so far established by Tsuchiya et al. (1974) and Fukazawa et al. (1976) are shown in Table 1. Previously, Tsuchiya et al. (1974) postulated that the standard strain of *C. albicans* M 1012 has a unique antigen 7, in addition to antigen 6. They tentatively named this strain type C. Later, however, difficulties in preparing factor 7 antibody made it desirable to combine serotype C with serotype A (Fukazawa et al., 1976).

In the precipitin reaction system, when mannans extracted from two serotypes of *C. albicans* were compared in immunodiffusion and quantitative precipitin reactions, other disparities were found that support the actual existence of type B-specific factors (Tsuchiya et al., 1974). Immunodiffusion results indicate that anti-A antiserum reacts weakly with type B mannan (Reiss et al., 1974; Lehmann and Reiss, 1980) with only 21% of anti-type A antibody nitrogen being precipitated by type B mannan as compared with precipitation by the homologous type A mannan. Moreover, subfractions of type B mannan enriched in a B-specific antigen were prepared by anion exchange chromatography

(Okubo et al., 1980). Anti-*S. cerevisiae* (baker's yeast) antiserum does not crossreact with type A mannan, but does so with that from type B. As a corollary, *S. cerevisiae* mannan was precipitated by anti-*C. albicans* type A antibody to less than 10% of the homologous *C. albicans* type A reaction (Summers et al., 1964).

Quantitative precipitin reactions (Sunayama, 1970; Summers et al., 1964) revealed that *C. tropicalis* mannan is closely related to type A and that anti-*C. albicans* type A antiserum adsorbed with type B yeast was still reactive with *C. tropicalis* mannan. In that sense, *C. tropicalis* is closer to *C. albicans* A than it is to *C. albicans* B. Reciprocal cross-adsorptions between *C. stellatoidea* (a synonym of *C. albicans*) and *C. albicans* type B revealed no disparities in agglutinin test. On the other hand, quantitative precipitin reactions of isolated *C. stellatoidea* mannan with anti-*C. albicans* type B antiserum precipitate less antibody nitrogen than homologous type B or type A mannan so that *C. stellatoidea* and *C. albicans* B may not be serologically analogous (Sunayama and Suzuki, 1970). Another difference between *C. albicans* B and *C. stellatoidea* mannan was obvious in their acetolysis fingerprints.

Specificity of Agglutinating Monoclonal Antibodies against *C. albicans* Serotypes A and B Strains and *C. tropicalis*

Two clones, CA4-2 and CA5-4, which produced agglutinating monoclonal immunoglobulin M (IgM) antibodies (MAbs) against mannan antigens of *C. albicans* serotype A, were recently established (Miyakawa et al., 1986). The specificity of each MAb was determined by slide agglutination tests for crossreactivity patterns against the homologous and the other six strains of *Candida* and a strain of *Torulopsis*. Cross-reactivity patterns as well as the competitive binding experiments by immunofluorescence staining with two MAbs and polyclonal factor sera (PAb factors) 5 and 6 suggested that the MAb from clone CA4-2 did not altogether correspond to PAb factor 6 and that the MAb from CA5-4 was distinct from PAb factor 5 in its mode of binding to determinants. Cross-reactivity patterns, nevertheless, furnished evidence that these two MAbs could replace the known PAb factors 6 and 5 for the rapid identification of strains of *C. albicans* and their serotypes in clinical laboratories.

Two other clones, CA1-2 and CB13-3 produced monoclonal immunoglobulin M (IgM) antibodies against *C. albicans* serotype A and B that demonstrated unique cross-reactivity patterns as shown in Table 2. These heterogeneous specificities of MAbs suggested that antibody recognition of epitopes exhibit extreme complexity.

More recently, Reiss et al. (1986) reported on the development of MAbs against *C. tropicalis*. Mice were immunized with *C. tropicalis* cell walls, and antibodies against mannan were detected by indirect enzyme immunoassay (EIA). Fusion of the high-responder BALB/c mouse with a plasmacytoma cell line re-

Table 2 Slide Agglutination Test with Four Monoclonal Antibodies in a Comparative Experiment

Species	No. of strains tested	% Positive agglutination with MAb			
		CA4-2[a]	CA5-4[b]	CA1-2	CB13-3
C. albicans (A)	112	100	100	100	96
C. albicans (B)	19	0	100	84	100
C. tropicalis	33	91	100	100	91
C. guilliermondii	15	0	80	100	34
C. krusei	15	0	0	0	0
C. parapsilosis	6	0	0	0	0
C. kefyr (C. pseudotropicalis)	9	0	0	0	0
C. glabrata (T. glabrata)	16	100	0	44	25

[a]Related to polyclonal factor 6.
[b]Related to polyclonal factor 5.

sulted in 41 clones secreting anti-mannan monoclonal antibodies (MAbs). Four clones selected for propagation included one IgM and one IgG MAb that reacted with mannans of *C. albicans* serotypes A and B and of *C. tropicalis* and two IgM MAbs specific for an epitope only in the mannans of *C. albicans* serotype A and *C. tropicalis*. There was 95.8% agreement in the results of serotyping using MAbs as reagents compared with PAb factors. Competitive inhibition in EIA between MAb CB6 and monospecific antisera against *C. albicans* factors 1, 4, and 6 indicated that MAb CB6 binds to an epitope, which is probably factor 6. A serological relationship between factor 4 and the binding site of MAb AC3 was also defined.

METHODS USED FOR THE IMMUNOCHEMISTRY OF YEAST MANNANS

Extraction and Purification

Mannans can be extracted from whole yeast cells or cell walls by several methods, depending on the purpose. The degree of polymerization of mannans may be influenced by exposure to extreme pH.

Hot Water Extraction (Peat et al., 1961)

Whole packed yeast cells are autoclaved in dilute neutral citrate buffer. The soluble extract is acidified to precipitate protein or deproteinized with chloroform-butanol (Okubo et al., 1980). The protein-depleted extract is neutralized and precipitated with ethanol. The polysaccharide is dissolved in water and mannan is separated with Fehling solution (see below).

Hot Alkali Extraction (Gorin and Spencer, 1968)

Whole yeast cells are suspended in 2% potassium hydroxide and boiled for 2 h. The cooled suspension is then neutralized with acetic acid and centrifuged. The supernatants are concentrated and precipitated with ethanol. Mannan is then selectively separated by complexation with Fehling solution.

Cold Alkali Extraction from Cell Walls (Reiss et al., 1974)

Cell walls prepared by mechanical disruption are cleaned of membranes by washing in sodium dodecyl sulfate solution followed by chloroform-methanol extraction. The cell walls are then treated with ice-cold dilute alkali for 24 h resulting in solubilization of 25% of the cell wall dry weight. Mannan and glucan are present in the extract. Mannan prepared in this manner is immunogenic, whereas mannan prepared from the same isolate by Fehling precipitation is serologically active but not immunogenic (Lehmann and Reiss, 1980).

Purification of Mannan as Copper Complex (Peat et al., 1961)

The polysaccharide is dissolved in water, and mannan is separated from glucan and glycogen with Fehling solution, in cold temperature, to yield the Cu^{2+}-mannan complex. The complex is recovered by centrifugation, resuspended in water and dissociated with Amberlite IR 120 (H^+). Mannan is lyophilized.

Cetyltrimethylammonium Bromide Complexation (Lloyd, 1970; Nakajima and Ballou, 1974a)

Mannan extracted from whole yeast cells by autoclaving in neutral buffer is precipitated with ethanol and dialyzed. Cetyltrimethylammonium bromide (CTAB) is added to precipitate acidic proteins and nucleic acids. After removal of precipitate, the supernate receives 1% boric acid and is titrated to pH 8.8, resulting in borate complexation of vicinal hydroxyls in the phosphomannan giving a net negative charge to the molecule, which then forms an insoluble CTAB complex. The complex is recovered and dissociated at acidic pH (Nakajima and Ballou, 1974a) or with concentrated NaCl (Cherniak et al., 1980). In this manner, mannan is prepared without exposure to the alkaline Fehling solution.

Properties of the Extracted Mannans

Mannan isolated from baker's yeast by CTAB precipitation, avoiding exposure to alkali, had a molecular weight of 133,000 da, as determined by sedimentation-equilibrium-ultracentrifugation (Nakajima and Ballou, 1974a). Further treatment of this mannan with 0.1 M NaOH reduced the molecular weight to 40,000 da. Boiling in 1 M NaOH leads to extensive degradation by breakage of *O*-phosphonomannan, *O*-mannosidic hydroxyamino acid, and *N*-glucosaminyl-aspartamide. Even mild alkali digestion is known to reduce the molecular weight through β-elimination of *O*-mannosidic hydroxyamino acids (Reiss, 1986).

Acetylation and Acetolysis of Mannan (Kocourek and Ballou, 1969; Suzuki and Fukazawa, 1982)

Controlled partial acetolysis is a fragmentation method that provides a fingerprint of the oligosaccharide chains removed from mannan depending on the acidic lability of (1→6) linkages in the mannan backbone. The method of Kocourek and Ballou (1969) is generally employed. Mannans (100 mg) are acetylated in a 1:1 mixture of anhydrous pyridine and acetic anhydride (10 ml) by heating in a steam bath for 8 h, after which the solvent is evaporated. The acetylated polysaccharide is dissolved in 10.5 ml of acetolysis medium [a mixture of acetic acid, acetic anhydride, and concentrated sulfuric acid (10:10:1, v/v)]. The solution is kept in an oil bath at 40°C. To cleave (1→6) linkages completely, acetolysis is carried out for 13 h. The reaction is terminated by adding two volumes of anhydrous pyridine. The solvent is evaporated, the oily residue is extracted with 50 ml of a mixture consisting of chloroform and water (1:1, v/v), and the chloroform extract is evaporated to dryness. The residue is dissolved in 2 ml of anhydrous methanol, then sodium methoxide in methanol is added dropwise until the solution becomes alkaline (deacetylation). After standing for 20 min, it is centrifuged 15 min at 3,000 rpm and the precipitate is washed with dried methanol and dissolved in 2 ml of distilled water. This solution is neutralized by the addition of Amberlite IR 120 (H^+) resin. The deacetylated acetolysis products of mannan are applied to a Bio-Gel P-2 column (−400 mesh, 2.5 × 150 cm). The column is then eluted with distilled water at the rate of 20 ml/h at room temperature. Total carbohydrate in the effluents is determined by the phenol-sulfuric acid method (Dubois et al., 1956).

Methylation Analysis (Hakomori, 1964, Suzuki and Fukazawa, 1982)

Methylation-fragmentation allows for the determination of glycosidic linkage arrangements and the degree of substitution in a polysaccharide, making it feasible to determine a minimal repeating unit. To convert the reducing end of the oligosaccharide into sugar alcohol, the oligosaccharide in deuterium oxide is reduced with $NaBD_4$ for 2 h at room temperature. Amberlite IR 120 (H^+) resin is added to destroy excess reducing agent, and the material is filtered and concentrated to dryness in a vacuum rotator at 40°C. Boric acid is removed by repeated evaporation with methanol. The reduced oligosaccharide is methylated by treatment with methylsulfinyl carbanion-methyl iodide in dimethyl sulfoxide according to the method of Hakomori (1964) as described by Hellerqvist et al. (1968). The partially methylated oligosaccharide is hydrolyzed with 90% formic acid at 100°C for 2 h and evaporated to dryness. The residue is hydrolyzed with 0.13 M sulfuric acid at 100°C for 12 h. The cooled solution is neutralized with barium carbonate, filtered, and concentrated to dryness. The residue is reduced with $NaBH_4$ in water for 2 h at room temperature. Amberlite IR 120 (H^+) resin is added to destroy excess reducing agent, and the material is filtered and concentrated to dryness in a vacuum rotator at 40°C. Boric acid is removed by repeated evaporation with methanol. The partially methylated alditols obtained are acetylated with a 1:1 mixture of anhydrous pyridine and acetic anhydride at 100°C for 2 h. The partially methylated alditol acetates are analyzed by gas chromatography–mass spectrometry at 180°C on a 3% ECNSS-M column (2 mm × 2 m) and identified based on their fragmentation patterns by comparing these with those reported by Bjorndal et al. (1970) and Reske and Jann (1972).

Immunological Methods

Preparation of Antisera and Anti-Factor Sera (Fukazawa et al., 1980a)

Yeast cells grown on Sabouraud medium for 48 h at 27°C are harvested with saline solution, heated for 2 h at 100°C, washed three times with formalinized saline solution, and adjusted to No. 9 of the McFarland scale. Rabbits are intravenously injected at 4-day intervals with 1.0, 2.0, 4.0, 4.0, and 4.0 ml of cell suspension. Factor serum is prepared from antiserum by appropriate adsorption with heated yeast cells according to the antigenic formula (see Table 1).

Agar Gel Diffusion Test

Precipitin tests are performed by the gel diffusion method with 0.85% noble agar (Difco) in borate buffered saline, pH 8.4. Samples (100 μg/ml) of purified polysaccharide and undiluted antiserum are placed in the upper and lower wells, respectively.

Quantitative Precipitin and Inhibition Tests (Raschke and Ballou, 1971)

Quantitative precipitin tests between rabbit antisera (0.1 ml) and mannan antigens are performed in physiologic saline solution (PSS); the final volume was

1.0 ml. The reaction mixture is incubated at 4°C for 48 h, and the precipitate is washed twice with ice-cold PSS by centrifugation. Protein in precipitate is measured by the method of Lowry et al. (1951). For inhibition of the precipitin reaction, the antiserum is incubated for 2 h at 37°C in the presence of an acetolysis oligosaccharide inhibitor, after which the antigen is added. Incubation (48 h, 4°C) and subsequent steps are as for the quantitative precipitin test.

Slide Agglutination-Inhibition Tests (Funayama et al., 1984)

Slide agglutination-inhibition tests are performed to determine the specificity of the determinants for agglutination. Factor serum (0.1 ml) is incubated for 2 h at 37°C in the presence of the acetolysis oligosaccharide inhibitor (0.1 ml). Then serial twofold dilutions of the mixture in PSS are made on a slide with 12 circles by using a micropipette. The same volumes of heated homologous yeast cell suspensions adjusted to No. 10 of the McFarland scale are added and the slide is rotated at 150 rpm for 5 min. The amounts of inhibitor required to reduce agglutination by one-fourth of that of the control factor serum are compared.

Proton Magnetic Resonance (^{1}H-NMR) Spectra of Mannans and Oligosaccharides (Gorin and Spencer, 1968)

^{1}H-NMR spectra are recorded with a 100–500 MHz nuclear magnetic resonance spectrometer for solutions, in deuterium oxide, at 70°C, of cell wall mannans and their acetolysis oligosaccharides sodium 4,4-dimethyl-4-silapentane-1-sulfonate is used as the internal standard.

IMMUNODOMINANT EPITOPES OF *Saccharomyces cerevisiae* MANNANS

Generalized Structure of *S. cerevisiae* Mannan

Mannans in the cell wall of *S. cerevisiae* are found to be covalently linked to proteins, hence the term "mannoprotein" is correctly more apropos to describe its macromolecular status. It has been postulated that the yeast cell envelope contains at least three classes of mannoprotein, i.e., inducible hydrolytic enzymes, sexual agglutinins, and structural mannoproteins (Ballou, 1976). The structural mannoproteins are distributed over the surface of the cell wall.

In early studies, mannan was isolated by extracting either whole yeast cells or isolated cell walls with dilute alkali followed by precipitation of the mannan moiety as a copper complex with Fehling solution. It is now an established fact that alkali-labile bonds, such as glycosylserine, glycosylthreonine, phosphodiester, and disulfide linkages are degraded by this procedure.

Studies on the fine structure of mannoproteins have been conducted over the past three decades, and fine chemical, enzymologic, immunochemical, and

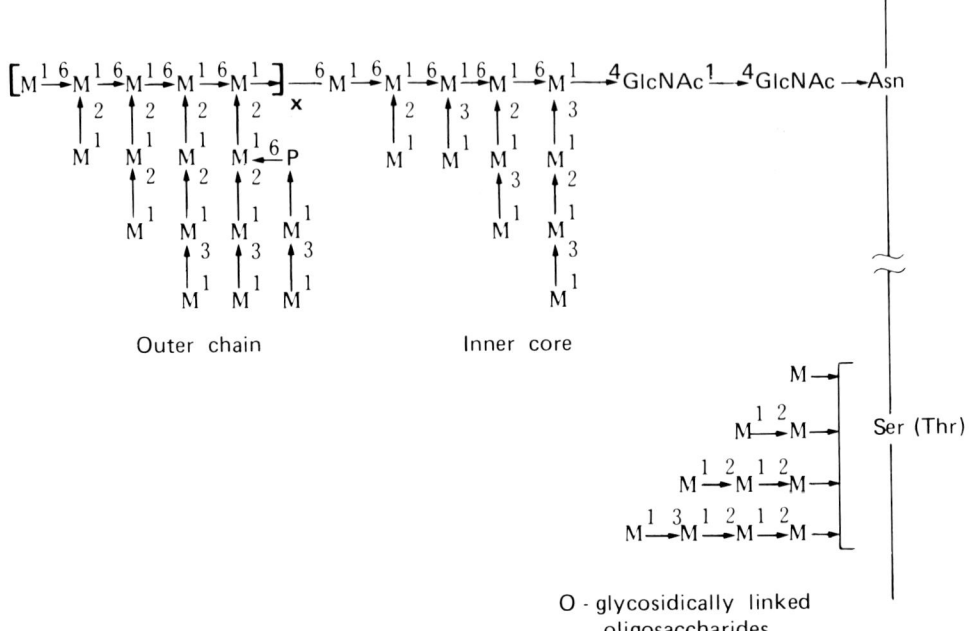

Figure 1 Generalized structure of yeast mannan: M, mannopyranose residue; P, orthophosphate; Asn, Ser, and Thr, amino acids asparagine, serine, and threonine, respectively (Nakajima and Ballou, 1974b).

genetic technologies have developed to elucidate its polysaccharide structures. The currently framed generalized structure of mannoprotein from *S. cerevisiae* proposed by Nakajima and Ballou (1974b) is shown in Figure 1. It consists of about 90% mannose, 10% protein, and modicum but varying amounts of phosphorus, depending on the strain. About 10% of the mannose occurs as short oligosaccharides that are linked to protein through serine and threonine residues. The remaining 90% of the mannose is connected by a double N-acetylglucosamine unit to asparagine in the protein. Outer chain structure demonstrated a linear (1→6)-linked backbone substituted with oligosaccharides having a degree of polymerization of 3 or 4 with a predominance of terminal (1→3)-linked mannose residues.

The fundamental techniques used to elucidate the structure in Figure 1 are briefly noted. First, acetolysis selectively cleaves the α-(1→6) linkages in the mannan backbone, thus liberating intact side chains that can be separated by gel filtration and quantitated by gas chromatography or high pressure liquid chroma-

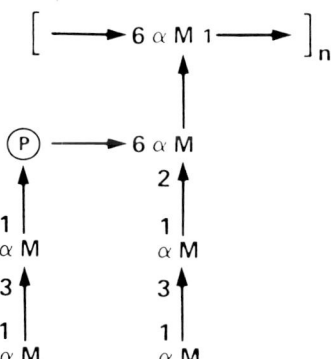

Figure 2 *O*-phosphonomannan in the outer chain region of *Saccharomyces cerevisiae* mannan (Ballou, 1976, 1980). ⓟ : orthophosphate.

tography. Second, selective enzymolysis with an exomannanase removes the unsubstituted mannosaccharide side chains, leaving essentially a linear backbone of α-(1→6)-linked residues. Third, mild alkali β-elimination selectively detaches the serine- and threonine-linked oligosaccharides. Finally, combined immunologic and genetic techniques have contributed substantially to an understanding of the fine structure of mannoprotein.

Epitopes of *S. cerevisiae* Mannan

A most salient observation has been the inhibition of homologous precipitin reactions by the oligosaccharide side chains isolated by the acetolysis procedure. Suzuki et al. (1968) have attempted analysis of the determinants of *S. cerevisiae* X2180 mannan by the precipitin inhibition assay using manno-oligosaccharides as inhibitory haptens. The results of this assay indicated that the tetraose, $M\underline{\alpha^3}M\underline{\alpha^2}M\underline{\alpha^2}M$, is the major antigenic determinant of this mannan, since this tetraose showed the strongest inhibition among the several haptens so far assayed. This finding was confirmed by Ballou (1970). In a similar fashion, it was established that the α-D-mannosyl-phosphate unit was also the major antigenic determinant in the side chains of some other strains of *S. cerevisiae* (Fig. 2) (Cawley and Ballou, 1972). On the other hand, Fukazawa et al. (1980b) found that the strains of *S. cerevisiae* and *S. uvarum* can be divided into three serotypes, Ia, Ib and II, and that these three serotypes were characterized by the presence or absence of antigenic factor 18. Immunochemical analysis of mannans of these three serotypes revealed that mannans from *S. cerevisiae* having antigenic factor 18 (serotypes Ia and Ib) contained side chains of manno-

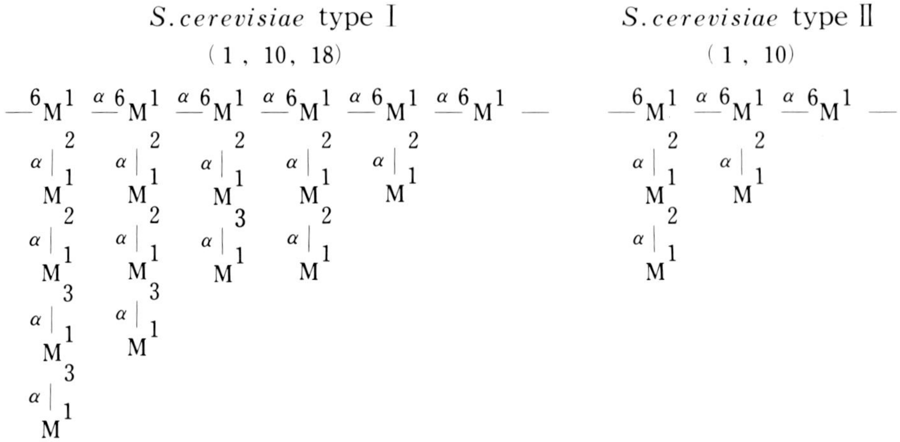

$$
\begin{array}{cc}
S.\,cerevisiae \text{ type I} & S.\,cerevisiae \text{ type II} \\
(1,\ 10,\ 18) & (1,\ 10)
\end{array}
$$

$$
-{}^6\!M^1 \xrightarrow{\alpha} {}^6\!M^1 \xrightarrow{\alpha} {}^6\!M^1 \xrightarrow{\alpha} {}^6\!M^1 \xrightarrow{\alpha} {}^6\!M^1 \xrightarrow{\alpha} {}^6\!M^1 - \qquad -{}^6\!M^1 \xrightarrow{\alpha} {}^6\!M^1 \xrightarrow{\alpha} {}^6\!M^1 -
$$

Structure for type I (left):
- Main chain: $-{}^6M^1 \xrightarrow{\alpha} {}^6M^1 \xrightarrow{\alpha} {}^6M^1 \xrightarrow{\alpha} {}^6M^1 \xrightarrow{\alpha} {}^6M^1 \xrightarrow{\alpha} {}^6M^1-$
- Side chains (α1→2 and α1→3 linked M residues) extending down from each backbone mannose, with branch positions at 2 and 3 as shown.

Structure for type II (right):
- Main chain: $-{}^6M^1 \xrightarrow{\alpha} {}^6M^1 \xrightarrow{\alpha} {}^6M^1-$
- Shorter α1→2 linked side chains.

Figure 3 Probable model for the immunodominant epitopes of *S. cerevisiae* serotypes I and II (Fukazawa et al., 1980a). ( ): antigenic structure.

pentaose and mannotetraose with two terminal α-(1→3) linkages and one terminal α-(1→3) linkage, respectively. Conversely, mannan from *S. cerevisiae* serotype II (factor 18 deficient strain) yielded acetolysis fragments of only mannotriose and mannobiose in which terminal α-(1→2) linkages were demonstrated (Fukazawa et al., 1980a). These results strongly suggest that serotype-specific determinants reside on the alkali-extracted mannan and that side chains with two terminal α-(1→3) linkages or one terminal α-(1→3) linkage are responsible for the serotype I specificity of *S. cerevisiae*. Probable models for epitopes of *S. cerevisiae* type I and type II, as well as ^{1}H-NMR spectra of whole mannans of these two serotypes, recently confirmed by Nishikawa and Shinoda (personal communication), are shown in Figures 3 and 4. It is therefore suggested that phosphate-containing side chains are not serotype-specific determinants, although phosphate-containing mannan was shown to possess strong reactivity (Shibata et al., 1983). As for the reason for the strong reactivity of phosphate-containing mannans, it appears likely that phosphate-containing mannans form larger molecules than non–phosphate-containing mannans and that the acidic phosphate-containing molecules may have the advantage of forming aggregates. The backbone structure of the mannoprotein is not an important antigenic determinant, since α-(1→6)-linked oligosaccharides are poor inhibitors of mannan precipitin reactions and a backbone devoid of side chains does not precipitate with homologous anti-mannan antisera (Ballou, 1970).

Immunochemical techniques have been used to select and characterize various mutants of *S. cerevisiae* that have altered mannoprotein structures. As might

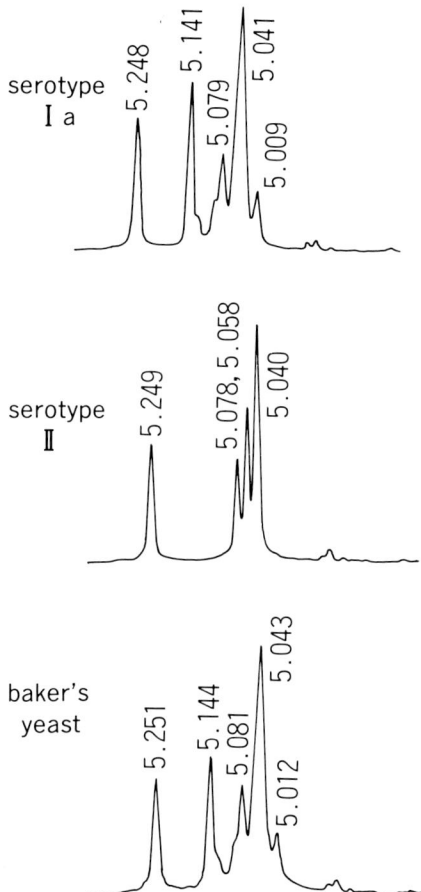

Figure 4 ^{1}H-NMR spectra of mannans from *S. cerevisiae* serotypes Ia and II and baker's yeast. Alkali-extracted mannans were determined by means of a JEOL 400 MHz nuclear magnetic resonance spectrometer (Japan Electron Optics, Tokyo) (obtained through the courtesy of Drs. Shinoda and Nishikawa).

be expected, these various mutations affect the immunochemical properties of the mannoprotein (Cohen and Ballou, 1981; Tsai et al., 1984).

IMMUNODOMINANT EPITOPES OF *CANDIDA* SPECIES MANNANS

C. albicans Serotype A and B Mannans

In view of the medical significance of candidosis (candidiasis), a number of serological and immunochemical studies have been made on *Candida* species. Some early studies of *C. albicans* mannan were, however, performed before the refinement of these techniques. Bishop and co-workers (1960), reporting on the results obtained from methylation and Smith degradation techniques, proposed that this mannan consists of α-(1→2)- and α-(1→6)-linked D-mannopyranosyl residues involving many side chains. Accordingly, they suggested that this mannan possesses a common structure with *S. cerevisiae*. Yu et al. (1967) subsequently reported amended structures for the mannans of *C. albicans* serotypes A and B, *C. tropicalis*, *C. stellatoidea*, and *C. parapsilosis*. A most significant finding is the presence of α-(1→3)-linked D-mannopyranosyl residues in all of these mannans.

Suzuki and Sunayama (1968) analyzed the antigenic determinants of the mannan of a *C. albicans* strain (probably serotype A) by means of precipitin-inhibition assay and revealed that a mannohexaose consisting of one α-(1→3) and four α-(1→2) linkages showed the strongest precipitin-inhibitory potency among more than 20 manno-oligosaccharides. Subsequently, Sunayama (1970) has analyzed the antigenic determinant groups of the mannan of *C. albicans* NIH A-207 (serotype A) and indicated that the manno-oligosaccharide that showed the most potent inhibitory action against the homologous precipitin reaction system was a mannoheptaose consisting solely of α-(1→2) linkages. Ananlyses of *C. albicans* NIH B-792 (serotype B) mannan indicated that a mannohexaose fraction consisting of α-(1→3) and α-(1→2) linkages in a ratio of 1:10 showed the strongest precipitin-inhibitory potency (Sunayama and Suzuki, 1970). Thus they concluded that, regardless of serotype-specificity of the parent mannan, the antigenic determinants of *C. albicans* correspond to the largest acetolysis products, namely hexaose or heptaose.

Recently Reiss et al. (1981) reported on the results obtained by analyzing methylated, fragmented, and peracetyl aldononitrile derivatives of mannans from *C. albicans* (serotypes A and B) and *Torulopsis (Candida) glabrata* by using gas-liquid chromatraphy–mass spectrometry. There was a greater amount of α-(1→2)-linked mannose in *C. albicans* serotype A mannan than in serotype B, indicating longer side chains in serotype A. This finding, however, does not prove that the only difference between serotypes is the length of the side chains.

A larger number of mannosyl branch points linked through C-1, C-3, and C-6 were found in serotype B mannan. The ratio of α-(1→2) to α-(1→3) linkages in serotype A mannan was 9.8:1 (Reiss et al., 1981), indicating two types of side chain: one type with five α-(1→2)-linked mannose residues, and the other with four α-(1→2) linkages and one α-(1→3) linkage, these findings being consistent with other reports (Suzuki and Sunayama, 1968; Fukazawa et al., 1980a).

The overview of the outer chain region of *C. albicans* mannan suggested by methylation analysis and compatible with serologic findings is of a linear α-(1→6)-linked backbone heavily substituted with oligosaccharides having a degree of polymerization of 5 or 6 in which α-(1→2) linkages are ubiquitous, but with terminal (1→3)-linked mannose residues in every other side chain. The weak cross-reactivity between serotypes A and B and the resistance of serotype A mannan to exo-α-mannan hydrolase imply that chain length alone cannot account for serotype specificity (Reiss, 1986).

The oligomannosides produced from *Saccharomyces kluyveri* lb cell wall mannan by either acetolysis or by β-elimination were compared (Zhang and Ballou, 1981). Although acetyolysis fragments have proven to be a valuable aid in fingerprinting yeast mannans, in this instance two immunochemically important structural features were labile to this treatment. Terminal β-(1→2)-linked mannose and single (1→6) branches in the oligomannosides were lost. The (1→6) branches were detected by the presence of the methylated fragment 2,4,-di-*O*-methylmannose. The β-linked mannose in the otherwise α-linked mannan was detected in the spectrum as a signal at σ = 4.75 ppm. The β linkage was labile to β-mannosidase, indicating that it occupied a terminal position. The removal of (1→6) branches and nonreducing β-(1→2)-mannosyl termini greatly reduced the hapten inhibitory power of oligomannosides from *S. kluyveri*. This study has important implications for the immunochemistry of *C. albicans* mannan. Antiserum to *S. kluyveri* lb cells cross-reacted strongly with *C. albicans* B311A (serotype A) mannan but not at all with mannan of *C. albicans* B792 (serotype B). Mannan of serotype A was previously shown to be resistant to exo-α-mannan (Reiss et al., 1981), thus β-linked mannosyl termini appear to play an immunodominant role in type A mannan (Reiss, 1986) (Fig. 5).

Fukazawa et al. (1980a) compared the structures of haptenic oligomannosides produced by acetolysis of bulk mannans from *C. albicans* serotypes A and B. The oligomannosides were studied using ^{1}H-NMR, methylation-fragmentation and quantitative precipitin analysis. Acetolysis fingerprints of serotypes A and B oligomannosides chromatographed on Bio-Gel P-2 were very similar. ^{1}H-NMR spectra of mannopentaoses (M5) were similar, containing (1→2)-linked residues and signals indicating (1→3) linkage. Serotype A mannohexaose (M6) yielded four anomeric proton signals, including one referrable to a terminal α-(1→3) linkage, whereas serotype BM6 had only three such signals due to the absence of a (1→3)-terminal residue. Methylation-fragmentation analysis showed that sero-

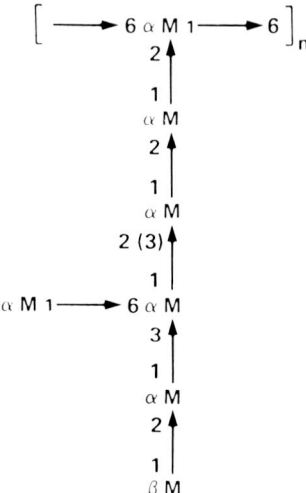

Figure 5 Probable model of the immunodominant epitopes of *Candida albicans* serotype A mannan. Zhang and Ballou (1981) have shown that nonreducing terminal β-(1→2) mannosyl residues and single (1→6) branches in the oligomannosides are labile to acetolysis in *Saccharomyces kluyveri* mannan. These linkages may also occur in *C. albicans* serotype A for the reason that antiserum to *S. kluyveri* reacts with *C. albicans* A but not with serotype B (Reiss, 1986).

type A M5 is a mixture of α (1→2)-linked pentaose and one with three (1→2) and one (1→3) linkages. Serotype A M6 was also a mixture of one purely (1→2)-linked hexaose and one with α (1→3) terminal. The major difference in serotype B M6 was the occurrence of the (1→3) linkage as an internal branch point. Inhibition of the homologous reaction between serotype A mannan and anti-*C. albicans* A serum was maximal with serotype A mannoheptaose (M7) but M6 was nearly as good an inhibitor. Oligomannosides of serotype B including M6 had comparatively less ability to inhibit the serotype A-anti-A reaction. The reaction between serotype B mannan and anti-*C. albicans* B serum was inhibited equally well by M6 and M7 of serotype B or by M6 of serotype A.

Taking into consideration all three types of analysis (acetolysis, methylation-fragmentation, hapten inhibition), the linkage of sugar residues in the mannohexaose from these two serotypes are responsible for serotype specificity. The immunodominant hapten of serotype A consisted of a straight chain of (1→2)-linked oligosaccharides with α (1→3) terminal. Type B M6 is more complex, lacking a terminal (1→3) linkage but having instead a single C-1, C-2, C-3 branch point and an additional internal (1→3) linkage.

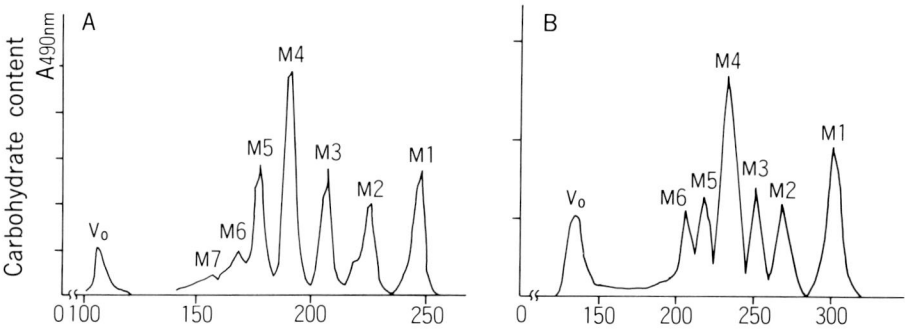

Figure 6 Gel filtration profiles of the acetolysis oligosaccharides of *C. albicans* serotype A and B mannans on a Bio-Gel P-2 column. (A) *C. albicans* serotype A mannan (Suzuki and Fukazawa, 1982). (B) *C. albicans* serotype B mannan (Funayama et al., 1983). M1, mannose; M2, mannobiose; M3, mannotriose; M4, mannotetraose; M5, mannopentaose; M6, mannohexaose; M7, mannoheptaose.

Later, Suzuki and Fukazawa (1982) further investigated the immunochemical structure of the specific epitope in the mannan of *C. albicans* M-1012 (serotype A) strain. Acetolysis of the mannan, obtained by alkali extraction and purified as a copper complex, yielded mannose and five oligosaccharide (from mannobiose to -hexaose) and a small amount of a mannoheptaose (Fig. 6). They examined the inhibition by these oligosaccharides of the precipitin reaction between anti-factor 6 serum specific for serotype A (see Table 1) and the homologous mannan, and found that the mannohexaose was the most effective in-

Table 3 Methylation Analysis of Reduced Acetolysis Fragments of *C. albicans* Serotype A Mannan (Suzuki and Fukazawa, 1982)

Partially acetylated mannitol derivative	Relative retention time	Molar ratios				
		M2	M3	M4	M5	M6
1,3,4,5,6-Penta-*O*-methyl	0.38	1.02	0.83	0.72	0.80	0.75
2,3,4,6-Tetra-*O*-methyl	1.00	1.00	1.00	1.00	1.00	1.00
3,4,6-Tri-*O*-methyl	1.84	–	0.78	1.89	1.62	2.32
2,4,6-Tri-*O*-methyl	1.98	–	0.12	–	0.57	0.52

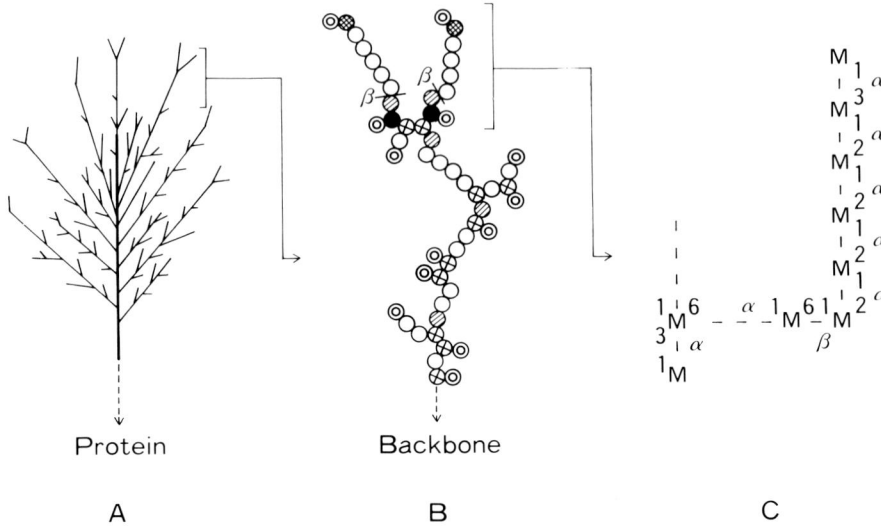

Figure 7 Probable model for the immunodominant epitope corresponding to specificity of factor 6 in relation to the possible overall structure of *C. albicans* serotype A mannan. (A) The overall structure of *C. albicans* serotype A mannan. The heavy line shows the protein-bound backbone to which is attached a side chain consisting of a primary and a secondary side chain. (B) Enlargement of the terminal portion with antigenic factor 6. ◉, nonreducing end; ◒, $-^6M^1$; ○, $-^2M^1-$; ◍, $-^6M_2{}^1$; ●, $-^3M^1-$; ●, $-^6M_3{}^1-$. (C) Enlargement of antigenic factor 6. M, mannopyranosyl residue (Suzuki and Fukazawa, 1982).

hibitor. These results, together with those obtained by ^{1}H-NMR, methylation analysis (Table 3), and other structural studies, suggested that the main component of this hexaose consists of one terminal α-(1→3) linkage in addition to four α-(1→2) linkages, and that this α-(1→3)-containing mannohexaose may be responsible for the specificity of antigenic factor 6. Additional results obtained by analyses of polarimetry, ^{1}H-NMR spectroscopy, and chromium trioxide oxidation-methylation of *C. albicans* M-1012 mannan and its acid-degraded product suggested that the mannan has a β linkage in addition to α linkages, and that the mode of the β linkage is mainly (1→6) linkage. Further data obtained by Smith degradation-methylation analysis and by quantitative precipitin reactions of intact and acid-degraded mannan suggested that the epitope of antigenic factor 6 may be bound, *via* the β-(1→6) linkage, to C-6 of mannose residues involved in oligosaccharide side chains of serotype A mannan. In order to account for these structural details, Suzuki and Fukazawa (1982) proposed an arboreal

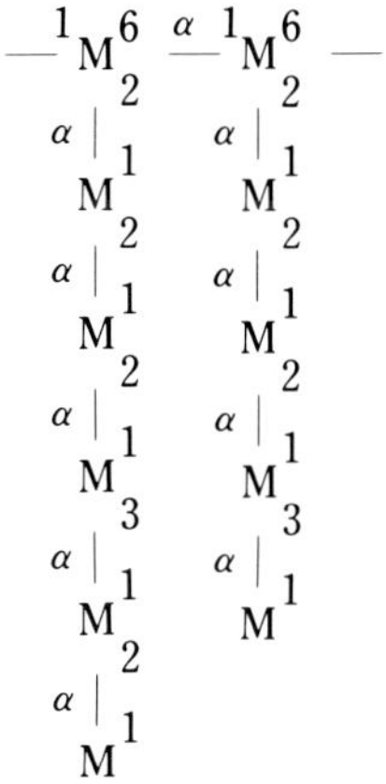

Serologic Specificity of Mannans in Yeasts

```
   1  6  α  1  6
 ─ M      ─ M    ─
    2        2
  α |  1   α |  1
    M        M
    2        2
  α |  1   α |  1
    M        M
    2        2
  α |  1   α |  1
    M        M
    3        3
  α |  1   α |  1
    M        M
    2
  α |  1
    M
```

Figure 8 Probable model for the immunodominant epitopes corresponding to specificities of antigen factors 13 and 13b in *C. albicans* serotype B and *C. parapsilosis* mannans (Funayama et al., 1984).

rather than a comblike structure for *C. albicans* type A mannan, i.e., composed of a protein trunk with a linear branch of alternating (1→6)-mannose residues two to three units long flanking (1→2)-linked mannobiose. Type specificity was considered to reside in sparse terminal subbranches of mannohexaose with four α-(1→2) linkages and one α-(1→3) linkage, joined to the main branch by β-(1→6)-linked mannose (Fig. 7).

Relationship Between *C. parapsilosis* and *C. albicans* Serotype B Mannans

Funayama et al. (1983) investigated the chemical structure of the specific epitope in the mannan of *C. parapsilosis* M1015 strain. Acetolysis of *C. parapsilosis* cell wall mannan, obtained by alkali extraction and purified as a copper complex, yielded six oligosaccharides by Bio-Gel P-2 filtration. Their chemical structure was examined by ^{1}H-NMR spectroscopy, methylation analysis, and partial acid hydrolysis. The proposed structures of penta- and hexasaccharides are shown in Figure 8. Inhibition of oligosaccharides with anti-*C. parapsilosis* serum and homologous mannan indicated that although M5 and M6 exhibited inhibitory activity, M6 was the more effective inhibitor, and that M6 may be the immunodominant or may be responsible for the specificity of *C. parapsilosis* mannan in the precipitin reaction system, or both.

Funayama et al. (1984) further examined the antigenic relationship between *C. parapsilosis* and *C. albicans* serotype B with respect to antigenic factors 13 and 13b (see Table 1), which are specific for the former species and common to both species, respectively. Acetolysis of *C. albicans* serotype B cell wall mannan yielded six oligosaccharides. Their chemical structure was determined by ^{1}H-NMR spectroscopy, methylation analysis, and partial acid hydrolysis. The structure of the hexasaccharide derived from *C. albicans* serotype B mannan was α-D-man-(1-2)-α-D-Man-(1-3)-α-D-Man-(1-2)-α-D-Man-(1-2)-α-D-Man-(1-2)-D-man (M6), which is identical to that from *C. parapsilosis* mannan. Inhibition of two precipitin reaction systems (anti-*C. albicans* serotype B serum and anti-*C. parapsilosis* serum to the respective homologous mannan) by oligosaccharides from homologous and heterologous mannans indicated that M6 from either *C. albicans* serotype B or *C. parapsilosis* was the most effective inhibitor. Moreover, inhibition of the agglutination reaction between factor serum containing anti-factors 13 and 13b and *C. albicans* serotype B or *C. parapsilosis* cells by oligosaccharides from both mannans also indicated that the M6s were the most effective inhibitors. These results suggest that the M6s derived from the two species are identical in their chemical structure, although the structures of the whole mannans of the two species are not identical and that M6s may be involved in the specificities of antigenic factors 13 and 13b. The amount of M6 is larger in *C. parapsilosis* cell wall mannan, suggesting that high repeating frequency of M6 fragment may induce the antibody specific for *C. parapsilosis* (Fig. 8).

C. krusei Mannan

Nishikawa et al. (1982) investigated the chemical structure of the antigenic determinant of *C. krusei* cell wall mannan. Complete acetolysis of the mannan, obtained by extraction with alkali and purified as a copper complex, yielded five oligosaccharides (from mono- to pentasaccharide) and a small amount of an octasaccharide. Partial acetolysis of the mannan yielded a large amount of mannooctaose. They examined the inhibition by these oligosaccharides of the precipitin reaction between anti-*C. krusei* serum and homologous mannan, and found that the mannooctaose was the most effective inhibitor. Results obtained by ^{1}H-NMR spectroscopy, methylation analysis, and other structural studies of *C. krusei* mannan suggested that the octasaccharide possesses six α-(1→2) linkages and one α-(1→6) linkage located in the middle of the chain, and that this mannooctaose may be responsible for the specificity of antigenic factor 11 (see Table 1). A hypothetical model for the type-specific epitope of *C. krusei* is shown in Figure 9.

Phosphonomannan of *Candida*

Phosphate content, which is expressed as the molar ratio of mannose/phosphate, is higher in *C. albicans* serotype B mannan (mannose/phosphate = 18) than in baker's yeast mannan (mannose/phosphate = 78–120) (Stewart and Ballou,

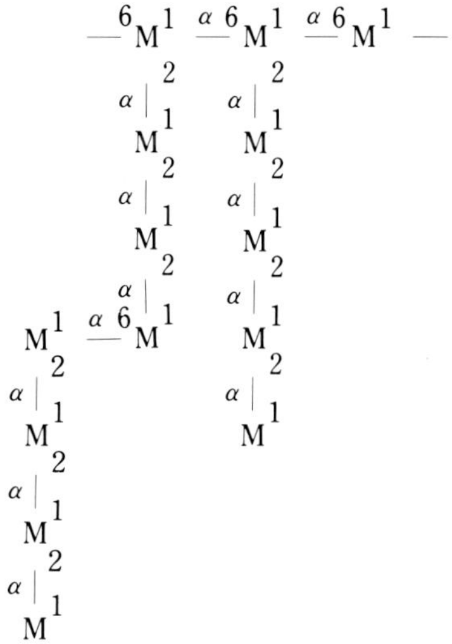

$$\begin{array}{c}
{-}{}^6M^1 \xrightarrow{\alpha\ 6} M^1 \xrightarrow{\alpha\ 6} M^1 {-} \\
{}_2{}_2 \\
\alpha\Big|_1\alpha\Big|_1 \\
MM \\
{}_2{}_2 \\
\alpha\Big|_1\alpha\Big|_1 \\
MM \\
{}_2{}_2 \\
\alpha\Big|\alpha\Big|_1 \\
M^1 \xrightarrow{\alpha\ 6} M^1 M \\
{}_2{}_2 \\
\alpha\Big|_1\alpha\Big|_1 \\
MM \\
{}_2 \\
\alpha\Big|_1 \\
M \\
{}_2 \\
\alpha\Big|_1 \\
M
\end{array}$$

Figure 9 Probable model for the immunodominant epitopes corresponding to specificity of antigenic factor 11 in *C. krusei* mannan (Nishikawa et al., 1982).

1968). The total phosphate content of bulk mannans has been measured for *C. albicans* B792, 0.86%; for *C. albicans* 311A, 0.25%; and for *S. cerevisiae*, 0.1% (Hasenclever and McAtee, 1977). Much information concerning the molecular organization of *O*-phosphonomannan of saprophytic yeasts has accumulated, thereby providing a model for analysis of *C. albicans* mannan.

In baker's yeast (Ballou, 1976) phosphodiesters link (1→3)-α-Man disaccharide to a mannotriose at the C-6 position of the reducing terminal mannose residue joined to the linear α-(1→6)-mannan backbone (see Fig. 2). The α-mannosyl phosphate groups of some *Saccharomyces* isolates are important antigenic determinants (Ballou, 1976).

Bulk mannan prepared from *C. albicans* by autoclaving and Fehling precipitation was separated into five fractions on DEAE-Sephadex column according to their phosphate content (Okubo et al., 1979, 1980). The three known serotypes of *C. albicans* were compared by this means. Serotype B mannan was the most highly phosphorylated, 33% of the bulk mannan was eluted at high ionic strength, and serotype C (belonging to A) mannan was the least phosphorylated. Quantitative precipitin reactions showed that the amount of antibody precipitated was proportional to the phosphate content of column-derived fractions.

Further studies with *C. albicans* mannan (Okubo et al., 1980) showed that the fractions highest in phosphate also contained oligomannosides with the longest chain length. Phosphate, protein, and longer oligomannosides are concentrated in the acidic fractions of mannan separated by anion exchange chromatography. The location of O-phosphonomannan of *C. albicans* with respect to the inner core–outer chain model of *S. cerevisiae* by Nakajima and Ballou (1974b) remains unknown.

Recently, Shibata et al. (1985) investigated the immunochemical properties of mannans of *C. albicans* serotypes A, B, and C (belonging to A) strains, prepared by cetyltrimethylammonium bromide (CTAB). They suggested that phosphate-bound, acid-labile oligomannosyl residues might exert extremely strong antibody-precipitating activity not only in strain B mannan but also in the mannans of the other two strains (A and C). Whether serotype specific epitopes reside in the mannooligosaccharides composed of β-D-mannopyranosyl residues (see Fig. 5) or in phosphate-bound oligomannosyl residues remains a subject of contention.

CONCLUDING REMARKS

Characteristics of outermost cell wall structures have been demonstrated indirectly by serological methods, such as agglutination test and fluorescent antibody staining of whole cells or precipitin reaction of extracts. Several medically important *Candida* species have been well defined in their antigenic relationships, e.g., proposed antigenic formula by mutual antibody-adsorption analysis. In the antigenic formula, species- or serotype-specific antigens have been shown as antigenic factors.

It has been firmly established that the serologic specificity of yeast cells resides in cell wall mannans, especially side chains consisting of oligomannosyl residues. However, molecular structures of epitopes (determinants) of *C. albicans* serotypes and other *Candida* species proposed by many investigators have not been equally consistent with each other. It has been suggested that outermost side chains of oligomannosyl residues of pathogenic yeast, especially of *C. albicans* serotype A, are branched arboreal rather than comblike in structure.

Participation of phosphate oligomannosyl residues and alkali-labile terminal mannose with β-$(1\rightarrow2)$ linkage in the serologic specificity, in addition to mannooligosaccharides composed of α-D-mannopyranosyl residues, is presently under argumentation.

The term "immunodominant" is not equivalent to "specificity," since specific epitopes may or may not reside in immunodominant structures. Even small oligosaccharide fragments may contain multiple epitopes, which are recognized by antibodies.

Studies on the specificity of monoclonal antibodies directed against *C. albicans* serotypes A and B mannans suggest that monoclonal antibodies recognize part of the known epitopes, hence, exhibiting heterogeneous cross-reactivities. These results may lead to further understanding of the mechanisms of antibody synthesis.

REFERENCES

Ballou, C. E. (1970). A study of the immunochemistry of three yeast mannans. *J. Biol. Chem. 245*:1197-1203.

Ballou, C. E. (1976). Structure and biosynthesis of the mannan component of the yeast cell envelope. *Adv. Microb. Physiol. 14*:93-158.

Ballou, C. E. (1980). Genetics of yeast mannoprotein biosynthesis. In *Fungal Polysaccharides*. Edited by P. A. Bandford, and K. Matsuda. American Chemical Society, Washington, D.C., pp. 1-14.

Biguet, J., Tran Van Ky, P., and Andrieu, S. (1962). Eteude electrophoretique et immunochimique compare des antigenes de quelques levures du genure *Candida*. *Mycopathol. Mycol. Appl. 17*:239-254.

Bishop, C. T., Blank, F., and Gardner, P. E. (1960). The cell wall polysaccharides of *Candida albicans*: glucan, mannan and chitin. *Can. J. Chem. 38*:869-881.

Bjorndal, H., Hellerqvist, C. G., Lindberg, B., Svensson, S. (1970). Gas-liquid chromatography and mass spectrometry in methylation analysis of polysaccharides. *Angew. Chem. Int. Ed. Engl. 9*:610-619.

Cassone, A., Mattia, E., and Boldrini, L. (1978). Agglutination of blastospores of *Candida albicans* by concanavalin A and its relationship with the distribution of mannan polymers and the ultrastructure of the cell wall. *J. Gen. Microbiol. 105*:263-273.

Cawley, T. N., and Ballou, C. E. (1972). Identification of two *Saccharomyces cerevisiae* cell wall mannan chemotypes. *J. Bacteriol. 111*:690-695.

Cherniak, R., Reiss, E., Slodki, M. E., Plattner, R. D., and Blumer, S. O. (1980). Structure and antigenic activity of the capsular polysaccharides of *Cryptococcus neoformans* serotype A. *Mol. Immunol. 17*:1025-1032.

Cohen, R. E., and Ballou, C. E. (1981). Mannoproteins: Structure. In *Plant Carbohydrates II. Extracellular Carbohydrates*, Encyclopedia of Plant Physiology, vol. 13B. Edited by W. Tanner and F. A. Loewus. Springer-Verlag, New York, pp. 441-458.

Dubois, M., Gilles, K. A., Hamilton, J. K., Rebers, P. A., and Smith, F. (1956). Colorimetric method for determination of sugars and related substances. *Anal. Chem. 28*:350-356.

Fukazawa, Y., Shinoda, T., Nishikawa, A., Kagaya, K., and Yonezawa, Y. (1976). Serological and immunological studies of the serotypes of *Candida albicans*. In *Yeasts and yeast-like microorganisms in medical science*. Edited by K. Iwata. Tokyo University Press, Tokyo, pp. 213-218.

Fukazawa, Y., Nishikawa, A., Suzuki, M., and Shinoda, T. (1980a). Immunochemical basis of serologic specificity of the yeast: Immunochemical determinants of several antigenic factors of yeasts. *Zentralbl. Bakteriol. Mikrobiol. Suppl. 8*:127-136.

Fukazawa, Y., Shinoda, T., Nishikawa, A., and Nakase, T. (1980b). Synonymy of *Saccharomyces cerevisiae* Hansen 1883 and *Saccharomyces uvarum* Beijerinck 1898: Significance of cell wall antigens in yeast classification. *Int. J. Syst. Bacteriol. 30*:196-205.

Funayama, M., Nishikawa, A., Shinoda, T., and Fukazawa, Y. (1983). Immunochemical determinant of *Candida parapsilosis. Carbohydr. Res. 117*:229-239.

Funayama, M., Nishikawa, A., Shinoda, T., Suzuki, M., and Fukazawa, Y. (1984). Antigenic relationship between *Candida parapsilosis* and *Candida albicans* serotype B. *Microbiol. Immunol. 28*:1359-1371.

Gorin, P. A. J., and Spencer, J. E. T. (1968). Galactomannans of *Trichosporon fermentans* and other yeasts. Proton magnetic resonance and chemical studies. *Can. J. Chem. 46*:2299-2304.

Hakomori, S. (1964). A rapid permethylation of glycolipid, and polysaccharide catalyzed by methylsulfinyl carbanion in dimethyl sulfoxide. *J. Biochem. 55*: 205-208.

Hasenclever, H. F., and McAtee, F. J. (1977). Antigenic relationships of *Candida albicans, Saccharomyces telluris* and *Saccharomyces cerevisae*. In *Host-Parasite Relationships in Systemic Mycoses*. Edited by A. M. Beemer, et al., S. Karger, Basel, pp. 126-137.

Hasenclever, H. F., and Mitchell, W. O. (1961). Antigenic studies of *Candida*. I. Observation of two antigenic groups in *Candida albicans. J. Bacteriol. 82*: 570-573.

Hasenclever, H. F., and Mitchell, W. O. (1964). Immunochemical studies on polysaccharides of yeasts. *J. Immunol. 93*:763-771.

Hellerqvist, C. G., Lindberg, B., and Svensson, S. (1968). Structural studies on the *O*-specific side-chains of the cell wall lipopolysaccharide from *Salmonella typhimurium* 396 MS. *Carbohydr. Res. 8*:43-55.

Ikeda, R., Shinoda, T., Fukazawa, Y., and Kaufman, L. (1982). Antigenic characterization of *Cryptococcus neoformans* serotypes and its application to serotyping of clinical isolates. *J. Clin. Microbiol. 16*:22-29.

Kocourek, J., and Ballou, C. E. (1969). Method for fingerprinting yeast cell wall mannans. *J. Bacteriol. 100*:1175-1181.

Lehmann, P. F., and Reiss, E. (1980). Detection of *Candida albicans* mannan by immunodiffusion, counterimmunoelectrophoresis and enzyme-linked immunoassay. *Mycopathologia 70*:83-88.

Lloyd, K. O. (1970). Isolation, characterization and partial structure of peptide-galactomannans from the yeast form of *Cladosporium werneckii*. *Biochemistry 9*:3446-3470.

Lowry, O. H., Rosebrough, N. J., Farr, A. L., and Randall, R. J. (1951). Protein measurement with the Folin phenol reagent. *J. Biol. Chem. 193*:265-275.

Martin, D. S. (1942). Studies on the immunologic relationships among various species of the genus *Candida* (*Monilia*). *Am. J. Trop. Med. 22*:295-303.

Miyakawa, Y., Kagaya, K., Fukazawa, Y., and Soe, G. (1986). Production and characterization of agglutinating monoclonal antibodies against predominant antigenic factors for *Candida albicans. J. Clin. Microbiol. 23*:881-886.

Müller, H. L., and Hirsch, M. (1967). Zur antigenstrukture von *Candida albicans* und *C. tropicalis. Zentralbl. Bakteriol. Parasitenkd. Infektionskr. Abt. 1. 202*:247-256.

Nakajima, T., and Ballou, C. E. (1974a). Characterization of the carbohydrate fragments obtained from *Saccharomyces cerevisiae* mannan by alkaline degradation. *J. Biol. Chem. 249*:7679-7684.

Nakajima, T., and Ballou, C. E. (1974b). Structure of the linkage region between the polysaccharide and protein parts of *Sacchromyces cerevisiae* mannan. *J. Biol. Chem. 249*:7685-7694.

Nishikawa, A., Shinoda, T., and Fukazawa, Y. (1982). Immunochemical determinant and serological specificity of *Candida krusei. Mol. Immunol. 19*:367-373.

Okubo, Y., Honma, Y., and Suzuki, S. (1979). Relationship between phosphate content and serological activities of the mannans of *Candida albicans* strains NIH A-207, NIH B-792, and J-1012. *J. Bacteriol. 137*:677-680.

Okubo, Y., Honman, Y., Ichikawa, T., and Suzuki, S. (1980). Immunochemistry of *Candida albicans* mannan. In *Chemistry and Biochemistry of Fungal Polysaccharides*, vol. 126. Edited by P. A. Sandford and K. Matsuda. American Chemical Society, Washington, D.C., pp. 95-112.

Peat, S., Whelan, W. J., and Edwards, T. E. (1961). Polysaccharides of baker's yeast. IV. Mannan. *J. Chem. Soc.* (London) *1*:29-34.

Pospisil, L. (1959). Agglutinations- Komplementbindungs-, and Hemmagglutinationsreaktion bei der Bestimmung von *Candida*-arten. *Dermatologia 118*: 65-73.

Raschke, W. C., and Ballou, C. E. (1971). Characterization of a yeast mannan containing *N*-acetyl-D-glucosamine as an immunochemical determinant. *Biochemistry 10*:4130-4135.

Reiss, E. (1986). *Molecular Immunology of Mycotic and Actinomycotic Infections*. Elsevier Science Publishers, New York.

Reiss, E., Stone, S. H., and Hasenclever, H. F. (1974). Serological and cellular immune activity of peptidoglucomannan fractions of *Candida albicans* cell walls. *Infect. Immun. 9*:881-890.

Reiss, E., Patterson, D. G., Yert, L. W., Holler, J. S., and Ibrahim, B. K. (1981). Structural analysis of mannans from *Candida albicans* serotypes A and B and from *Torulopsis galbrata* by methylation gas chromatography mass spectrometry and exo-α-mannanase. *Biomed. Mass Spectrom. 8*:252-255.

Reiss, E., De Repentigny, L., Kuykendall, R. J., Carter, A. W., Galindo, R., Auger, P., Bragg, S. L., and Kaufman, L. (1986). Monoclonal antibodies against *Candida tropicalis* mannan: Antigen detection by enzyme immunoassay and immunofluorescence. *J. Clin. Microbiol. 24*:796-802.

Reske, K., and Jann, K. (1972). The O8 antigen of *Escherichia coli* structure of the polysaccharide chain. *Eur. J. Biochem. 31*:320-328.

Seeliger, H. P. R. (1958). *Mycologische Serodiagnostik*. Johan Ambrosius Barth, Leipzig.

Shibata, N., Mizugami, K., Takano, K., and Suzuki, S. (1983). Isolation of mannan-protein complexes from viable cells of *Saccharomyces cerevisiae* X2180-1A wild type and *Saccharomyces cerevisiae* X2180-1A-5 mutant strains by the action of zymolyase-60,000. *J. Bacteriol. 156*:552-558.

Shibata, N., Ichikawa, T., Tojo, M., Takahashi, M., Ito, N., Okubo, Y., and Suzuki, S. (1985). Immunochemical study on the mannans of *Candida albicans* NIH A-207, NIH B-792, and J-1012 strains prepared by fractionated precipitation with cetyltrimethylammonium bromide. *Arch. Biochem. Biophys. 243*:338-348.

Shibata, N., Kobayashi, H., Tojo, M., and Suzuki, S. (1986). Characterization of phosphomannan-protein complexes isolated from viable cells of yeast and mycelial forms of *Candida albicans* NIH B-792 strain by the action of Zymolyase-100T. *Arch. Biochem. Biophys. 251*:697-708.

Stewart, T. S., and Ballou, C. E. (1968). A comparison of yeast mannans and phosphomannans by acetolysis. *Biochemistry 7*:1855-1863.

Summers, D. F., Grollman, A. P., and Hasenclever, H. F. (1964). Polysaccharides antigens of the *Candida* cell wall. *J. Immunol. 92*:491-499.

Sunayama, H. (1970). Studies on the antigenic activities of yeasts. IV. Analysis of the antigenic determinant groups of the mannan of *Candida albicans* serotype A. *Jpn. J. Microbiol. 14*:27-39.

Sunayama, H., and Suzuki, S. (1970). Studies on the antigenic activities of yeasts. VI. Analysis of the antigenic determinant groups of the mannan of *Candida albicans* serotype B-792. *Jpn. J. Microbiol. 14*:371-379.

Suzuki, M., and Fukazawa, Y. (1982). Immunochemical characterization of *Candida albicans* cell wall antigens: specific determinant of *Candida albicans* serotype A mannan. *Microbiol. Immunol. 26*:387-402.

Suzuki, S., and Sunayama, H. (1968). Studies on the antigenic activities of yeasts II. Isolation and inhibition assay of the oligosaccharides from acetolysate of mannan of *Candida albicans*. *Jpn. J. Microbiol. 12*:413-422.

Suzuki, S., Sunayama, H., and Saito, T. (1968). Studies on the antigenic activities of yeasts I. Analysis of determinant groups of the mannan of *Saccharomyces cerevisiae*. *Jpn. J. Microbiol. 12*:19-24.

Sweet, C. E., and Kaufman, L. (1970). Application of agglutinins for the rapid and accurate identification of medically important *Candida* species. *Appl. Microbiol. 19*:830-836.

Tsai, P-K., Frevert, J., and Ballou, C. E. (1984). Carbohydrate structure of *Saccharomyces cerevisiae* mnn9. *J. Biol. Chem. 259*:3805-3811.

Tsuchiya, T., Fukazawa, Y., Miyazaki, F., and Kawakita, S. (1955). Studies on the classification of the genus *Candida*. Thermo-stable and thermo-labile antigens of seven species of genus *Candida*. *Jpn. J. Exp. Med. 25*:75-83.

Tsuchiya, T., Fukazawa, Y., Taguchi, M., Nakase, T., and Shinoda, T. (1974). Serologic aspects of yeast classification. *Mycopathol. Mycol. Appl. 53*:77-91.

Yu, R. J., Biship, C. T., Cooper, F. P., Hasenclever, H. F., and Blank, F. (1967). Structural studies of mannans from *Candida tropicalis* (serotypes A and B), *Candida parapsilosis, Candida stellatoidea* and *Candida tropicalis*. *Can. J. Chem. 45*:2205-2211.

Zhang, W. J., and Balou, C. E. (1981). *Saccharomyces kluyveri* cell wall mannoprotein. *J. Biol. Chem. 256*:10073-10079.

3
Antigenic Structure of *Cryptococcus neoformans* Capsular Polysaccharides

THOMAS R. KOZEL
University of Nevada School of Medicine, Reno, Nevada

INTRODUCTION

Cryptococcus neoformans, the etiological agent of cryptococcosis, is an encapsulated yeast that reproduces by budding. Cultures of the yeast exhibit wide variation in individual cell size, with the diameter ranging from 4-20 μm. The capsule diameter varies from inapparent to 5-10 μm in width. Capsule size is influenced by the strain, culture medium, growth conditions, and age of culture. Kwon-Chung established that *C. neoformans* is a Basidiomycete with the perfect state *Filobasidiella neoformans* (Kwon-Chung, 1975). Two varieties of the yeast have been identified. In both cases, variety is closely linked to the serotype of the capsular polysaccharide that is produced by the yeast. *C. neoformans* var. *neoformans* is serotype A or D, whereas *C. neoformans* var. *gattii* is serotype B or C. Cross-hybridization between the two varieties shows only 55-63% homology (Aulakh et al., 1981), and numerous biochemical differences have been demonstrated between var. *neoformans* and var. *gattii*. The corresponding perfect states produced by compatible mating types are *F. neoformans* var. *neoformans* (*C. neoformans* var. *neoformans*) and *F. neoformans* var. *gattii* (*C. neoformans* var. *gattii*) (Kwon-Chung et al., 1982).

Cryptococcosis may occur as a pulmonary infection, but serious infection most often involves the central nervous system. Cryptococcosis occurs most frequently as an opportunistic infection. Individuals found at risk for infection include those with malignant neoplasms and kidney transplant recipients. More recently, cryptococcosis has emerged as a common life-threatening infection in patients with AIDS (Kovacs et al., 1985). Cryptococcosis takes a particularly difficult course in the AIDS patient; dissemination occurs frequently, and the infection responds poorly to conventional treatment (Kovacs et al., 1985; Eng et al., 1986; Rinaldi et al., 1986).

The capsular polysaccharide is the most important antigen produced by the yeast. The structure, immunochemistry, and biological properties of cryptococcal polysaccharide have been studied extensively. There are several reasons for this interest in the capsule. First, the capsule is an essential virulence factor. Second, as noted above, capsular serotype is closely tied to the taxonomy of the yeast. Finally, cryptococcal polysaccharide is released by the yeast into body fluids of patients with cryptococcosis. Serological assays for cryptococcal polysaccharide have become an important procedure for diagnosis and evaluation of the prognosis of cryptococcosis.

EXTRACELLULAR POLYSACCHARIDES OF
C. neoformans

Several early studies of *C. neoformans* noted the presence of polysaccharides in culture filtrates of the yeast (Kligman, 1947; Mager and Aschner, 1947; Neill et al., 1949). The first definitive analysis of the composition of these extracellular polysaccharides was reported by Evans and Mehl (1951). This study found an ethanol precipitable polysaccharide that contained galactose, mannose, xylose, and glucuronic acid. Subsequent reports from several laboratories invariably noted the presence of mannose, xylose, and glucuronic acid; however, galactose was a constituent also reported by some groups (Rebers et al., 1958; Blandamer and Danishefsky, 1966; Kozel and Cazin, 1971) but not others (Bhattacharjee et al., 1978, 1979I, 1979b, 1980). Evans and Theriault (1953) found that the ethanol precipitable polysaccharide could be fractionated with heavy metal salts. The crude polysaccharide contained galactose. Galactose was not found in a fraction precipitated with lead acetate, but galactose was found in polysaccharide in the supernatant fluid. These results suggested the presence of at least two extracellular polysaccharides in culture filtrates of *C. neoformans*. Unfortunately, this is an issue that was not pursued by other investigators, and many of the early studies of cryptococcal polysaccharide were undoubtedly done with preparations that contained more than one component.

The number of polysaccharide components in *C. neoformans* culture filtrates was readdressed by Cherniak and co-workers (Cherniak et al., 1980). The polycationic detergent cetyl trimethyl ammonium bromide (CTAB) was used to fractionate crude ethanol precipitates. This procedure produced a CTAB precipitable fraction and a nonprecipitable fraction. The CTAB precipitable fraction comprised the bulk of the extracellular polysaccharide and was composed of mannose, xylose, and glucuronic acid. This fraction, termed the glucuronoxylomannan (GXM), is the major type-specific polysaccharide produced by the yeast.

Precipitation of the GXM by CTAB is due to an ionic interaction between the CTAB and the charged uronic acid found in the acidic cryptococcal polysaccharide. The nonprecipitated fraction was found to contain galactose as a major constituent as well as xylose and mannose. This fraction was termed galactoxylomannan (GalXM). Subsequent fractionation of GalXM by affinity chromatography on concanavalin A (Con A) demonstrated the presence of a fraction that was adherent to Con A and a fraction that did not bind to Con A (Turner et al., 1984). The fraction binding to Con A was found to be a mannoprotein. The unbound fraction contained little protein and was composed of xylose, mannose, galactose, and a small amount of uronic acid.

Accumulated evidence has shown that ethanol precipitates of culture filtrates contain a mixture of at least three heteroglycans. As a consequence, studies on the immunochemistry and composition of unfractionated preparations must be interpreted with caution. The relative contribution of each of these polysaccharides to the immunogenicity of the yeast has not been determined, although it appears certain that most, if not all, of the serotype specificity resides in the major acidic polysaccharide, the GXM. As a consequence, this review will deal with the immunochemistry of the GXM.

BIOLOGICAL PROPERTIES OF CRYPTOCOCCAL POLYSACCHARIDE

Cryptococcal polysaccharides have several biological properties that undoubtedly contribute to the pathogenesis of cryptococcosis. Several studies have shown that acapsular variants of the yeast are markedly deficient in their ability to produce disease in mice. Bulmer et al. (1967) used UV irradiation to induce seven nonencapsulated variants of the yeast. Initially, these mutants were avirulent for mice while the parent strain was virulent. Six of the seven isolates reverted to the encapsulated state and regained varying degrees of virulence. These results suggest that a strong correlation exists between the presence of a capsule and virulence of *C. neoformans*. Kozel and Cazin (1971) found a stable acapsular variant that was similarly avirulent for mice. The specific genetic lesion(s) responsible for defects in capsule production in each of these reports were not known, consequently, the presence of defects in addition to capsule production cannot be excluded. More recent studies by Fromtling et al. (1982) demonstrated that six acapsular strains obtained by chemical mutagenesis were avirulent in mice.

One of the most important biological properties of the cryptococcal capsule is the ability of the capsular polysaccharide to inhibit phagocytosis of the yeast. Bulmer and Sans (1967) found that encapsulated cryptococci were highly resistant to phagocytosis by human neutrophils, whereas phagocytosis was three times more effective with nonencapsulated mutants. The GXM was shown to be the capsular component responsible for inhibiting phagocytosis. Addition of microgram amounts of GXM to nonencapsulated variants completely inhibits phagocytosis by human neutrophils (Bulmer and Sans, 1968) and mouse peritoneal macrophages (Kozel, 1977). Inhibition of phagocytosis was shown to be a consequence of binding of the polysaccharide to the nonencapsulated variants rather than a direct effect of the polysaccharide on the phagocytes (Bulmer and Sans, 1968; Kozel, 1977; Kozel and Hermerath, 1984; Small and Mitchell, 1986).

One or more components of the cryptococcal capsule are potent activators of the alternative complement pathway. Incubation of encapsulated cryptococci in normal human serum leads to deposition of C3 in the cryptococcal capsule (Goren and Warren, 1968; Diamond et al., 1974). Ultrastructural studies using peroxidase-labeled antiserum specific for C3 demonstrated that the capsular surface was the site for deposition of the C3. Further studies have shown that the C3 fragments are largely in the form of iC3b, and the number of bound C3 fragments is $5-10 \times 10^6$ molecules/yeast (Kozel and Pfrommer, 1986). The specific component of the capsule that is responsible for complement activation has not been clearly determined. Studies by Gadebusch (1961) and Diamond et al. (1974) found that incubation of normal human serum with purified GXM produced measurable activation of the complement cascade. Subsequent studies in our laboratory (Laxalt and Kozel, 1979) were unable to demonstrate complement activation by serotype D GXM in concentrations as high as 1 mg/ml. These differences may be due to differences in cryptococcal serotype or to the presence of contaminating polysaccharides in the GXM preparations.

Cryptococcal GXM is poorly immunogeneic in humans and experimental animals. GXM-specific antibody occurs in only a fraction of patients with cryptococcosis, and titers are usually quite low (Diamond and Bennett, 1972; Kozel and Cazin, 1972). Most successful attempts at immunization of experimental animals have used protein conjugates of the polysaccharides as immunogens. High titers of antibody have been produced by immunization with polysaccharide coupled to bovine serum albumin (Kozel and Follette, 1981), bovine gamma globulin (Goren, 1967; Kozel and Cazin, 1974; Shinoda et al., 1980), and sheep erythrocytes (Eckert and Kozel, 1987). At least one study has suggested that cryptococcal polysaccharide is a T-independent antigen (Breen et al., 1982). This may account in part for the poor immunogenicity.

In addition to being poorly immunogenic, cryptococcal polysaccharide will induce immunological unresponsiveness in mice and humans. Mice treated with 100-800 μg of cryptococcal polysaccharide had a reduced ability to produce antibody in response to a normally immunogenic challenge immunization (Kozel et al., 1977). This immunological unresponsiveness was shown to be due to a suppression in the number of antibody producing cells (Murphy and Cozad, 1972; Breen et al., 1982). Henderson et al. (1982) demonstrated that immunological unresponsiveness also occurs in humans.

Cryptococcal GXM persists for extended periods of time in host tissue. This is probably due to the poor immune response to the polysaccharide and to the absence of hydrolytic enzymes capable of degrading the polysaccharide. Studies in mice have shown that GXM will persist in serum for one week. There was a marked accumulation of antigen in the kidney where the antigen could be detected for as long as 14 weeks after intraperitoneal injection of 100 μg polysaccharide (Kozel et al., 1977). Studies by Eng et al. (1986) have suggested that patients with AIDS are particularly deficient in their ability to clear the polysaccharide.

SEROTYPES OF CRYPTOCOCCAL POLYSACCHARIDE

Development of a serotyping scheme for *C. neoformans* has been a difficult task because experimental animals have a poor immune response to the polysaccharide. Typing of *C. neoformans* was first reported by Evans in 1949 (Evans, 1949). Twelve strains isolated from human infections were shown on the basis of reciprocal agglutination to fall into three distinct serotypes (serotypes A, B, and C). Evans examined a larger number of strains in a later report (Evans, 1950) and used agglutination, cross-adsorption of rabbit antisera, and a microscopic "capsular reaction" to confirm the presence of three serotypes and to demonstrate that the capsule is the microbial structure concerned with type specificity. The corresponding role of a soluble polysaccharide in type specificity of *C. neoformans* was demonstrated when an ethanol precipitable extracellular polysaccharide gave the same pattern of reactivity in a precipitin reaction. It was fortuitous that these early studies were done at the University of Southern California because subsequent epidemiological studies have shown that serotypes B and C are found almost exclusively in southern California and in tropical areas of the world (Bennett et al., 1977; Kwon-Chung and Bennett, 1984).

Further studies of serological groupings of *C. neoformans* (Wilson et al., 1968) demonstrated the presence of a fourth serotype (serotype D). Antisera in this study were cross-adsorbed to produce monospecific typing sera. Of the 106 strains examined in this study, 102 isolates were reactive with only one type-specific antiserum. Four were reactive with monospecific antisera specific for serotype A and serotype D, suggesting the presence of a variant serotype. The presence of isolates reactive with both serotype A- and serotype D-specific typing sera was confirmed by Ikeda et al. (1982), who found three serotype A-D isolates among 58 Japanese strains of *C. neoformans*. These studies demonstrate that serotypes A, B, C, and D comprise the major serotypes of *C. neoformans*, and A-D strains represent a variant that is intermediate between serotypes A and D.

Table 1 Structures of the Glucuronoxylomannans of *Cryptococcus neoformans*

Serotype	Structure	References

D	→3)-α-D-Man$p$(1→3)-α-D-Man$p$(1→3)-α-D-Man$p$(1→
	       2 2
	       ↑ ↑
	       1 1
	     β-D-Xyl$p$ β-D-Glc$p$A
		Bhattacharjee et al., 1979

A	→3)-α-D-Man$p$(1→3)-α-D-Man$p$(1→3)-α-D-Man$p$(1→
	      2 2 β-D-Glc$p$A
	      ↑ ↑
	      1 1
	    β-D-Xyl$p$ β-D-Xyl$p$ β-D-Xyl$p$
	                1
	                ↓
	                4
		Merrifield and Stephen, 1980
		Bhattacharjee et al., 1981

B

$$\begin{array}{c}
\rightarrow 3)\text{-}\alpha\text{-D-Man}_p(1\rightarrow 3)\text{-}\alpha\text{ D-Man}_p(1\rightarrow 3)\text{-}\alpha\text{-D-Man}_p(1\rightarrow 3)\text{-}\alpha\text{-D-Man}_p(1\rightarrow \\
\begin{array}{ccc} 2 & 2 & 2 \\ \uparrow & \uparrow & \uparrow \\ 1 & 1 & 1 \end{array} \\
\beta\text{-D-Xyl}_p \quad \beta\text{-D-Xyl}_p \quad \beta\text{-D-Glc}_p\text{A} \\
\begin{array}{c} \beta\text{-D-Xyl}_p \\ 1 \\ \downarrow \\ 4 \end{array} \quad \begin{array}{c} \beta\text{-D-Xyl}_p \\ 1 \\ \downarrow \\ 4 \end{array}
\end{array}$$

Bhattacharjee et al., 1980

C

$$\begin{array}{c}
\rightarrow 3)\text{-}\alpha\text{-D-Man}_p(1\rightarrow 3)\text{-}\alpha\text{ D-Man}_p(1\rightarrow 3)\text{-}\alpha\text{-D-Man}_p(1\rightarrow \\
\begin{array}{ccc} 2 & 2 & 2 \\ \uparrow & \uparrow & \uparrow \\ 1 & 1 & 1 \end{array} \\
\beta\text{-D-Xyl}_p \quad \beta\text{-D-Xyl}_p \quad \beta\text{-D-Glc}_p\text{A}
\end{array}$$

Bhattacharjee et al., 1978

STRUCTURES OF CRYPTOCOCCAL POLYSACCHARIDES

The structures of the major acidic polysaccharides of cryptococcal serotypes A, B, C, and D have been extensively studied (Bhattacharjee et al., 1978, 1979a, 1979b, 1980, 1981; Cherniak et al., 1980; Merrifield and Stephen, 1980). Polysaccharides of all four serotypes contain a backbone of α-(1→3)-linked D-mannopyranosyl residues. Side chains of single xylopyranosyl and glucuronopyranosyl residues are attached to the mannan backbone. Regardless of serotype, the glucuronosyl residue is linked β-(1→2) to the backbone. Xylose residues are linked β-(1→2) to the mannan backbone, and, depending on the serotype, also may be linked β-(1→4) to the backbone. All sugars occur in the pyranose form. Serotypes D and A are the simplest of the polysaccharides, with the mannosyl residues substituted only a O-2. Serotypes B and C are more complex because the mannopyranosyl residues are also substituted with xylopyranosyl residues at O-4 as well as O-2. The four serotypes also differ with regard to the extent of O-acetylation. Serotype D is the most heavily O-acetylated, while serotype C is the least O-acetylated.

The structure of serotype D polysaccharide is shown in Table 1. It can be seen that each repeating unit of three mannosyl residues in the backbone contains one mannopyranosyl residue substituted β-(1→2) with a xylopyranosyl residue, one mannopyranosyl residue substituted β-(1→2) with a glucuronopyranosyl residue, and one unsubstituted mannopyranosyl residue. Serotype D is the most heavily O-acetylated (Table 2). The location of the O-acetyl groups is not known, but the mannose backbone is the most likely possibility (Bhattacharjee et al., 1979).

Two studies of serotype A have shown that the mannan backbone of this polysaccharide is slightly more substituted than serotype D polysaccharide. Like serotype D, serotype A has a repeating unit of three mannopyranosyl residues in the backbone, which contains one mannopyranosyl residue substituted β-(1→2) with a xylopyranosyl residue and one mannopyranosyl residue substituted β-(1→2) with a glucuronopyranosyl residue. The third mannose residue is also substituted β-(1→2) with a xylopyranosyl residue. Serotype A polysaccharide con-

Table 2 O-acetyl Content of Cryptococcal Polysaccharides of the Four Major Serotypes

Serotype	O-acetyl content
A	9.6%, 12% (Kozel and Hermerath, 1984; Merrifield and Stephen, 1980)
B	7.9%, 10.4% (Kozel and Hermerath, 1984; Bhattacharjee et al., 1980)
C	4.0%, 3% (Kozel and Hermerath, 1984; Bhattacharjee et al., 1978)
D	16.5%, 10.3% (Kozel and Hermerath, 1984; Bhattacharjee et al., 1979)

tains a large number of O-acetyl groups, although the precise amount of O-acetylation varies in different reports (Table 2).

Serotype B differs from serotype A because one of the three mannopyranosyl residues in the mannan backbone is doubly substituted, with a glucuronosyl side chain bound β-(1→2) and a xylopyranosyl side chain bound β-(1→4). Available evidence suggests that serotype B polysaccharide is slightly less O-acetylated than serotypes A and D (Table 2).

Serotype C polysaccharide is the most structurally complex of the type-specific polysaccharides and contains the largest number of xylopyranosyl residues. Like serotype B polysaccharide, one mannopyranosyl residue in the backbone is doubly substituted with a glucuronosyl side chain bound to O-2 and a xylopyranosyl residue bound to O-4. A second mannopyranosyl residue in the backbone is doubly substituted with xylopyranosyl side chains bound to O-2 and O-4. The third mannopyranosyl residue is singly substituted with a xylopyranosyl residue bound at O-2. Serotype C polysaccharide contains the least amount of O-acetyl groups. The location of the O-acetyl groups has not been determined, but the mannose backbone is a likely location. It is attractive to suggest that the extensive xylopyranosyl substitution on the backbone limits the amount of O-acetylation that would be possible on the backbone, thus accounting for the apparent inverse reaction between xylose substitution and O-acetylation that is observed with the different serotypes.

The possibility must be raised that cryptococcal polysaccharides may occur as structures other than the four discrete serotypes shown in Table 1. There is increasing evidence for intermediate forms between the four serotypes. Clear evidence exists for an A,D or A/D serotype, which is reactive with polyclonal antisera specific for both serotype A and serotype D. This continuum in structures between the discrete serotypes may account for discrepancies that have occurred in the reported structures of serotype A polysaccharide. Cherniak et al. (1980) examined the structure of the strain NIH 371, a large capsule isolate presumed to be of serotype A. Structural studies showed this strain to have a repeating unit containing five mannopyranosyl residues in the backbone (Table 3). One mannopyranosyl residue was singly substituted β-(1→2) with a glucu-

Table 3 Structure of a Variant Polysaccharide of Serotype A or D

$$\begin{array}{c}
\rightarrow 3)\text{-}\alpha\text{-D-Man}_p(1\rightarrow 3)\text{-}\alpha\text{ D-Man}_p(1\rightarrow 3)\text{-}\alpha\text{-D-Man}_p(1\rightarrow 3)\text{-}\alpha\text{-D-Man}_p(1\rightarrow 3)\text{-}\alpha\text{-D-Man}_p \\
\quad\quad 2 \quad\quad\quad\quad\quad\quad\quad\quad\quad\quad 2 \quad\quad\quad\quad\quad\quad\quad\quad\quad\quad\quad 2 \\
\quad\quad \uparrow \quad\quad\quad\quad\quad\quad\quad\quad\quad\quad \uparrow \quad\quad\quad\quad\quad\quad\quad\quad\quad\quad\quad \uparrow \\
\quad\quad 1 \quad\quad\quad\quad\quad\quad\quad\quad\quad\quad 1 \quad\quad\quad\quad\quad\quad\quad\quad\quad\quad\quad 1 \\
\beta\text{-D-Xyl}_p \quad\quad\quad\quad\quad\quad \beta\text{-D-Xyl}_p \quad\quad\quad\quad\quad\quad\quad \beta\text{-D-Glc}_p\text{A}
\end{array}$$

ronosyl residue; two mannopyranosyl residues were substituted β-$(1\rightarrow2)$ with xylopyranosyl residues; and two mannopyranosyl residues in the backbone were unsubstituted. A comparison of this structure with the structures of serotypes A and D in Table 1 suggests that this variant structure represents an intermediate form between the structures of serotypes A and D.

The apparent anomaly between reported structures of serotype A polysaccharide prompted Bhattacharjee et al. (1981) to reexamine the structure of the acidic polysaccharide produced by strain NIH 371. It was noted that on subculture, the strain produced two distinct colonies that had different characteristics. Subcloning produced two stable populations of cells. Structural analysis of the acidic polysaccharides produced by the two subcultures showed that one isolate, designated strain 371-3, was agglutinated by antiserum specific for serotype A and produced a polysaccharide identical to the serotype A polysaccharide reported by Merrifield and Stephen (1980). The other culture obtained from strain NIH 371, designated 371-a, reacted with antisera specific for serotype D and produced an acidic polysaccharide that was quite similar to the structure reported by Cherniak et al. (1980) for serotype A polysaccharide. The close similarity between serotype D polysaccharide and the structure of 371-a and the reactivity of strain 371-a with serotype D antiserum led the authors to conclude that strain 371-a was an isolate of serotype D. However, the lack of identity between polysaccharide of strain 371-a and the reported structure of serotype D (Bhattacharjee et al., 1979b) raises the possibility that strain 371-a is an intermediate strain between serotypes D and A.

IMMUNOCHEMICAL RELATIONSHIP BETWEEN CRYPTOCOCCAL POLYSACCHARIDES

The immunochemical relationship between the various serotypes closely parallels the known varietal status of *C. neoformans*. Polysaccharides of serotypes A or D are produced by *C. neoformans* var. *neoformans*, whereas isolates of var. *gattii* produce polysaccharides of serotypes B or C. Polysaccharides of serotypes A and D show a close immunochemical relationship to one another, and serotypes B and C are closely related. This immunochemical relationship also correlates well with the known structures of the polysaccharides discussed above. The structures shown in Table 1 predict that the relative degree of cross-reactivity will occur in order from D to A to B to C, i.e., serotype D is most closely related to serotype A and least related to serotype C. Conversely, serotype C is most closely related to serotype B and least related to serotype D. Similarly, serotype B is more closely related to serotypes A and C than it is to serotype D.

The immunochemical relationship between the polysaccharides has been demonstrated by a variety of immunochemical techniques. We used double immuno-

C. neoformans Capsular Polysaccharides

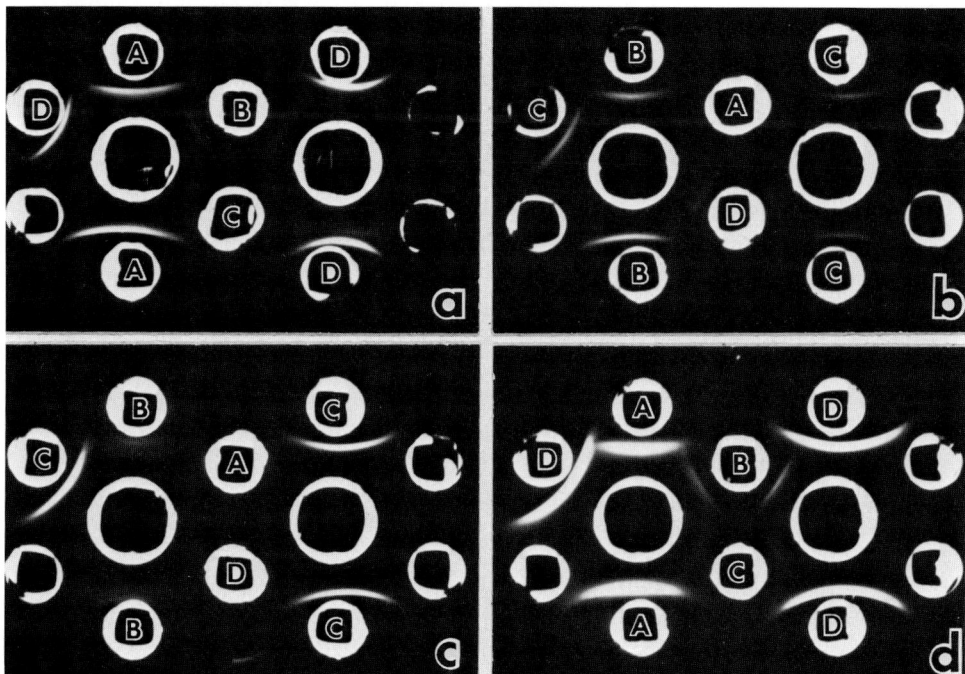

Figure 1 Ouchterlony immunodiffusion reactions. Center wells contain (a) antiserum raised against a methylated bovine serum albumin (MBSA) conjugate of serotype A polysaccharide; (b) antiserum raised against a MBSA conjugate of serotype B polysaccharide; (c) antiserum raised against a MBSA conjugate of serotype C polysaccharide; and (d) antiserum raised against a MBSA conjugate of serotype D polysaccharide. Outer wells contained polysaccharides of the indicated serotypes.

diffusion in gel to examine this relationship. Results using purified cryptococcal polysaccharides and antisera raised against methylated bovine serum albumin (MBSA) conjugates of the four serotypes are shown in Figure 1. Antiserum raised against serotype D polysaccharide showed strong reactivity with polysaccharides of serotypes A and D and weak reactivity with serotypes B and C. Spur formation was evident when the reactivity of the anti-D serum with serotype D polysaccharide was compared with precipitation of serotype A, B, and C polysaccharides. This pattern suggests that type D contains epitopes not found on the remaining serotypes. Antiserum raised against type A polysaccharide showed strong reactivity with polysaccharides of serotypes A and D and weak reactivity with serotypes B and C. Lines of identity were evident between polysaccharides of A and D, indicating a close relationship between these polysaccharides. The lack of spur formation between serotypes A and D using type A antiserum is surprising considering numerous reports that serotype A contains determinants not found on serotype D (Wilson et al., 1968; Ikeda et al., 1982). This is undoubtedly due to our use of MBSA conjugates of the polysaccharide as immunogens rather than whole cells.

Antiserum raised against a MBSA conjugate of serotype B polysaccharide showed strong reactivity with polysaccharides of serotypes B and C, but only very weak reactivity was observed with serotype D and A polysaccharides. A line of identity was observed between serotypes B and C, indicating that antiserum raised against the MBSA conjugate of serotype B could not distinguish between these two serotypes. Finally, antiserum raised against serotype C polysaccharide showed strong reactivity with polysaccharide of serotype C, weak reactivity with serotype B, and no reactivity with serotype A and D polysaccharides. Spur formation between polysaccharides of serotypes C and B indicates that antisera raised against the MBSA conjugate of serotype C recognizes epitopes on type C polysaccharides that are not present on serotype B. Once again, these results differ from results obtained when whole cells are used as immunogens. Immunization with whole serotype B or C cells produces antibodies that are reactive with epitopes unique to each respective serotype.

Ikeda et al. (1982) reported a pattern for the distribution of epitopes on the four major serotypes that accounts for both the uniqueness of each serotype as well as the observed cross-reactivity between the serotypes. Polyclonal antiserum was produced in this study by immunization of rabbits with whole cryptococcal cells. Cross-adsorption of the antisera with cells of the four serotypes demonstrated the presence of at least eight distinct epitopes. Some of these epitopes were unique to a single serotype, others were found on two or more serotypes. One putative determinant was common to all four serotypes. An examination of the distribution of these epitopes (Table 4) shows a close correlation between the various factors and the observed cross-reactivity between the polysaccharides. For example, serotype D shares three factors with serotype A, two factors with serotype B, and one factor with serotype C. Thus, it is not surprising that antiserum raised against serotype D shows strong reactivity with serotype A polysaccharide and only weak reactivity with polysaccharides of serotypes B or C (Fig. 1). Similar comparisons can be made between the remaining serotypes in which the presence of epitopes on polysaccharides of each serotype (Table 4) can explain the cross-reactivity observed between the serotypes (Fig. 1).

The results reported by Ikeda et al. (1982) would suggest that it should be possible to produce monoclonal antibodies that have reactivity patterns similar to the reactivity of the various factor sera. Initial studies by Dromer et al. (1987) and studies from our laboratory (Eckert and Kozel, 1987) indicate that this is indeed the case. One monoclonal antibody, designated MAb E_1 (Dromer et al.,

Table 4 Association of Specific Antigenic Determinants with Polysaccharides of the Four Major Cryptococcal Serotypes

Antigenic factor or determinant (Ikeda et al., 1982)	Serotype				Possible MAb reactive with each antigenic factor
	D	A	B	C	
Factor 1	+	+	+	+	MAb 439 (Eckert and Kozel, 1987)
Factor 2	+	+	+	-	MAb 386 and MAb 1255 (Eckert and Kozel, 1987)
Factor 3	+	+	-	-	MAb 302 (Eckert and Kozel, 1987)
Factor 4	-	-	+	+	None reported
Factor 5	-	-	+	-	None reported
Factor 6	-	-	-	+	None reported
Factor 7	-	+	-	-	MAb E_1 (Dromer et al., 1987)
Factor 8	+	-	-	-	None reported

1987), was reactive primarily with serotype A, corresponding with the distribution of the putative factor 7 determinant. Similarly, we have reported monoclonal antibodies that have reactivities that closely parallel the distribution of the factor 1 determinant (MAb 439), the factor 2 determinant (MAb 386 and MAb 1255), and the factor 3 determinant (MAb 302). It is notable that one of the monoclonal antibodies reactive with serotypes A, B, and D was produced from mice immunized with serotype A polysaccharide (MAb 1255), whereas, another monoclonal antibody with the same reactivity pattern (MAb 386) was produced from mice immunized with serotype D polysaccharide. This is a result that could be predicted by the distribution of epitopes suggested by Ikeda et al. (1982). These results also demonstrate the utility of monoclonal antibodies for epitope mapping of cryptococcal polysaccharides.

A caveat must be applied to interpretation of data generated from studies of polyclonal antisera. Each of the "factor" sera described by Ikeda et al. (1982) may contain polyclonal antibodies reactive with more than one epitope. That is, there could be more than one epitope that is unique only, for example, with serotype A (factor 7). Similarly, there could be more than one epitope that is shared by all four serotypes (factor 1). Thus, a monoclonal antibody that has a reactivity that parallels one of the factor sera may recognize one, but not necessarily all, of the epitopes recognized by the factor sera. A clear correlation between the epitope specificities of monoclonal antibodies and the apparent specificity of sera produced by cross-adsorption of polyclonal antisera (factor sera) will require competitive binding studies. To date, such studies have not been reported. Nevertheless, it is attractive to compare the observed reactivity of these "factor sera" and their corresponding monoclonal antibodies.

Results from several studies suggest that the antigen selected for use as an immunogen may markedly influence the repertoire of epitopes recognized by the antisera that are produced. Whole cells appear to induce antibodies reactive with epitopes that are unique to each serotype (Wilson et al., 1968; Ikeda et al., 1982). Our results indicate that immunization of rabbits with MBSA conjugates of serotype D and C polysaccharides will induce antibodies that are uniquely reactive with each respective serotype (Fig. 1). In contrast, immunization with MBSA conjugates of serotypes A or B did not induce antibodies that are reactive only with serotypes A and B. Immunization of mice with purified polysaccharide of serotype A allowed the production of a hybridoma secreting antibodies with strong reactivity for serotype A polysaccharide and very weak reactivity with the remaining serotypes. Immunization of mice with serotype A polysaccharide coupled to sheep erythrocytes allowed the production of hybridomas that secreted antibodies reactive with several serotypes, but hybridomas reactive only with serotype A were not observed (Eckert and Kozel, 1982). These results suggest that the manner in which the polysaccharides are presented will markedly influence the specificity of the antibodies that are produced. In particular, it appears that conjugation of polysaccharide to carriers diminishes the ability to produce antibodies reactive with epitopes found only on serotypes A and B.

STRUCTURES OF IMMUNODETERMINANTS ON CRYPTOCOCCAL POLYSACCHARIDES

Few studies have attempted to assign specific structures to individual epitopes. As a consequence, one can only speculate as to the structures responsible for the mosaic of epitopes on cryptococcal polysaccharides. Most available evidence indicates that the immunodeterminants occur only on the side chains, with little or no reactivity of antibody with the mannan backbone. Cherniak et al. (1980) produced a soluble polyalcohol of serotype A polysaccharide by use of periodate oxidation followed by reduction with $NaBH_4$. This procedure will cleave the xylopyranosyl and glucuropyranosyl side chains. In addition, the alkaline $NaBH_4$ is likely to hydrolyze the O-acetyl groups. The resulting polyalcohol product was not reactive with latex particles coated with anticryptococcal antibody. Selective removal of the xylopyranosyl and glucuronopyranosyl side chains on serotype D polysaccharide by Smith degradation generates the mannan backbone without the attached side chains. This product was not reactive with polyclonal antiserum to either serotype C or D polysaccharides (Bhattacharjee et al., 1979). Once again, the Smith degradation involves exposure of the polysaccharide to $NaBH_4$, consequently, O-acetyl groups are probably absent from the mannan backbone. The lack of reactivity of polyclonal antiserum with the mannose backbone was further confirmed when Bhattacharjee et al. (1983) demonstrated that methyl α-D-mannopyranoside would not appreciably inhibit the reaction between radiolabeled serotype D polysaccharide and polyclonal antiserum. Our studies with monoclonal antibodies have produced results similar to those reported for polyclonal antibodies. In no instance have we been able to identify a monoclonal antibody that is reactive with the mannose backbone produced by Smith degradation of serotype A or D polysaccharides (Eckert and Kozel, 1987).

The O-acetyl group is a significant epitope on the polysaccharide. Removal of O-acetyl groups from serotype D polysaccharide effected a partial reduction in the reactivity of the polysaccharide with rabbit antiserum to type D cells (Bhattacharjee et al., 1979). Studies in our laboratory have shown that it is possible to isolate antibodies that are specific for the O-acetyl group from polyclonal antiserum raised against a MBSA conjugate of serotype D polysaccharide (Kozel and Gotschlich, 1982). Whole cells of serotype D were de-O-acetylated by alkaline hydrolysis and used to adsorb polyclonal antiserum. This procedure removed all antibodies other than those reactive with O-acetyl groups. This adsorbed antiserum was reactive with untreated type D polysaccharide and carboxyl-reduced polysaccharide, but not de-O-acetylated type D polysaccharide. We have also reported the production of a monoclonal antibody (MAb 302; Eckert and Kozel, 1987) that was raised in response to immunization with serotype D polysaccharide. This monoclonal antibody was reactive with polysaccharides of serotypes A and D. The lack of reactivity with de-O-acetylated polysaccharide suggested that MAb 302 was specific for one of the O-acetyl groups on these polysaccharides. The lack of reactivity of MAb 302 with serotypes B and C is not surprising considering the limited level of O-acetylation on these serotypes (Kozel and Hermerath, 1984). Additional evidence for the importance of O-acetylation in the antigenicity of serotype A polysaccharide was provided by Cherniak et al. (1980), who found that de-O-acetylated polysaccharide had a 50% reduced ability to inhibit agglutination of latex particles coated with agglutinating antibody.

Few studies have examined the role of the xylosyl side chain in the antigenicity of cryptococcal polysaccharides. Bhattacharjee et al. (1983) found that rabbit antiserum to serotype D polysaccharide contained antibodies reactive with the xylopyranosyl residue. An affinity column was prepared in which oligosaccharides of serotype D polysaccharide were coupled to a gel. Antibodies were found which bound to the column and were eluted with methyl-β-D-xyloside. Antibody isolated in this manner was not able to produce precipitin lines in Ouchterlony diffusion against purified cryptococcal polysaccharide; however, the affinity-purified anti-xylosyl antibody was shown to bind to cryptococcal polysaccharide in a Farr assay. The presence of anti-xylosyl antibody was further demonstrated when methyl-β-D-xyloside was shown to effect a 40% inhibition in binding between cryptococcal polysaccharide and anti-cryptococcal antibody. The authors suggested that the inability of the anti-xylopyranosyl antibody to precipitate polysaccharide might occur because the affinity of antibody for a single neutral sugar might not be sufficient to cause precipitation. To date, no monoclonal antibodies have been reported which are specifically reactive with the xylopyranosyl residue. Studies of anti-xylopyranosyl monoclonal antibodies would substantially improve our understanding of the role of this epitope in the immunochemistry of cryptococcal polysaccharides. It should also be noted that serotypes B and C have a high density of xylosyl substitution, and xylose side chains might play a much larger role in the antigenicity of these serotypes.

Several studies have shown the importance of the glucuronopyranosyl side chains in the antigenicity of cryptococcal polysaccharides. Reduction of the carboxyl group has been shown to reduce the reactivity of serotype D polysaccharide with rabbit antiserum (Bhattacharjee et al., 1979). In addition, the reactivity of type C antiserum with serotype D polysaccharide was markedly inhibited by carboxyl reduction of serotype D polysaccharide. This suggests that the putative epitope that is shared by different cryptococcal serotypes either includes the carboxyl group or is dependent on the presence of the carboxyl group for its conformation. The importance of the carboxyl group for the reactivity of serotype A polysaccharide was demonstrated by Cherniak et al. (1980), who found that carboxyl-reduced polysaccharide had a 75% reduced ability to agglutinate latex particles coated with agglutinating antibody.

Two complementary procedures have been used to isolate antibodies reactive with the glucuronopyranosyl residues. Bhattacharjee et al. (1983) prepared an affinity column in which β-D-glucuronopyranoside was coupled to an agarose gel. Rabbit antiserum to serotype D was passed over the column, and the bound antibody was eluted with ammonium methyl β-D-glucopyranosido uronate. Antibody that was affinity purified in this manner was reactive in Ouchterlony diffusion with polysaccharides from serotypes D and A. As an alternative approach to isolation of antibodies specific for the glucouronopyranosyl side chain, we prepared carboxyl-reduced whole yeast cells by chemical modification (Taylor and Conrad, 1972) of the carboxyl group (Kozel and Gotschlich, 1982). These modified cells were used as a solid phase adsorbent to remove all antibodies other than those reactive with the carboxyl group. The unabsorbed antibodies were reactive with unmodified and de-O-acetylated cryptococcal polysaccharide but not carboxyl-reduced polysaccharide.

An intact carboxyl group on cryptococcal polysaccharide was essential for the reactivity of three monoclonal antibodies produced against serotype A or D polysaccharides (Eckert and Kozel, 1987). This may have been due to the presence of the carboxyl group in the reactive epitope, or it may indicate a key role for the carboxyl group in maintaining the conformation of the polysaccharide. This latter explanation may be a key factor in the reactivity of MAb 302. The reactivity of this antibody with polysaccharide was dependent on the presence of the O-acetyl group; however, carboxyl-reduced polysaccharide also failed to react withMAb 302. The fact that all three antibodies (MAb 302, 439, and 1255) were reactive with epitopes shared by at least two serotypes strengthens the argument that the carboxyl group is an important factor in cross-reactivity.

ASSOCIATION BETWEEN SEROTYPE AND EPIDEMIOLOGY OF CRYPTOCOCCOSIS

The A and D serotypes differ markedly from the B and C sterotypes with regard to their natural habitat and geographic distribution (Bennett et al., 1977; Kwon-Chung and Bennett, 1984). The most prevalent serotype in the United States is serotype A. Serotypes B and C (*C. neoformans* var. *gattii*) are relatively infrequent causes of human disease in North America. Most North American isolates of serotypes B and C have been from Southern California. Indeed, it appears that serotypes B and C are found only in tropical and subtropical climates. Serotype D is rarely isolated in the United States, but it appears to be relatively common in European countries such as Italy, Denmark, Germany, and Switzerland.

Early epidemiological studies found that *C. neoformans* was isolated with some frequency from pigeon excreta or from soil contaminated with avian feces, and this was believed to be the primary source of infection. More recent studies have shown that *C. neoformans* var. *gattii* is not found in pigeon droppings, and, to date, this variety has been cultured only from clinical samples. The environmental niche for var. *gattii* has yet to be determined. Recent studies have suggested that serotype may play some role in the occurrence of cryptococcosis in AIDS. To date, most, if not all, isolates of *C. neoformans* from AIDS patients have been var. *neoformans* (Rinaldi et al., 1986; Bottone et al., 1987). Even geographic areas such as Southern California, where the incidence of var. *gattii* (serotypes B and C) may be as high as 40%, are reporting that all isolates of *C. neoformans* from patients with AIDS are var. *neoformans* (Shimizu et al., 1986). Similar results have been reported from Central Africa where all isolates from patients with AIDS were shown to be var. *neoformans* (Swinne et al., 1986). Studies of cultures collected before the occurrence of AIDS found that 100% of clinical isolates from Central Africa were var. *gattii* (Kwon-Chung and Bennett, 1984). These results suggest that a dramatic shift in the epidemiology of cryptococcis has occurred in association with the advent of AIDS. The reason(s) for this exclusive occurrence of var. *neoformans* in patients with AIDS is not known.

ROLE OF CRYPTOCOCCAL POLYSACCHARIDES IN DIAGNOSIS OF CRYPTOCOCCOSIS

Assay of body fluids for the presence of cryptococcal polysaccharide has become a reliable diagnostic tool for cryptococcal meningitis. As with the many encapsulated bacteria that cause meningitis, cryptococci shed capsular material into the cerebral spinal fluid (CSF). This excess antigen can be detected by a variety of procedures including immunoprecipitation (Neill et al., 1951), complement fixation (Neill et al.,.1951; Bennett et al., 1964), passive hemagglutination inhibition (Kozel and Cazin, 1972), counterimmunoelectrophoresis (Maccani, 1977), latex agglutination (Bloomfield et al., 1963), and enzyme immunoassay (Scott et al., 1980). High sensitivity, ease of use, and commercial availability have made latex agglutination a routine procedure for the diagnosis of cryptococcosis. Cryptococcal antigen can be detected in CSF in patients with negative India-ink preparations and cultures (Goodman et al., 1971; Gordon and Vedder, 1966). A high antigen titer is presumed to be caused by a large antigen load and has been correlated with a poor prognosis (Diamond and Bennett, 1974). In contrast, a reduction in antigen titer with treatment has been correlated with a favorable prognosis (Gordon and Vedder, 1966; Diamond and Bennett, 1974).

Few problems have been encountered in the use of the latex agglutination assay. The primary difficulty has been the occurrence of false positive results due to the presence of interference factors, primarily rheumatoid factor. A study of 9001 CSF samples and 18,510 sera tested for cryptococcal antigen yielded a 1% false positive rate for CSF and a 2.4% false positive rate in sera (Stockman and Roberts, 1982). Treatment of clinical samples with dithiothreitol (Gordon and Lapa, 1974), EDTA-heat extraction (Eng and Person, 1981), or pronase (Stockman and Roberts, 1982) have been used to eliminate interference factors. The latex agglutination assay has proven to be remarkably specific for cryptococcal polysaccharide, however, sera from patients with infection by *Trichosporon beigelii* have been shown to give positive results in the assay (McManus et al., 1985).

Cryptococcosis in the AIDS patient may pose new problems for interpretation of serological assays for cryptococcal polysaccharide. Dissemination occurs much more frequently in the patient with AIDS. The frequency of disseminated infection in patients with AIDS was found to be 50% in one retrospective study (Eng et al., 1986). This suggests that the ability to reliably detect cryptococcal polysaccharide in body fluids other than CSF may become increasingly important. It is noteworthy that remarkably high titers of cryptococcal polysaccharide have been observed in sera from some patients with AIDS who have disseminated cryptococcosis. Titers as high as 1:2,000,000 have been reported (Eng et al., 1986). Titers of this magnitude were not observed with any frequency prior to the advent of AIDS. It has been suggested that patients with AIDS have high serum antigen levels due to a defect in antigen elimination rather than due to continued multiplication of the yeast during therapy (Eng et al., 1986). This is a hypothesis that will require further study. If this is in fact the case, it will be difficult to use serum antigen levels as indicators of patient prognosis.

AREAS FOR FUTURE STUDY

Cryptococcal polysaccharides offer rich opportunities for future study. Structural features of the polysaccharide that account for the biological properties of the molecule are not known. The molecular mechanism(s) for assembly of the polysaccharide into the macromolecular capsule offer an opportunity to address the entire subject of microbial capsular architecture. The genetics of capsule synthesis is in the early stages of study. Knowledge of the regulation of capsule synthesis at a genetic level will contribute greatly to our understanding of strain variation in capsule formation and phenotypic variation in capsule size. Further studies are needed to elucidate the fine structure of the various serotypes. In particular, the sites for O-acetylation have yet to be determined. A clear understanding of the structural determinants of antigenicity will allow improvements in the construction of serological assays for cryptococcal polysaccharides.

The unique aspects of cryptococcosis in the patients with AIDS provide several new and unexpected areas for study. The exclusive occurrence of var. *neoformans* in the AIDS patient requires explanation. A corollary to this question is a determination of the environmental niche of var. *gattii*. Diagnosis of cryptococcosis in the patient with AIDS poses new problems in the serodiagnosis of cryptococcosis. Development of new and sensitive assays for the detection of cryptococcal polysaccharide in body fluids other than CSF will become increasingly important. The occurrence of unusually high levels of antigenemia requires further study. Finally, improved serological methods are needed for assessing the effectiveness of antimicrobial therapy in the AIDS patient with cryptococcosis.

ACKNOWLEDGMENTS

Portions of this work were supported by Public Health Service grants AI-14209 and AI-24357 from the National Institute for Allergy and Infectious Diseases.

REFERENCES

Aulakh, H. S., Straus, S. E., and Kwon-Chung, K. J. (1981). Genetic relatedness of *Filobasidiella neoformans* (*Cryptococcus neoformans*) and *Filobasidiella bacillispora* (*Cryptococcus bacillisporus*) as determined by deoxyribonucleic acid base composition and sequence homology studies. *Int. J. Syst. Bacteriol.* *31*:97-103.

Bennett, J. E., Hasenclever, H. F., and Tynes, B. S. (1964). Detection of cryptococcal polysaccharide in serum and spinal fluid: value in diagnosis and prognosis. *Trans. Assoc. Amer. Physicians 77*:145-150.

Bennett, J. E., Kwon-Chung, K. J., and Howard, D. H. (1977). Epidemiologic differences among serotypes of *Cryptococcus neoformans*. *Amer. J. Epidemiol. 105*:582-586.

Bhattacharjee, A. K., Bennett, J. E., Bundle, D. R., and Glaudemans, C. P. J. (1983). Anticryptococcal type D antibodies raised in rabbits. *Mol. Immunol. 20*:351-359.

Bhattacharjee, A. K., Bennett, J. E., and Glaudemans, C. P. J. (1984). Capsular polysaccharides of *Cryptococcus neoformans*. *Rev. Infect. Dis.* 6:619-624.

Bhattacharjee, A. K., Kwon-Chung, K. J., and Glaudemans, C. P. J. (1978). On the structure of the capsular polysaccharide from *Cryptococcus neoformans* serotype C. *Immunochemistry 15*:673-679.

Bhattacharjee, A. K., Kwon-Chung, K. J., and Glaudemans, C. P. J. (1979a). On the structure of the capsular polysaccharide from *Cryptococcus neoformans* serotype C-II. *Mol. Immunol. 16*:531-532.

Bhattacharjee, A. K., Kwon-Chung, K. J., and Glaudemans, C. P. J. (1979b). The structure of the capsular polysaccharide from *Cryptococcus neoformans* serotype D. *Carb. Res. 73*:183-192.

Bhattacharjee, A. K., Kwon-Chung, K. J., and Glaudemans, C. P. J. (1980). Structural studies on the major, capsular polysaccharide from *Cryptococcus bacillisporus* serotyp B. *Carb. Res. 82*:103-111.

Bhattacharjee, A. K., Kwon-Chung, K. J., and Glaudemans, C. P. J. (1981). Capsular polysaccharides from a parent strain and from a possible mutant strain of *Cryptococcus neoformans* serotype A. *Carb. Res. 95*:237-248.

Blandamer, A., and Danishefsky, I. (1966). Investigations on the structure of the capsular polysaccharides from *Cryptococcus neoformans* type B. *Biochim. Biophys. Acta 117*:305-313.

Bloomfield, N., Gordon, M. A., and Elmendorf, D. F., Jr. (1963). Detection of *Cryptococcus neoformans* antigen in body fluids by latex particle agglutination. *Proc. Soc. Exp. Biol. Med. 114*:64-67.

Bottone, E. J., Salkin, I. F., Hurd, N. J., and Wormser, G. P. (1987). Serogroup distribution of *Cryptococcus neoformans* in patients with AIDS. *J. Infect. Dis. 156*:242.

Breen, J. F., Lee, I. C., Vogel, F. R., and Friedman, H. (1982). Cryptococcal capsular polysaccharide-induced modulation of murine immune responses. *Infect. Immun. 36*:47-51.

Bulmer, G. S., Sans, M. D., and Gunn, C. M. (1967). *Cryptococcus neoformans* I. Nonencapsulated mutants. *J. Bacteriol. 94*:1475-1479.

Bulmer, G. S., and Sans, M. D. (1967). *Cryptococcus neoformans* II. Phagocytosis by human leukocytes. *J. Bacteriol. 94*:1480-1483.

Bulmer, G. S., and Sans, M. D. (1968). *Cryptococcus neoformans* III. Inhibition of phagocytosis. *J. Bacteriol. 95*:5-8.

Cherniak, R., Reiss, E., Slodki, M. E., Plattner, R. D., and Blumer, S. O. (1980). Structure and antigenic activity of the capsular polysaccharide of *Cryptococcus neoformans* serotype A. *Mol. Immunol. 17*:1025-1032.

Cherniak, R., Reiss, E., and Turner, S. H. (1982). A galactoxylomannan antigen of *Cryptococcus neoformans* serotype A. *Carb. Res. 103*:239-250.

Diamond, R. D., and Bennett, J. E. (1974). Prognostic factors in cryptococcal meningitis. *Ann. Inter. Med. 80*:176-181.

Diamond, R. D., May, J. E., Kane, M. A., Frank, M. M., and Bennett, J. E. (1974). The role of the classical and alternate complement pathways in host defenses against *Cryptococcus neoformans* infection. *J. Immunol. 112*:2260-2270.

Diamond, R. D., Root, R. K., and Bennett, J. E. (1972). Factors influencing killing of *Cryptococcus neoformans* by human leukocytes in vitro. *J. Infect. Dis. 125*:367-376.

Dromer, F., Salamero, J., Contrepois, A., Carbon, C., and Yeni, P. (1987). Production, characterization, and antibody specificity of a mouse monoclonal antibody reactive with *Cryptococcus neoformans* capsular polysaccharide. *Infect. Immun. 55*(3):742-748.

Eng, R. H. K., Bishburg, E., Smith, S. M., and Kapila, R. (1986). Cryptococcal infections in patients with acquired immune deficiency syndrome. *Amer. J. Med. 81*:19-23.

Eng, R. H. K., and Person, A. (1981). Serum cryptococcal antigen determination in the presence of rheumatoid factor. *J. Clin. Microbiol. 14*:700-702.

Evans, E. E. (1950). The antigenic composition of *Cryptococcus neoformans*. I. A serologic classification by means of the capsular and agglutination reactions. *J. Immunol. 64*:423-430.

Evans, E. E., and Mehl, J. W. (1951). A qualitative analysis of capsular polysaccharides from *Cryptococcus neoformans* by filter paper chromatography. *Science 114*:10-11.

Evans, E. E., and Theriault, R. J. (1953). The antigenic composition of *Cryptococcus neoformans* IV. The use of paper chromatography for following purification of the capsular polysaccharide. *J. Bacteriol. 65*:571-577.

Fromtling, R. A., Shadomy, H. J., and Jacobson, E. S. (1982). Decreased virulence in stable, acapsular mutants of *Cryptococcus neoformans*. *Mycopathologia 79*:23-29.

Gadebusch, H. H. (1961). Natural host resistance to infection with *Cryptococcus neoformans*. I. The effect of the properdin system on the experimental disease. *J. Infect. Dis. 109*:147-153.

Goodman, J. S., Kaufman, L., and Koenig, M. G. (1971). Diagnosis of cryptococcal meningitis. Value of immunologic detection of cryptococcal antigen. *N. Eng. J. Med. 285*:434-436.

Gordon, M. A., and Lapa, E. W. (1974). Elimination of rheumatoid factor in the latex test for cryptococcosis. *Amer. J. Clin. Pathol. 61*:488-494.

Gordon, M. A., and Vedder, D. K. (1966). Serologic tests in diagnosis and prognosis of cryptococcosis. *J. Amer. Med. Assoc. 197*:961-967.
Goren, M. B. (1967). Experimental murine cryptococcosis: effect of hyperimmunization to capsular polysaccharide. *J. Immunol. 98*:914-922.
Goren, M. B., and Warren, J. (1968). Immunofluorescence studies of reactions at the cryptococcal capsule. *J. Infect. Dis. 118*:215-229.
Henderson, D. K., Bennett, J. E., and Huber, M. A. (1982). Long-lasting specific immunologic unresponsiveness associated with cryptococcal meningitis. *J. Clin. Invest. 69*:1185-1190.
Ikeda, R., Shinoda, T., Fukazawa, Y., and Kaufman, L. (1982). Antigenic characterization of *Cryptococcus neoformans* serotypes and its application to serotyping of clinical isolates. *J. Clin. Microbiol. 16*:22-29.
Kligman, A. M. (1947). Studies of the capsular substance of *Torula histolytica* and the immunologic properties of Torula cells. *J. Immunol. 57*:395-401.
Kovacs, J. A., Kovacs, A. A., Polis, M., Wright, W. C., Gill, V. J., Tuazon, C. U., Gelmann, E. P., Lane, H. C., Longfield, R., Overturf, G., Macher, A. M., Fauci, A. S., Parrillo, J. E., Bennett, J. E., Masur, H. (1985). Cryptococcosis in the acquired immunodeficiency syndrome. *Ann. Intern. Med. 103*:533-538.
Kozel, T. R. (1977). Non-encapsulated variant of *Cryptococcus neoformans*. II. Surface receptors for cryptococcal polysaccharide and their role in inhibition of phagocytosis by polysaccharide. *Infect. Immun. 16*:99-106.
Kozel, T. R., and Cazin, J., Jr. (1971). Non-encapsulated variant of *Cryptococcus neoformans*. I. Virulence studies and characterization of soluble polysaccharide. *Infect. Immun. 3*:287-294.
Kozel, T. R., and Cazin, J., Jr. (1972). Immune response to *Cryptococcus neoformans* soluble polysaccharide. I. Serological assay for antigen and antibody. *Infect. Immun. 5*:35-41.
Kozel, T. R., and Cazin, J., Jr. (1974). Induction of humoral antibody response by soluble polysaccharide of *Cryptococcus neoformans: Mycopathol. Mycol. Appl. 54*:21-30.
Kozel, T. R., and Follette, J. L. (1981). Opsonization of encapsulated *Cryptococcus neoformans* by specific anti-capsular antibody. *Infect. Immun. 31*: 978-984.
Kozel, T. R., and Gotschlich, E. C. (1982). The capsule of *Cryptococcus neoformans* passively inhibits phagocytosis of the yeast macrophages. *J. Immunol. 129*:1675-1680.
Kozel, T. R., Gulley, W. F., and Cazin, J., Jr. (1977). Immune response to *Cryptococcus neoformans* soluble polysaccharide: immunological unresponsiveness. *Infect. Immun. 18*:701-707.
Kozel, T. R., and Hermerath, C. A. (1984). Binding of cryptococcal polysaccharide to *Cryptococcus neoformans. Infect. Immun. 43*:879-886.
Kozel, T. R., Highison, B., and Stratton, C. J. (1984). Localization on encapsulated *Cryptococcus neoformans* of serum components opsonic for phagocytosis by macrophages and neutrophils. *Infect. Immun. 43*:574-579.

Kozel, T. R., and Pfrommer, G. S. T. (1986). Activation of the complement system by *Cryptococcus neoformans* leads to binding of iC3b to the yeast. *Infect. Immun.* 52:1-5.

Kwon-Chung, K. J. (1975). Description of a new genus, *Filobasidiella*, the perfect state of *Cryptococcus neoformans*. *Mycologia* 67:1197-1200.

Kwon-Chung, K. J., and Bennett, J. E. (1984). Epidemiologic differences between the two varieties of *Cryptococcus neoformans*. *Amer. J. Epidemiol.* 120:123-130.

Kwon-Chung, K. J., Polacheck, I., and Bennett, J. E. (1982). Improved diagnostic medium for separation of *Cryptococcus neoformans* var. *neoformans* (serotypes A and D) and *Cryptococcus neoformans* var. *gattii* (serotypes B and C). *J. Clin. Microbiol.* 15:535-537.

Laxalt, K. A., and Kozel, T. R. (1979). Chemotaxigenesis and activation of the alternative complement pathway by encapsulated and non-encapsulated *Cryptococcus neoformans*. *Infect. Immun.* 26:435-440.

Maccani, J. E. (1977). Detection of cryptococcal polysaccharide using counter-immunoelectrophoresis. *Amer. J. Clin. Pathol.* 68:39-44.

Mager, J., and Aschner, M. (1947). Biological studies on capsulated yeasts. *J. Bacteriol.* 53:283-295.

McManus, E. J., Bozdech, M. J., and Jones, J. M. (1985). Role of the latex agglutination test for cryptococcal antigen in diagnosing disseminated infections with *Trichosporon beigelli*. *J. Infect. Dis.* 151:1167-1169.

Murphy, J. W., and Cozad, G. C. (1972). Immunological unresponsiveness induced by cryptococcal capsular polysaccharide assayed by the hemolytic plaque technique. *Infect. Immun.* 5:896-901.

Neill, J. M., Castillo, C. G., Smith, R. H., and Kapros, C. E. (1949). Capsular reactions and soluble antigens of *Torula histolytica* and of *Sporotrichum schenckii*. *J. Exp. Med.* 89:93-106.

Neill, J. M., Sugg, J. Y., and McCauley, D. W. (1951). Serologically reactive material in spinal fluid, blood, and urine from a human case of cryptococcosis (torulosis). *Proc. Soc. Exp. Biol. Med.* 77:775-778.

Rebers, P. A., Barker, S. A., Heidelberger, M., Dische, Z., and Evans, E. E. (1958). Precipitation of the specific polysaccharide of *Cryptococcus neoformans* A by type II and XIV antipneumococcal sera. *J. Amer. Chem. Soc.* 80:1135-1137.

Rinaldi, M. G., Drutz, D. J., Howell, A., Sande, M. A., Wofsy, C. B., and Haldey, W. K. (1986). Serotypes of *Cryptococcus neoformans* in patients with AIDS. *J. Infect. Dis.* 153:642.

Scott, E. N., Muchmore, H. G., and Felton, F. G. (1980). Comparison of enzyme immunoassay and latex agglutination methods for detection of *Cryptococcus neoformans* antigen. *Amer. J. Clin. Pathol.* 73:790-794.

Shimizu, R. Y., Howard, D. H., and Clancy, M. N. (1986). The variety of *Cryptococcus neoformans* in patients with AIDS. *J. Infect. Dis.* 154:1042.

Shinoda, T., Ikeda, R., Nishikawa, A., and Fukazawa, Y. (1980). The serological, chemical and physicochemical analyses of cryptococcal capsular polysaccharides. *Jpn. J. Med. Mycol.* 21:230-238.

Small, J. M., and Mitchell, T. G. (1986). Binding of purified and radioiodinated capsular polysaccharides from *Cryptococcus neoformans* serotype A strains to capsule-free mutants. *Infect. Immun. 54*:742-750.

Stockman, L., and Roberts, G. D. (1982). Specificity of the latex text for cryptococcal antigen: a rapid, simple method for eliminating interference factors. *J. Clin. Microbiol. 16*:965-967.

Swinne, D., Nkurikiyinfura, J. B., and Muyembe, T. L. (1986). Clinical isolates of *Cryptococcus neoformans* from Zaire. *Eur. J. Clin. Microbiol. 5*:50-51.

Taylor, R. L., and Conrad, H. E. (1972). Stoichiometric depolymerization of polyuronides and glycosaminoglycuronans to monosaccharides following reduction of their carbodiimide-activated carboxyl groups. *Biochemistry 11*: 1383-1388.

Turner, S. H., Cherniak, R., and Reiss, E. (1984). Fractionation and characterization of galactoxylomannan from *Cryptococcus neoformans. Carb. Res. 125*:343-349.

Wilson, D. E., Bennett, J. E., and Bailey, J. W. (1968). Serologic grouping of *Cryptococcus neoformans. Proc. Soc. Exp. Biol. Med. 127*:820-823.

4
Antigenic Structure of *Aspergillus* Species

VERONICA M. HEARN and DONALD W. MACKENZIE
Mycological Reference Laboratory, Central Public Health Laboratory, London, England

INTRODUCTION

Species of *Aspergillus* synthesize a wide range of complex antigenic molecules. Some serve as functional elements of the organism; others enter the growth medium by active or passive transfer during the maturation process (Kim and Chaparas, 1978; Kauffman and de Vries, 1980a).

Some investigations have focused on wall structure and antigenically active components of conidia (Scholer, 1974; Barreto-Bergter et al., 1981a; Cole and Pope, 1981; Cole et al., 1982). While shared components exist, conidial antigens often differ considerably from those of the mycelium (Amos, 1970; Kauffman et al., 1984). However, most diagnostic and immunological studies have centered on antigens that are integral to the fungal mycelium or on the extracellular products that demonstrate IgG and IgE-binding properties.

Hyphal wall polysaccharides are composed of glucans (which may be β-(1,3)-, α-(1,3)- and α-(1,4)-linked) and chitin (a homopolymer of β-(1,4)-N-acetylglucosamine units) at concentrations between 50 and 80% of the dry weight (Johnson, 1965; Bull, 1970; Zonneveld, 1971; Stagg and Feather, 1973). Galactomannan (GM) and galactosaminogalactan occur as minor constituents (Bardalaye and Nordin, 1977). Small amounts of protein and lipid are also present. Studies of tissues invaded by *Aspergillus fumigatus* when probed, in thin section, with fluorescein-conjugated lectins show strong hyphal staining with lectins specific for N-acetylgalactosamine and β-D-galactose, but only moderate staining with a lectin specific for α-D-glucose and α-D-mannose residues (Stoddart and Herbertson, 1978).

It is important to bear in mind that the conditions of growth in vitro and the method of preparation are likely to influence cell wall analysis. Nevertheless, preliminary data on the basic chemical composition of wall material indicates considerable structural differences among *Aspergillus* species (Johnson, 1965; Bull, 1970; Hearn and Mackenzie, 1979). This finding is reflected at the macromolecular level where nigeran, a glucan with alternating α-1,3 and α-1,4 linkages, is found in *Aspergillus niger* but not in *A. fumigatus* (Bobbitt and Nordin, 1978). Differences in susceptibility of intact organisms to the hydrolytic action of a variety of proteinases and polysaccharidases is also indicative of interspecies differences in structural organization of the cell wall (Hearn, 1984).

In only a few instances has an antigenic role been ascribed to defined chemical structures isolated from aspergilli. Of these the GMs of *A. fumigatus* have been most completely analyzed, either as a pure polysaccharide or complexed with peptides (Sakaguchi et al., 1968; Azuma et al., 1971; Reiss and Lehmann, 1979; Bennett et al., 1985). GM identical or closely related to that of cellular origin has also been isolated from culture fluid (Sakaguchi et al., 1968). A glucan, which in many reports was coextracted with GM, was found, on separation, to have no identifiable immunological reactivity (Azuma et al., 1971; Reiss and Lehmann, 1979).

Water-soluble constituents of *Aspergillus*, obtained by physical rupture of the mycelium, have proved a very rich source of diagnostic antigens (Tran Van Ky et al., 1968; Bardana et al., 1972a; Proctor, 1976). In young cultures, the adverse effects of prolonged exposure of antigens to autolytic enzymes is avoided. As many as 40 detectable bands have been demonstrated in preparations from young mycelia by fused-rocket immunoelectrophoresis (Kim and Chaparas, 1978). Mycelial extracts have been fractionated, and some of the component molecules have been partially or extensively purified and chemically and immunologically characterized (Kim et al., 1978; Hearn et al., 1980; Calvanico et al., 1981; Piechura et al., 1983; Schønheyder and Andersen, 1983, 1984a,b; Wilson et al., 1984).

Because of the ease of preparation and variety of components, antigens derived from culture filtrates have long been popular as diagnostic antigens (Longbottom and Pepys, 1964; Coleman and Kaufman, 1972; Dee, 1975). Such preparations are still much used (Kauffman and de Vries, 1980a; Kurup et al., 1983 and 1986; Kauffman et al., 1985). They contain antigens which are found only in the culture medium as well as many components shared with the parent mycelium (Biguet et al., 1964; Piechura et al., 1983; De Magaldi and Mackenzie, 1984; Schønheyder et al., 1985a,b; Harvey and Longbottom, 1986). *Aspergillus*, in common with many other fungi, actively secretes protein and glycoprotein complexes, some with enzymatic activity, into the culture medium. The pioneering work in this field was done by Tran Van Ky et al. (1966a,b; 1968). More recent work has centered on antigens which possess catalase or a chymotryptic-like protease activity (Girault et al., 1977; Schønheyder et al., 1984a,b). These have been isolated from both cellular material and culture medium.

ANTIGENS LOCATED AT THE CELL SURFACE OR IN THE WALL OF *ASPERGILLUS*

Antigens in situ on hyphal surfaces have been recognized by the binding of specific anti-*Aspergillus* antibodies to fragments, thin sections, or intact hyphae (Drouhet et al., 1972; Warnock, 1974; Schønheyder et al., 1982). Germinating conidia were reported by Gordon et al. (1977) to provide more specific binding to antibodies in immunofluorescence tests. In such germlings, staining tended to concentrate at the growing tips of hyphae. However, little or nothing is known about the structure of these reactive surface antigens.

Evidence for a layered arrangement of the hyphal cell wall, where pretreatment with an exo-β-(1,3)-glucanase is required before chitin release is achieved with chitinase, has been reported for *Aspergillus oryzae* (Horikoshi and Arima, 1962) and for *Aspergillus nidulans* (Bull, 1970). Changes in the chemical composition of many fungal surfaces are known to occur with age (Hunsley and Burnett, 1970; Barkai-Golan et al., 1978). This is also the case with *A. fumigatus*, where age-related differences have been noted when binding of specific IgG to the hyphal surface was monitored (John et al., 1984). Incubation of intact germ tubes with rabbit antiserum to an alkaline extract of *A. fumigatus* mycelium resulted in negligible immunoglobulin uptake. The extract used to prepare this antiserum (a generous gift from Dr. E. Reiss, Centers for Disease Control, Atlanta, GA) is considered to contain predominantly GMs and α-(1,3)-glucan moieties (Dr. E. Reiss, personal communication). However, when this antiserum was incubated with mature hyphae, binding values approximately 50% above those of negative control sera were obtained (Hearn, 1984). Additional evidence for age-related structural differences was also provided by the differential susceptibles of *A. fumigatus* organisms, germlings and mature hyphae, to the action of hydrolytic enzymes, including α-mannosidase and chitinase (Hearn, 1984; John et al., 1984).

Analysis of detergent extracts of *A. fumigatus* hyphal wall material by polyacrylamide gel electrophoresis in sodium dodecyl sulfate (SDS-PAGE) has shown a complex array of molecules varying in unit size from >100 kD to <10 kD mw (Hearn et al., in press). Most of these moieties appear to be glycoprotein, staining for protein with a combined Coomassie Blue/silver stain and for carbohydrate with a variety of lectins with specificities for different sugar residues. All extracts proved highly immunogenic and reacted with specific anti-*Aspergillus* antisera. Exposure of antigens to *Cytophaga* lytic enzyme or α-amylase considerably reduced antigen-antibody binding and indicated that (1,3)-β-, (1,4)-β-, and (1,4)-α-D-glucans were implicated in these reactions.

GALACTOMANNANS

Evidence that a chemical structure of *A. fumigatus* defined as GM has an immunological role originated with the studies reported by Sakaguchi et al. (1967) and Azuma et al. (1967). More recently, GMs have been characterized following alkaline extraction of the cell walls of *A. niger* and *A. fumigatus* (Bardalaye and Nordin, 1977; Reiss and Lehmann, 1979). A preparation has also been obtained by mechanical breakage of *A. fumigatus* mycelium followed by purification from the cell sap (Bennett et al., 1985). Most of this work has been reported by Reiss in his comprehensive review (1986) and is only briefly summarized in Table 1. Detailed below is the method used by Barreto-Bergter et al. (1981a), which, although no antigenic activity has been assigned to it, represents an exhaustive chemical analysis of a GM preparation.

Following extraction of wall material with alkali, the polysaccharides were precipitated with excess alcohol and purified by passage through a mixed bed of cation and anion exchange resins (see Sakaguchi et al., 1968). Monosaccharides were characterized as their derived alditol acetates (Sawardeker et al., 1965). Polysaccharides were per-o-methylated (Haworth, 1915; Kuhn et al., 1955), hydrolyzed with 10% aqueous H_2SO_4, reduced with sodium borohydride, and acetylated. Partially o-methylated alditol acetates were identified by use of a gas-liquid chromatography unit equipped with a capillary column filled with either OV-225 or OV-17 and coupled to a mass spectrometer. Partial acetolysis (Lee and Ballou, 1965) of the polysaccharide provided fragments that were converted into free sugars. On a paper chromatogram they had mobilities corresponding to those of galactose, mannose, and α-D-(1,2)-linked mannobiose, mannotriose, and mannotetraose, which is consistent with consecutive (1,2)-links between α-D-mannopyranosyl residues. The composite structure proposed by Barreto-Bergter et al. for their GM preparation is given in Table 1.

In the analysis conducted by Reiss and Lehmann (1979), who also isolated GM from wall by alkali extraction, monosaccharides were characterized as their trimethylsilyl ether derivatives (Sweeley et al., 1966). GM was permethylated with dimethylsulfinyl-sodium and methyl iodide (Hakomori, 1964). The methylated polysaccharide was formalyzed, hydrolyzed (with 0.3 N trifluoroacetic acid), and reduced with sodium borohydride. Alditol acetates of the methylated monosaccharides were prepared (Lindberg, 1972) and separated by gas-liquid chromatography on a column of 3% ECNSS-M (ethylene succinate-phenyl-silicone copolymer) coated on 100/120 mesh Gas Chrom Q. The structure that Reiss and Lehmann assigned to GM is shown in Table 1.

Table 1 Galactomannans Isolated from *Aspergillus fumigatus* Mycelium

Extraction from mycelium	G/M ratio	Mol. wt.	Mannan core	Side chains	References
45% Aqueous[a] phenol	1.0:1.0		D-man-p	D-gal-f	Sakaguchi et al., 1968
Ether-ethanol (1:1, v/v), refluxed	1.0:1.5	—	D-man-p(α1,2) D-man-p(α1,6)	D-gal-f	Azuma et al., 1967, 1971
Alkaline borohydride, 4°C	1.0:1.2	25–75 kD	D-man-p(α1,6) D-man-p ?	D-gal-f	Reiss and Lehmann, 1979
6% Aqueous KOH, 100°C	1.6:1.0	—	D-man-p(α1,6) D-man-p(α1,2)	D-gal-f(β1,5)	Barreto-Bergter et al., 1981a
Mechanical breakage, soluble supernatant	1.9:1.0	21–38 kD	D-man-p(α1,2)	D-gal-f	Bennett et al., 1985

[a]Also prepared from culture filtrate (S_{20} = 1.8); Mycelial GM: S_{20} = 2.1.

Bennett et al. (1985) prepared *A. fumigatus* GM from the ammonium sulfate-soluble material from the water-soluble fraction of mechanically disrupted mycelium. It was partially freed from contaminants by diethyaminoethyl (DEAE) cellulose column chromatography and purified by its affinity binding to concanavalin A (ConA). Monosaccharide composition was established on the alditol acetate derivatives. The authors used the Hakomori method (1964) to prepare partially methylated alditol acetates (see Reiss and Lehmann, 1979). Gas chromatography was performed using a capillary column of OV-225.

Some of the reported evidence supports the model of a core of D-mannopyranosyl residues, which are α-(1,6)-linked (Reiss and Lehmann, 1979; Barreto-Bergter et al., 1981a). Mannosyl residues linked α-(1,2) are also present and have been assigned a side chain position (Barreto-Bergter et al., 1981a). The opposite arrangement has also been proposed, namely, that α-1,2-mannose units constitute the backbone with α-1,6-mannose units as side chains (Azuma et al., 1971). A high degree of branching (branch points at C6 and either C2 or C4) with oligogalactose side chains of 3–6 units has been postulated. Terminal galactose occurs in the furanose form (gal-f).

Where α-1,6-linked mannopyranosyl units have not been detected, the problem may be one of column efficiency. It has been reported that 2,3,4-tri-o-methylmannitol and 2,3,6-tri-o-methylgalactitol are nonresolvable on either OV-225 or ECNSS-M liquid phases (Barreto-Bergter et al., 1981b). The composition of the GM being analyzed will, in any case, depend on the source material and the methods employed in its separation. The macromolecular structures present in the cell wall will vary in size and composition depending on the mode of extraction. Methods used range from strong alkali under harsh extraction conditions (Barreto-Bergter et al., 1981a), to ether-ethanol extraction under reflux conditions (Azuma et al., 1967), a weak acid, phenol (Sakaguchi et al., 1968), and controlled alkaline extraction in the presence of borohydride (Reiss and Lehmann, 1979). Variation in molecular size may also be a feature of GM location. That found in a water-soluble form in the cell sap or extracellular fluid will, of necessity, be of relatively small size and may, in the first instance, represent a precursor of wall-form GM and, in the latter, represent a breakdown product. Such macromolecules are unlikely to be homogeneous but will occur as a family of closely related molecules varying in their degree of branching and in the length of their side chains. Composition appears also to be age related. The data presented here refer to 5-day-old mycelium (Barreto-Bergter et al., 1981a); in a 10-day-old culture differences occurred with time, especially in the proportion of D-gal, though not of the β-D-gal-f residues. A similar finding was reported for the cell wall composition of *A. niger* (Johnson, 1965).

GMs are regarded as among the most widely distributed polysaccharides found in fungi (Gorin and Spencer, 1968) and are responsible for a degree of intra- and interspecies cross-reactivity (Suzuki et al., 1967; Suzuki and Takeda,

1975). They share common immunological determinants in the form of gal-f residues. Sakaguchi et al. (1968) showed that controlled hydrolysis with dilute acid liberated gal-f from *A. fumigatus* GM with a concomitant loss in precipitin activity of the core mannan. Bennett et al. (1985), who studied inhibition of GM-antibody binding by different monosaccharides, have confirmed the immunodominant role of gal-f residues.

GLYCOPROTEIN ANTIGENS WITH CONCANAVALIN A-BINDING PROPERTIES

Pyridine extraction of a galactomannan-protein (APSK-66) from the defatted mycelium of *A. fumigatus* has been reported (Azuma et al., 1968). The protein moiety was responsible for eliciting a delayed-type skin reactivity in both sensitized guinea pigs and rabbits, while the carbohydrate moiety elicited an Arthus-type reaction (see Suzuki and Hayashi, 1975). Analysis of the carbohydrate fraction of APSK-66 and GM, obtained by alcohol extraction, fractional precipitation with acetone, and ion-exchange chromatography (Azuma et al., 1967), demonstrated immunological identity between the two components (Azuma et al., 1971).

Weiner and Coats-Stephen (1979) fractionated an *A. fumigatus* homogenate-supernatant on Sephacryl S-200 to yield an acidic glycoprotein with MW = 150–180 kD. When analyzed on polyacrylamide gel-isoelectric focusing (PAG-IEF), it stained for protein with Coomassie Blue and for sugars with PAS: it also bound to ConA (see Table 2). Precipitinogen reactivity showed identity with a cell wall extract, which was sensitive to periodate oxidation and partially resistant to tryptic digestion. Schønheyder and Andersen (1984b) fractionated the crude cell sap material of *A. fumigatus* by hydrophobic interaction chromatography on Phenyl-Sepharose CL-4B. Five peaks were separated by desorption from the support medium by stepwise reduction in the salt content of the elution buffer. The final fraction eluted with water. This material seemed most discriminatory in detecting patients with aspergillosis and was further fractionated by gel filtration on Sephadex G-200 into three major components. Gel filtration of the material which eluted with 2 M NaCl from Phenyl-Sepharose gave one major protein peak of MW = 470 kD. All preparations showed high reactivity in binding specific *Aspergillus* antibodies from patient's sera when tested in an enzyme-linked immunosorbent assay (ELISA) system (Schønheyder and Andersen, 1984a). The precipitinogens were analyzed by crossed immunoelectrophoresis (XIE). The glycoprotein nature of some of these components was established by their affinity for ConA-Sepharose in an intermediate gel system (Table 3). The same authors have also shown that the 470 kD antigen cross-reacts with antigens in *A. flavus* and *A. nidulans* but not in *A. niger* or *A. terreus* (Schønheyder et al., 1985a).

Table 2 Characterization of Glycoprotein Antigens with ConA-binding Components

Antigen fraction	Preparation	MW (kD) peak	MW (kD) subunits	pI value	Specific binding CB[a]	Specific binding PAS	Specific binding WGA	Reference
myc,G.P.[b]	gel filtration, affinity, chromatography	150–180	—	acidic	p	p	—	Weiner and Coats-Stephen, 1979
met,IIb	preparative-IEF ConA and antibody affinity chromatography	—	20, 40, and 80	6–6.5	p	p	—	Kurup et al., 1983 and 1986
myc,pk.a	preparative-IEF	—	—	3.5–4.5	p	n	p	Wilson et al., 1984
myc,BF	ConA-affinity chromatography	—	—	—	p	n	p	Wilson and Hearn, 1983
met,Ag 7	salt precipitation, gel filtration, affinity antibody chromatography	150–200	36	acidic	p	p	—	Harvey and Longbottom, 1986

[a]Reactivity of components separated on SDS-PAGE to Coomassie Blue, periodic acid Schiff, and wheat germ agglutinin; p = positive; n = negative.
[b]myc = mycelial; met = metabolic; G.P. = glycoprotein; IIb, etc. = antigen abbreviations as used by the authors (see text); subunit size was determined by SDS-PAGE.

Table 3. *A. fumigatus* Antigen Fractions of MW 470,000, 250,000 and 25,000–50,000 Investigated by SDS-PAGE and CIE

Technique	*A. fumigatus* antigen fractions		
	MW 470,000	MW 250,000	MW 25,000–50,000
Number of protein bands	SDS-PAGE	7	6	14
Number of precipitating antigens	CIE[a]	3	2	10
Reaction of identity	CIE/intermediate gel	x———0———x		
			x———1———x	
		x———2———x		
Number of glycoprotein antigens	CIE/ConA	3	2	3

CIE was carried out with hyperimmune rabbit sera, and reactions of identity between antigens of the three fractions were ascertained by intermediate gel technique. The glycoprotein nature of some precipitating antigens was determined by affinity to ConA-Sepharose 4-B added to the intermediate gel.

[a] The number of precipitin lines obtained with the most polyvalent antiserum for each fraction: for the MW 470,000 antigen fraction it was antiserum to a partially purified high MW antigen fraction (fraction III), and for MW 250,000 and 25,000–50,000 antigen fractions antiserum to crude *A. fumigatus* antigen.

Source: Shønheyder and Andersen, 1984, p. 265. Reprinted from *Int. Archs. Allergy Appl. Immunol.* with the permission of the publisher. S. Karger AG, Basel.

Kurup et al. (1983) separated extracellular components of a young *A. fumigatus* culture by preparative IEF in a Sephadex G-75 glat-bed system utilizing a gradient of pH 4-6.5. A pooled fraction with an isoelectric point (pI) of 6-6.5, shown on bidimensional electrophoresis to have subunits of MW = 20, 40, and 80 kD, was further purified by affinity chromatography on ConA-Sepharose. The fraction elutable with α-methyl-D-mannoside was used to raise monospecific antiserum in rabbits. From this an IgG fraction was coupled to cyanogen bromide-activated (CNBr) sepharose and components, which eluted with 8 M urea, were used in antibody assays (Kurup et al., 1986). Preparative IEF has also been used to fractionate mycelial extracts by Wilson et al. (1984). Fractions, pooled on the basis of their strong and reproducible reactivity in an ELISA system (pk. a) had a pI value of 3.5-4.5 and were subsequently shown to have ConA-binding properties (Hearn, unpublished results).

Harvey and Longbottom (1986) isolated a specific antigen, designated Ag7, from both cell sap and culture filtrate of *A. fumigatus*. It was recovered by a combination of gel filtration on Sephacryl S-200 and affinity-antibody chromatography. An IgG fraction from monospecific antisera (raised in rabbits to the relevant peak excised from XIE gels) was attached to CNBr-activated Sepharose 4B. That portion of the partially purified antigenic mixture which bound was eluted with 1 M propionic acid. The antigen eluted from molecular sieving with a MW = 150-200 kD. On SDS-PAGE it gave a single diffuse band with subunit MW = 36 kD (with or without prior reduction) and stained for protein and carbohydrate with Coomassie Blue and PAS, respectively. It bound to ConA in an intermediate gel-XIE test, indicating the presence of α-D-man-p and/or α-D-glu-p end residues. The antigenic binding site to specific IgG and IgE antibodies was heat stable at 100°C for 5 min (see Reiss and Lehmann, 1979). Another antigen, designated Ag 13, was prepared using a similar procedure. It eluted from molecular sieving with an approximate MW = 70 kD. While this antigen also bound to ConA, it differed from Ag 7 in being heat sensitive and in reacting primarily with IgG class antibodies (Harvey and Longbottom, 1987).

The affinity of specific sugar residues for ConA has been used as a basic step in purifying immunologically reactive components from *A. fumigatus* (Kurup et al., 1983; Wilson and Hearn, 1983; Le Pape and Deunff, 1987). A large number of precipitinogens can be seen when such a fraction is reacted in XIE with specific antisera raised in rabbits and with sera from patients with proven aspergillosis. High specific binding to IgG in ELISA can be observed (Wilson and Hearn, 1983). On SDS-PAGE it can be separated into approximately 25 bands covering a range of MW = 12-98 kD; most of these constituent molecules bound ConA: many also bound wheat germ agglutinin (with specificity for N-acetyl-glucosamine) following Western blotting onto cellulose nitrate membranes (Hearn, unpublished results).

Le Pape and Deunff (1987) obtained a ConA-binding fraction from a mixture of cellular and extracellular antigens of *Aspergillus*. Analysis of *A. fumigatus* extracts on SDS-PAGE showed that the number of Coomassie Blue-staining components varied from 19 to 25, depending on the strain used. When the separated components were blotted onto cellulose nitrate, exposed to ConA labeled with peroxidase, and visualized with a suitable substrate, a maximum of 12 glycoprotein moieties were seen (Le Pape and Morin, 1986). A similar technique was used to compare culture filtrate antigens from four *A. fumigatus* strains, one of *A. fisheri* and one of *A. flavus*. Six ConA-binding glycoprotein antigens were found to be common to the four strains of *A. fumigatus*. The authors suggest that these entities may be helpful in the investigation of the immunological system in patients with aspergillosis. In a system analagous to that proposed by Mishra (1984), the percentage of common bands was 60% for *A. fumigatus*, 44% for *A. fisheri*, and only 16% for *A. flavus*.

GLYCOPROTEIN ANTIGENS OF UNDEFINED CARBOHYDRATE COMPOSITION

Bardana et al. (1972a) obtained a trichloroacetic acid–soluble, carbohydrate-enriched antigenic preparation from the supernatant of a mycelial homogenate. The mixture was radiolabeled with ^{125}I-NaI. PAGE revealed an anionic radiolabeled component which bound, but did not form precipitin arcs with, specific anti-*A. fumigatus* antisera. It was identified as a mucopolysaccharide through positive staining with alcian blue. In addition, it was deduced from its extreme sensitivity to periodate oxidation and relative insensitivity to pepsin that immunodominance resided in the glycosidic moiety. This fraction was later shown to produce a line of identity with the large MW antigenic fraction (470 kD) of Schønheyder et al. (1985a) on line immunoelectrophoresis.

Components of the homogenate-supernate from 4-day-old *A. fumigatus* mycelium were subjected to salt fractionation by Kim et al. (1978). That portion insoluble in 75% ammonium sulfate was separated by molecular sieving on Sephadex G-75 into four fractions. The fraction eluting early from gel filtration (APIFA) was immunologically very reactive in tests for both humoral and cell-mediated immune responses. So, too, was an ammonium sulfate soluble, DEAE-binding fraction. All antigenically active fractions contained both Coomassie Blue and PAS-staining components. The second fraction to elute from gel filtration (APIFB) was the most skin-reactive substance in both immediate and delayed-type hypersensitivity tests. However, this fraction proved a potent inhibitor of lymphocyte transformation in vitro (Chaparas et al., 1986).

A cell sap component (CS2) with MW = 150–200 kD (established by molecular sieving) was characterized as an acidic glycoprotein with an isoelectric point of 5.2–5.6 and a neutral hexose content of 12.5% (Calvanico et al., 1981). It

Table 4 Effect of Polysaccharidases on Antibody Binding to *Aspergillus* Mycelial Surfaces

Species	T. harzianum culture filtrate	Cytophaga lytic enzyme	Exo-(1,3)-β-glucanase	α-Amylase	β-Amylase	α-Mannosidase	Chitinase
A. fumigatus	121	75	NS	65	NS	75	NS
A. flavus	NS	NS	134	67	NS	NS	NS
A. terreus	77	dis	NS	NS	55	57	124
A. niger	NS	dis	NS	64	NS	NS	76
A. nidulans	NS	66	NS	62	64	NS	NS

Results are expressed as percentage binding of antibody relative to control values. Hydrolytic enzyme control, β-mannosidase, gave a mean binding value of 93% ± 4.2; NS (not significant) (set an arbitrary value of 100 ± 20% with a permissible coefficient of variation among replicates of 10%); dis = mycelial disintegration.

Source: Hearn, 1984. Reprinted from *J. Gen. Microbiol.* with the permission of the copyright holder, the Society for General Microbiology.

reacted in double immunodiffusion tests with 75% of sera from patients with aspergilloma and allergic bronchopulmonary aspergillosis (ABPA). In later work from the same group (Piechura et al., 1983), three major entities with MW = 41, 53, and 83 kD were detected, after reduction, on SDS-PAGE. This indicated that the major antigenic component was a protein of 177 kD composed of three subunits. Also considered was the possibility that, in view of their similar pI values, the 41 kD moiety was a degradation product of the 53 kD component. If this were so, it would give a major antigenic component consisting of two disulfide-linked polypeptide chains, one of $\sim$ 83 kD and the other 53 kD to give a MW = 136 kD to the unreduced protein. The authors subsequently described another fraction obtained from cell sap material of *A. fumigatus*, consisting of mainly low molecular weight components and designated CS3. When examined by bidimensional electrophoresis, *Aspergillus* species (including *A. flavus, A. fischeri, A. terreus*, and *A. niger*) appeared to share certain CS3 components, but to a lesser extent than the CS2 components (Piechura et al., 1985).

Antigenic material that binds to ConA has α-D-man-p and/or α-D-glu-p terminal residues. It is known that GM and glucan can be isolated by their ability to bind to ConA-Sepharose (Reiss and Lehmann, 1979; Bennett et al., 1985). It is thus possible that some of the glycoprotein antigens described here contain GM or GM-like structures (see Azuma et al., 1968). In this connection the observation of Reiss and Lehmann (1979) is of interest. Their serum antigen with MW $>$ 125 kD may constitute a complex of GM to protein through alkali-labile o-glycosidically linked serine and threonine residues. What may also be of significance is that a polyvalent antiserum raised to *A. fumigatus* binds with equal facility to surface immunodominant antigens of the homologous fungus when compared with heterologous species such as *A. flavus, A. terreus, A. niger, A. nidulans*. It is nevertheless known that different species exhibit major differences in surface architecture (Table 4).

ASPERGILLUS FUMIGATUS ANTIGENS IN VIVO

Antigenic metabolites have been detected in the serum of immunosuppressed animals infected with *A. fumigatus* (White et al., 1977; Rippon and Anderson, 1978). This may be due in part to hypogammaglobulinemia resulting in the circulation in body fluids of free antigen or immune complexes in antigen excess. It has been recognized that sensitive methods such as radioimmunoassays (RIA) and ELISA are necessary to demonstrate *Aspergillus* antigenemia in rabbit model systems (Shaffer et al., 1979a; Weiner and Coats-Stephen, 1979; Sabetta et al., 1985). In each case, carbohydrate antigens were used in competition assays. On analysis, one antigen stained for protein with Coomassie Blue and for carbohydrate with PAS; it was periodate sensitive and bound to ConA (Weiner and Coats-Stephen, 1979). Direct iodination of the alkali-soluble antigen used by Shaffer et al. (1979a) yielded a product, only 7% of which bound to rabbit antiserum to *Aspergillus* at a dilution of 1:1000. However, when the extract was tyramine-conjugated prior to iodination, 72% of the product bound rabbit antibody at a dilution of 1:20,000. This indicates that antibody is directed primarily towards the carbohydrate moiety of the antigen. The methods have been successfully applied to the detection of circulating antigen in humans with invasive infections (Shaffer et al., 1979b; Weiner, 1980; Sabetta et al., 1985; Talbot et al., 1987).

A constructive attempt to define an in vivo antigen of *A. fumigatus* is represented by the rabbit model devised by Lehmann and Reiss (1978). Rabbits, immunosuppressed with cortisone and cyclophosphamide, developed invasive aspergillosis following injection of *A. fumigatus* conidia. Serum containing circulating antigen produced in vivo was used to produce antiserum by injection into a second group of normal rabbits. This diagnostic antiserum detected circulating antigen by counterimmunoelectrophoresis in a patient with invasive aspergillosis. Absorption of this antiserum with chemically characterized GM (Reiss and Lehmann, 1979) removed the activity specific for circulating antigen.

Purified GM has now been used as the basis for antigenemia detection in rabbit model systems and in patients with invasive aspergillosis (deRepentigny et al., 1987). While a relatively high detection rate of antigen was reported in patients' urine (7/13), serum antigen was found in only 17% of cases (2/12) (Dupont et al., 1987).

Recently, ELISA inhibition methods which utilize a ConA-binding, glycoprotein fraction as detector antigen have been described (Le Pape and Deunff, 1987; Wilson et al., 1987). While it is probable that these partially purified ConA-binding fractions contain GM, free and/or proteinbound, it is worth noting the resemblance that exists between this product and the serum antigen described by Reiss and Lehmann (1979). This latter antigen has a MW > 125 kD, while that of purified GM lies in range 25-75 kD. Reiss (1986) has suggested that during antigenemia, GM is complexed in serum, probably to immunoglobulin. In some studies, heat dissociation of inferred immune complexes has been used successfully to increase the incidence of antigen detection (Weiner and Coats-Stephen, 1979; de Repentigny et al., 1987; Wilson et al., 1987). It is also possible that in serum, GM exists as a complex with protein through alkali-labile o-glycosyl-serine or -threonine bonds (Reiss and Lehmann, 1979). It should be kept in mind that in one instance the ConA-binding material was not only used as a detector antigen but also served to coat the plastic support (Le Pape and Deunff, 1987). Given the competition for binding sites, it is to be expected that protein molecules, rather than carbohydrate, will preferentially bind to the plastic (Scott et al., 1981; Kenny and Dunsmoor, 1983). The serum antigen described by Lehmann and Reiss (1978) was able to elicit specific antibodies in rabbits. In contrast, injection of chemical extracts of *Aspergillus* mycelium containing GM were unable to do so. It may be concluded that there is considerable support for the existence of a GM-protein complex (see Azuma et al., 1968), with the likelihood that the carbohydrate moiety plays an immunodominant role.

Difficulties in the detection of serum GM have been explained on the basis of its clearance from the bloodstream by macrophages through binding of the antigenic moieties to mannose receptor sites (Bennett et al., 1987). Reported levels of unsubstituted mannosyl units in GM vary from 53% (Bennett et al., 1985) to 20% (Barreto-Bergter et al., 1981a) to 34% (Reiss and Lehmann, 1979). It is tempting to speculate that the rate of removal from the bloodstream may depend in part on the physical composition of the GM, i.e., the number of unsubstituted mannosyl residues available for binding to macrophages.

ANTIGENS WITH ENZYMIC FUNCTIONS

Antigens possessing enzymatic activity have been demonstrated for *A. fumigatus* and other *Aspergillus* species in immunoelectrophoresis precipitin arcs visualized with appropriate substrates. A total of 18 enzymatically active antigens were identified including phosphatases, esterases, dehydrogenases, glucuronidases, proteases, catalase, and amylase (Tran Van Ky et al., 1966a,b, 1968). While several of these antigen-enzymes were found to react with specific *Aspergillus* antibodies, the ones which have proved most consistently useful as diagnostic tools are antigens with protease or catalase activity. Biguet et al. (1967) identified two precipitating antigens with protease activity; one with chymotryptic activity reacted with sera from each of 100 patients with proven aspergillosis. In a separate study, trypsinlike protease activity was located in precipitin in arcs nos. 13 and 18 when rabbit antiserum to *A. fumigatus* was reacted with homologous culture filtrate antigens on XIE. Precipitate no. 18, which displayed two maxima, also showed chymotryptic activity (Kauffman and de Vries, 1980b).

Chymotryptic antigen from *A. fumigatus* has now been purified by affinity chromatography on ϵ-amino-caproyltryptophanmethylester coupled to agarose (Bout et al., 1973). This fraction has been used successfully to detect specific anti-*Aspergillus* antibodies in a passive hemagglutination test (Senet et al., 1978a). *A. fumigatus* catalase antigen has also been purified by a combination of DEAE-ion exchange chromatography and gel filtration (Girault et al., 1977). Both antigens gave four components on SDS-PAGE with MW ~ 250 kD (relative to a reference catalase).

More recently, constituents from ruptured mycelium and culture filtrate of *A. fumigatus*, when separated by a combination of hydrophobic interaction chromatography and gel filtration, have yielded highly reactive specific antigens with protease and catalase activities (Schønheyder and Andersen, 1984a, 1985a). The protease fraction separated with MW = 25-50 kD and the catalase antigenic fraction eluted with MW = 250 kD (Table 3). The catalase antigen detected antibodies in 13 of 14 sera from patients with aspergilloma, while it reacted with only one of 9 sera from proven ABPA patients. Whether this low rate of positivity is due to the relative insensitivity of the detection system used (immunoelectrophoresis) or to a lack of specific antibodies to this antigen in cases of allergic aspergillosis is unresolved (Schønheyder et al., 1985b). In contrast, the C (chymotryptic) antigen described by Tran van Ky (1966), when isolated and tested in an ELISA system by De Magaldi and Mackenzie (1984), gave positive results with 15/17 cases of aspergilloma and 12/14 cases of ABPA. Similarly encouraging results were obtained with Ag 13, found to be identical to the chymotryptic antigen (Harvey and Longbottom, 1987). In a sandwich ELISA, all ABPA sera (25/25) had significantly raised levels of specific IgG. Reactions given by 2/5 aspergilloma sera showed very low IgG levels as did 5/12 sera from farmer's lung patients.

Cross-reactivity has been shown to occur when antiserum raised in rabbits to the catalase fraction from *A. fumigatus* was tested on XIE with isolates of *A. flavus*, *A. terreus*, and *A. nidulans*. No reaction was obtained with *A. niger* extracts (Schønheyder et al., 1985a). These results are in agreement with the earlier work of Tran Van Ky et al. (1968). However, *A. fumigatus* catalase antigen was found to cross-react with antisera raised in rabbits in the order *A. flavus* > *A. niger* > *A. nidulans*. Using the same passive hemagglutination system, the same result was obtained with *A. fumigatus* chymotryptic antigen (Senet et al., 1978a,b).

CONCLUSION

The differences and similarities found in the chemical composition of the aspergilli are reflected in their antigenic reactivities. *Aspergillus* has been shown to possess both specific antigens and to share a limited number of common antigens among the different species (Biguet et al., 1964; Kim and Chaparas, 1979; De Magaldi and Mackenzie, 1984). A degree of cross-reactivity is seen among mycelial antigens (Hearn et al., 1980; Piechura et al., 1985) and culture filtrate aspergillins (Gerber and Jones, 1973). However, in general, common antigens are less frequent than specific ones. Chaparas et al. (1980) found that an antigen-rich mycelial extract of *A. fumigatus* detected less than 50% of cases of aspergillosis where *A. flavus* or *A. niger* was the infecting agent. The necessity for using antigenic preparations from a range of *Aspergillus* species for diagnostic purposes has been reported by Kurup et al. (1979) and Froudist et al. (1986). There have also been cases of aspergillosis detectable only by an antiserum raised to the *Aspergillus* strain isolated from the patient's sputum (Kim, personal communication).

An additional problem appears to be the inherent variability of the antigenic preparations (Philpot and Mackenzie, 1976; Kurup et al., 1978; Reed, 1978; Wallenbeck et al., 1984). The problem remains despite the use of defined media and controlled conditions (Young and Bennett, 1971; Kim and Chaparas, 1978; Kurup et al., 1978; Kauffman and de Vries, 1980a). Variability of defined antigens has been reported as greater in the culture filtrate than in the mycelium (Piechura et al., 1983). Nevertheless, batch-to-batch differences occur routinely, irrespective of antigen source (Wilson et al., 1984; Martinez and Torres, 1985; Schønheyder et al., 1985a; Longbottom and Austwick, 1986).

The methods used to demonstrate the antigenicity of fungal preparations include precipitin tests, agglutination reactions, complement fixation, and indirect fluorescent antibody tests. More sensitive methods include ELISA and RIA. Detection of antigenic activity will depend in part on the system used to evaluate it (Kauffman et al., 1983). The importance of measuring primary antigen-antibody interaction has been stressed by Bardana (1972a). Some reports have suggested that the presence of antigen be assessed by more than one detection method, e.g., with antigens that give weak precipitation reactions but may be readily identified in a Farr-type RIA (Weiner and Coats-Stephen, 1979; Bennett et al., 1985). Similarly, antigens that show high IgE-binding properties but are either only weakly visible or undetectable as precipitating components have been noted (Longbottom, 1983; Kauffman et al., 1985). The advantages of using several different methods to demonstrate serum antibodies and to avoid false negative and false positive results has been remarked by Kurup et al. (1984). Similarly it has been shown that some antigens that give recognizable precipitates with *Aspergillus* antisera may either fail to detect, or show limited ability to detect, specific-IgG in patients serum by ELISA (Wilson and Hearn, 1982, 1983) or by SDS-PAGE and Western blotting techniques (Hearn, unpublished observations). It should also be emphazied that antibodies to *Aspergillus* occur among the normal population (Bardana et al., 1972b; Malo et al., 1977; Schønheyder and Andersen, 1982; Richardson et al., 1984) and must be allowed for when calculating what constitutes a significant concentration of specific immunoglobulin.

REFERENCES

Amos, W. M. G. (1970). The extraction of fungal antigens and their use in serological tests as an aid to the diagnosis of bronchial disorders. *J. Med. Lab. Technol.* 27:18-32.

Azuma, I., Kimura, H., Hirao, F., Tsubura, E., and Yamamura, Y. (1967). Biochemical and immunological studies on *Aspergillus* I. Chemical and biological investigations of lipopolysaccharide, protein and polysaccharide fractions isolated from *Aspergillus fumigatus*. *Jap. J. Med. Mycol.* 8:210-220.

Azuma, I., Kimura, H., and Yamamura, Y. (1968). Purification and characterization of an immunologically active glycoprotein from *Aspergillus fumigatus*. *J. Bact. 96*:272-273.

Azuma, I., Kimura, H., Hirao, F., Tsubura, E., Yamamura, Y., and Misaki, A. (1971). Biochemical and immunological studies on *Aspergillus III*. Chemical and immunological properties of glycopeptide obtained from *Aspergillus fumigatus*. *Jap. J. Microbiol. 15*:237-246.

Bardalaye, P. C., and Nordin, J. H. (1977). Chemical structure of the galactomannan from the cell wall of *Aspergillus niger*. *J. Biol. Chem. 252*:2584-2591.

Bardana, E. J., McClatchy, J. K., Farr, R. S., and Minden, P. (1972a). The primary interaction of antibody to components of aspergilli. I. Immunologic and chemical characteristics of a non-precipitating antigen. *J. Allergy Clin. Immunol. 50*:208-221.

Bardana, E. J., McClatchy, J. K., Farr, R. S., and Minden, P. (1972b). The primary interaction of antibody to components of aspergilli. II. Antibodies in sera from normal persons and from patients with aspergillosis. *J. Allergy Clin. Immunol. 50*:222-234.

Barkai-Golan, R., Mirelman, D., and Sharon, N. (1978). Studies on growth inhibition by lectins of Penicillia and Aspergilli. *Arch. Microbiol. 116*:119-124.

Barreto-Bergter, E. M., Gorin, P. A., and Travassos, L. R. (1981a). Cell constituents of mycelia and conidia of *Aspergillus fumigatus*. *Carbohydr. Res. 95*:205-218.

Barreto-Bergter, F. M., Hogge, L., and Gorin, P. A. (1981b). Gas-liquid chromatography of partially methylated alditol acetates on capillary columns of OV-17 and OV-225. *Carbohydr. Res. 97*:147-150.

Bennett, J. E., Bhattacharjee, A. K., and Glaudemans, C. P. J. (1985). Galactofuranosyl groups are immunodominant in *Aspergillus fumigatus* galactomannan. *Mol. Immunol. 22*:251-254.

Bennett, J. E., Freidman, M. M., and Dupont, B. (1987). Receptor-mediated clearance of *Aspergillus* galactomannan. *J. Infect. Dis. 155*:1005-1010.

Biguet, J., Tran Van Ky, P., Andrieu, S., and Fruit, J. (1964). Analyse immunoélectrophorétique d'extraits cellulaires et de milieux de culture d'*Aspergillus fumigatus* par des immunsérums expérimentaux et des sérums de malades atteints d'aspergillome bronchopulmonaire. *Ann. Inst. Pasteur (Paris) 107*:72-97.

Biguet, J., Tran Van Ky, P., Fruit, J., and Andrieu, S. (1967). Identification d'une activité chymotrypsique au niveau de fractions remarquables de l'extrait antigénique *d'Aspergillus fumigatus*. Répercussions sur le diagnostic immunologique de l'aspergillose. *Rev. Immunol. 31*:317-328.

Bobbitt, T. F., and Nordin, J. H. (1978). Hyphal nigeran as a potential phylogenetic marker for *Aspergillus* and *Penicillium* species. *Mycologia 70*:1201-1211.

Bout, D., Fruit, J., and Capron, A. (1973). Application de la chromatographie d'affinité à l'isolement des fractions antigéniques *d'Aspergillus fumigatus* supportant une activité chymotrypsique. *CR Acad. Sci. Paris 276*:2341-2344.

Bull, A. T. (1970). Chemical composition of wild-type and mutant *Aspergillus nidulans* cell walls. The nature of polysaccharide and melanin constituents, *J. Gen. Microbiol. 63*:75-94.

Calvanico, N. J., Du Pont, B. L., Huang, C. J., Patterson, R., Fink, J. N., and Kurup, V. P. (1981). Antigens of *Aspergillus fumigatus*. I. Purification of a cytoplasmic antigen reactive with sera of patients with *Aspergillus*-related disease. *Clin. Exp. Immunol. 45*:662-671.

Chaparas, S. D., Kaufman, L., Kim, S. J., and McLaughlin, D. W. (1980). Characterization of antigens from *Aspergillus fumigatus*. V. Reactivity in immunodiffusion tests with serums from patients with aspergillosis caused by *Aspergillus flavus, A. niger*, and *A. fumigatus. Am. Rev. Resp. Dis. 122*:647-649.

Chaparas, S. D., Morgan, P. A., Holobaugh, P., and Kim, S. J. (1986). Inhibition of cellular immunity by products of *Aspergillus fumigatus. J. Med. Vet. Mycol. 24*:67-76.

Cole, G. T., and Pope, L. M. (1981). Surface wall components of *Aspergillus niger* conidia. In *The Fungal Spore: Morphogenetic Controls*. Edited by G. Turian and H. Hohl. Academic Press, London, pp. 195-215.

Cole, G. T., Sun, S. H., and Huppert, M. (1982). Isolation and ultrastructural examination of conidial wall components of *Coccidioides* and *Aspergillus. Scanning Electron Microsc. Part. IV*:1677-1685.

Coleman, R. M., and Kaufman, L. (1972). Use of the immunodiffusion test in the serodiagnosis of aspergillosis. *Appl. Microbiol. 23*:301-308.

Dee, T. H. (1975). Detection of *Aspergillus fumigatus* serum precipitins by counter immunoelectrophoresis. *J. Clin. Immunol. 2*:482-485.

De Magaldi, S. W., and Mackenzie, D. W. R. (1984). Specificity of antigens from pathogenic *Aspergillus* species. I. Studies with ELISA and immunofluorescence. *Sabouraudia 22*:381-394.

Drouhet, E., Camey, L., and Segrétain, G. (1972). Valeur de l'immunoprécipitation et de l'immunofluorescence indirecte dans les aspergilloses bronchopulmonaires. *Ann. Inst. Pasteur (Paris) 123*:379-395.

Dupont, B., Huber, M., Kim, S. J., and Bennett, J. E. (1987). Galactomannan antigenemia and antigenuria in aspergillosis: Studies in patients and experimentally infected rabbits. *J. Infect. Dis. 155*:1-11.

Froudist, J. H., McAleer, R., and Cherian, G. (1986). Serology for the detection of antibody to *Aspergillus*: use of extracts from four species. *Australian J. Med. Lab. Sc. 7*:57-60.

Gerber, J. D., and Jones, R. D. (1973). Immunologic significance of Aspergillin antigens of six species of *Aspergillus* in the serodiagnosis of aspergillosis. *Am. Rev. Resp. Dis. 108*:1124-1129.

Girault, A., Boyer, J. P., Senet, J. M., and Girault, M. (1977). Isolement et purification de protéines d'*Aspergillus fumigatus* présentant respectivement une activité chymotrypsique et catalasique. *Bull. Soc. Fr. Mycol. Med. 6*:75-79.

Gordon, M. A., Lapa, E. W., and Kane, J. (1977). Modified indirect fluorescent antibody test for aspergillosis. *J. Clin. Microbiol. 6*:161-165.

Gorin, P. A. J., and Spencer, J. F. T. (1968). Galactomannans of *Trichosporon fermentans* and other yeasts; proton magnetic resonance and chemical studies. *Can. J. Chem. 46*:2299-2304.

Hakomori, S. I. (1964). A rapid permethylation of glycolipid and polysaccharide catalysed by methyl sulfinyl carbanion in dimethyl sulfoxide. *J. Biochem. (Tokyo) 55*:205-208.

Harvey, C., and Longbottom, J. L. (1986). Characterization of a major antigenic component of *Aspergillus fumigatus*. *Clin. Exp. Immunol. 65*:206-214.

Harvey, C., and Longbottom, J. L. (1987). Characterization of a second major antigen, Ag 13 (antigen C) of *Aspergillus fumigatus* and investigation of its immunological reactivity. *Clin. Exp. Immunol. 70*:247-254.

Haworth, W. N. (1915). A new method of preparing alkylated sugars. *J. Chem. Soc. 107*:8-16.

Hearn, V. M. (1984). Surface antigens of intact *Aspergillus fumigatus* mycelium: their localization using radiolabeled Protein A as marker. *J. Gen. Microbiol. 130*:907-917.

Hearn, V. M., and Mackenzie, D. W. R. (1979). The preparation and chemical composition of fractions from *Aspergillus fumigatus* wall and protoplasts possessing antigenic activity. *J. Gen. Microbiol. 112*:35-44.

Hearn, V. M., Proctor, A. G., and Mackenzie, D. W. R. (1980). The preparation and partial characterization of antigenic fractions obtained from the mycelial walls of several *Aspergillus* species. *J. Gen. Microbiol. 119*:41-49.

Hearn, V. M., Wilson, E. V., and Mackenzie, D. W. R. (in press). Characterization of purified wall antigens obtained from *Aspergillus* species. In *Fungal Antigens*, 1st Int. Symp. Edited by E. Drouhet, G. T. Cole, L. de Repentigny, J. P. Latgé and B. Dupont. Plenum Press, New York.

Horikoshi, K., and Arima, K. (1962). X-ray diffraction patterns of the cell wall of *Aspergillus oryzae*. *Biochim. Biophys. Acta 57*:392-394.

Hunsley, D., and Burnett, J. H. (1970). The ultrastructural architecture of the walls of some hyphal fungi. *J. Gen. Microbiol. 62*:203-218.

John, J., Wilson, E. V., and Hearn, V. M. (1984). Analysis of *Aspergillus fumigatus* germ tube surface structures by an immunofluorescent labelling technique. *Mykosen 27*:485-497.

Johnson, I. R. (1965). The composition of the cell wall of *Aspergillus niger*. *Biochem. J. 96*:651-658.

Kauffman, H. F., and de Vries, K. (1980a). Antibodies against *Aspergillus fumigatus*. I. Standardization of antigenic composition. *Int. Archs. Allergy Appl. Immunol. 62*:252-264.

Kauffman, H. F., and de Vries, K. (1980b). Antibodies against *Aspergillus fumigatus* II. Identification and quantification by means of crossed immunoelectrophoresis. *Int. Archs. Allergy Appl. Immunol. 62*:265-275.

Kauffman, H. F., Beaumont, F., Meurs, H., van der Heide, S., and de Vries, K. (1983). Comparison of antibody measurements against *Aspergillus fumigatus* by means of double-diffusion and enzyme-linked immunosorbent assay (ELISA). *J. Allergy Clin. Immunol. 72*:255-261.

Kauffman, H. F., van der Heide, S., Beaumont, F., de Monchy, J. G. R., and de Vries, K. (1984). The allergenic and antigenic properties of spore extracts of *Aspergillus fumigatus*: a comparative study of spore extracts with mycelium and culture filtrate extracts. *J. Allergy Clin. Immunol.* 73:567-573.

Kauffman, H. F., van der Laan, S., van der Heyden, P. J., van der Heide, S., and de Vries, K. (1985). Antibody determination against *Aspergillus fumigatus* by means of the enzyme-linked immunosorbent assay. I. Liberation of polystyrene binding components during cultivation. *Int. Archs. Allergy Appl. Immunol.* 77:343-348.

Kenny, G. E., and Dunsmoor, C. L. (1983). Principles, problems and strategies in the use of antigenic mixtures for the enzyme-linked immunosorbent assay. *J. Clin. Microbiol.* 17:655-665.

Kim, S. J., and Chaparas, S. D. (1978). Characterization of antigens from *Aspergillus fumigatus* I. Preparation of antigens from organisms grown in completely synthetic medium. *Am. Rev. Resp. Dis.* 118:547-551.

Kim, S. J., and Chaparas, S. D. (1979). Characterization of antigens from *Aspergillus fumigatus* III. Comparison of antigenic relationships of clinically important Aspergilli. *Am. Rev. Resp. Dis.* 120:1297-1303.

Kim, S. J., Chaparas, S. D., Brown, T. M., and Anderson, M. C. (1978). Characterization of antigens from *Aspergillus fumigatus* II. Fractionation and electrophoretic, immunologic and biologic activity. *Am. Rev. Resp. Dis.* 118:553-560.

Kuhn, R., Trischmann, H., and Löw, I. (1955). Zur Permethylierung von Zuckern und Glykosiden. *Angew. Chem.* 67:32.

Kurup, V. P., Fink, J. N., Scribner, G. H., and Falk, M. J. (1978). Antigenic variability of *Aspergillus fumigatus* strains. *Microbios* 19:191-204.

Kurup, V. P., Fink, J. N., Barboriak, J. J., and Scribner, G. (1979). The detection of circulating antibodies against antigens from three strains of *Aspergillus fumigatus*. *Mykosen* 23:368-372.

Kurup, V. P., Ting, E. Y., and Fink, J. N. (1983). Immunochemical characterization of *Aspergillus fumigatus* antigens. *Infect. Immun.* 41:698-701.

Kurup, V. P., Resnick, A., Scribner, G. H., Kalbfleisch, J. H., and Fink, J. N. (1984). Comparison of antigens and serological methods in *Aspergillus fumigatus* antibody detection. *Mykosen* 27:43-50.

Kurup, V. P., John, K. V., Resnick, A., and Fink, J. N. (1986). A partially purified glycoprotein antigen from *Aspergillus fumigatus*. *Int. Archs. Allergy Appl. Immunol.* 79:263-269.

Lee, Y.-C., and Ballou, C. E. (1965). Preparation of mannobiose, mannotriose and a new mannotetraose from *Saccharomyces cerevisiae* mannan. *Biochemistry* 4:257-264.

Lehmann, P. F., and Reiss, E. (1978). Invasive aspergillosis: antiserum for circulating antigen produced after immunization with serum from infected rabbits. *Infect. Immun.* 20:570-572.

Le Pape, P., and Morin, O. (1986). Electrophorèse en gel de polyacrylamide et immunoempreinte dans l'étude des *Aspergillus*. *Bull. Soc. Fr. Mycol. Med.* 15:485-490.

Le Pape, P., and Deunff, J. (1987). Antigène glycoprotéique circulant d'*Aspergillus fumigatus*: détection dans le sérum de souris par une technique ELISA. *Bull. Soc. Fr. Mycol. Med. 16*:169-172.

Lindberg, B. (1972). Methylation analysis of polysaccharides. *Methods Enzymol. 28*:178-195.

Longbottom, J. L. (1983). Antigens/allergens of *Aspergillus fumigatus*. Identification of antigenic components reacting with both IgG and IgE antibodies of patients with allergic bronchopulmonary aspergillosis. *Clin. Exp. Immunol. 53*:354-362.

Longbottom, J. L., and Pepys, J. (1964). Pulmonary aspergillosis: diagnostic and immunological significance of antigens and C-substance in *Aspergillus fumigatus*. *J. Path. Bact. 88*:141-152.

Longbottom, J. L., and Austwick, P. K. C. (1986). Antigens and allergens of *Aspergillus fumigatus*. I. Characterization by quantitative immunoelectrophoretic techniques. *J. Allergy Clin. Immunol. 78*:9-17.

Malo, J. L., Longbottom, J. L., Mitchell, J., Hawkins, R., and Pepys, J. (1977). Studies in chronic allergic bronchopulmonary aspergillosis. *Thorax 32*:269-274.

Martinez, J., and Torres, J. M. (1985). Caracterización de exoantígenos de *Aspergillus fumigatus*. Estudio de factores que influyen en su calidad. *Allergol. Immunopathol. 13*:501-508.

Mishra, S. K. (1984). Antigenic profile of some typical and septate phimide-strains of *Aspergillus fumigatus*. *Sabouraudia 22*:91-100.

Philpot, C. M., and Mackenzie, D. W. R. (1976). A comparison of antibodies to *Aspergillus fumigatus* in agar gel with different antigens and immunodiffusion patterns. *J. Biol. Stand. 4*:73-79.

Piechura, J. E., Huang, C., Cohen, S. H., Kidd, J. M., Kurup, V. P., and Calvanico, N. J. (1983). Antigens of *Aspergillus fumigatus* II. Electrophoretic and clinical studies. *Immunology 49*:657-665.

Piechura, J. E., Kurup, V. P., Fink, J. N., and Calvanico, N. J. (1985). Antigens of *Aspergillus fumigatus* III. Comparative immunochemical analyses of clinically relevant aspergilli and related fungal taxa. *Clin. Exp. Immunol. 59*:716-724.

Proctor, A. G. J. (1976). *Mycological Methods*. In *Microbiological Methods*, 4th ed. Edited by C. H. Collins and P. M. Lyne. Butterworths, London.

Reed, C. (1978). Variability of antigenicity of *Aspergillus fumigatus*. *J. Allergy Clin. Immunol. 61*:227-229.

Reiss, E. (1986). *Aspergillus fumigatus*. In *Molecular Immunology of Mycotic and Actinomycotic Infections*. Elsevier, New York.

Reiss, E., and Lehmann, P. F. (1979). Galactomannan antigenemia in invasive aspergillosis. *Infect. Immun. 25*:357-365.

de Repentigny, L., Bonshira, M., Ste-Marie, L., and Bosisio, G. (1987). Detection of galactomannan antigenemia by enzyme immunoassay in experimental invasive aspergillosis. *J. Clin. Microbiol. 25*:863-867.

Richardson, M. D., Llewellyn, P. A., and Warnock, D. W. (1984). Antibody to *Aspergillus fumigatus* antigens in normal sera: influence on positive-negative discrimination in ELISA. *J. Immunoassay 5*:205-220.

Rippon, J. W., and Anderson, D. N. (1978). Experimental mycosis in immunosuppressed rabbits. II. Acute and chronic aspergillosis. *Mycopathologia 64*: 97-100.

Sabetta, J. R., Miniter, P., and Andriole, V. T. (1985). The diagnosis of invasive aspergillosis by an enzyme-linked immunosorbent assay for circulating antigen. *J. Infect. Dis. 152*:946-953.

Sakaguchi, O., Yokota, K., and Suzuki, M. (1967). Biochemical and immunochemical studies on fungi. XII. On the galactomannans obtained from culture filtrate and cells of *Aspergillus fumigatus*. *Yakugaku Zasshi 87*:1268-1272.

Sakaguchi, O., Suzuki, M., and Yokota, K. (1968). Effect of partial acid hydrolysis on precipitin activity of *Aspergillus fumigatus* galactomannan. *Jap. J. Microbiol. 12*:123-124.

Sawardeker, J. S., Sloneker, J. H., and Jeanes, A. (1965). Quantitative determination of monosaccharides as their alditol acetates by gas-liquid chromatography. *Anal. Chem. 37*:1602-1604.

Schønheyder, H., and Andersen, P. (1982). An indirect immunofluorescence study of antibodies to *Aspergillus fumigatus* in sera from children and adults without aspergillosis. *Sabouraudia 20*:41-50.

Schønheyder, H., Anderson, P., and Stenderup, A. (1982). Serum antibodies to *Aspergillus fumigatus* in patients with pulmonary aspergillosis detected by immunofluorescence. *Acta Pathol. Microbiol. Immunol. Scand. Sect. B 90*:273-279.

Schønheyder, H., and Andersen, P. (1983). Determination of antibodies to partially purified *Aspergillus* antigens by an enzyme-linked immunosorbent assay. *Int. Arch. Allergy Appl. Immunol. 70*:108-111.

Schønheyder, H., and Andersen, P. (1984a). IgG antibodies to purified *Aspergillus fumigatus* antigens determined by enzyme-linked immunosorbent assay. *Int. Arch. Allergy Appl. Immunol. 74*:262-269.

Schønhyeder, H., and Andersen, P. (1984b). Fractionation of *Aspergillus fumigatus* antigens by hydrophobic interaction chromatography and gel filtration. *Int. Arch. Allergy Appl. Immunol. 73*:231-236.

Schønheyder, H., Storgaard, L., and Andersen, P. (1985a). Variation of a 470000 daltons antigen complex and catalase antigen in clinical isolates of *Aspergillus fumigatus*. *Sabouraudia 23*:339-349.

Schønheyder, H., Andersen, P., and Munck Petersen, J. C. (1985b). Rapid immunoelectrophoretic assay for detection of serum antibodies to *Aspergillus fumigatus* catalase in patients with pulmonary aspergillosis. *Eur. J. Clin. Microbiol. 4*:299-303.

Scholer, H. J. (1974). Specific mycelial and conidial antigens from *Aspergillus fumigatus*. In *Aspergillosis and Farmer's Lung in Man and Animal*. Edited by R. Haller and F. Suter. Hans Huber, Bern, pp. 45-54.

Scott, E. N., Muchmore, H. G., and Felton, F. G. (1981). Enzyme-linked immunosorbent assays in murine cryptococcosis. *Sabouraudia 19*:257-265.

Senet, J. M., Girault, A., Robert, R., and Girault, M. (1978a). Diagnostic de l'aspergillose par hémagglutination indirecte. 1. Utilisation d'une fraction chymotrypsique purifiée *d'Aspergillus fumigatus. Bull. Soc. Fr. Mycol. Med. 7*:225-228.

Senet, J. M., Girault, A., Robert, R., and Girault, M. (1978b). 2. Utilisation d'une fraction catalasique purifiée *d'Aspergillus fumigatus. Bull. Soc. Fr. Mycol. Med. 7*:229-232.

Shaffer, P. J., Medoff, G., and Kobayashi, G. S. (1979a). Demonstration of antigenemia by radioimmunoassay in rabbits experimentally infected with *Aspergillus. J. Infect. Dis. 139*:313-319.

Shaffer, P. J., Kobayashi, G. S., and Medoff, G. (1979b). Demonstration of antigenemia in patients with invasive aspergillosis by solid phase (Protein A-rich *Staphylococcus aureus*) radioimmunoassay. *Am. J. Med. 67*:627-630.

Stagg, C. M., and Feather, M. S. (1973). The characterization of a chitin-associated D-glucan from the cell walls of *Aspergillus niger. Biochim. Biophys. Acta 320*:64-72.

Stoddart, R. W., and Herbertson, B. M. (1978). The use of fluorescein-labelled lectins in the detection and identification of fungi pathogenic for man: a preliminary study. *J. Med. Microbiol. 11*:315-324.

Suzuki, M., and Hayashi, Y. (1975). Skin reaction and macrophage migration inhibition tests for polysaccharides from *Aspergillus fumigatus* and *Candida albicans. Jap. J. Microbiol. 19*:335-362.

Suzuki, S., and Takeda, N. (1975). Serologic cross-reactivity of the D-galacto-D-mannans isolated from several pathogenic fungi against anti-*Hormodendrum pedrosoi* serum. *Carbohydr. Res. 40*:193-197.

Suzuki, S., Suzuki, M., Yokota, K., Sunayama, H., and Sakaguchi, O. (1967). On the immunochemical and biochemical studies of fungi. XI. Cross-reaction of the polysaccharides of *Aspergillus fumigatus, Candida albicans, Saccharomyces cerevisiae* and *Trichophyton rubrum* against *Candida albicans* and *Saccharomyces cerevisiae* antisera. *Jap. J. Microbiol. 11*:269-273.

Sweeley, C. C., Wells, W. W., and Bentley, R. (1966). Gas chromatography of carbohydrates. *Methods Enzymol. 8*:95-108.

Talbot, G. H., Weiner, M. H., Gerson, S. L., Provencher, M., and Hurwitz, S. (1987). Serodiagnosis of invasive aspergillosis in patients with hematologic malignancy: validation of the *Aspergillus fumigatus* antigen radio-immunoassay. *J. Infect. Dis. 155*:12-27.

Tran Van Ky, P., Biguet, J., and Fruit, J. (1966a). Localisation et fréquence des arcs des immunoélectrophoregrammes produits par le sérum des malades atteints de mycétomes aspergillaires appliqués contre l'antigène *Aspergillus fumigatus. Rev. Immunol. 30*:13-20.

Tran Van Ky, P., Uriel, J., and Rose, F. (1966b). Caractérisation de types d'activités enzymatiques dans des extraits antigéniques *d'Aspergillus fumigatus* après électrophorèse et immunoélectrophorèse en agarose. *Ann. Inst. Pasteur 111*:161-170.

Tran Van Ky, P., Biguet, J., and Vaucelle, T. (1968). Étude d'une fraction antigénique d'*Aspergillus fumigatus* support d'une activité catalasique. Conséquence sur le diagnostic immunologique de l'aspergillose. *Rev. Immunol. 32*: 37-52.
Wallenbeck, I., Ankrust, L., and Einarsson, R. (1984). Antigenic variability of different strains of *Aspergillus fumigatus*. *Int. Arch. Allergy Appl. Immunol. 73*:166-172.
Warnock, D. W. (1974). Indirect immunofluorescence test for the detection of *Aspergillus fumigatus* antibodies. *J. Clin. Pathol. 27*:911-912.
Weiner, M. H. (1980). Antigenemia detected by radioimmunoassay in systemic aspergillosis. *Ann. Int. Med. 92*:793-796.
Weiner, M. H., and Coats-Stephen, M. (1979). Immunodiagnosis of systemic aspergillosis. I. Antigenemia detected by radioimmunoassay in experimental infection. *J. Lab. Clin. Med. 93*:111-119.
White, L. O., Richardson, M. D., Newham, H. C., Gibb, E., and Warren, R. C. (1977). Circulating antigen of *Aspergillus fumigatus* in cortisone-treated mice challenged with conidia: detection by counterimmunoelectrophoresis. *FEMS Microbiol. Letts. 2*:153-156.
Wilson, E. V., and Hearn, V. M. (1982). A comparison of surface and cytoplasmic antigens of *Aspergillus fumigatus* in an enzyme-linked immunosorbent assay (ELISA). *Mykosen 25*:653-661.
Wilson, E. V., and Hearn, V. M. (1983). Use of *Aspergillus fumigatus* mycelial antigens in enzyme-linked immunosorbent assay and counterimmunoelectrophoresis. *J. Med. Microbiol. 16*:97-105.
Wilson, E. V., De Magaldi, S. W., and Hearn, V. M. (1984). Preparative isoelectric focussing of immunologically reactive components of *Aspergillus fumigatus* mycelium. *J. Gen. Microbiol. 130*:919-925.
Wilson, E. V., Hearn, V. M., and Mackenzie, D. W. R. (1987). Evaluation of a test to detect circulating *Aspergillus fumigatus* antigen in a survey of immunocompromised patients with proven or suspected invasive disease. *J. Med. Vet. Mycol. 25*:365-374.
Young, R. C., and Bennett, J. E. (1971). Invasive aspergillosis. Absence of detectable antibody response. *Am. Rev, Resp. Dis. 104*:710-716.
Zonneveld, B. J. M. (1971). Biochemical analysis of the cell wall of *Aspergillus nidulans*. *Biochim. Biophys. Acta 249*:506-514.

5
Antigens of Dermatophytes and Their Characterization Using Monoclonal Antibodies

PETER DE HAAN, JIM R. WIKLER, ELISABETH M. H. VAN DER RAAY-HELMER, and DICK M. BOORSMA
Academic Hospital of the Free University, Amsterdam, The Netherlands

INTRODUCTION

Dermatophytes are a homogeneous group of keratinophilic fungi which can cause dermatomycosis of the human skin. Dermatophytes succeed in breaking the first line of host defense by way of infestation and subsistence of the cornified skin. The infection remains limited to the cornified layers of the skin; the dermatophytes do not penetrate deeper in the human skin, which probably reflects the effectiveness of the host defense system. The array of defense mechanisms thought active against dermatophytes consists of $\alpha 2$ macroglobulin keratinase inhibitor, unsaturated transferrin, epidermal desquamation, lymphocytes and lymphokines, and antifungal antibody. In 1972, Yu et al. (1972) described an $\alpha 2$ macroglobulin present in normal human serum, which inhibited keratinase. Keratinase is an extrinsic enzyme of the dermatophytes which breaks down keratin. However the exact role it plays in the host defense has not been clarified so far. Iron is an essential nutrient for dermatophytes. In serum its level is regulated by transferrin. Unsaturated transferrin binds avidly iron so that this metal becomes unavailable for fungal nutrition (King et al., 1975). Large quantities of iron have been detected in normal human stratum corneum (Monroe et al., 1976). The fact that dermatophytes can colonize the stratum corneum is indicative for the ample supply of iron.

Lateral invasion of the dermatophytes into adjacent stratum corneum does usually not occur. The mechanisms involved in prohibiting lateral invasion are unknown, but it is tempting to suggest that unsaturated transferrin is involved. It might act in concert with inflammatory damage caused by lymphocyte infiltration. Another defense mechanism against dermatophyte infection is epidermal desquamation. Dermatophytes can be removed from infected skin by accelerated epidermal turnover (Berk et al., 1976). Lymphocytes and lymphokines play a major role in the pathogenesis of the inflammation of the infected skin and help to eliminate the infection. After an initial infection with dermatophytes, a group of lymphocytes is generated, which is activated by trichophytin glycopeptides. Such a cell-mediated immunity helps to control and eliminate the infection. Artis and Jones (1976) suggested that lymphokines initiate epidermal cell damage, whereafter release of unsaturated transferrin, $\alpha 2$ macroglobulin keratinase inhibitor, and possibly other (soluble) factors can occur. These agents can then act on the fungus and inhibit further proliferation.

The humoral immune response to dermatophyte antigens (trichophytin) is generally very weak. Specific circulating antibodies against dermatophytes could be demonstrated in serum of subjects suffering from dermatomycosis (Kaaman et al., 1981).

Although the above is true for acute infections that heal spontaneously after acquired dermatophyte infection can also be chronic. These chronic infections, which can persist for a long time, are mainly caused by *Trichophyton rubrum*. They are characterized by a deficient specific cellular immunity and by the presence of a significant humoral immune response. The latter can be demonstrated in vivo by a positive immediate-type reaction to dermatophyte antigens and in vitro by the demonstration of specific circulating antibodies.

There are three genera of dermatophytes: *Trichophyton, Epidermophyton* and *Microsporum*. They comprise 39 species, 11 of which are important in human diseases. There is a wide variety of antigens originating from dermatophytes, as is well documented in literature (Grappel et al., 1974). The composition of antigenic material can vary widely under the influence of environmental conditions as is described by Jones and Artis (1981); trichophytin produced from one single strain of *Trichophyton mentagrophytes* varied in chemical composition and quantity. Therefore it is reasonable to assume that the immune response is not evoked against one single antigen but to a mixture of antigens present in varying amounts. Trichophytins obtained from the three different dermatophyte genera show structural similarities. Antibodies raised against trichophytin from one species react also with trichophytins from different species. They also react with trichophytins from other dermatophyte genera and even with antigenic material from unrelated fungi such as *Penicillium* or *Homodendrum*. Antisera against trichophytin can be obtained by immunizing rabbits with the antigenic material. A disadvantage of such antisera is that they are unable to detect one separate antigen but instead detect several antigenic determinants. Also, they cannot be used infinitely. On the other hand, mono-

Characterization of Dermatophyte Antigens

clonal antibodies (MAbs) can detect one specific antigenic determinant, and they can be used infinitely. With such monoclonal antibodies, possible similarities in antigenic material of different fungi can be detected. As host defense to chronic dermatophyte infections consists mainly of humoral-immune response against dermatophyte antigens, we produced MAbs against antigens of *Trichophyton rubrum*, a dermatophyte which is often associated with chronic dermatophyte infections. With these monoclonal antibodies two culture filtrates of *Trichophyton rubrum* were screened for the presence of antigens. We investigated the relationship between the antigenic determinant and the molecular weight of the antigens. These MAbs were also used for the detection of antigenic similarities between the different dermatophytes and unrelated fungi (cross-reaction).

MATERIALS AND METHODS

Isolation of Dermatophyte Antigens

Many different methods for the preparation of dermatophyte antigens have been developed. We have used the following methods (Cruickshank et al., 1960): Dermatophytes were obtained from human skin scrapings, which were cultured on Sabouraud's glucose. After identification of the dermatophyte according to col-

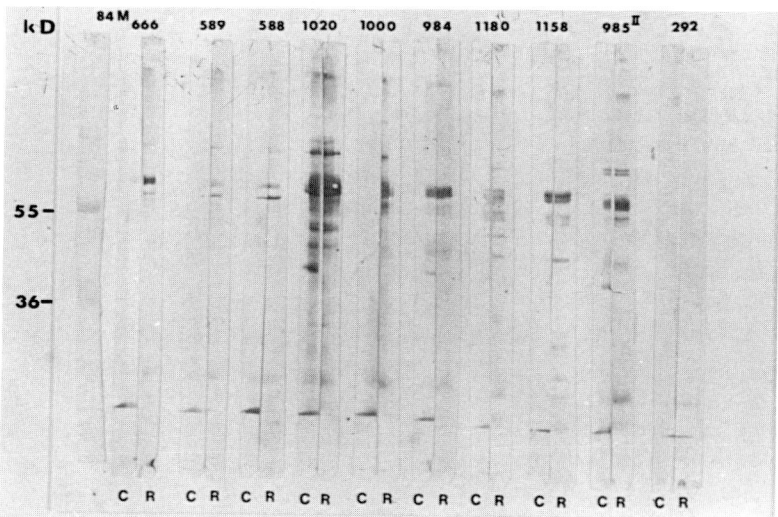

Figure 1 Results of 10 culture filtrates from *Trichophyton rubrum* in the immunoblot assay using Cult (C) and Rubrac (R) as antiserum.

ony morphology and microscopic appearance, the dermatophyte was inoculated in growth medium. The medium consisting of glucose, pepton containing streptomycin, and penicillin was shaken during 14 days at room temperature. Thereafter the mycelium, consisting of hyphae, was isolated by filtration and separated from the culture filtrate. Antigens were extracted either from the mycelium or from culture filtrates. For antigen isolation from the culture filtrate, the filtrate was concentrated in an Amicon concentration cell (YM 10 filter) and was rinsed three times with aquadest. The glycoproteins were isolated by alcohol (ethanol) precipitation. This precipitate was spun off, redissolved in water, lyophylized, and coded CF. Glycoproteins were isolated from *Trichophyton rubrum, Trichophyton tonsurans*, and *Epidermophyton floccosum*.

Selection of Antigen to Be Used for Immunization of Mice

In order to obtain MAbs, a suitable antigen preparation has to be selected which can be used for the immunization of mice. Several culture filtrates were screened for the presence of different antigens. This was done by immunoblot analysis using rabbit antisera raised against dermatophyte antigens (*Trichophyton rubrum*).

These rabbit antisera were raised as follows: Antigens from seven culture filtrates were isolated, mixed, and used for immunization of three rabbits (antisera called Cult). Another three rabbits were immunized with acetone-dried material from mycelia (antiserum called Rubrac).

Diversity of antigenic extracts was demonstrated by these antisera (de Haan, et al., 1987). With the help of these antisera a suitable antigen preparation was selected out of a number of culture filtrate extracts. These extracts were analyzed using immunoblotting after polyacrylamide gel electrophoresis (PAGE). The extracts were placed on a 10% separating polyacrylamide gel. Each sample applied consisted of 20 µl extract solution (20 mg/ml) and 10 µl sample buffer (buffered glycerol/SDS). The antigens were electroblotted onto nitrocellulose paper. After rinsing with PBS containing Tween 20 (0.05%), the paper was dried and stored. The nitrocellulose paper was cut into strips, each strip containing all blotted glycoproteins or polysaccharides of a sample. The strips were cut into two identical halves and incubated with a 1:10 dilution of either Cult or Rubrac in PBS/Tween 20, followed by an incubation with sheep anti-rabbit IgG conjugated with horseradish peroxidase (SaR-HRP, 20 µg/ml). The enzyme activity was revealed using aminoethylcarbazole (AEC)/H_2O_2. As can be seen from Figure 1, the 10 culture filtrates showed different patterns after incubation with the antisera Rubrac and Cult.

Antigens with low molecular weight are mostly found with Cult. In two culture filtrates antigens could be demonstrated with Rubrac, which were not

Table 1 Immunization Scheme

	Day	Dose (μg)	Route
Priming	−135	50	IP + CFA
Booster	−99	150	IP + CFA
Final booster	−3	1000 + 1000	IV + IP
Splenectomy	0		

CFA = Complete Freund's adjuvant; IP = intraperitoneal injection; IV = intravenous injection.

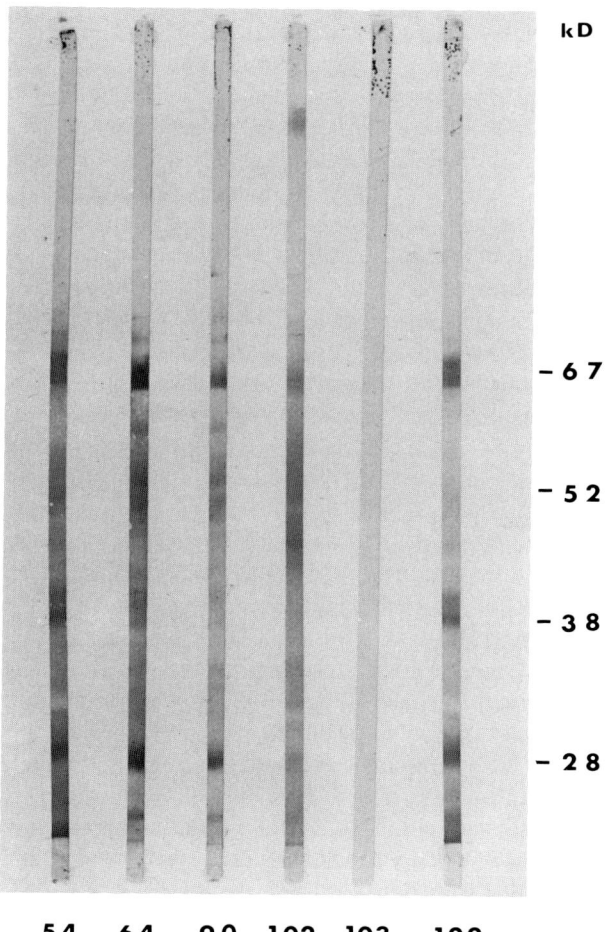

Figure 2 Results of immunoblot assay using 84 M1020 as antigen. Strips were incubated with MAbs 54, 64, 90, 102, 103, and 122.

detected by Cult (666, 1000). The different culture filtrates displayed a diversity in antigen composition. We selected culture filtrate 84M 1020 as antigen preparation for the immunization of mice, as this is the antigen preparation with the broadest spectrum of antigenic determinants, which is reflected in the diversity of molecular weights, in order to obtain monoclonal antibodies. This culture filtrate also shows the strongest staining pattern.

Preparation of Monoclonal Antibodies

Female BALB/c mice (10 weeks old) were immunized with *Trichophyton rubrum* antigen 84M 1020 according to the scheme shown in Table 1. Two weeks after the booster injection, a small sample was taken from each animal, and the sera were screened for a specific anti-*Trichophyton rubrum* immune response using ELISA with *Trichophyton rubrum* 84M 1020 as antigen. After the final booster injection, the mice were splenectomized after neck dislocation. For both somatic cell hybridization and culture of hybridoma cells, the methods described by Kenneth (1980), though in a modified way, were used (de Haan et al., 1985). Culture medium was RPMI 1640 (Flow), with added glutamine and the antibiotics penicillin and streptomycin at 100E and 100 μg/ml, respectively. Fetal calf serum (10%) was added to this culture medium. Half of the splenocyte population obtained after splenectomyzation was mixed with 10^8 mouse plasmacytoma cells X63-Ag8.653 (cell ratio approximately 1:1) and fused using slightly alkaline 50% PEG 4000 5% DMSO in RPMI medium. The remainder of the spleen cells were processed for cryopreservation in RPMI + 10% DMSO. After culture in bulk overnight, half of the fused cell population was transferred to RPMI+HAT medium and distributed at 10^5 cells/well into 96-well multiplates with at least 10^5 peritoneal macrophages/well in RPMI+ HAT, seeded 2 days previously. The other half of the fused cell population was processed for cryopreservation. At 15 days after fusion, a complete medium change was performed, and 3 days later colony supernatants were screened using ELISA for antibodies to *Trichophyton rubrum* antigen 84M 1020 (de Haan and Kalsbeek, 1983). Positive supernatants were also tested in the immunoblot system with the 84M 1020 antigen. Hybridoma cell cultures producing antibodies with the desired specificity were subcloned twice by limiting dilution (0.5 cell/well with 10^5 peritoneal macrophages in RPMI) and the supernatants were screened as described above. Secondary subclones with a stable production of monoclonal antifungal antibodies were expanded and processed for cryopreservation. Ascites was induced by interperitoneal injection of $2-5 \times 10^6$ washed hybridoma cells per female BALB/c mouse and harvested 2-4 weeks later. The results of immunoblot of the antigen 84M 1020 with monoclonal antibodies, which had been subcloned twice, are shown in Figure 2. Of the six monoclonal antibodies (54, 60, 90, 102, 103, 122), the most reactive with the

Table 2 Molecular Weight of Antigens in Six Culture Filtrates Demonstrated with MAb 54

MW (kD)	Filtrate					
	A	B	C	D	E	F
110						
96						
86						
76						
72						
70	+					
67	++					
61		+				
60						
58						
54	+					+
52	+					
50						
48	+	+				
46		+	+			
44						
42			+			
38	+					+
36	+					+
35						
34						
32		+				
30						
28	++					
26						
23	++	+				

A = *Trichophyton rubrum* 84 M1020; B = *Trichophyton rubrum* (HAL); C = *Trichophyton mentagrophytes* (HAL); D = *Trichophyton tonsurans*; E = *Penicillium brevicompactum*; F = *Aspergillus fumigatus*; + = positive band; ++ = strong positive band.

Table 3 Molecular Weight of Antigens in Six Culture Filtrates Demonstrated with MAb 64

MW (kD)	Filtrate					
	A	B	C	D	E	F
110			++			
96		++	++		+	
86	+					
76	+					
72						
70	+				+	
67	++		+		+	
61		++				
60	++	++	++		++	
58	++			++	++	++
54	+	++			++	
52	+	++	+	++		++
50					++	
48	++	++	+			
46		++	++			
44				+		
42		++	++			+
38	+		+	+	++	++
36	+	+	+		++	++
35					++	
34					++	
32	+	++			++	
30	+	++	+		++	
28	++	++	++			+
26						
23	++	++		++		

A = *Trichophyton rubrum* 84 M1020; B = *Trichophyton rubrum* (HAL); C = *Trichophyton mentagrophytes* (HAL); D = *Trichophyton tonsurans*; E = *Penicillium brevicompactum*; F = *Aspergillus fumigatus*; + = positive band; ++ = strong positive band.

Table 4 Molecular Weight of Antigens in Six Culture Filtrates Demonstrated with MAb 90

MW (kD)	Filtrate					
	A	B	C	D	E	F
110						
96						
86	+					
76	+					
72						
70	+					
67	++					
61		++				
60	+	++				
58	+					
54		++				
52	+	++				
50				ND	ND	ND
48	++	++	+			
46	+	++	+			
44						
42			++			
38						
36						
35						
34						
32		+				
30						
28	+		++			
26						
23	+	++				

A = *Trichophyton rubrum* 84 M1020; B = *Trichophyton rubrum* (HAL); C = *Trichophyton mentagrophytes* (HAL); D = *Trichophyton tonsurans*; E = *Penicillium brevicompactum*; F = *Aspergillus fumigatus*; + = positive band; ++ = strong band. ND = not done.

Table 5 Molecular Weight of Antigens in Six Culture Filtrates Demonstrated with MAb 102

MW (kD)	Filtrate					
	A	B	C	D	E	F
110	++					
96						
86	+					
76	+					
72						
70	+					
67	++					
61		+				
60	+					
58						+
54	+					
52	+	+				
50						
48	+	+				
46	+	+				
44	+					
42	+					
38	+				+	
36	+				+	
35						
34						
32	+					
30	+					
28	++		+			
26						
23	++	+				

A = *Trichophyton rubrum* 84 M1020; B = *Trichophyton rubrum* (HAL); C = *Trichophyton mentagrophytes* (HAL); D = *Trichophyton tonsurans*; E = *Penicillium brevicompactum*; F = *Aspergillus fumigatus*; + = positive band; ++ = strong positive band.

Table 6 Molecular Weight of Antigens in Six Culture Filtrates Demonstrated with MAb 103

MW (kD)	A	B	C	D	E	F
110						
96						
86					+	
76						
72						
70						
67						
61		+				
60					+	
58						+
54						
52						
50						
48		+				
46		+	+			
44						
42			+			
38			++			+
36						+
35					+	
34					+	
32					+	
30						
28	±	+	+			
26						
23		+				

A = *Trichophyton rubrum* 84 M1020; B = *Trichophyton rubrum* (HAL); C = *Trichophyton mentagrophytes* (HAL); D = *Trichophyton tonsurans*; E = *Penicillium brevicompactum*; F = *Aspergillus fumigatus*; + = positive band; ++ = strong positive band.

Table 7 Molecular Weight of Antigens in Six Culture Filtrates Demonstrated with MAb 122

MW (kD)	Filtrate					
	A	B	C	D	E	F
110						
96						
86						
76						
72						
70						
67	++					
61		+				
60				+		
58				+		
54						+
52		+				
50				+		
48		++	++			
46		+	++			
44						
42			+			
38	++		++		+	+
36	++		+		+	+
35						+
34						+
32	+	+	+			+
30			+			
28	++		++			
26						
23	++	+				

A = *Trichophyton rubrum* 84 M1020; B = *Trichophyton rubrum* (HAL); C = *Trichophyton mentagrophytes* (HAL); D = *Trichophyton tonsurans*; E = *Penicillium brevicompactum*; F = *Aspergillus fumigatus*; + = positive band; ++ = strong positive band.

immunizing culture filtrate were 54 and 64. These monoclonal antibodies appeared to detect a wide range of antigens with different molecular weights. As can be seen from Tables 2-7, MAb 102 detects the greatest number of antigens, reflected by the molecular weights, but MAb 64 shows the strongest reaction. MAb 103 detects only one antigen (28 kD band) weakly.

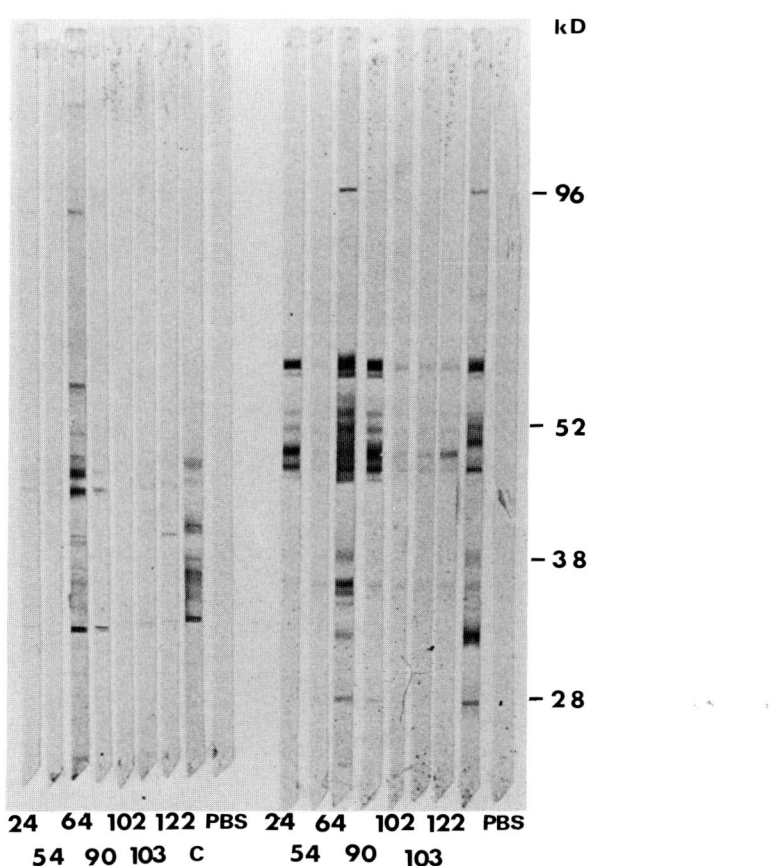

Figure 3 Results in the immunoblot assay using *Trichophyton rubrum* (right) and *Trichophyton mentagrophytes* (left) as antigen. Strips were incubated with MAbs 54, 64, 90, 102, 103, and 122. C = Cult; PBS = phosphate-buffered saline.

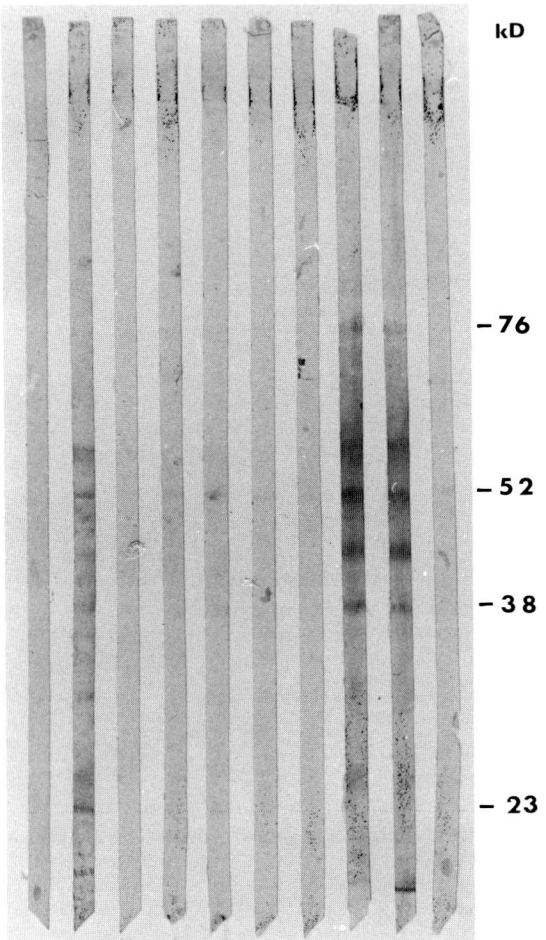

Figure 4 Results in the immunoblot assay using *Trichophyton tonsurans* as antigen. Strips were incubated with MAbs 54, 64, 102, 103, and 122. MS = normal mouse serum; R = Rubrac; C = Cult; PBS = phosphate-buffered saline.

RESULTS

Monoclonal Antibody Reaction with Commercial *Trichophyton rubrum* and *Trichophyton mentagrophytes* Antigens

Two antigen extracts of *Trichophyton rubrum* (84M 1020 and one commercially obtained, HAL Haarlem, The Netherlands) and an extract of *Trichophyton mentagrophytes* (HAL, The Netherlands) were electrophorized and blotted onto nitrocellulose. The nitrocellulose strips with these antigens were cut and incubated with MAbs 54, 64, 90, 102, 103, and 122. Using these MAbs a different staining pattern was obtained with the commerical antigens from that obtained with the 84M 1020 antigen, which was used for immunization of the mice (Fig. 2). It is remarkable that the MAbs react for the greater part with other antigens (Tables 2-7), reflected in the molecular weights. MAb 64 can detect the most common antigenic determinants both in the 84M 1020 antigen and in the commercial antigen extracts. As can be seen from Figure 3 and Tables 3, 4, and 7, the MAbs 64, 90, and 122 react with both the *Trichophyton rubrum* antigen extract and the *Trichophyton mentagrophytes* extract. The other MAbs do not react, or react only slightly, with the Trichophyton Mentagrophytes antigen extract. Apparently the MAbs 64, 90, and 122 cross-react with *Trichophyton mentagrophytes* antigen extracts, moreover, for a number of components they react with antigens having the same molecular weight (kD) as those of *Trichophyton rubrum*.

Monoclonal Antibody Reaction with Other Dermatophytes—Cross-Reactivity

As MAbs against *Trichophyton rubrum* cross-react with *Trichophyton mentagrophytes* antigens, it is of interest to know if they also cross-react with other dermatophytes. Nitrocellulose strips with, respectively, the antigens of *Epidermophyton floccosum* and *Trichophyton tonsurans* were incubated with MAbs 54, 64, 102, 103, and 122. These MAbs reacted only with components in the antigen extract of *Trichophyton tonsurans* (Fig. 4). MAb 64 was mainly accountable for the reaction with the *Trichophyton tonsurans* antigens.

The antisera Rubrac and Cult, which were produced by immunization of rabbits (see selection of antigen to be used for immunization of mice), also reacted with *Trichophyton tonsurans* antigens.

Monoclonal Antibody Reaction with Nondermatophytes

The MAbs 54, 64, 102, 103, and 122 were also tested for reactivity with antigen extracts from nondermatophyte fungi. From the fungi *Penicillium brevicompactum* and *Aspergillus fumigatus* (HAL, The Netherlands), antigen extracts

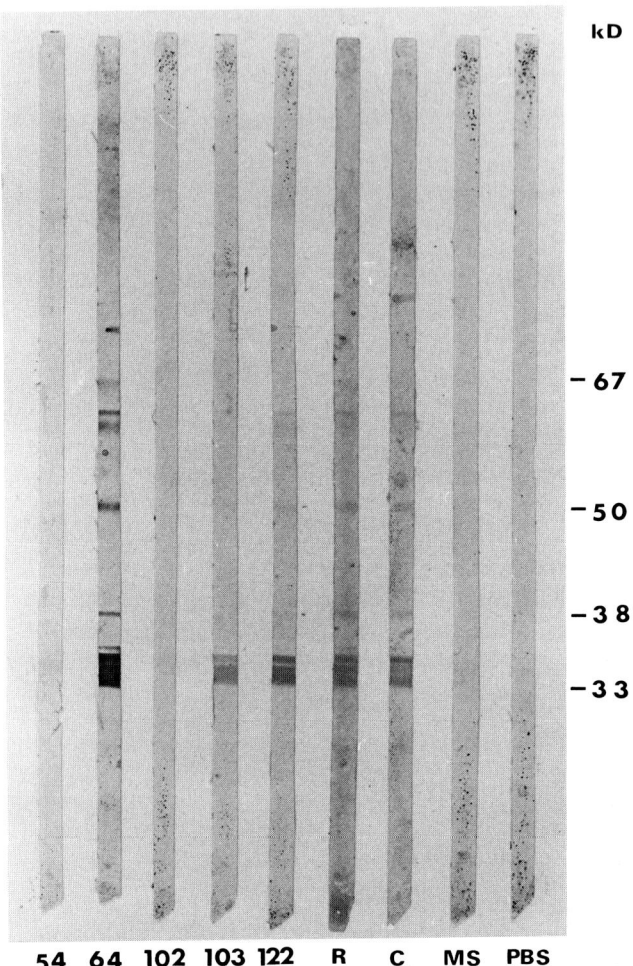

Figure 5 Results in the immunoblot assay using *Penicillium brevicompactum* as antigen. Strips were incubated with MAbs 54, 64, 102, 103, and 122. MS = normal mouse serum; C = Cult; R = Rubrac; PBS = phosphate-buffered saline.

Characterization of Dermatophyte Antigens

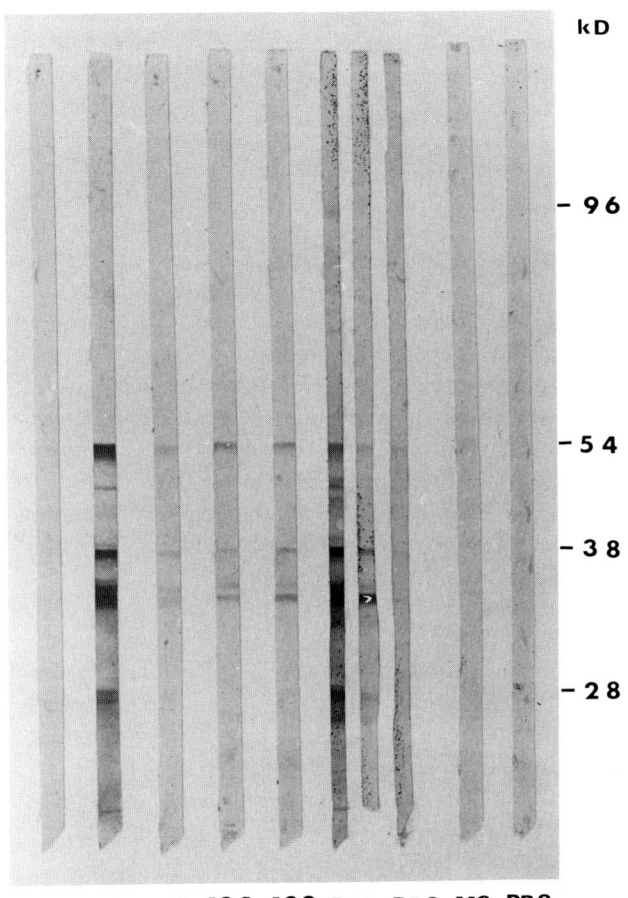

Figure 6 Results in the immunoblot assay using *Aspergillus fumigatus* as antigen. Strips were incubated with MAbs 54, 64, 102, 103, and 122. MS = normal mouse serum; C = Cult; R = Rubrac; PBS = phosphate-buffered saline.

were prepared and blotted and thereafter incubated with the MAbs. For MAb 64 there was a similarity in the staining pattern for *Penicillium brevicompactum* antigens and 84M 1020 *Trichophyton rubrum* antigens (Table 3). The other MAbs did not react, or reacted slightly, with the *Penicillium brevicompactum* antigen extract (Figure 5). With MAbs 64, 103, and 122 antigenic components were also demonstrated in the *Penicillium brevicompactum* extract, which were not detected in the *Trichophyton rubrum* extract. Rubrac and Cult detected the same antigenic components as did MAb 64. MAb 64 reacted also with the *Aspergillus fumigatus* (Fig. 6) antigen extract, although not intensively. The other MAbs reacted only slightly with the same small number of antigenic components of *Aspergillus fumigatus*.

DISCUSSION

This study shows that the MAbs raised against the fungus *Trichophyton rubrum* are suitable for characterization of dermatophyte antigens. In order to produce these MAbs, an appropriate antigen preparation was selected out of a number of culture filtrate extracts, with the help of the antisera Rubrac and Cult. The antigen preparation showing the broadest reactivity with both Rubrac and Cult was chosen for the production of MAbs with as many different specificities as possible.

The MAbs produced (54, 64, 90, 102, 103, 122) detect different antigenic determinants in the culture filtrate (84M 1020), which was used for the immunization of mice. This was apparent from the different staining patterns in the immunoblot tests. It is remarkable that all MAbs detect a range of antigenic components with different molecular weights. These different components probably do not reflect degradation of material, as a diffuse staining pattern would be expected. A detection of subunits is unlikely because a clear relationship between the molecular weight of the different components does not exist. Therefore, one may conclude that the MAbs detect one single determinant, which is located on a number of antigenic components with different molecular weights. It is well known that culture filtrate glycoproteins are comprised of about 80% polysaccharides. Therefore, it is reasonable to assume that the antigenic determinants recognized by the MAbs are oligosaccharides.

When reactivity patterns from two different *Trichophyton rubrum* preparations with the MAbs are compared, though a partial overlap can be detected, a considerable difference is apparent. This confirms earlier results (de Haan et al., 1987) documenting diversity in dermatophyte antigens from different isolations of *Trichophyton rubrum*. In what way this diversity in combination with the host response plays a role in the development of chronic *Trichophyton rubrum* infection is as yet unclear.

Until now, the commercially available trichophytin preparations were not standardized. The present MAbs make it possible to analyze, characterize, and standardize these preparations, thus creating uniformity by which the results of investigations can be compared more accurately.

Another interesting feature of the present panel of MAbs is their reactivity pattern toward other dermatophytes and toward unrelated fungi. Culture filtrate extracts from *Trichophyton mentagrophytes*, *Trichophyton tonsurans*, *Epidermophyton floccosum*, *Penicillium brevicompactum*, and *Aspergillus fumigatus* were studied. It appeared that MAbs 54 and 102 reacted almost exclusively with *Trichophyton rubrum* antigens. The MAbs 64 and 122 reacted more or less with all the investigated extracts. From these data it can be concluded that MAbs 64 and 122 recognize a very common determinant present on many related and unrelated fungi. Several authors (Blank, 1953; Bishop et al., 1965; 1966) demonstrated the presence of similar polysaccharide moieties from different dermatophytes. These results are actually confirmed by the data obtained in the present study using MAbs. The presence of identical polysaccharide moieties from dermatophytes and from these unrelated fungi has not been documented so far. Our results clearly demonstrate that such moieties can be found. Patients with diseases caused by these unrelated fungi sometimes show a positive reaction to an intracutaneous test (humoral immune response) with *Trichophyton* (Jones et al., 1973). A possible relationship between those diseases caused by the unrelated fungi and the immune response to dermatophytes could exist.

Further characterization of the antigens involved will help to clarify pathological mechanisms of dermatomycoses.

REFERENCES

Artis, W. M., and Jones, H. E. (1976). The effect of a mammalian-cell active lymphokine on fungal growth. *Fed. Proc. 35*:2120.

Berk, S. H., Penneys, N. S., and Weinstein, G. D. (1976). Epidermal activity in annular dermatophytosis. *Arch. Dermatol. 112*:485-488.

Bishop, C. T., Perry, M. B., and Blank, F. (1966). The water-soluble polysaccharides of dermatophytes. V. Galactomannans II from *Trichophyton granulosum, Trichophyton interdigitale, Microsporum quinckeanum, Trichophyton rubrum*, and *Trichophyton schönleinii*. *Can. J. Chem. 44*:2291-2297.

Bishop, C. T., Perry, M. B., Blank, F., and Cooper, F. P. (1965). The water-soluble polysaccharide of dermatophytes. IV. Galactomannans I from *Trichophyton granulosum, Trichophyton interdigitale, Microsporum quinckeanum, Trichophyton rubrum*, and *Trichophyton schönleinii*. *Can. J. Chem. 43*:30-39.

Blank, F. (1953). The chemical composition of the cell walls of dermatophytes. *Biochim. Biophys. Acta 10*:110-113.

Cruickshank, C. N. D., Trotter, M. D., and Wood, S. R. (1960). Studies on trichophytin hypersensitivity. *J. Invest. Derm. 35*:219-223;

De Haan, P., and Kalsbeek, G. L. (1983). Humoral immune response to some benzylpenicillin preparations. *Allergy 39*:319-324.

De Haan, P., de Jonge, A. J. R., Vergrugge, T., and Boorsma, D. M. (1985). Three epitope-specific monoclonal antibodies against the hapten penicillin. *Int. Archs. Allergy Appl. Immun. 76*:42-46.

De Haan, P., Van der Raay-Helmer, E. M. H., and Boorsma, D. M. (1987). Diversity of antigenic extracts from the dermatophyte *Trichophyton rubrum*. *Mykosen 30*:427-433.

Grappel, S. F., Bishop, C. T., and Blank, F. (1974). Immunology of dermatophytes and dermatophytosis. *Bacteriol. Rev. 38*:222–250.

Jones, H. E., and Artis, W. M. (1981). Dermatophytosis. In *Immunodermatology*. Edited by Safai, B., and Good, R. A. Plenum Medical Book Company, London.

Jones, H. E., Rinaldi, M. G., Chai, H., and Kahn, G. (1973). Apparent cross-reactivity of airborn moulds and the dermatophytic fungi. *J. All. Clin. Immunol. 52*:346–351.

Kaaman, T., von Stedingk, L., von Stedingk, M., and Wasserman, J. (1981). ELISA-determined serological reactivity against purified trichophytin in dermatophytosis. *Acta Derm. Venerol. 61*:313–317.

Kenneth, R. H. (1980). Hybridomas: a new dimension in biological analysis. In *Monoclonal antibodies*. Plenum Press, New York.

King, R. D., Kahn, H. A., Foye, J. C., Greenberg, J. H., and Jones, H. E. (1975). Transferrin, iron, and dermatophyte inhibitory component definitively identified as unsaturated transferrin. *J. Lab. Clin. Med. 86*:204–212.

Monroe, E. W., Wade, T. R., Artis, W. M., Young, J. C., and Jones, H. E. (1976). The transition metal content of human plantar stratum corneum. *Clin. Res. 25*:284A.

Yu, R. J., Grappel, S. F., and Blank, F. (1972). Inhibition of keratinases by $\alpha 2$-macroglobulin. *Experientia 28*:886.

6
Antigenic Structure of *Coccidioides immitis*

REBECCA A. COX
San Antonio State Chest Hospital, San Antonio, Texas

INTRODUCTION

Coccidioides immitis, the etiologic agent of coccidioidomycosis, is a dimorphic fungus which exists in the soil in the arid and semiarid regions of the Western Hemisphere (Pappagianis, 1980). In its natural habitat, the fungus propagates in a mycelial form, which gives rise to enterothallic arthroconidia. Primary infection is acquired by inhalation of the arthroconidia, which undergo a morphogenetic conversion into a spherule/endospore phase within host tissue.

The spectrum of disease caused by *C. immitis* ranges from an asymptomatic infection to a severe, disseminated process, which commonly involves the skin, bones and joints, meninges, and genitourinary tract (Drutz and Catanzaro, 1978b). Early epidemiological studies by Smith and colleagues (1946) established that fully 60% of primary infections are asymptomatic, evidenced only by skin test reactivity to coccidioidal antigen. Symptomatic disease occurs in about 40% of primary infections and, although most of these are resolved without complication, approximately 5% develop into a persistent pulmonary disease and about 1% progress to disseminated coccidioidomycosis. Exceptions to the incidence of disease involvement occur in persons exposed to heavy inocular of airborne arthroconidia (Werner et al., 1972; Larsen et al., 1985), in pregnant women (Smale and Birsner, 1949; Vaughan and Ramirez, 1951; Harvey, 1980), in immunocompromised persons (Deresinski and Stevens, 1975; Rowland et al., 1977), including those with the acquired immunodeficiency syndrome (Roberts 1984; Wolf et al., 1986; Prichard et al., 1987), and in persons of certain ancestry, notably Filipinos and blacks (Drutz and Catanzaro, 1987a; Pappagianis, 1980; Johnson, 1982).

Host response to primary infection with *C. immitis* is manifested by the acquisition of delayed-type hypersensitivity and the production of antibodies to coccidioidal antigens. The profile of these responses as they relate to the course of disease was delineated in a series of epoch studies by Smith and co-workers (1948, 1950, 1956). Approximately 83% of persons with nonprogressive, pulmonary coccidioidomycosis acquire skin test reactivity to coccidioidin (CDN) during the first week of clinical illness, 93% are skin test positive by the second week, and virtually all persons are reactive by the third week. Skin test reactivity persists in most persons who recover from their primary coccidioidal infection, and, with few exceptions, such individuals are endowed with life-long immunity to the disease. In contrast to the strong T-lymphocyte reactivity manifested in self-limited disease, progressive coccidioidomycosis is typically associated with depressed or nondemonstrable T-cell reactivity to CDN (Catanzaro et al., 1975; Cox and Vivas, 1977; Harvey and Stevens, 1981).

Antibody responses to coccidioidal antigens have been detected within the IgM, IgG, IgA, and IgE classes (Pappagianis et al., 1965; Sawaki et al., 1966; Cox and Arnold, 1979). Primary infection is accompanied by the production of tube precipitin (TP) antibody, which is predominantly of the IgM isotype (Pappagianis et al., 1965; Sawaki et al., 1966). The TP antibody response is short-lived, being demonstrable in 90% or more of patients within the first 4 weeks of clinical onset and decreasing to less than 10% of patients within the next 4 months (Smith et al., 1950, 1956). Complement-fixing (CF) antibody to CDN is detected later during the course of coccidioidomycosis, generally within 3 to 4 months of clinical onset, and persists in most patients throughout active disease (Smith et al., 1950, 1956). This antibody response is primarily of the IgG isotype (Pappagianis et al., 1965; Sawaki et al., 1966).

Detection of the TP or CF antibody responses can be accomplished using the classical TP and CF assays described by Smith and co-workers (1950) or by the immunodiffusion (ID) assays developed by Huppert and colleagues (1965a,b), and designated the IDTP and IDCF assays, respectively. It has not been proved that the IDTP and IDCF assays detect the same antigen/antibody reactions as the TP and CF tests; however, the high correlation between the results are consistent with this presumption (Huppert et al., 1965a,b; Wood et al., 1982).

MORPHOGENESIS AND ULTRASTRUCTURE

Morphogenetic Cycle

Although *C. immitis* is classified as a dimorphic fungus, it exhibits a complex and unique multiphasic life cycle (Fig. 1). As a soil saprophyte and on most common laboratory media, the fungus propagates in a mycelial phase which

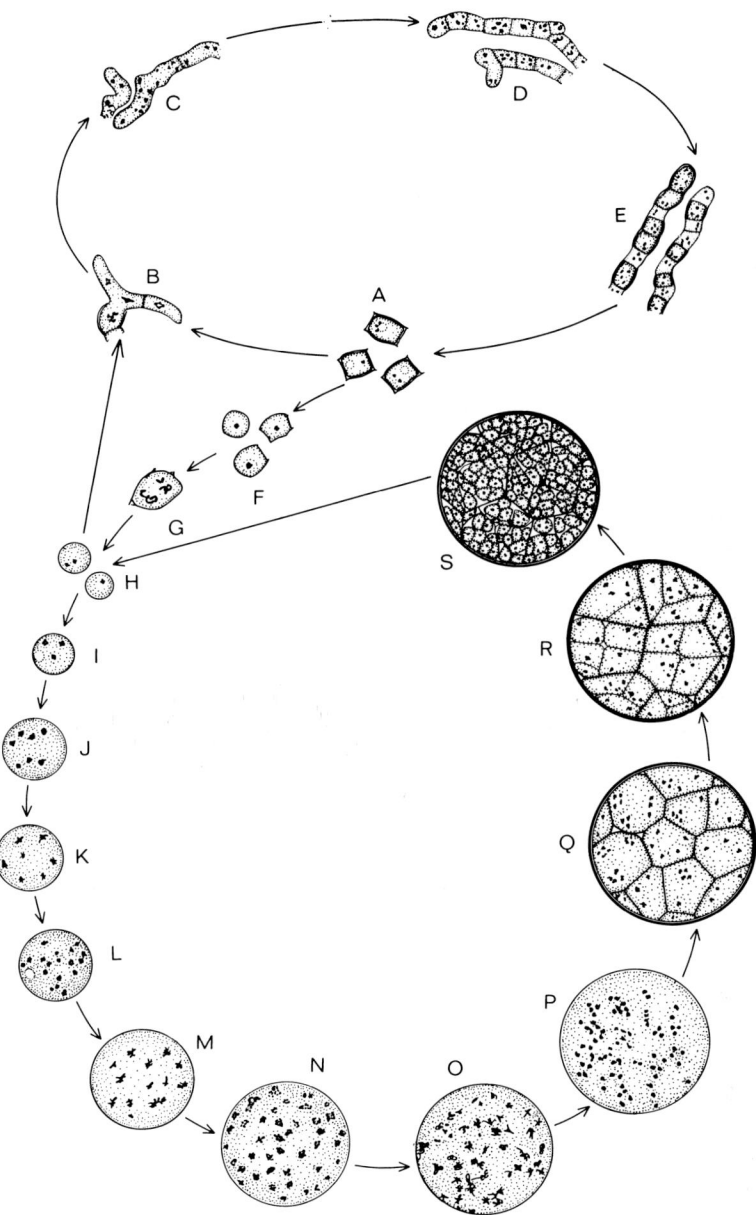

Figure 1 Saprobic (A–E) and parasitic (F–S) life cycles of *C. immitis*. (From Sun and Huppert, 1976.)

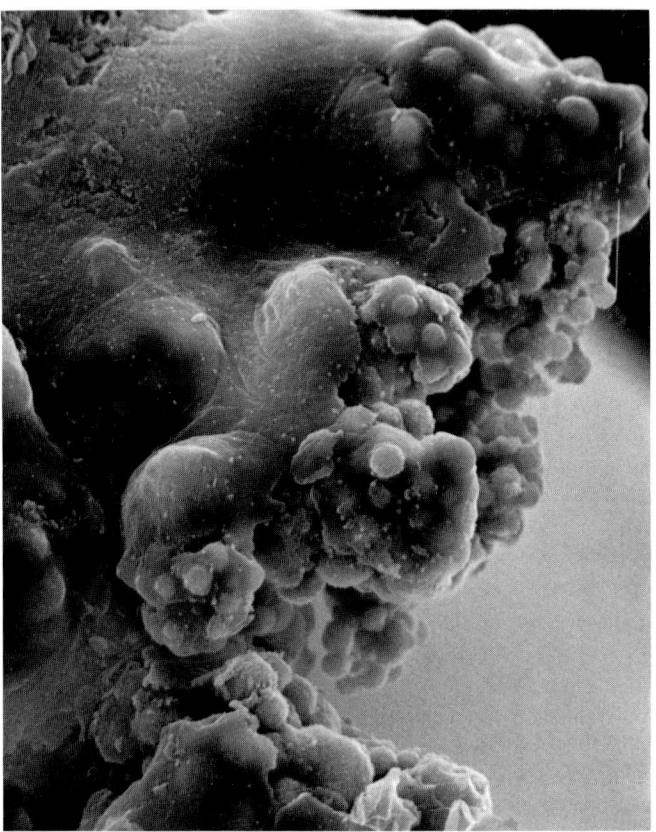

Figure 2 Ruptured spherules showing newly released endospores encased in a fibrillar membrane. (From Huppert, Sun, and Harrison, 1982.)

gives rise to arthroconidia by a process of enterothallic development (Huppert et al., 1982; Cole and Sun, 1985a). At the initiation of this process, the hyphae cease apical extension and undergo septation, forming chains of multinucleated compartments. Cells that are destined to become arthroconidia show wall synthesis and condensed cytoplasm, while the adjacent cells become nonrefractile and eventually autolyze, resulting in the disarticulation of the arthroconidia. The newly released arthroconidia measure 3-6 μm in length and 2.5-4 μm in diameter and are characteristically thick-walled, barrel-shaped cells.

Conversion of the arthroconidia into the parasitic phase occurs within tissues of a susceptible host or in vitro by incubating the mycelial-phase cells in

The Antigenic Structure of Coccidioides immitis

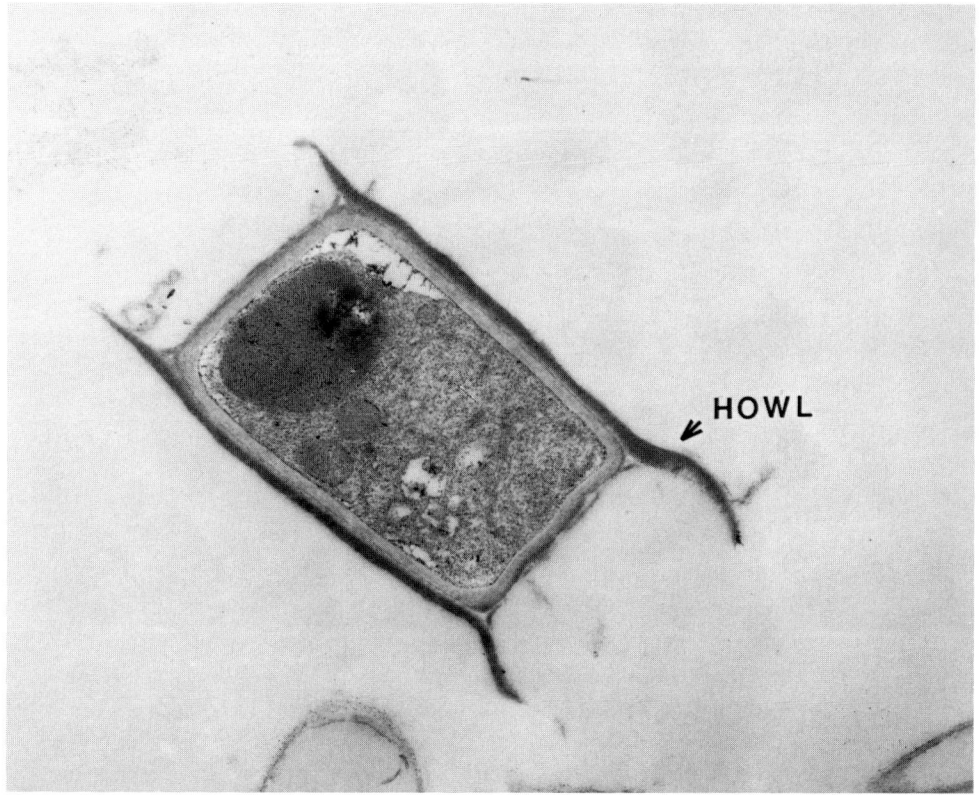

Figure 3 Arthroconidium showing remnants of the hyphal outer wall layer (HOWL). (From Huppert, Sun, and Harrison, 1982.)

Converse medium (1955, 1956, 1957) at 37–40°C to prevent mycelial growth and under a 20% CO_2 atmosphere. Under these conditions, the arthroconidia begin to swell and round up and then undergo multiple and apparently synchronous nuclear divisions. Nuclear division ceases and is followed by a progressive segmentation of the protoplasm, eventually giving rise to uninucleated endospores. At maturity the spherules (30–80 μm in diameter) rupture, releasing numerous endospores, which are initially encased in a fibrillar membrane (Fig. 2). Each endospore (2–4 μm in diameter) has the potential of developing into a mature, endosporulating spherule, thus repeating the parasitic cycle. The

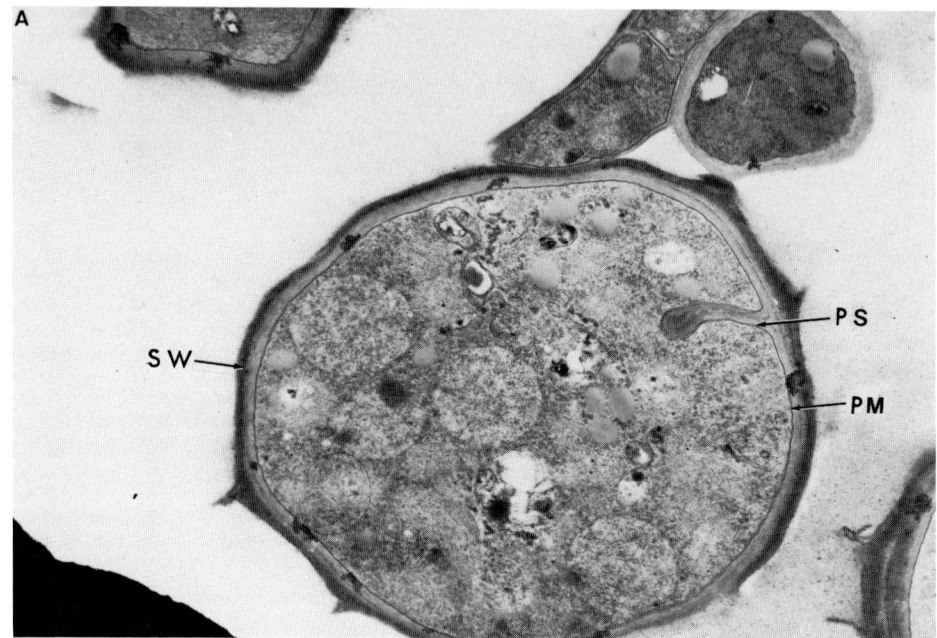

Figure 4 (A) Young immature spherule showing initiation of primary segmentation (PS) from inner wall layer of young spherule wall (SW). (PM, plasma membrane.) (B) Progressive segmentation of protoplasm. (PS,SS, and TS are primary, secondary and tertiary segmentation planes, respectively.) (C) Mature, endosporulating spherule. (E, endospore; N, nucleus; SW, spherule wall.) (From Sun, Sekhon, and Huppert, 1979.)

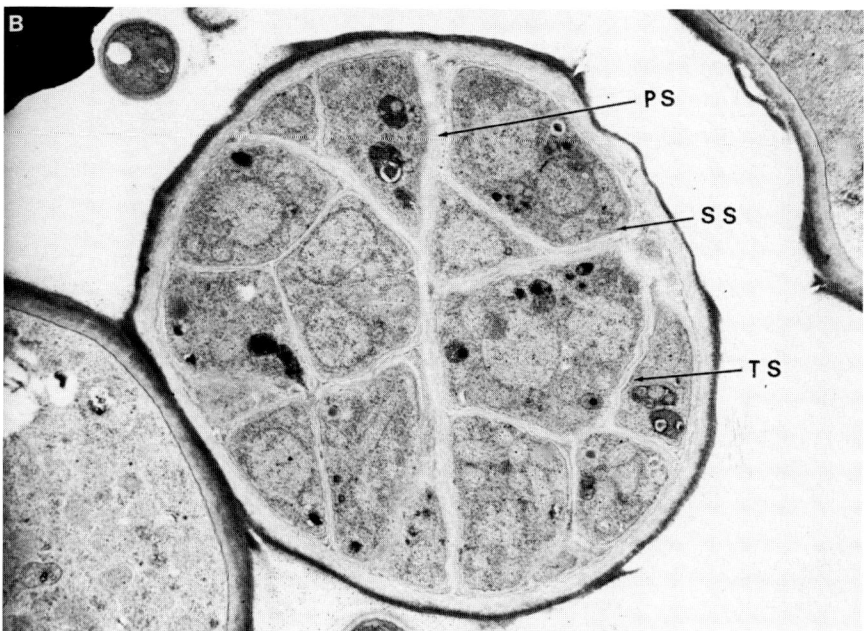

Figure 4

developmental stages of endosporulating spherules observed in vitro appear to be representative of their development in host tissues (Donnelly and Yunis, 1974; Sun and Huppert, 1976; Gabal, 1985; Miyaji et al., 1985; Sun et al., 1986).

Ultrastructure

By electron microscopy the cell wall of the arthroconidium is shown to be comprised of an electron-opaque outer layer that measures about 0.1 μm in thickness and an inner electron translucent layer that measures approximately 0.2 μm in thickness (Fig. 3). The outer wall layer is a remnant from the parent hyphal wall (HOWL) and, in newly released arthroconidia, is observed to extend

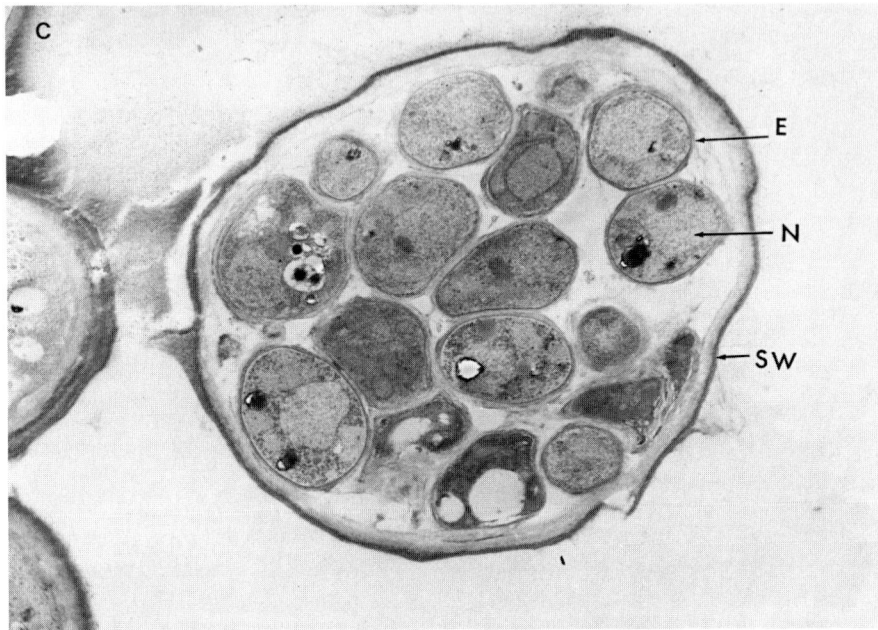

Figure 4

beyond the cell as tags from the walls of the degenerated intercalary compartments. Scanning electron microscopy reveals that this hydrophobic wall layer occurs as irregular folds over the arthroconidium.

Electron micrographs of successive stages in the in vitro conversion of an arthroconidium into a mature, endosporulating spherule are shown in Figure 4. The cell wall of the young, immature spherule resembles that of the arthroconidium in having an outer electron-dense wall layer and a multilayered, electron-transparent inner wall. The primary segmentation process appears to begin with an invagination of the two innermost layers of the inner wall (Huppert et al., 1982). This wall ingrowth continues, resulting in the segmentation of the cell into multinucleated compartments and eventually uninucleated endospores.

Limited information has accrued on the nature of the components that comprise the cell walls of *C. immitis* (reviewed in Reiss, 1986). Hector and Pappagianis (1982) reported that treatment of immature spherules with chitinase digested the inner cell wall layer adjacent to the cytoplasmic membrane without a detectable effect on the outer cell wall. Pronase treatment resulted in a buckling of the wall and detachment of the wall from the cytoplasmic membrane. A similar effect was observed following treatment with mannosidase. The morphological alterations noted in spherules treated with pronase or mannosidase, alone and sequentially, suggest that a mannoprotein complex is distributed throughout the inner cell wall layer.

Immature and mature spherules have an electron-dense, fibrillarlike matrix on their surface (Frey and Drutz, 1986). This matrix material has cytochemical staining characteristics of a glycoprotein and is believed to impede contact between spherule-phase cells and phagocytes, a feature that would favor survival of the fungus in vivo.

IMMUNOREACTIVE ANTIGENS

Crude Lysate Antigens

The classical antigen preparation used to evaluate cell-mediated and humoral responses in coccidioidomycosis has been coccidioidin (CDN), prepared as the broth culture filtrate of mycelial-phase cells grown for several weeks in a synthetic asparagine medium (Smith et al., 1948). CDN preparations of proven efficacy have also been produced from 5-day-old mycelia suspended in 3% aqueous toluene and agitated on a gyratory shaker to facilitate enzymatic autolysis (Pappagianis et al., 1961b). With the development of a medium for the cultivation of spherules in vitro (Converse, 1955, 1956, 1957), Levine and co-workers (1969) prepared a biologically active aqueous extract from lysed spherules and designated this extract spherulin (SPH). In a series of studies, SPH was reported to exhibit a greater degree of sensitivity than CDN in eliciting cell-mediated immune responses (Levine et al., 1969; Stevens et al., 1974, 1975; Deresenski et al., 1974, 1977) and in detecting CF antibody (Scalarone et al., 1974). In other investigations, SPH was comparable to or less reactive than CDN (Huppert et al., 1977; Cox et al., 1977a, 1981; Gifford and Catanzaro, 1981). The specificity of SPH approximates that of CDN in skin tests of persons with noncoccidioidal mycoses, notably histoplasmosis and paracoccidioidomycosis (Levine et al., 1975), but has been reported to be significantly more cross-reactive than CDN in assays for CF antibody (Huppert et al., 1977).

Although CDN and SPH have an acceptable range of sensitivity and specificity (when compared to other fungal antigens), their heterogeneous composition limits their utility in immunological investigations. In an effort to circumvent

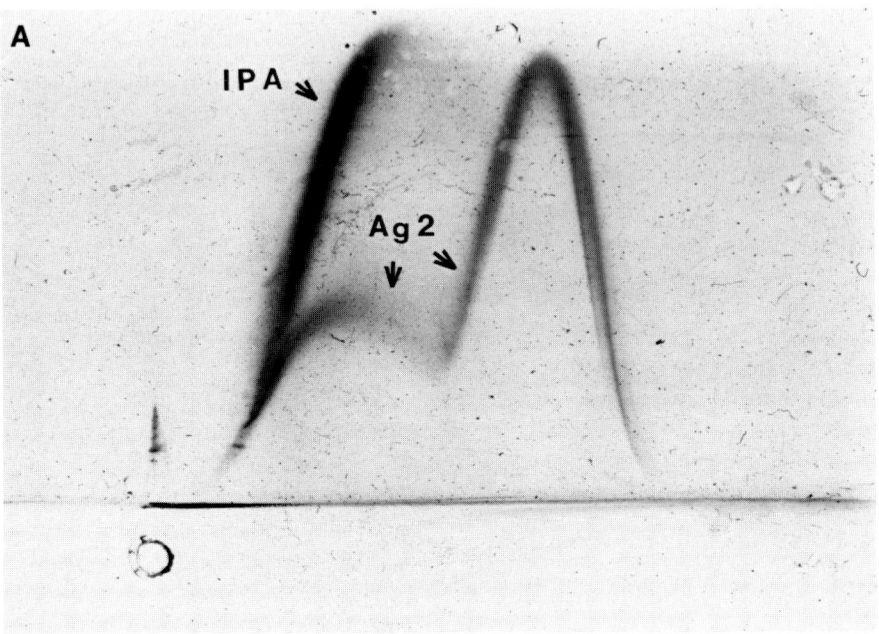

Figure 5 Detection of the incomplete precipitating antigen (IPA) and Antigen 2 (Ag2) in (A) C-ASWS and (B) CDN by 2D-IEP against goat antiserum to C-ASWS. (From Cox and Britt, 1985.)

this problem, investigators have conducted studies to isolate pure antigens from CDN or SPH or, as an alternative approach, directly from intact cells or cell walls. The results obtained in these studies are reviewed in the following sections.

TP Antigen

Early work established that the TP antigen is remarkably stable to heat treatment, with no significant decrease in precipitinogen activity even after autoclaving (Smith et al., 1950). The antigen was also shown to be stable at an acid pH and insoluble in 80% ethanol (Pappagianis et al., 1961).

Cumulative evidence suggests that the TP antigen is a cell wall component of *C. immitis*. Collins and co-workers (1977) showed that enzymatic digestion of intact spherules or spherule-derived cell walls with chitinase or lysozyme solubilized the TP antigen, as assessed by the reactivity of the fraction in the IDTP assay. Cox and associates (1984) reported that extraction of trypsin-treated cell

The Antigenic Structure of *Coccidioides immitis* 143

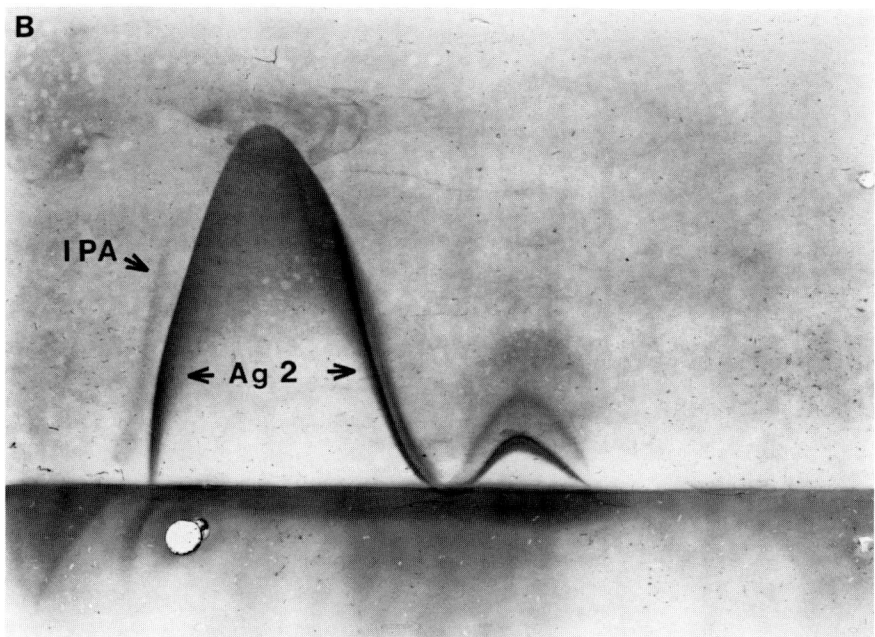

Figure 5B

walls with alkali yielded an alkali-soluble, water-soluble fraction (designated C-ASWS), which was reactive both in the IDTP and TP tests. More recently, Cole and co-workers (1987) isolated cell walls of arthroconidia by subjecting the cells to light sonication followed by fractionation in a Ribi apparatus. The soluble supernatant fraction was reactive with IDTP antibody.

Antigenic analysis of the cell wall extracts obtained by enzymatic digestion or mechanical dissection established the presence of other components, in addition to the IDTP antigen (Collins et al., 1977; Cole et al., 1987). Our initial analyses of C-ASWS were performed by 2D-IEP of C-ASWS in tandem with CDN against burro anti-CDN (Cox et al., 1984). The results indicated that this fraction was antigenically homogeneous, comprised of a polymeric antigen having antigenic activity with a CDN precipitinogen previously designated Antigen 2 (Ag2) by Huppert and co-workers (1978a). However, subsequent studies using goat antiserum prepared against C-ASWS (Cox and Britt, 1985) revealed the presence of a second component, characterized as an incomplete precipitating antigen (IPA) in 2D-IEP by having a complete cathodal precipitin leg and an

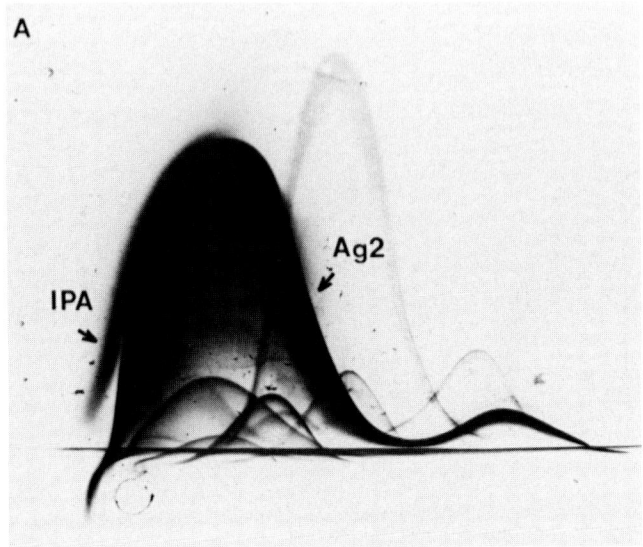

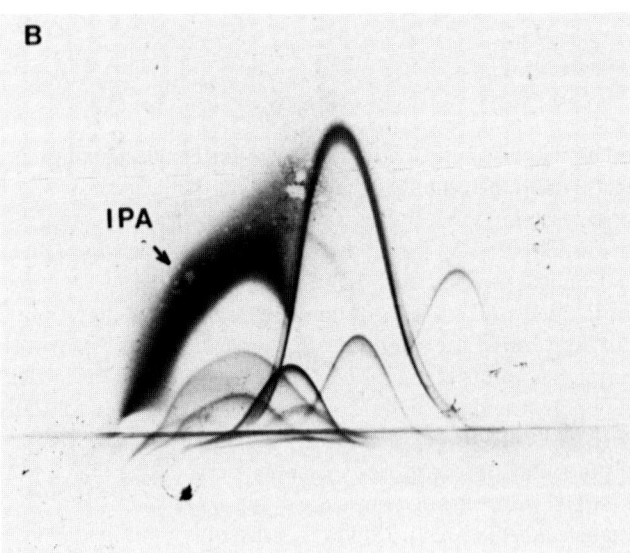

Figure 6 Antigenic analyses of (A) unfractionated CDN and (B) column effluent and (C) eluate fractions obtained in immunoaffinity chromatography of CDN with goat antiserum to C-ASWS, as assessed by 2D-IEP against goat anti-CDN. (From Cox and Britt, 1986b.)

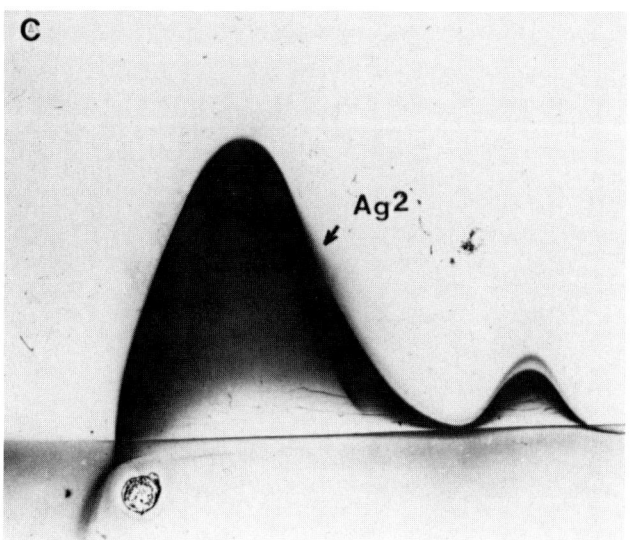

Figure 6C

incomplete anodal leg (Fig. 5A). The latter study also established that although the C-ASWS Ag2 retained antigenic activity, the immunoelectrophoretic profile of this polymeric component differed from CDN Ag2. Whereas the CDN Ag2 polymer showed a predominant cathodal peak (Fig. 5B), the C-ASWS polymer showed a predominant anodal peak. This difference suggests that the conformation and/or configuration of the latter was altered during physicochemical extraction.

The finding that C-ASWS was comprised of Ag2 and the IPA and was reactive in the IDTP assay prompted studies to isolate Ag2 and the IPA directly from CDN by immunoaffinity chromatography with goat antiserum to C-ASWS (Cox and Britt, 1986b). Analyses of the resulting immunosorbent effluent (nonadsorbed) and column eluate (desorbed) fractions by 2D-IEP established that Ag2 was completely adsorbed by goat anti-CASWS, whereas only trace amounts of the IPA were adsorbed (Fig. 6). Reactivity in the IDTP assay was demonstrable with both the column eluate and effluent, which, on the basis that the IPA was the only precipitinogen common to the two fractions, indicated that this component was the IDTP antigen. Attempts were made to purify the IPA from Ag2 by chromatography of the column effluent immunoadsorbent fraction on concanavalin A (ConA), Sephadex G-200, and DEAE-Sephacel. Both the IPA and Ag2 bound to the lectin and coeluted on molecular-sieve and ion-

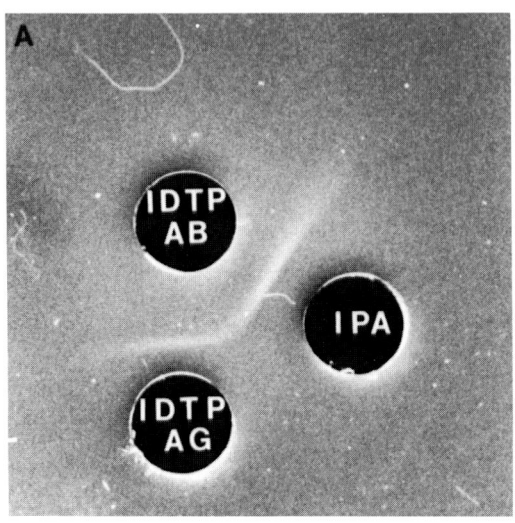

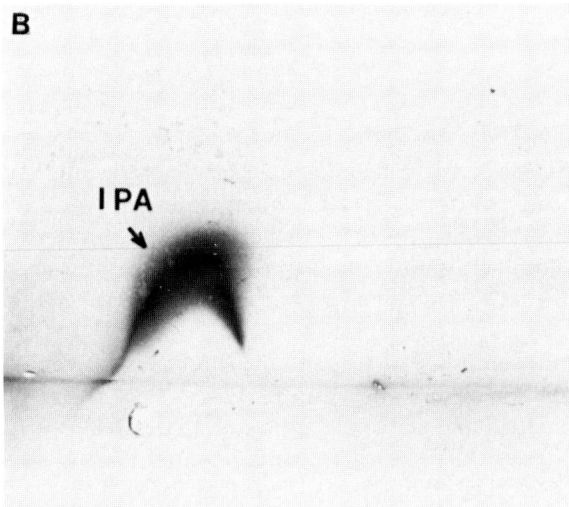

Figure 7 Identification of the IPA as the IDTP antigen. Reactivity of the DEAE-purified IPA in (A) 2D-IEP against goat anti-CDN and (B) in the IDTP assay. (From Cox and Britt, 1986b.)

The Antigenic Structure of *Coccidioides immitis*

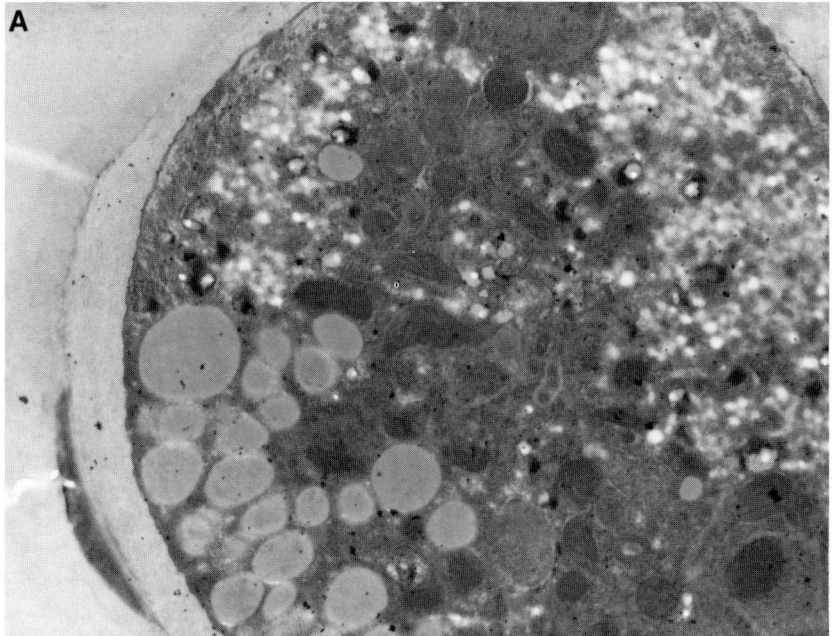

Figure 8 Immunoelectron microscopy of immature spherules with (A) normal goat antiserum or (B) goat antiserum to the purified IDTP antigen. The thin sections were reacted with the goat antisera and, after removal of nonbound antibodies, colloidal gold–labeled rabbit anti-goat was added. (From Harrison, Sun, and Cox, 1988.)

exchange columns. Purification of the IPA was achieved, however, by ion-exchange chromatography of the solid-phase immunoadsorbent column effluent fraction, which was devoid of Ag2 (Cox and Britt, 1986b). Of the multiple components present in the DEAE effluent fraction, only the IPA eluted from the column before initiation of the salt gradient (Fig. 7a). The DEAE-purified IPA was reactive in the IDTP assay (Fig. 7B) and in the classical TP assay. We also prepared goat antiserum to the IPA and showed that this antiserum yielded a line of total identity with the precipitin band obtained between reference IDTP antiserum and IDTP antigen, and by 2D-IEP detected only the IPA (Cox and Britt, 1987). These collective results established that the IPA is antigenically identical to the IDTP antigen, and, on this basis, the IPA will hereafter be designated the IDTP antigen.

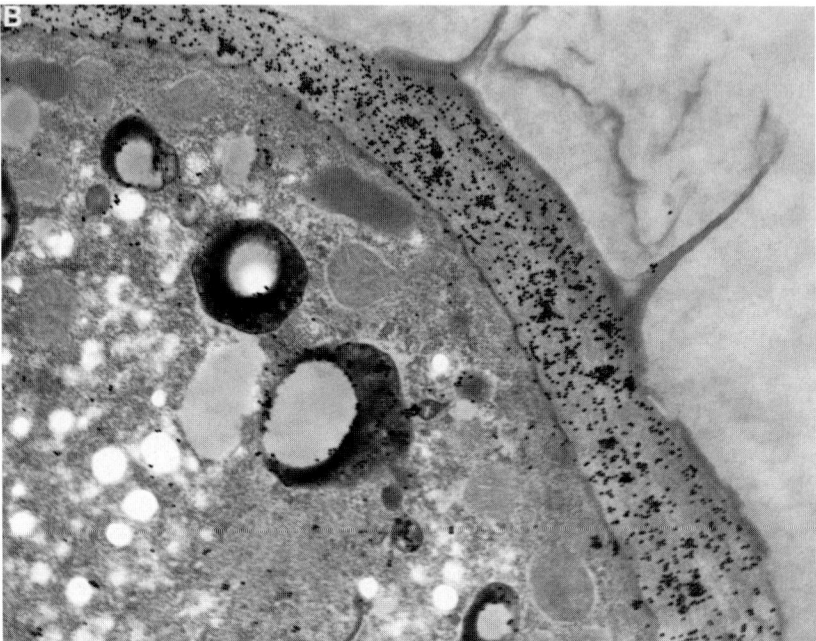

Figure 8B

The cellular location of the IDTP antigen in *C. immitis* has been examined by immunoelectron microscopy with goat anti-IDTP (Harrison et al., 1988). Reactivity was observed throughout the inner cell wall layer of *C. immitis* spherules (Fig. 8) and mycelia, with an increased concentration in the cell walls of the parasitic phase. In addition, the IDTP antigen was detected within cytoplasmic organelles having morphological and staining characteristics consistent with those of lysosomes. These results, taken together, indicate that the IPA is synthesized or modified within the lysosomes (or lysosomelike vesicles) and subsequently transported to the cell wall.

Our more recent work has been directed toward producing monoclonal antibody (MAb) against the IDTP antigen (Dolan et al., 1989). We have obtained an IgG1 MAb by fusing spleen cells of SPH-immunized mice with SP2/0 Ag14 myeloma cells. This MAb is detectable as a precipitin and, in the IDTP assay, yields a band of partial identity with the reaction between reference IDTP antigen and antiserum (Fig. 9). This partial identity can be explained on the basis of the diverse epitope–antibody interactions with the polyclonal anti-IDTP serum as

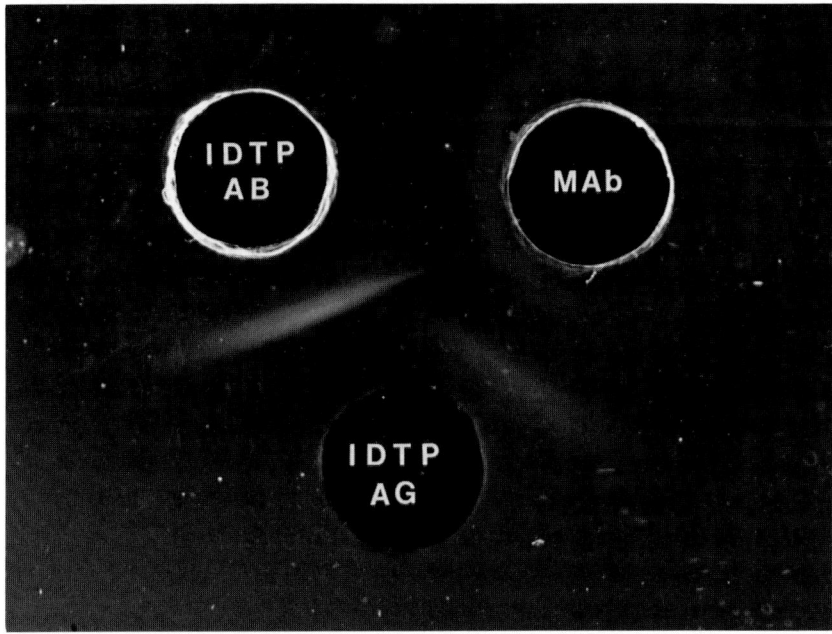

Figure 9 Reactivity of MAb in the IDTP assay. Note the reaction of partial identity. (From Dolan, Cox, Williams, and Wooley, 1989.)

opposed to the single epitope-MAb interaction. The MAb is also reactive with CDN in the classical TP assay, yielding the typical button-type precipitate described by Smith and associates (1950).

When the MAb was assayed by 2D-IEP against CDN (or SPH), only the IDTP antigen was precipitated (Fig. 10). However, incorporation of the MAb in a gel interposed between the first-dimension electrophoresis of CDN (or SPH) and homologous goat antiserum resulted in a slight reduction of the height of the cathodal peak of the Ag2 polymer as well as the precipitin peak of the IDTP antigen. These results would be consistent with the MAb recognizing an epitope which is common to the IDTP antigen and Ag2 but present in lower concentrations on the latter antigen. To more fully define the reactivity of the MAb with Ag2, we combined the techniques of autoradiography and 2D-IEP as described by Harboe and Ivanyi (1987). For this procedure, CDN was subjected to 2D-IEP against ^{125}I-labeled MAb admixed with goat anti-CDN. As shown in Figure 11, the radiolabeled MAb was reactive with the IDTP antigen

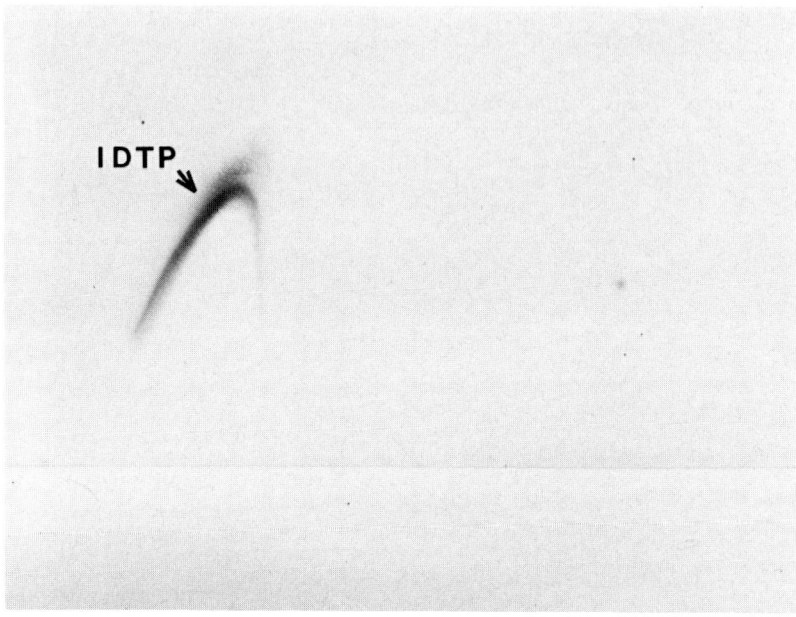

Figure 10 Reactivity of MAb with Ag2 in CDN as assessed by 2D-IEP. (From Colan, Cox, Williams, and Woolley, 1989.)

and with Ag2. Hence, the MAb recognized an epitope present on both of these antigens.

Immunoblots of the purified IDTP antigen probed with the MAb yielded a diffusely staining band with a molecular weight distribution of 130–330 kD. Treatment of the IDTP antigen with sodium metaperiodate, but not protease, abolished its reactivity with the MAb; hence, the epitope recognized by this antibody has a carbohydrate composition. It can also be concluded that this epitope is repetitive, since the MAb is able to precipitate the IDTP antigen.

The diffusely reactive band obtained in immunoblots of the IDTP antigen against MAb would suggest that the antigen itself is predominantly polysaccharide. This interpretation is supported by the fact that sodium dodecyl sulfate–polyacrylamide gel electrophoresis (SDS-PAGE) of the purified antigen yields a broad band that is strongly reactive with periodate acid Schiff, but only weakly reactive with the silver stain (unpublished data).

In recent investigations, Calhoun and co-workers (1986) and Zimmer and Pappagianis (1986) probed immunoblots of spherule- and mycelial-phase anti-

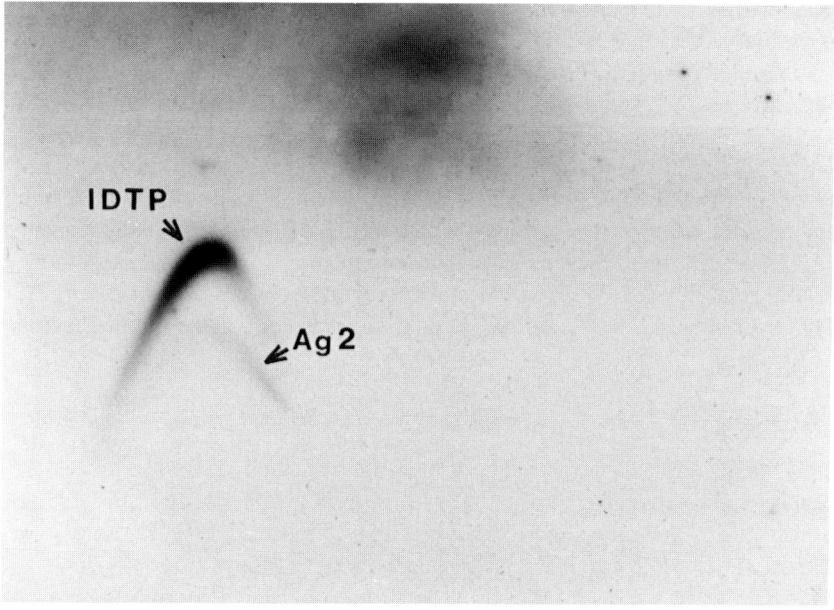

Figure 11 Reactivity of ^{125}I-labeled MAb with Ag2 and the IDTP antigen in CDN as assessed by autoradiography. Second-dimension gel contained radiolabeled MAb admixed with goat anti-CDN. (From Dolan, Cox, Williams, and Woolley, 1989.)

gens with sera of coccidioidomycosis patients who were positive for IDTP antibody. Calhoun and colleagues (1986) demonstrated reactivity of IgM antibody(s) with a broad band having a molecular weight distribution of 80-120 kD. At variance with this finding and with our results using MAb to the IDTP antigen, Zimmer and Pappagianis (1986) demonstrated IgM reactivity with double bands in the 50- to 65-kD range. These divergent data may be attributed to the differences in antigen preparations or to the methodologies employed for immunoblotting.

In early studies, Smith and co-workers (1956) reported that the TP assay yielded cross-reactivity with sera of patients with primary histoplasmosis. We have also observed cross-reactivity with sera of histoplasmosis patients in the IDTP assay (unpublished data). The precipitin band obtained is, however, of partial identity rather than total identity with the reference IDTP antigen/antibody band. In contrast to the partial identity observed with *Histoplasma capsulatum*, extracts from arthroconidia-forming gymnoascaceous saprophytes, such as *Malbranchea, Arachniotus*, and *Auxarthron* species, were shown to contain an antigen of total identity with the IDTP antigen (Kaufman et al., 1985).

One other aspect of the IDTP/TP antigen that merits discussion is the possibility that this polysaccharide is tolerogenic. This supposition is based on the fact that humoral response to the TP antigen characteristically follows a course of early rise in antibody(s) of the IgM isotype, which decreases to nondetectable levels within 4 months of clinical onset (Smith et al., 1950, 1956). Subsequent IgG response is modest to absent. This decreased antibody response, together with the resistance of polysaccharides to degradation, would suggest that the IDTP/TP antigen might be a suitable candidate for detection by an antigenemia assay.

CF Antigen

The diagnostic and prognostic value of measuring antibody response to the CF antigen is well established (Smith et al., 1950, 1956). Titers of 1:16 or less indicate localized pulmonary disease, whereas titers of 1:32 or above are consistent with disseminated disease. A rise in the antibody response to this antigen, as measured by the CF test or IDCF assay, denotes progressive coccidioidomycosis, and, conversely, a decrease in CF antibody denotes disease stabilization or regression. The antigen that induces the CF antibody response differs from the TP antigen in being labile to heat treatment at 56°C for 30 min, labile to an acid pH, and nonprecipitable in (or inactivated by) 80% ethanol (Huppert et al., 1967; Pappagianis et al., 1961).

Enzymatic treatment of spherules or spherule cell walls with chitinase or lysozyme solubilizes the CF antigen, as measured by the reactivity of the fraction in the IDCF assay (Collins et al., 1977). Similarly, the cell wall fraction obtained by subjecting arthroconidia to light sonication followed by fractionation in a Ribi apparatus contains a low but detectable amount of the IDCF antigen (Cole et al., 1987). As noted earlier, these fractions are comprised of other antigens as well, including the IDTP antigen. In contrast to the reactivity of wall fractions obtained by enzymatic digestion or mechanical treatment, the C-ASWS cell wall extract, which as discussed in the preceding section is comprised of Ag2 and the IDTP antigen, is not reactive in the IDCF or CF assay. This lack of reactivity is not attributable to the treatment of the walls with trypsin prior to alkaline extraction, i.e., C-ASWS from native walls is also nonreactive (unpublished data).

We have recently reported the isolation of the IDCF antigen from CDN (Cox et al., 1987). The experimental approach used was to fractionate CDN using conventional methods and then monitor the fractions for reactivity in the IDCF assay and for antigenic heterogeneity by 2D-IEP. This strategy exploited

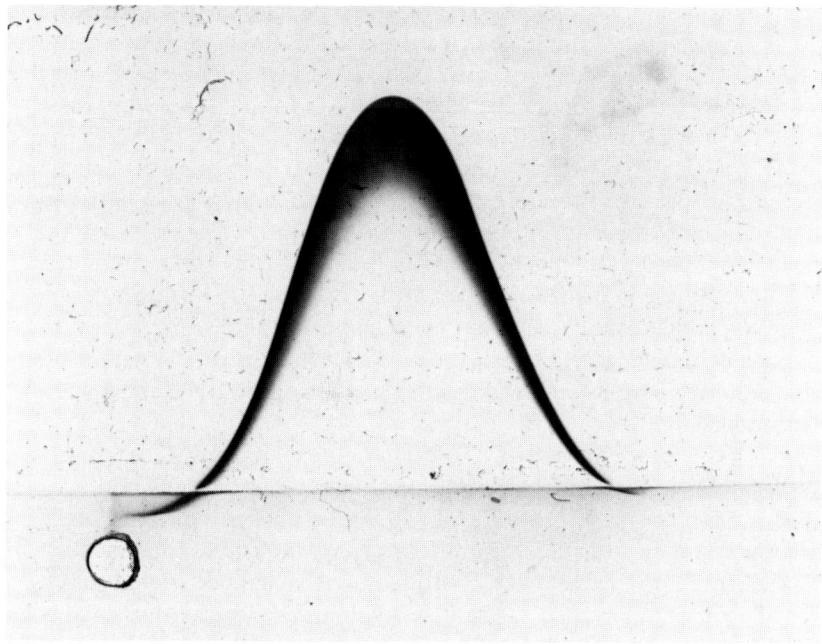

Figure 12 Specificity of goat antiserum to the purified IDCF antigen, as assessed by 2D-IEP against CDN. (From Cox, Britt, and Michael, 1987.)

the fact that, as a precipitinogen, the IDCF antigen would be detectable by 2D-IEP and, second, that heat treatment of the antigen would abolish its precipitinogen activity. The CDN preparation used was prepared as a broth culture filtrate (Huppert and Bailey, 1965a) and, by 2D-IEP against hyperimmune goat anti-CDN, contained seven precipitinogens. Lectin-affinity chromatography of this preparation on ConA yielded an effluent fraction which was reactive in the IDCF assay and was comprised of three antigens by 2D-IEP. One of these was labile to heat treatment. By using goat antiserum prepared against the heat-treated ConA effluent fraction as a ligand in immunoaffinity chromatography, the two heat-stable antigens were adsorbed from the ConA effluent, leaving a heat-labile precipitinogen which was reactive in the IDCF assay. The identification of this antigen as the IDCF precipitinogen was confirmed in studies using goat antiserum prepared against the purified antigen (Cox et al., 1987). This antiserum detected a single antigen in CDN by 2D-IEP (Fig. 12) and, when used as a ligand, specifically adsorbed the IDCF antigen from CDN.

Western blotting of CDN against goat anti-IDCF established that the IDCF antigen has a molecular weight range of 43–46 kD. This finding is concordant with studies by Zimmer and Pappagianis (1986) and Calhoun and co-workers (1986), who reported that sera of coccidioidomycosis patients who were positive by the IDCF assay demonstrated IgG reactivity to a 48-kD band and a band between 45 and 48 kD, respectively, in immunoblots of *C. immitis* extracts.

Preliminary analyses of the composition of the IDCF antigen have been performed by subjecting the antigen to various treatments and then evaluating the effects of these procedures on its reactivity in the IDCF assay and its immuno-electrophoretic profile in 2D-IEP (unpublished data). The IDCF antigen is susceptible to periodate oxidation and to enzymatic digestion with protease, glucose oxidase, neuramidase, and lipase. It is resistant to treatment with mannosidase and chitinase. These combined results suggest that the IDCF antigen is a glycolipoprotein complex.

Extensive cross-reactivity is known to occur in the CF assay with sera from patients with other mycoses, in particular histoplasmosis or blastomycosis (Campbell and Binkley, 1953; Smith et al., 1956). By contrast, the IDCF assay is specific for antibody to *C. immitis* in that a precipitin band of total identity with the reference IDCF antigen/IDCF antibody system has been detected only with sera of coccidioidomycosis patients (Huppert and Bailey, 1965a; Wood et al., 1982). Precipitin bands of partial identity have been observed, however, with sera of patients with noncoccidioidal mycoses (unpublished data). The presence of a cross-reactive epitope(s) poses a major problem in the development of an ELISA or immunoblot procedure for specifically detecting anti-*Coccidioides* IDCF antibody. It does not preclude, however, the possibility that the cross-reactive epitope might be removed by physicochemical means or, alternatively, that a specific peptide might be produced using recombinant technology.

Heat-Stable Exoantigen

Identification of a mycelial colony as *C. immitis* is firmly established by demonstrating the production of a heat-stable (HS) exoantigen designated as the HS antigen (Standard and Kaufman, 1977; Kaufman and Standard, 1978; Huppert et al., 1978b; DiSalvo et al., 1980). This antigen is detected by the IDHS assay (Standard and Kaufman, 1977) and, although originally believed to be the same as the IDTP antigen, has since been shown to be antigenically distinct from the latter (Kaufman et al., 1985).

During the course of studies to isolate the IDCF antigen, we assayed the CDN-derived fractions for the presence of other relevant antigens, including the HS exoantigen. We found that the column effluent fraction obtained from ConA affinity chromatography exhibited strong reactivity in the IDHS assay (Cox and Britt, 1986a). Antigenic analysis of this fraction by 2D-IEP revealed two precipitinogens, one heat labile at 56°C for 30 min (i.e., the IDCF antigen) and the other stable to heat treatment. Polyclonal antiserum was produced to the latter antigen by immunizing goats with the precipitin arc obtained in 2D-IEP of the heat-treated ConA effluent fraction against goat anti-CDN (Cox and Britt, 1987). This antiserum was reactive in the IDHS assay (Fig. 13A) and detected a single CDN precipitinogen when assayed by 2D-IEP (Fig. 13B). When used as a ligand in solid-phase immunoaffinity chromatography, the antiserum effectively adsorbed the HS antigen from CDN and the desorbed antigen retained antigenic reactivity in the IDHS assay.

Although the HS antigen is produced in abundance by *C. immitis*, it is poorly immunogenic. Kaufman and associates (1985) analyzed sera from coccidioidomycosis patients for anti-HS antibody using the IDHS assay. Less than 8% of the 27 sera assayed demonstrated precipitin reactivity. We have observed a similar low level of reactivity using an ELISA with the HS antigen, i.e., less than 29% of patients with active disease had detectable levels of antibody to this antigen.

Antigens Reactive in Cell-Mediated Immune Assays

Several investigators have attempted to isolate the skin test active component(s) of CDN. Pappagianis (1961a) reported that skin test activity was associated with a fraction that precipitated in 95% ethanol. This fraction yielded a single peak by ultracentrifugation and, when analyzed chemically, was reported to contain a polysaccharide comprised of mannose, galactose, and a third unidentified reducing sugar. Isolation of an ethanol-insoluble, skin test active polysaccharide had been reported earlier by Hirsch and D'Andrea (1927) and Hassid and associates (1943). The unidentified reducing sugar described by Hassid and co-workers (1943) and later by Pappagianis and his colleagues (1961a) was subsequently shown to be 3-*O*-methylmannose (Scheer et al., 1970, 1971).

Anderson and colleagues (1971) fractionated CDN by ultrafiltration and showed that skin test activity was demonstrable with the ultrafiltrate material from an XM 50, but not PM 10 membrane. Molecular-sieve chromatography on Sephadex G-25 yielded a fraction which was active in skin tests and was comprised of 3-*O*-methylmannose, mannose, glucose, galactose, and a peptide representing approximately 6% as dry weight.

In analysis of SPH, Levine and Scalarone (1974) reported that the skin test active component was nondialyzable, resistant to heat, and unaffected by treatment with trypsin. Sucrose gradient ultracentrifugation of SPH indicated that the skin test active antigen(s) was heterogeneous in molecular weight, distributed throughout the 7 to 17S fractions. This disperse distribution of skin test active fractions resembles the results obtained when CDN was fractionated by ethanol precipitation or phenol extraction, i.e., virtually all fractions elicited DTH responses (Wheat and Chung, 1977). These data suggest that either

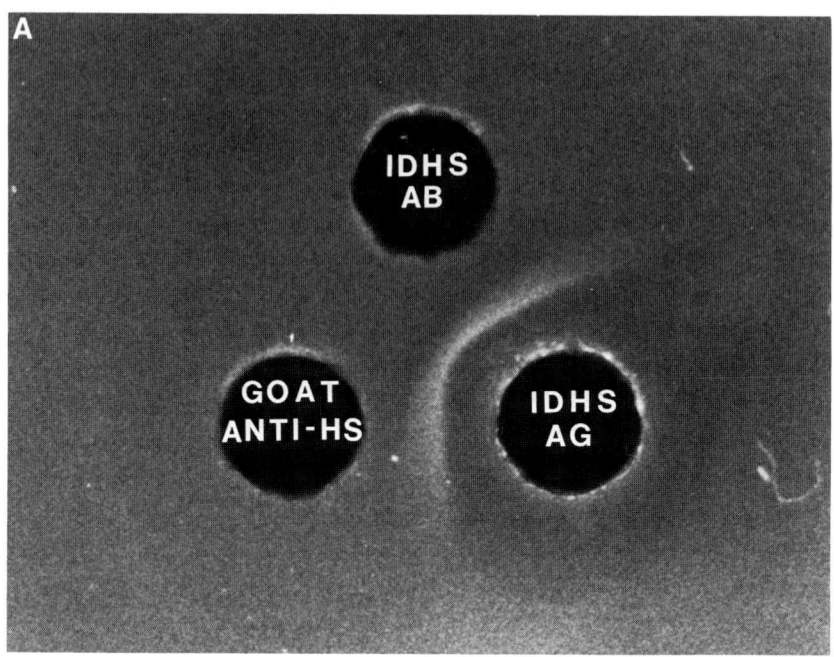

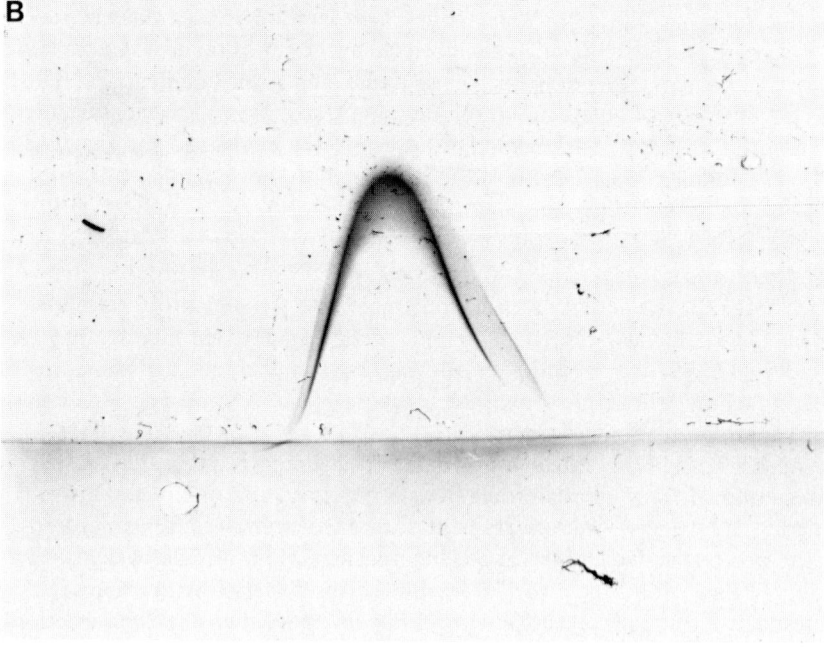

Figure 13 Reactivity of goat anti-HS antigen in (A) the IDHS assay and (B) 2D-IEP against CDN. (From Cox and Britt, 1987.)

multiple sensitins are present or that the active component occurs in polymeric form.

As an alternative approach to isolating the skin test active antigen of *C. immitis*, we and others assayed fractions extracted from cells or cell walls of *C. immitis* (Ward et al., 1975; Cox et al., 1977a; and Wheat et al., 1978, 1983). Early studies showed that the C-ASWS extract from mycelial-phase cells was active in eliciting skin test responses in sensitized guinea pigs and delayed-type footpad hypersensitivity in experimentally infected mice (Ward et al., 1975). The sensitivity of the C-ASWS extract was comparable to that of CDN and, when assayed in skin tests of *Histoplasma*-immunized guinea pigs, appeared to be specific for *C. immitis*. Subsequent investigations established that C-ASWS was also effective in eliciting in vitro lymphocyte transformation and the production of macrophage migration inhibitory factor by peripheral blood mononuclear cells from healthy, CDN skin test positive persons and patients with active coccidioidomycosis (Cox et al., 1977a; Cox and Vivas, 1977b). Cross-reactivity was observed, however, in in vitro lymphocyte transformation assays of healthy subjects who were skin test positive to histoplasmin (Cox, 1979). A similar level of cross-reactivity was also observed with CDN and SPH.

On the basis that the C-ASWS extract (of mycelial and spherule cell walls) is reactive in eliciting cell-mediated immune responses and is comprised of Ag2 and the IDTP antigen, we have attempted to determine which of these two components elicits T-lymphocyte reactivity (unpublished data). Footpad hypersensitivity tests of *Coccidioides*-infected mice with the purified IDTP antigen failed to reveal any reactivity. Parallel studies were performed with the column eluate obtained in solid-phase immunoadsorption of CDN with goat anti-CASWS, a fraction which is comprised of Ag2 and trace amounts of the IDTP antigen. Significant footpad hypersensitivity responses were observed in mice tested with this Ag2-enriched fraction to a level comparable to those elicited by unfractionated CDN. On the presumption that the DTH-eliciting component can be detected as a precipitinogen, these data suggest that Ag2 is the component that elicits DTH response.

Preliminary analyses have been performed to assess the composition of Ag2 by examining its susceptibility to various enzymes and its affinity for plant lectins. As shown in Figure 14, treatment of CDN with protease significantly altered the immunoelectrophoretic profile of Ag2, resulting in a complete reduction in the cathodal peak and a concomitant increase in the height of the anodal peak. Mannosidase also effected an increase in the anodal peak, but to a much lesser extent. The affinity of Ag2 for lectins was assessed using the technique of immunoaffinity electrophoresis (Owen and Smyth, 1977). The incorporation of lentil lectin (from *Lens culinaris*) in a gel interposed between the first-dimension electrophoresis of CDN and a second-dimension gel containing goat anti-CDN effected the in situ adsorption of Ag2. Identical results were obtained with con-

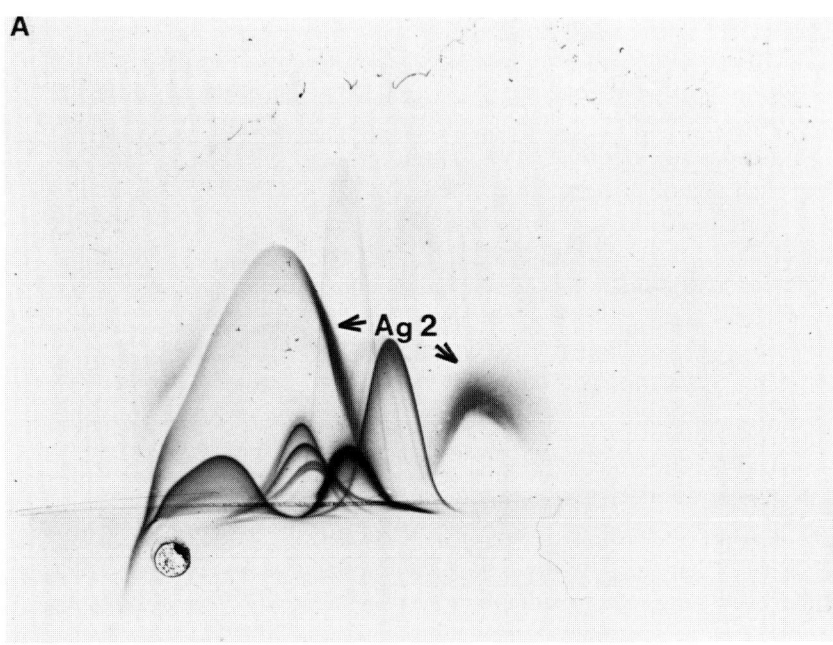

Figure 14 Effect of protease treatment on the immunoelectrophoretic profile of Ag2 as assessed by 2D-IEP. (A) Untreated CDN. (B) CDN treated with protease. Note increase in the peak height of the anodal precipitin peak of Ag2. (From Cox, R. A., unpublished data.)

canavalin A, which, like lentil lectin, binds α-D-mannosyl and α-D-mannosyl residues (Sharon and Lis, 1972). By contrast, lectins from *Ricinus communis* and *PTiLota plumosa* which bind α-D-glycosyl residues, respectively, did not bind Ag2. These collective results provide evidence that this polymeric antigen has a mannoprotein composition.

Relevant to these studies is the report by Wheat and colleagues (1983) that C-ASWS contains 3-O-methylmannose. This methylated carbohydrate has previously been shown to be present in skin test active fractions of CDN (Anderson et al., 1971), and, although it was thought to be specific to *C. immitis*, more recent studies have demonstrated that the ASWS extract of cell walls from *Malbranchea dendritica* also contains 3-O-methylmannose (Cole et al., 1985c). It is emphasized that at the time these studies were performed, it was not known that C-ASWS was comprised of two components, i.e., Ag2 and the IDTP antigen. Hence, additional work is needed to determine if 3-O-methylmannose is a constituent of Ag2, the IDTP antigen, or both.

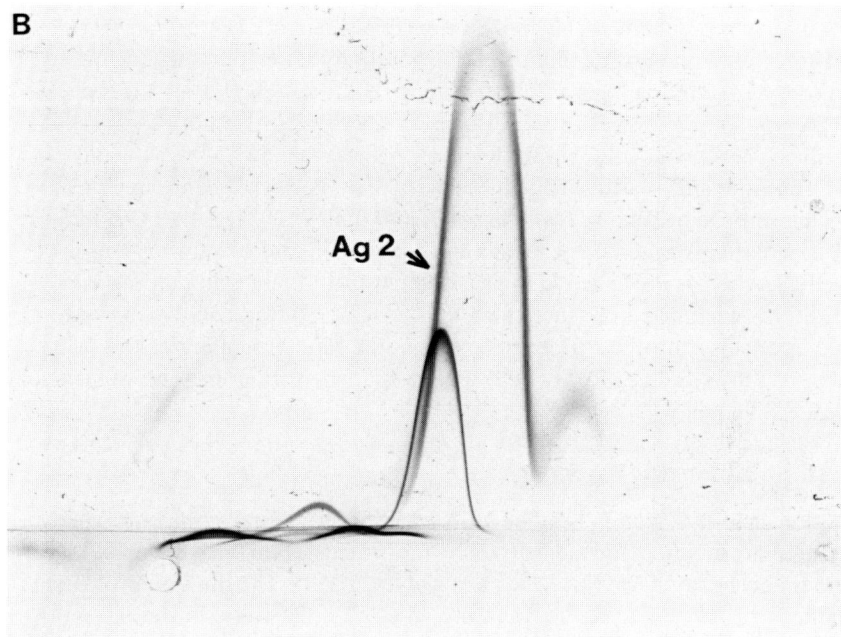

Figure 14B

Using the technique of advanced-line IEP, Cole and co-workers (1985c) reported that the ASWS extracts from cell walls of *M. dendritica, H. capsulatum, B. dermatitidis,* and *Aspergillus fumigatus* contained a component of antigenic identity with CDN Ag2. Although we have corroborated their finding that Ag2 is present in the ASWS extract from *M. dendritica* cell walls, we have not been able to confirm the presence of this antigen in extracts from *H. capsulatum, B. dermatitidis,* or *A. fumigatus*. One possible explanation for these discordant results is that the advancing line identified by Cole and his colleagues (1985c) as Ag2 consists of the IDTP antigen as well as Ag2, in which case cross-reactivity with either antigen would have resulted in a disruption of the advancing line. If this interpretation is correct and if Ag2 is the C-ASWS component which contains 3-O-methylmannose, then it would provide an explanation for detection of this carbohydrate in the ASWS extract of *C. immitis* and *M. dendritica* but not in *H. capsulatum, B. dermatitidis,* or *A. fumigatus* (Cole et al., 1985c).

Wheat and co-workers (1978) reported the extraction of a skin test active component from intact, defatted mycelia using aqueous phenol. In a subsequent study, these investigators (Wheat et al., 1983) compared the composition and skin test activity of the phenol-soluble, water-soluble (PSWS) extract with C-ASWS. The two preparations elicited comparable skin test responses in sensitized guinea pigs, and both exhibited similar carbohydrate composition. When analyzed by Ouchterlony gel diffusion against rabbit anti-SPH, PSWS was of antigenic nonidentity with C-ASWS. However, alkali extraction of PSWS yielded an alkali-soluble, water-soluble fraction (i.e., ASWS from PSWS) which contained a component of antigenic identity with one of the two bands in C-ASWS. One other important finding that emerged from their study was that C-ASWS could be isolated from crude cell walls and from walls pretreated with trypsin or 2% SDS, whereas PSWS could not be isolated from these wall preparations. The authors concluded from these data that C-ASWS, but not PSWS, is a cell wall component of mycelia. The similarities in the biological activity and composition of PSWS and C-ASWS would suggest that the former may be a precursor to the C-ASWS cell wall fraction.

Immunomodulatory Antigens

Protection against *C. immitis* can be induced in experimental animal models by intramuscular, subcutaneous, or intraperitoneal vaccination with formalin-killed spherules (Levine et al., 1960, 1965; Kong et al., 1963). The immunoprotective component was shown to reside primarily, but not exclusively, in the cell walls of this fungus (Kong et al., 1963). Whereas the insoluble fraction obtained from mechanically disrupted spherules afforded strong protection, the soluble cytosol fraction was weakly immunogenic.

Pappagianis and co-workers (1979) reported that the component(s) released from spherule cell walls incubated for 5 days in phosphate-buffered saline (with 2% chloroform as preservative) induced protection in mice when administered with adjuvant. The level of protection was comparable to that obtained with killed spherules. When chitinase was added during the extraction procedure, the immunogenicity of the wall-derived fraction was reduced. As noted by these investigators, chitinase preparations are commonly contaminated with β-(1-3)-glucanase, thus the latter enzyme may have been responsible for this effect.

Lecara and co-workers (1983) reported that subcutaneous vaccination of mice with C-ASWS, admixed with complete Freund adjuvant, induced protective immunity against intranasal challenge. Unlike the extract studied by Pappagianis and co-workers (1979), the immunoprotective effect of C-ASWS (from mycelia or spherules) was less than that of intact spherules (unpublished data). This reduced efficacy may be attributed to a denaturation of the immunogenic component(s) during alkali extraction or to incomplete extraction of the immunogen(s). It has not yet been determined whether the protection induced by C-ASWS is attributable to the TP antigen or Ag2. We have found that mice immunized with the purified TP antigen are not protected against challenge (unpublished data) which, by inference, suggests that Ag2 is the immunoprotective component.

In marked contrast to the protection afforded by vaccination with low doses of coccidioidal immunogens, high doses of antigen or administration of antigen by an intravenous route suppresses immune response (Levine et al., 1966; Scalarone and Levine, 1971; Ibrahim and Pappagianis, 1973; Cox and Kennell, 1988). Evidence that antigen-induced suppression of host response occurs during the course of coccidioidomycosis has been obtained in a murine model (Cox et al., 1988). *Coccidioides*-infected BALB/c mice were shown to mount a T-lymphocyte reactivity to CDN early during the course of coccidioidal disease, but then become anergic thereafter. The development of anergy was coincident with the detection of high levels of circulating Ag2 and/or the TP antigen, as measured by an ELISA with goat antibodies to C-ASWS (Cox and Kennell, 1988). Sera from infected mice, when transferred intravenously to syngeneic mice, suppressed the T-lymphocyte response of recipients to immunization with coccidioidal antigens. This suppressive activity was removed by solid-phase immunoadsorption of the sera with goat antibodies to C-ASWS, a procedure that would remove Ag2 and the TP antigen. Considering the critical role of T cells in host resistance to *C. immitis* (Beaman et al., 1977, 1979), it is reasonable to predict that suppression of cell-mediated immune responses by antigen would adversely modulate the course of the disease.

Antigenic Comparisons of Morphological Forms of *C. immitis*

Numerous studies have documented differences in the biological activities of cells (and derived extracts) from the saprobic and parasitic forms of *C. immitis*. In a direct comparison of the vaccine potential of killed mycelia and spherules, Kong and Levine (1967) reported that spherules were significantly more protective than mycelia. We have obtained similar results in vaccine studies of the C-ASWS extracts, i.e., C-ASWS from spherule cell walls was more protective than the mycelial-phase C-ASWS extract (unpublished data). Comparative analyses of SPH and CDN have also shown that spherule-derived extracts are more potent than mycelial-derived antigens. This increased immunological activity of SPH was demonstrable in tests for delayed-type dermal hypersensitivity (Levine et al., 1969, 1973; Stevens et al., 1974, 1975), in lymphocyte transformation assays (Deresinski et al., 1974, 1977), and in CF antibody tests (Scalarone et al., 1974). However, in some studies, using different preparations of CDN and SPH, other investigators have found that the mycelial-phase extract was more sensitive in detecting T-lymphocyte responses (Cox et al., 1977a, 1981; Gifford and Catanzaro, 1981).

It has not yet been established whether the differences observed in the biological activities of spherule and mycelial phase antigens of *C. immitis* are attributed to qualitative or quantitative differences in their antigenic composition. Antigenic comparisons by 2D-IEP and Western blotting have revealed that some antigens appear to be unique to CDN or SPH, suggesting qualitative differences (Huppert et al., 1978a; Karu et al., 1985; Calhoun et al., 1986; Zimmer and Pappagianis, 1986). The possibility that these reflect quantitative differences has not been excluded. For example, Huppert and co-workers (1978a) identified two precipitinogens in SPH by 2D-IEP which were not demonstrable in CDN. However, as noted by these authors, these two SPH antigens were detected using antiserum from a burro hyperimmunized with CDN; hence, they are present in CDN but at insufficient concentrations for detection by 2D-IEP.

In recent studies, we employed polyclonal antisera prepared against the purified IDTP, IDCF, and IDHS derived from CDN to assess the identity of these precipitinogens with components in SPH by 2D-IEP (Cox and Britt, 1987). All three antigens were demonstrable in SPH and, although quantitative differences were observed, the extent of these quantitative differences varied with successive lots of SPH and CDN. This latter observation is consistent with earlier reports that the biological activities of *C. immitis* extracts vary from lot to lot, even when prepared from the same strain(s) and using the same culture conditions (Smith et al., 1950; Pappagianis et al., 1961). Comparative studies of extracts from the various morphological forms should, therefore, be made using several lots, and preferably lots prepared from different strains of *C. immitis*.

SUMMARY

Coccidioides immitis is a unique fungal pathogen, characterized by a saprobic mycelial phase that gives rise to infectious arthroconidia which, in host tissue, convert into a morphologically distinct spherule-endospore phase. This multiphasic cycle presents a formidable challenge to the host.

The cell wall of *C. immitis* is a major reservoir of immunoreactive antigens and thus figures prominently in host–parasite interaction in this disease. Extracts of cell walls have yielded antigens reactive with the TP and CF antibodies, antigens that elicit in vivo and in vitro T-cell responses, and antigens that augment or suppress host response.

By using techniques of 2D-IEP and Western blotting, investigators have provided evidence that the TP antigen has a predominant polysaccharide composition, with a molecular weight range of 130–330 kD. This antigen may have tolerogenic activity, as suggested by the induction of IgM antibody early during the disease, which then wanes and is followed by a modest or nondemonstrable IgG antibody response. In contrast, the CF antigen is highly immunogenic, inducing IgG levels within 1 to 3 months of clinical illness, which then persist throughout the course of disease. Compositional analyses of the CF antigen indicate that this heat-labile precipitinogen is a glycolipoprotein complex with a molecular weight range of 43-48 kD.

The skin test active component(s) of *C. immitis* has not been purified, but preliminary evidence suggests that it may be a polymeric antigen designated as Ag2. This polymeric antigen is susceptible to enzymatic treatment with mannosidase and protease, indicating a mannoprotein composition, although other constituents may be present as well. Despite differences in the chemical composition of Ag2 and the TP antigen, these two components coelute on molecular sieve, ion-exchange, and lectin affinity columns and both antigens are solubilized from cell walls with alkali. These similar chromatographic profiles suggest that Ag2 and the TP antigen are intimately associated and perhaps present as a complex.

Comparative analyses of morphological forms of *C. immitis* have provided evidence that extracts from spherules are more potent than those from the mycelial phase in eliciting immune responses. These differences may, however, be attributed to quantitative rather than qualitative differences in the antigenic composition of these cell types.

In summary, many advances have been made in the isolation and characterization of antigens from this pathogenic fungus. However, additional work is required to fully delineate their production and expression during the morphogenetic phases of *C. immitis* and to assess their role in the pathophysiology of coccidioidomycosis.

ACKNOWLEDGMENTS

Portions of the work from this laboratory were supported by Public Health Service grant AI-21431.

REFERENCES

Anderson, K. L., Wheat, R. W., and Conant, N. I. (1971). Fractionation and composition studies of skin test-active components of sensitins from *Coccidioides immitis*. *Appl. Microbiol. 22*:294-299.

Beaman, L., Benjamini, E., Pappagianis, D. (1977). Significance of T cells in resistance to experimental murine coccidioidomycosis. *Infect. Immun. 17*: 580-585.

Beaman, L., Pappagianis, D., and Benjamini, E. (1979). Mechanisms of resistance to infection with *Coccidioides immitis* in mice. *Infect. Immun. 23*: 681-685.

Calhoun, D. L., Osir, E. O., Dugger, K. O., Galgiani, J. N., and Law, J. H. (1986). Humoral antibody responses to specific antigens of *Coccidioides immitis*. *J. Infect. Dis. 154*:265-272.

Campbell, C. C., and Binkley, G. E. (1953). Serologic diagnosis with respect to histoplasmosis, coccidioidomycosis, and blastomycosis and the problem of cross reactions. *J. Lab. Clin. Med. 42*: 896–906.

Catanzaro, A., Spitler, L. E., and Moser, K. M. (1975). Cellular immune response in coccidioidomycosis. *Cell. Immunol. 15*:360–371.

Cole, G. T., Sun, S. H., and Huppert, M. (1982). Isolation and ultrastructural examination of conidia wall components of *Coccidioides* and *Aspergillus*. *Scanning Electron Microsc.* (IV):1677–1685.

Cole, G. T., Pope, L. M., Huppert, M., Sun, S. H., and Starr, P. (1983). Ultrastructure and composition of conidial wall fractions of *Coccidioides immitis*. *Exp. Mycol. 7*:297–318.

Cole, G. T., and Sun, S. H. (1985a). Arthroconidium-spherule-endospore transformation in *Coccidioides immitis*. In *Fungal Dimorphism*. Edited by P. J. Szaniszlo. Plenum Press, New York, pp. 281–333.

Cole, G. T., Pope, L. M., Huppert, M., Sun, S. H., and Starr, P. (1985b). Wall composition of different cell types of *Coccidioides immitis*. In *Proceedings of The Fourth International Conference on Coccidioidomycosis*. Edited by H. E. Einstein and A. Catanzaro. The National Foundation for Infectious Diseases, pp. 112–119.

Cole, G. T., Chinn, J. W., Jr., Pope, L. M., and Starr, P. (1985c). Characterization and distribution of 3-O-methylmannose in *Coccidioides immitis*. In *Proceedings of The Fourth International Conference on Coccidioidomycosis*. Edited by H. E. Einstein and A. Catanzaro. The National Foundation for Infectious Disease, pp. 130–145.

Cole, G. T. (1986). Models of cell differentiation in conidial fungi. *Microbiol. Rev. 50*:95–132.

Cole, G. T., Kirkland, T. N., and Sun, S. H. (1987). An immunoreactive, water-soluble conidial wall fraction of *Coccidioides immitis*. *Infect. Immun. 55*: 657–667.

Cole, G. T., Sun, S. H., Dominguez, J., Yuan, L., Franco, M., and Kirkland, T. N. (1989). Wall-associated antigens of *Coccidioides immitis*. In *Isolation, Purification, and Characterization of Fungal Antigens*. Edited by E. Drouhet and J. P. Latge. Plenum Press, New York (in press).

Collins, M., and Pappagianis, D. (1973). Effects of lysozyme and chitinase on the spherules of *Coccidioides immitis in vitro*. *Infect. Immun. 7*:817–822.

Collins, M. S., Pappagianis, D., and Yee, J. (1977). Enzymatic solubilization of precipitin and complement fixing antigen from endospores, spherules, and spherule fraction of *Coccidioides immitis*. In *Coccidioidomycosis: Current Clinical and Diagnostic Status*. Edited by L. Ajello. Symposia Specialists, pp. 429–444.

Converse, J. L. (1955). Growth of spherules of *Coccidioides immitis* in a chemically defined liquid medium. *Proc. Soc. Exp. Biol. Med. 90*:709–711.

Converse, J. L. (1956). Effect of physio-chemical environment on spherulation of *Coccidioides immitis* in a chemically defined medium. *J. Bacteriol. 72*: 106–107.

Converse, J. L. (1957). Effect of surface active agents on endosporulation of *Coccidioides immitis* in a chemically defined medium. *J. Bacteriol. 74*:784-792.

Cox, R. A., Brummer, E., and Lecara, G. (1977a). In vitro lymphocyte responses of coccidioidin skin-test positive and -negative persons to coccidioidin, spherulin, and a *Coccidioides* cell wall antigen. *Infect. Immun. 15*:751-755.

Cox, R. A., and Vivas, J. R. (1977). Spectrum of *in vivo* and *in vitro* cell mediated immune responses in coccidioidomycosis. *Cell. Immunol. 31*:130-141.

Cox, R. A., and Arnold, D. R. (1979). Immunoglobulin E in coccidioidomycosis. *J. Immunol. 123*:194-200.

Cox, R. A. (1979). Cross-reactivity between antigens of *Coccidioides immitis*, *Histoplasma capsulatum*, and *Blastomyces dermatitidis* in lymphocyte transformation assay. *Infect. Immun. 25*:932-938.

Cox, R. A., Mead, C. G., and Pavey, E. F. (1981). Comparisons of mycelia- and spherule-derived antigens in cellular immune assays of *Coccidioides immitis*-infected guinea pigs. *Infect. Immun. 31*:687-692.

Cox, R. A., Huppert, M., Starr, P., and Britt, L. A. (1984). Reactivity of an alkali-soluble, water-soluble antigen of *Coccidioides immitis* with anti-*Coccidioides* immunoglobulin M precipitin antibody. *Infect. Immun. 43*: 502-507.

Cox, R. A., and Britt, L. A. (1985). Antigenic heterogeneity of an alkali-soluble, water-soluble cell wall extract of *Coccidioides immitis. Infect. Immun. 50*: 365-369.

Cox, R. A., and Britt, L. A. (1986a). Isolation and identification of an exoantigen specific for *Coccidioides immitis. Infect. Immun. 52*:138-143.

Cox, R. A., and Britt, L. A. (1986b). Isolation of a coccidioidin component that reacts with immunoglobulin M precipitin antibody. *Infect. Immun. 53*: 449-453.

Cox, R. A., Britt, L. A., and Michael, R. A. (1987). Isolation of *Coccidioides immitis* F antigen by immunoaffinity chromatography with monospecific antiserum. *Infect. Immun. 55*:227-232.

Cox, R. A., and Britt, L. A. (1987). Antigenic identity of biologically active antigens in coccidioidin and spherulin. *Infect. Immun. 55*:2590-2596.

Cox, R. A., Kennell, W., Boncyk, L., and Murphy, J. W. (1988). Induction and expression of cell-mediated immune responses in inbred mice infected with *Coccidioides immitis. Infect. Immun. 56*:13-17.

Cox, R. A., and Kennell, W. (1988). Suppression of T-lymphocyte response by *Coccidioides immitis* antigens. *Infect. Immun. 56*:1424-1429.

Deresinski, S. C., and Stevens, D. A. (1975). Coccidioidomycosis in compromised hosts. Experience at Stanford University Hospital, *Medicine 54*:377-395.

Deresinski, S. C., Levine, H. B., and Stevens, D. A. (1974). Soluble antigens of mycelia and spherules in the *in vitro* detection of immunity to *Coccidioides immitis. Infect. Immun. 10*:700-704.

Deresinski, S. C., Applegate, R. J., Levine, H. B., and Stevens, D. A. (1977). Cellular immunity to *Coccidioides immitis*: in vitro lymphocyte response to spherules, arthrospores, and endospores. *Cell. Immunol. 32*:110-119.

DiSalvo, A. F., Sekhon, A. S., Land, G. A., and Fleming, W. H. (1980). Evaluation of the exoantigen test for identification of *Histoplasma* species and *Coccidioides immitis* culture. *J. Clin. Microbiol. 11*:238-241.

Dolan, M. J., Cox, R. A., Williams, V., and Woolley, S. (1989). Development and characterization of a monoclonal antibody against the tube precipitin antigen of *Coccidioides immitis. Infect. Immun. 57*: (in press, April issue, #4).

Donnelly, W. H., and Yunis, E. J. (1974). The ultrastructure of *Coccidioides immitis*. Study of a human infection. *Arch. Pathol. 94*:227-232.

Drutz, D., and Catanzaro, A. (1978a). Coccidioidomycosis. State of the art. Part I, *Am. Rev. Respir. Dis. 117*:559-585.

Drutz, D., and Catanzaro, A. (1978b). Coccidioidomycosis. State of the art. Part II, *Am. Rev. Respir. Dis. 117*:727-771.

Drutz, D. J., and Huppert, M. (1983). Coccidioidomycosis: factors affecting the host-parasite interaction. *J. Infect. Dis. 147*:372-390.

Frey, C. L., and Drutz, D. J. (1986). Influence of fungal surface components on the interaction of *Coccidioides immitis* with polymorphonuclear neutrophils. *J. Infect. Dis. 153*:933-943.

Gabal, M. A. (1985). Pathogenesis and electron microscopic changes of spherulogenesis of *Coccidioides immitis* (valley fever). *Am. J. Vet. Res. 46*:671-675.

Gifford, J., and Catanzaro, A. (1981). A comparison of coccidioidin and spherulin skin testing in the diagnosis of coccidioidomycosis. *Am. Rev. Respir. Dis. 124*:440-444.

Harboe, M., and Ivanyi, J. (1987). Analysis of monoclonal antibodies to *Mycobacterium leprae* by crossed immunoelectrophoresis. *Scand. J. Immunol. 25*: 133-138.

Harrison, J., Sun, S. H., and Cox, R. A. (1988). Cytological studies of *Coccidioides immitis* tube precipitin antigen by immunoelectron microscopy. Presented at the Annual American Society for Microbiology meeting, Miami Beach, Abst. F-53.

Harvey, R. P. (1980). Coccidioidomycosis in pregnancy. In *Coccidioidomycosis*. Edited by D. A. Stevens. Plenum Press, New York, pp. 241-244.

Harvey, R. P., and Stevens, D. A. (1981). *In vitro* assays of cellular immunity in progressive coccidioidomycosis. *Am. Rev. Respir. Dis. 123*:665-669.

Hassid, W. Z., Baker, E. E., and McCready, R. M. (1943). An immunologically active polysaccharide produced by *Coccidioides immitis* Rixford and Gilchrist. *J. Biol. Chem. 149*:303-311.

Hector, R. F., and Pappagianis, D. (1982). Enzymatic degradation of the walls of spherules of *Coccidioides immitis. Exp. Mycol. 6*:136-152.

Hirsch, E. F., and D'Andrea, D. (1927). The specific substance of *Coccidioides immitis. J. Infect. Dis. 40*:634-637.

Huppert, M., and Bailey, J. W. (1965). The use of immunodiffusion tests in coccidioidomycosis. I. The accuracy and reproducibility of the immunodiffusion tests which correlates with complement fixation. *Am. J. Clin. Pathol. 44*:364-368.

Huppert, M., and Bailey, J. W. (1965b). The use of immunodiffusion tests in coccidioidomycosis. II. An immunodiffusion test as a substitute for the tube precipitin test. *Am. J. Clin. Pathol. 44*:369-373.

Huppert, M., Bailey, J. W., and Chitjian, P. (1967). Immunodiffusion as a substitute for complement fixation and tube precipitin tests in coccidioidomycosis. In *Coccidioidomycosis: Current Clinical and Diagnostic Status*. Edited by L. Ajello. Symposia Specialists, pp. 221-225.

Huppert, M., and Sun, S. H. (1980). Overview of mycology, and the mycology of *Coccidioides immitis*. In *Coccidioidomycosis*. Edited by D. A. Stevens. Plenum Press, New York, pp. 21-46.

Huppert, M., Krasnow, I., Vukovich, K. R., Sun, S. H., Rice, E. H., and Kutner, L. J. (1977). Comparison of coccidioidin and spherulin in complement fixation tests for coccidioidomycosis. *J. Clin. Microbiol. 6*:33-41.

Huppert, M., Spratt, N. S., Vukovich, K. R., Sun, S. H., and Rice, E. H. (1978a). Antigenic analysis of coccidioidin and spherulin determined by two-dimensional immunoelectrophoresis. *Infect. Immun. 20*:542-551.

Huppert, M., Sun, S. H., and Rice, E. H. (1978b). Specificity of exoantigens for identifying cultures of *Coccidioides immitis*. *J. Clin. Microbiol. 8*:346-348.

Huppert, M., Sun, S. H., and Harrison, J. L. (1982). Morphogenesis throughout saprobic and parasitic cycles of *Coccidioides immitis*. *Mycopathol. 78*:107-122.

Ibrahim, A. B., and Pappagianis, D. (1973). Experimental induction of anergy to coccidioidin by antigens of *Coccidioides immitis*. *Infect. Immun. 7*:786-794.

Johnson, W. M. (1982). Racial factors in coccidioidomycosis: mortality experience in Arizona. *Arizona Med. 39*:18-24.

Karu, A. E., Gennevois, D. J. P., Hoffman, J. W., and Levine, H. B. (1985). Preliminary characterization of monoclonal antibodies to *Coccidioides immitis* antigens. In *Proceedings of the Fourth International Symposium on Coccidioidomycosis*. Edited by H. E. Einstein and A. Catanzaro, pp. 227-238.

Kaufman, L., and Standard, P. (1978). Immuno-identification of cultures of fungi pathogenic to man. *Curr. Microbiol. 1*:135-140.

Kaufman, L., Standard, P. G., Huppert, M., and Pappagianis, D. (1985). Comparison and diagnostic value of the coccidioidin heat-stable (HS) and tube precipitin antigens in immunodiffusion. *J. Clin. Microbiol. 22*:515-518.

Kong, Y. M., Levine, H. B., and Smith, C. E. (1963). Immunogenic properties of nondisrupted and disrupted spherules of *Coccidioides immitis* in mice. *Sabouraudia 2*:131-142.

Kong, Y. M., and Levine, H. B. (1967). Experimentally induced immunity in the mycoses. *Bacteriol. Rev. 31*:35-53.

Landay, M. E., Wheat, R. W., Conant, N. F., and Lowe, E. P. (1967). Serological comparison of spherules and arthrospores of *Coccidioides immitis*. *J. Bacteriol. 94*:1400-1405.

Landay, M. E. (1973). Spherules in the serology of *Coccidioides immitis*. II. Complement fixation tests with human sera. *Mycopathologia 49*:45-52.

Larsen, R. A., Jacobson, J. A., Morris, A. H., and Benowitz, B. A. (1985). Acute respiratory failure caused by primary pulmonary coccidioidomycosis. Two case reports and a review of the literature. *Am. Rev. Respir. Dis. 131*:797–799.

Lecara, G., Cox, R. A., and Simpson, R. B. (1983). *Coccidioides immitis* vaccine: potential of an alkali-soluble, water-soluble cell wall antigen. *Infect. Immun. 39*:473–475.

Levine, H. B., Cobb, J. M., and Smith, C. E. (1960). Immunity to coccidioidomycosis induced in mice by purified spherule, arthrospore, and mycelial vaccines. *Trans. NY Acad. Sci. 22*:436–449.

Levine, H. B., Kong, Y. M., and Smith, C. E. (1965). Immunization of mice to *Coccidioides immitis*: dose, regimen and spherulation stage of killed spherule vaccines. *J. Immunol. 94*:132–142.

Levine, H. B., and Kong, Y. M. (1966). Immunologic impairment in mice treated with killed *Coccidioides immitis* spherules: suppressed response to intramuscular doses. *J. Immunol. 97*:297–305.

Levine, H. B., Cobb, J. M., and Scalarone, G. M. (1969). Spherule coccidioidin in delayed dermal sensitivity reactions of experimental animals. *Sabouraudia 7*:20–32.

Levine, H. B., Ochoa, A. G., and Ten Eyck, D. R. (1973). Dermal sensitivity to *Coccidioides immitis*. A comparison of responses elicited in man by spherulin and coccidioidin. *Am. Rev. Respir. Dis. 107*:379–386.

Levine, H. B., and Scalarone, G. M. (1974). Properties of spherulin, a skin-test reagent in coccidioidomycosis. Proceedings of the Third International Symposium on the Mycoses. *Pan Am. Health Organ. Sci. Publ. 304*:101–110.

Levine, H. B., Restrepo, A. M., Ten Eyck, D. R., and Stevens, D. A. (1975). Spherulin and coccidioidin: cross-reactions in dermal sensitivity to histoplasmin and paracoccidioidin. *Am. J. Epidemiol. 101*:512–516.

Miyaji, M., Nishimura, K., and Ajello, L. (1985). Scanning electron microscope studies on the parasitic cycle of *Coccidioides immitis*. *Mycopathologia 89*:51–57.

Owen, P., and Smyth, C. J. (1977). Enzyme analysis by quantitative immunoelectrophoresis. In *Immunochemistry of Enzymes and Their Antibodies*. Edited by M. R. J. Salton. John Wiley and Sons, New York, pp. 148–202.

Pappagianis, D., Putman, E. W., and Kobayashi, G. S. (1961a). Polysaccharide of *Coccidioides immitis*. *J. Bacteriol. 82*:714–723.

Pappagianis, D., Smith, C. E., Kobayashi, G. S., and Saito, M. T. (1961b). Studies of antigens from young mycelia of *Coccidioides immitis*. *J. Infect. Dis. 108*:35–44.

Pappagianis, D., Lindsey, N. J., Smith, C. E., and Saito, M. T. (1965). Antibodies in human coccidioidomycosis: immunoelectrophoretic properties. *Proc. Soc. Exp. Biol. Med. 118*:118–122.

Pappagianis, D. H., Hector, R., Levine, H. B., and Collins, M. S. (1979). Immunization of mice against coccidioidomycosis with a subcellular vaccine. *Infect. Immun. 25*:440–445.

Pappagianis, D. (1980). Epidemiology of coccidioidomycosis. In *Coccidioidomycosis*. Edited by D. A. Stevens, Plenum Press, New York, pp. 63-85.
Prichard, J. G., Sorotzkin, R. A., and James, R. E., III (1987). Cutaneous manifestations of disseminated coccidioidomycosis in the acquired immunodeficiency syndrome. *Cutis 39*:203-205.
Reiss, E. (1986). *Coccidioides immitis*. In *Molecular Immunology of Mycotic and Actinomycotic Infections*. Edited by E. Reiss. Elsevier, New York, pp. 53-76.
Roberts, C. J. (1984). Coccidioidomycosis in acquired immune deficiency syndrome. *Am. J. Med. 76*:734-736.
Rowland, V. S., Westfall, R. E., and Hinchcliffe, W. A. (1977). Fatal Coccidioidomycosis, analysis of host factors. In *Coccidioidomycosis*. Edited by D. A. Stevens. Plenum Press, New York, pp. 91-106.
Sawaki, Y., Huppert, M., Bailey, J. W., and Yagi, Y. (1966). Patterns of human antibody reactions in coccidioidomycosis. *J. Bacteriol. 91*:422-427.
Scalarone, G. M., and Levine, H. B. (1971). Attributes of deficient immunity in mice receiving *Coccidioides immitis* spherule vaccine by the intravenous route. *Sabouraudia 7*:169-177.
Scalarone, G. M., Levine, H. B., Pappagianis, D., and Chaparas, S. D. (1974). Spherulin as a complement-fixing antigen in human coccidioidomycosis. *Am. Rev. Respir. Dis. 110*:324-328.
Scheer, E. R., and Wheat, R. W. (1971). Skin test active extracts from *Coccidioides immitis* mycelia. *Bacteriol. Proceed.* Mm. 13. p. 130.
Scheer, E., Terai, T., Kulkarni, S., Conant, N. F., Wheat, R. W., and Lowe, E. P. (1970). Unusual reducing sugar from *Coccidioides immitis*. *J. Bacteriol. 103*:525-526.
Sharon, N., and Lis, H. (1972). Lectins: cell-agglutinating and sugar-specific proteins. *Science 177*:949-959.
Smale, L. E., and Birsner, J. W. (1949). Maternal deaths from coccidioidomycosis. *J. Am. Med. Assoc. 140*:1152-1154.
Smith, C. E., Whiting, E. G., Baker, E. E., Rosenberger, H. G., Beard, R. R., and Saito, M. T. (1948). The use of coccidioidin. *Am. Rev. Tuberc. 57*: 330-360.
Smith, C. E., Saito, M. T., Beard, R. R., Kepp, R. M., Clark, R. W., and Eddie, B. U. (1950). Serological tests in the diagnosis and prognosis of coccidioidomycosis. *Am. J. Hyg. 52*:1-21.
Smith, C. E., Saito, M. T., and Simons, S. A. (1956). Pattern of 39,500 serologic tests in coccidioidomycosis. *J. Am. Med. Assoc. 160*:546-552.
Standard, P. G., and Kaufman, L. (1977). Immunological procedures for the rapid and specific identification of *Coccidioides immitis* cultures. *J. Clin. Microbiol. 5*:149-153.
Stevens, D. A., Levine, H. B., and Ten Eyck, D. R. (1974). Dermal sensitivity to different doses of spherulin and coccidioidin. *Chest 65*:530-533.
Stevens, D. A., Levine, H. B., Deresinski, S. C., and Blaine, L. J. (1975). Spherulin in clinical coccidioidomycosis: comparison with coccidioidin. *Chest 68*:5-9.

Sun, S. H., and Huppert, M. (1976). A cytological study of morphogenesis in *Coccidioides immitis*. *Sabouraudia 14*:185–198.

Sun, S. H., Sekhon, S. S., and Huppert, M. (1979). Electron microscopic studies of saprobic and parasitic forms of *Coccidioides immitis*. *Sabouraudia 17*:265–273.

Sun, S. H., Cole, G. T., Drutz, D. J., and Harrison, J. L. (1986). Electron-microscopic observations of the *Coccidioides immitis* parasitic cycle *in vivo*. *J. Med. Vet. Mycology 24*:183–192.

Vaughan, J. E., and Ramirez, H. (1951). Coccidioidomycosis as a complication of pregnancy. *Calif. Med. 74*:121–123.

Ward, E. R., Jr., Cox, R. A., Smitt, J. A., Jr., Huppert, M., and Sun, S. H. (1975). Delayed-type hypersensitivity responses to a cell wall fraction of the mycelial phase of *Coccidioides immitis*. *Infect. Immun. 12*:1093–1097.

Werner, S. B., Pappagianis, D., Heindl, I., and Michel, A. (1972). An epidemic of coccidioidomycosis among archeology students in northern California. *N. Engl. J. Med. 286*:507–512.

Wheat, R. W., and Su Chung, K. S. (1977). Antigenic fractions of *Coccidioides immitis*. In *Coccidioidomycosis: Current Clinical and Diagnostic Status*. Edited by L. Ajello. Symposia Specialists, pp. 453–460.

Wheat, R. W., Chung, K. S. S., Ornellas, E. P., and Scheer, E. R. (1978). Extraction of skin test activity from *Coccidioides immitis* mycelial by water, perchloric acid, and aqueous phenol extraction. *Infect. Immun. 19*:152–159.

Wheat, R. W., Woodruff, W. W., III, and Haltiwanger, R. S. (1983). Occurrence of antigenic (species-specific?) partially 3-O-methylated heteromannans in cell wall and soluble cellular (nonwall) components of *Coccidioides immitis* mycelia. *Infect. Immun. 41*:728–734.

Wolf, J. E., Little, J. R., Pappagianis, D., and Kobayashi, G. S. (1986). Disseminated coccidioidomycosis in a patient with the acquired immune deficiency syndrome. *Diagn. Microbiol. Infect. Dis. 5*:331–336.

Wood, J. C., Friedly, G., Zartarian, M., Aarnaes, S., and Maza, L. M. (1982). Alternatives to the standardized laboratory branch complement fixation test for detection of antibodies to *Coccidioides immitis*. *J. Clin. Microbiol. 16*:1030–1033.

Zimmer, B. L., and Pappagianis, D. (1986). Comparison of immunoblot analyses of spherule-endospore-phase extracellular protein and mycelial-phase antigen of *Coccidioides immitis*. *Infect. Immun. 53*:64–70.

7
Antigenic Structure of *Paracoccidiodes brasiliensis*

GIOCONDA SAN-BLAS and FELIPE SAN-BLAS
Instituto Venezolano de Investigaciones Científicas, Caracas, Venezuela

INTRODUCTION

Paracoccidioidomycosis is the most frequent systemic mycosis in Latin America, where it is geographically confined from Central America to Argentina. Humans are apparently the only hosts susceptible to natural infection by its etiological agent, *Paracoccidioides brasiliensis*.

The dimorphic nature of *P. brasiliensis* was first described by Lutz (1908) 80 years ago. This fungus displays a yeastlike morphology in lesions and in cultures at 35-37°C and a mycelial phase at temperatures lower than 25°C. Transition from the yeast to the mycelial phase, or vice versa, depends exclusively on the temperature of incubation (Nickerson and Edwards, 1949), and it is independent of the nature and composition of the culture medium (Inlow, 1979). The ability of several auxotrophic mutants to undergo phase transition on minimal media that do not allow growth (San-Blas and Centeno, 1977) supports the role played by temperature as the only factor responsible for inducing dimorphism in *P. brasiliensis*.

Clinical diagnosis of paracoccidioidomycosis is based on demonstration of characteristic multiple budding yeast cells in biologic fluids or in tissue sections. Serological tests in conjunction with clinical and histological findings help greatly in establishing an unequivocal diagnosis. Specific and standardized preparations must be used to ensure diagnostic reliability and reproducibility within the same laboratory and to compare results from different laboratories. However, the efficacy of crude antigen preparations from *P. brasiliensis* varies greatly for several reasons. The first is the diversity of experimental conditions used for extraction: strain of choice, morphological phase, growth conditions, cell fractionation, recovery techniques, etc. Another reason is the variety of tests: skin tests, serological tests such as complement fixation, immunodiffusion, electrophoresis, ELISA, MELISA, etc. Finally, the numerous antigenic substances (polysaccharides, polypeptides, glycopeptides, lipids) synthesized by *P. brasiliensis* introduce an important factor of heterogeneity when comparing results from different laboratories. According to their origin, *P. brasiliensis* antigens have been classified as (1) cell wall antigens, (2) intracellular (somatic) antigens, and (3) exocellular (metabolic) antigens (Yarzabal, 1982).

CELL WALL ANTIGENS

Through its cell wall, a microorganism makes its first contact with the environment. Therefore, this structure becomes of primary importance when a pathogen invades an organism, since the survival of the host will depend in large measure on its abilities to react immunologically and to lyse the microbial cell wall.

The *P. brasiliensis* cell wall, like other fungal walls, is composed mainly of polysaccharides (50–80% dry weight) plus some proteins and lipids (Kanetsuna et al., 1969). These components are arranged as a single layer 80–150 nm wide in the mycelial form and as a two-layered structure 200–600 nm wide in the yeast form (Carbonell and Rodriguez, 1968). Through alkaline separation, four polysaccharides are obtained. Three of them (chitin, α-glucan, and β-glucan) are structural polysaccharides, important not only in giving shape to the yeast and the mycelial phases of the fungus (chitin and β-glucan) (Kanetsuna et al., 1972), but also in regulating host–parasite relationships (α-glucan) (San-Blas and San-Blas, 1977) and dimorphism (α- and β-glucans) (San-Blas and San-Blas, 1985). None of them elicit immunological responses of any kind (Kanetsuna et al., 1972). A fourth polysaccharide, an alkali- and acid-soluble galactomannan, is present in *P. brasiliensis* walls, particularly in the mycelial form where it constitutes about 6% of the total wall. This polysaccharide (invisible to the electron microscope) is responsible for the immunogenicity of the cell wall of *P. brasiliensis*. Its chemical structure was studied in detail by Azuma et al. (1974), together with galactomannans from the pathogenic dimorphic fungi *Histoplasma capsulatum, H. duboisii*, and *Blastomyces dermatitidis*, in one of the few reports known to date to deal specifically with the elucidation of the detailed chemical structure of a given antigen from *P. brasiliensis*. The experimental results suggest that mannose:galactose ratios varied from 3.8:1.0 in *P. brasiliensis* to 1.5:1.0 in *H. capsulatum*. In all cases, mannopyranose units were linked together as a main chain through (1→6) linkages, with branching of galactofuranosyl residues occurring at the second or third position of the main chain (Fig. 1). Serological activities of galactomannans purified from whole cells of the four

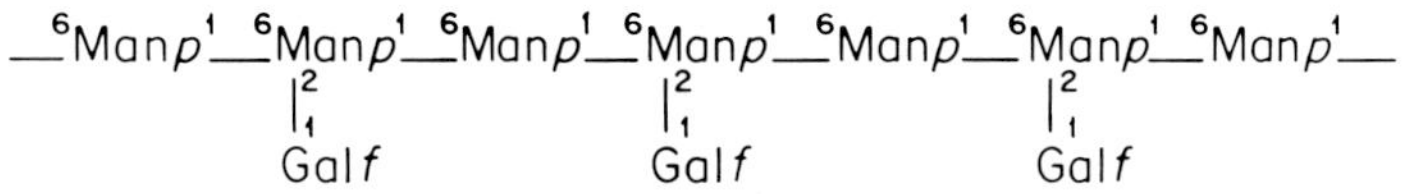

$$\text{—}^6\text{Man}p\underline{}^1\text{—}^6\text{Man}p\underline{}^1\text{—}^6\text{Man}p\underline{}^1\text{—}^6\text{Man}p\underline{}^1\text{—}^6\text{Man}p\underline{}^1\text{—}^6\text{Man}p\underline{}^1\text{—}^6\text{Man}p\underline{}^1\text{—}$$

$$|^2_1|^2_1|^2_1$$

$$\text{Gal}f\text{Gal}f\text{Gal}f$$

Figure 1 Chemical structure of the galactomannan from *P. brasiliensis* (Kanetsuna et al., 1972).

fungi listed above were determined by both the quantitative precipitation test and immunodiffusion analysis, with rabbit antisera against *P. brasiliensis, H. capsulatum*, and *Alternaria kikuchiana*. Mannan from *A. kikuchiana* was also included in the tests. With rabbit serum against whole cells of *H. capsulatum*, both galactomannans of *H. capsulatum* and *H. duboisii* showed higher reactivity than galactomannans of *P. brasiliensis* and *B. dermatitidis*, while *H. duboisii* galactomannan was less reactive with rabbit antiserum against *P. brasiliensis* whole cells. Mannan of *A. kikuchiana* did not give precipitin reaction with either antisera against *P. brasiliensis* or *H. capsulatum* (Fig. 2). Immunodiffusion patterns of galactomannans from *H. capsulatum, H. duboisii, P. brasiliensis*, and *B. dermatitidis* with rabbit antiserum against whole cells of *H. capsulatum* produced one single precipitin line. Upon immunoelectrophoretic analysis, all the galactomannans gave a single arc with rabbit antiserum against *H. capsulatum* cells. These results suggest that galactomannans tested in these experiments were immunologically related, having identical immunological determinant groups (Azuma et al., 1974). Since *A. kikuchiana* mannan did not cross-react, it was assumed that the nonreducing galactofuranosyl residue played an important role as an immunological determinant of these galactomannans. Intradermal injection of purified galactomannans into rabbits immunized with *H. capsulatum* elicited type III hypersensibility but not delayed-type skin reactions.

These results suggested that galactomannans were common antigens of these pathogenic fungi and, therefore, were of limited usefulness for skin and serological testing (Azuma et al., 1974).

INTRACELLULAR (SOMATIC) ANTIGENS

The search for specific antigens of *P. brasiliensis* has led some authors to explore the possibility of using intracellular substances, released after breakage of the cells by physical procedures. In these studies only the yeast phase of the fungus has been used so far. In fact, the first contributions to the serological study of paracoccidioidomycosis (Moses, 1916; Gomes and Assumpcao, 1924) were done with somatic antigens obtained from broken cells of *P. brasiliensis*. Using the complement fixation test, this antigen demonstrated antibodies against the fungus in 8 out of 10 paracoccidioidomycosis patients (Moses, 1916) and in 2

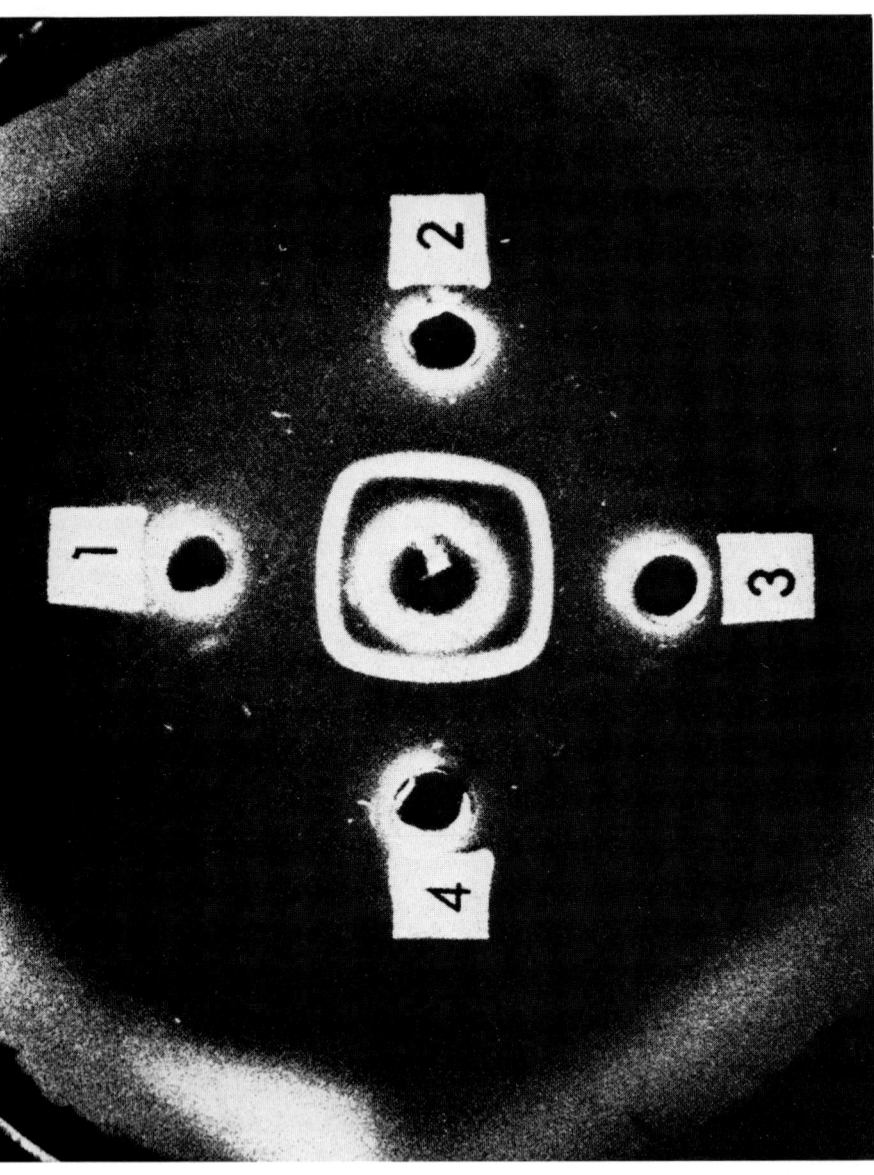

Figure 2 Immunodiffusion analysis of galactomannans with rabbit antiserum against *H. capsulatum*. Antigens used: (1) *H. capsulatum*; (2) *H. duboisii*; (3) *B. dermatiditidis*; (4) *P. brasiliensis*. Undiluted rabbit antiserum against *H. capsulatum* was placed in the center well (Azuma et al., 1974; reproduced by permission)

additional subjects (Gomes and Assumpcao, 1924). With this approach, Fava-Netto (1955) and Fava-Netto and Raphael (1961) reported the use of the so-called polysaccharide fraction employed for complement fixation tests for skin testing (Fava-Netto and Raphael, 1961), and for precipitation reactions (Fava-Netto, 1961). This material was derived from yeastlike cells grown for 60 days or later on for 20 days (Fava-Netto et al., 1969). Cells were collected by centrifugation, extracted with acetone and ethyl ether, and autoclaved, after which the pellet was discarded and the supernatant taken as antigen. No analytical methods were employed to assure the chemical homogeneity of this polysaccharide. Actually, an important proportion of this material must have been cell wall galactomannan, which is partially solubilized by autoclaving (San-Blas, G., unpublished results), as are mannans from *Saccharomyces cerevisiae* (Peat et al., 1961). When using it for skin tests (Fava-Netto and Raphael, 1961), positive results were obtained in very sick patients (67%) and in normal controls (87%). False positives were reported in 20 to 26% of controls. Expanding his studies on the "polysaccharide" antigen, Fava-Netto (1961) prepared 21 lots from eight strains of the fungus. Differences in the amount of polysaccharides released into the medium were detected; nevertheless, the author considered these paracoccidioidins as having the same composition. With them, he carried out serological studies in 314 paracoccidioidomycosis patients, 220 of them performing simultaneous complement fixation and precipitin tests. The results revealed two distinct clinicoimmunological forms, one mild with localized lesions and low levels of circulating antibodies and the other severe with disseminated lesions and high titer in serology. The precipitin test was positive in all patients having symptoms of the disease for less than a year and in 60% of those with symptoms for more than a year. The complement fixation test, on the other hand, was positive in two thirds of the patients with symptoms for less than a year and in 95% of those in the group with more prolonged evolution. In a later work, Fava-Netto (1965) concluded that precipitins appeared first in the course of paracoccidioidomycosis, whereas complement fixation antibodies appeared later and continued to be present after clinical cure. He also indicated that complement fixation antibody titers are low in localized forms of the disease and higher in disseminated paracoccidioidomycosis and decline as the patient's health improves with treatment or increase with relapses (Fig. 3).

Negroni and Negroni (1968) used as antigens both the supernatant and the pellet separated with Fava-Netto's technique besides an exocellular antigen released by the fungus after one month in culture. Their results indicated higher titers with exocellular antigens than with Fava-Netto's polysaccharidic antigen when using complement fixation tests. Cross-reactions with histoplasmosis and coccidioidomycosis sera were observed. No immunodiffusion reactions were detected with any of the preparations.

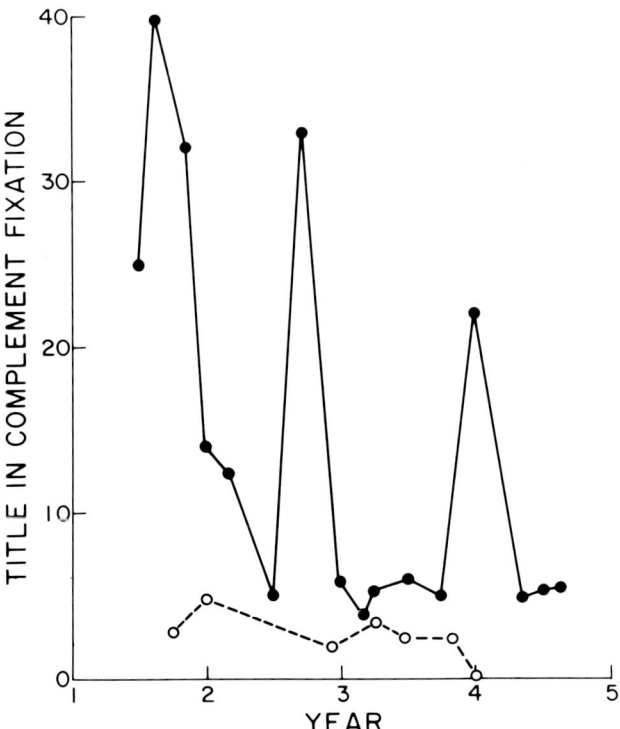

Figure 3 In the localized forms of the disease, complement fixation titers were low (o - - - - o), being higher in disseminated paracoccidioidomycosis (● ———— ●). Titers fell with clinical improvement and rose in relapses (Fava-Netto, 1965).

Franco et al. (1973) applied the same approach reported by Negroni and Negroni (1968) and used the yeast-insoluble fraction for indirect immunofluorescence while leaving the supernatant (equivalent to Fava-Netto's polysaccharide antigen) for complement fixation tests. The authors reported that all sera from 50 patients with paracoccidiodomycosis gave positive results in both tests while 50 sera from controls were negative. No sera from patients with other mycoses were checked for cross-reactivity.

Negroni et al. (1976) used two strains of *P. brasiliensis*, cultured in a complex medium at 37°C for 4 weeks, as the source of intracellular antigens. After collecting the cells by centrifugation, they were suspended in distilled water and incubated at 37°C with continuous agitation for 20 days. This was followed by breakage of cells in a homogenizer and centrifugation at 5000 rev/min for 15 min. The supernatant from this last step was lyophilized and used as antigen. Chemical analyses suggested the presence of 7.8 μg protein and 55 μg polysaccharides per mg dry weight (0.78 and 5.5%, respectively), the rest being of an unknown nature.

Skin testing with this material was done in conjunction with Fava-Netto antigen. In both cases, tests were negative in 10 out of 32 patients with paracoccidioidomycosis, while two out of 15 patients with histoplasmosis reacted positively to Negroni antigen. In a group of 70 patients with nonmycotic pulmonary lesions, 7 gave false positive reactions to this preparation. In general, Negroni paracoccidioidin was slightly more sensitive than Fava-Netto antigen. In both techniques, it is likely that prolonged culture times led to deterioration of cells, resulting in the solubilization of wall components after lytic action by cytoplasmic proteinases, and in the release of intracellular soluble material. This could be one of the reasons for the difficulties in reproducing results with different batches of such paracoccidioidins, even when produced under apparently identical conditions in the same laboratory.

Focusing perhaps on these limitations, Restrepo et al. (1983) prepared a yeast cytoplasmic antigen from strain LA, and later compared it with a mycelial filtrate from strain B339 (Restrepo et al., 1985). Results suggested that the yeast paracoccidioidin was superior to the mycelial preparation (62 vs. 24%) in patients with active paracoccidioidomycosis. However, the fact that both antigens were extracted from different strains introduces a bias in the comparisons established by the authors, since one of the factors to be considered when comparing results is the origin of strains (Restrepo, 1970). It was also pointed out in their more recent report that the reason for not using the same strain was the difficulty in handling the yeastlike phase of strain B339 for the production of cytoplasmic antigen. In spite of this unresolved issue, strain B339 was used in a following report (Restrepo et al., 1985) and subsequently by many other investigators to obtain cytoplasmic antigens, being nowadays a reference strain for the production of antigens in order to compare results among the various laboratories. Antigens are produced according to the following protocol: cells are grown in a chemically defined medium at 36°C for 10 days and used for the preparation of antigens only if their viability is in excess of 85% (Restrepo et al., 1985). Once cells are centrifuged and before mechanical rupture, phenylmethylsulfonylfluoride (PMSF) is added in order to inhibit serine proteinases. No other proteinase inhibitors are used. Addition of PMSF is a crucial step in the process since its absence left serine proteinases free to act and, as a consequence the antigen preparations obtained were weakly potent. Furthermore, it was observed that PMSF had to be present in high quantities and repeatedly throughout the experiment to ensure a relative stability of the antigens. Even so, the preparations were incompletely stable on storage (A. Restrepo, personal communication). Protein content varied from 8.5 to 10.3%, and carbohydrates ranged between 0.6 and 0.8%. Agar gel immunodiffusion showed that this extract was highly reactive (over 95%) with sera from patients with paracoccidioidomycosis. With the exception of histoplasmosis sera, which produced certain cross-reactivity (10%), the antigens proved specific, and none of the patients with other systemic mycoses, tuberculosis, or controls, reacted in the tests (Restrepo et al., 1985).

The preparation of cytoplasmic antigens for use as serological diagnostic tools in paracoccidioidomycosis has also been attempted by McGowan and Buckley (1985). A modification of the Restrepo et al. (1985) procedure combined with ELISA and immunodiffusion techniques was used. Their antigen preparation demonstrated a sensitivity of 100% and a specificity of 99% when tested against sera from 176 cases of paracoccidioidomycosis. Among 166 sera from patients with other fungal diseases tested, only two cases of blastomycosis gave positive reactions (ELISA titers $1 > 160$) (Fig. 4). The antigenic components of the cytoplasmic extract were separated by Concanavalin A-Sepharose 4B affinity chromatography. The fraction that remained bound to the column was reported to be identical to antigen 1 from Restrepo and Moncada (1974).

With regard to somatic antigens, more sophisticated techniques have been introduced recently to test for antibodies to *P. brasiliensis*, such as enzyme-linked immunosorbent assay (ELISA) (Cano et al., 1986) and magnetic enzyme-linked immunosorbent assay (MELISA) (Camargo et al., 1984c). These methods, based on immunoenzymatic reactions, may provide higher sensitivity, economy of reagents, precision in results, and can be easily automated. Using a yeast cytoplasmic antigen in an ELISA assay, Cano et al. (1986) found that 66% of the sera from patients with the disease proved reactive at titers greater than 1:128. At this dilution, sera from healthy blood donors and tuberculosis patients showed minimal reactivity (5%), and no positive reactions were observed with sera from patients with candidiasis, cryptococcosis, aspergillosis, or coccidioidomycosis. One third of the sera from patients with histoplasmosis reacted with *P. brasiliensis* antigens. The method lacked specificity at low serum dilutions. Camargo et al. (1984c) used several *P. brasiliensis* isolates to test the MELISA procedure, including strains B-339 and C-81. Both somatic and exocellular antigens were prepared according to Restrepo and Drouhet (1970). The exocellular (or metabolic) antigen had 1.9 mg protein per ml, and the somatic preparation, 0.5 mg per ml. The sensitivity of the MELISA test was highest when concanavalin A-activated beads and somatic antigen were used. When a comparative evaluation of MELISA, counterimmunoelectrophoresis and erythroimmunoassay methods was carried out with sera from 33 cases of paracoccidioidomycosis, 14 cases of histoplasmosis, and 20 normal human sera, MELISA was found to distinguish antibody levels in paracoccidioidomycosis from those of normal sera; however, two

Antigenic Structure of *Paracoccidioides brasiliensis* 179

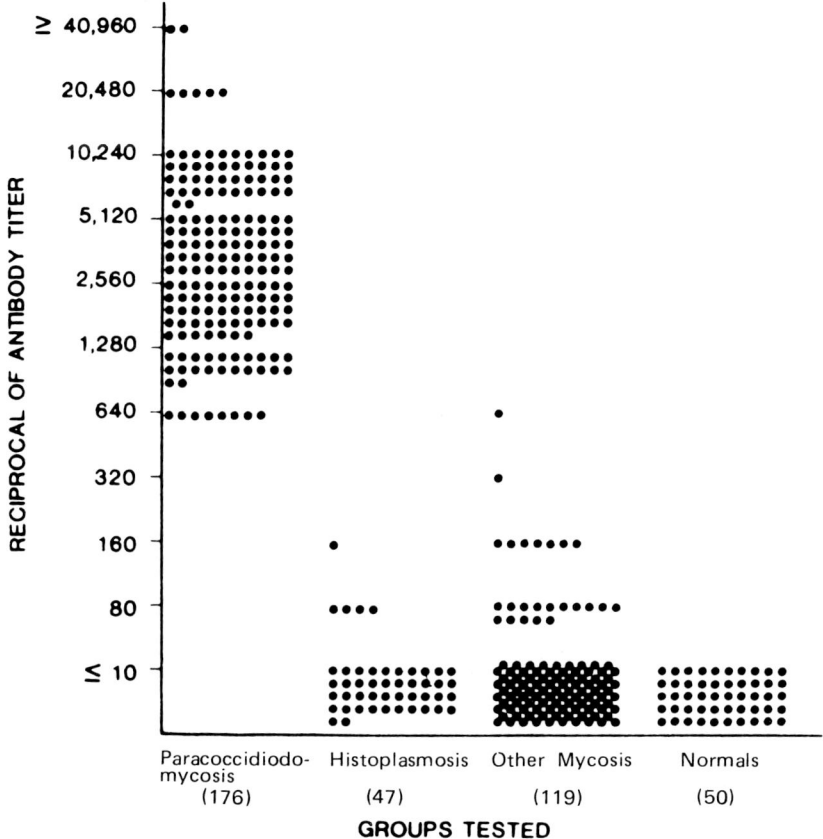

Figure 4 ELISA antibody titers of 392 sera tested with a cytoplasmic antigen of *P. brasiliensis*. The titers are expressed as the reciprocal of the highest dilution of serum to give positive results for antibodies to the cytoplasmic antigen. Groups tested included 50 normal sera, 47 sera from cases of histoplasmosis, 176 cases of paracoccidioidomycosis, and 119 sera from cases of other mycotic infections (McGowan and Buckley, 1985; reproduced with permission).

sera (14%) from histoplasmosis cases cross-reacted. MELISA titers seemed to be significantly higher in untreated patients than in those undergoing antimycotic treatment, indicating a potential usefulness of this method for follow-up studies in clinics (Camargo et al., 1986).

As suggested by Restrepo and Schneidau (1970), the binding of the specific somatic antigen to concanavalin A-activated polyacrylamide-agarose beads in

MELISA tests (Camargo et al., 1984c) is a clear indication of the glycopeptide nature of the *P. brasiliensis* main somatic antigen.

EXOCELLULAR (METABOLIC) ANTIGENS

The most widely used antigens of *P. brasiliensis* have been those released into the medium after prolonged (1 to 4 months) growth of the culture. This fact already suggests that while some of the components of such paracoccidioidins may be considered as truly excretion products from an active metabolic process, a great part of them must be passively freed into the medium as a consequence of an autolytic process, and therefore may correspond to all kinds of soluble cellular products, from wall polysaccharides like galactomannans to cytoplasmic proteins or glycoproteins. In fact, studies done in strain IVIC Pb9 (San-Blas et al., 1982) showed that *P. brasiliensis*, grown at 37°C, released an autoinhibitory factor of phenolic nature into the medium from day 9 onwards. Once this substance had accumulated in the medium, the fungus died out and cells disintegrated rapidly. Hence, the heterogeneity of paracoccidioidins prepared by prolonged culture must be one of the reasons for the variability of the responses observed in epidemiological surveys, even when using paracoccidioidins prepared under the same experimental protocol (Table 1).

Many investigators have employed either mycelial phase (Almeida et al., 1945; Lacaz, 1948, 1949; Carvalho, 1953, 1958; Douat and Dias, 1958) or yeast phase culture filtrates (Almeida and Lacaz, 1941; Lacaz, 1948, 1951; Mackinnon et al., 1953); Honnie and Artagaveitia-Allende, 1957; Lacaz et al., 1959) of *P. brasiliensis* to study skin reactivity in infected patients and in the normal population. Because of the heterogeneity of these antigenic materials, results were not comparable among different laboratories or even among batches prepared in a given laboratory (Restrepo and Schneidau, 1967). Mackinnon et al. (1953) observed that culture filtrates of the fungal yeastlike phase were reasonably specific and sensitive in a 1:100 dilution when tested in groups of guinea pigs incubated with other pathogenic fungi, whereas similar dilutions of culture filtrates of the mycelial phase were less sensitive and less specific.

Using paracoccidioidin obtained from supernatants of mycelial cultures grown for three months, Lacaz (1949) reported positive complement fixation tests in 87.5 to 93.5% of sera from paracoccidioidomycosis patients. The fact that seroconversion was seen with worsening of the clinical symptoms in those that initially tested negative made Lacaz suggest that complement fixation reaction should be periodically used throughout the treatment period to follow the therapeutic response of the patient. Some years later, Lacaz et al. (1959) performed skin tests on 529 apparently healthy persons using 1:10 and 1:100 dilutions of a paracoccidioidin, derived this time from a filtrate of yeastlike phase of *P. brasiliensis* grown on a liquid medium. They found positive reactions in

Table 1 Skin-test Surveys with *P. brasiliensis* Antigens

Author(s)	Number of subjects	Positivity (%)	Antigen
Lacaz, 1951 (Brazil)	330	7.8	MF
Negroni et al. (1952) (Argentina)	64	4.6	MF 1:10
Mackinnon et al. (1953) (Uruguay)	537	2.0	YF 1:10
Carvalho, 1953 (Brazil)	475	4.2	MF 1:10
Hounie and Artagaveitia (1957) (Uruguay)	30	0.0	YF 1:10
	12	25.0	YF 1:10
	12	41.6	YF 1:10 and 1:100
Aguirre (1957) (Chile)	137	0.0	YF 1:10
Douat and Dias (1958) (Brazil)	300	1.6	MF 1:10
Lacaz et al. (1959) (Brazil)	529	4.7	YF 1:10 and 1:100
Fava-Netto and Raphael (1961) (Brazil)	372	26.0	PA 1:10
	66	20.0	
	44	66.0	
Restrepo et al. (1968) (Colombia)	3,938	9.6	MFE
Albornoz et al. (1971) (Venezuela)	686	52.0	PA 1:10
	418	6.1	PA 1:10

MF = mycelial filtrate; YF = yeast filtrate; PA = polysaccharide antigen; MFE = mycelial filtrate extract.
Source: Conti-Diaz, 1972.

4.7% of the cases, with a maximum frequency in the 30 to 50 age group. No positive reactions were observed in those persons under 19 years of age. The complement fixation test was positive in five cases, and precipitin titers of over 1:5 were observed in six cases. These findings were considered suggestive of the existence of an asymptomatic or subclinical form of paracoccidioidomycosis.

At variance with the report of Mackinnon et al. (1953), Restrepo et al. (1968) concluded that the mycelial antigen was better after conducting skin test surveys in rural populations using paracoccidioidins prepared from both the mycelial and the yeastlike phases. However, a previous work by Restrepo and Schneidau (1967) carried out with three strains of *P. brasiliensis* indicated that 4-week-old culture filtrates of either M or Y phases, partially purified through precipitation with ethyl alcohol and extraction with chloroform-amyl alcohol, were able to elicit similar reactivities in skin tests done on animals infected with *P. brasiliensis*. Animals infected with *H. capsulatum* cross-reacted with the M or the Y metabolic antigen of *P. brasiliensis*, whereas in animals infected with *Sporothrix schenckii*, cross-reaction was limited to the M phase. Chemical analyses on both materials suggested the presence of hexoses (29% in yeasts; 37% in mycelium), and nitrogenous compounds (3.9% in yeasts; 2.7% in mycelium). The following amino acids were detected: leucine, methionine, alanine, glutamic acid, serine, threonine, and glycine. Both extracts had glucose, galactose, arabinose, and glucosamine. The fact that galactose was three times higher in the mycelial than in the yeast preparation, may suggest that this saccharide was present as a result of autolysis and not as a real exocellular product, since this proportion is found in the soluble galactomannan of *P. brasiliensis* cell walls (Kanetsuna et al., 1969). Restrepo and Schneidau (1967) suggested that the antigenically active substance might be a glycopeptide. Further studies by Restrepo-Moreno (1970) in six strains of *P. brasiliensis* grown for 4 weeks at 37°C indicated that supernatants contained variable amounts of nitrogen (0.3-1.9%) and carbohydrates (20-94%). This antigen was prepared according to Restrepo and Schneidau (1967) though omitting the ethyl alcohol and chloroform extractions. The individual mycelial antigens seemed to share a common epitope while yeast phase culture filtrates behaved as antigenically heterogeneous mixtures. It is interesting to observe the differences in reactivity of various sera towards the individual antigens. Five of the sera tested had antibodies reactive with each one of the six yeast antigens employed. The remaining three sera reacted with some, but not all, of the antigens. This suggested that individual strains, vary in the production or liberation in vitro of antigens capable of combining with antibodies present in the serum of patients with the disease; therefore, the authors recommend the use of a pool of antigens when serological tests were applied for diagnostic purposes. Electrophoretic analysis of this material (Restrepo and Drouhet, 1970) revealed five antigenic fractions in the yeast phase of *P. brasiliensis* (labeled from A to E) when tested against sera of paracoccidioidomycosis patients, arc A being found in every serum with precipitating activity. Specific precipitins were found in 28 out of 30 cases with paracoccidioidomycosis, as measured by the Ouchterlony technique or by immunoelectrophoresis analysis, methods which proved more sensitive than the complement fixation test, which was positive in only 27 cases.

In order to characterize the precipitin bands detected in the immunodiffusion test for paracoccidioidomycosis, a study was undertaken in 54 patients with the disease (Restrepo and Moncada, 1974), using paracoccidioidin prepared according to Restrepo and Schneidau (1967). On the basis of the reactivity of known control sera, three precipitation lines were identified by this technique. They were designated 1, 2, and 3 according to their location in the immunodiffusion plate, band 1 being located near the antigen well and band 3, closer to the serum well. At the time of diagnosis, 28 patients exhibited all three bands, 16 gave two bands, and 10 showed only one precipitin line. Over 50% of the sera with three bands had high complement fixation titers (above 1:512), whereas those with one band exhibited lower titers. A similar picture was obtained with the quantitative agar gel techniques, whereas titers of 1:64 and above were more commonly observed in sera with three precipitin lines. Followup studies carried out in 18 patients under treatment revealed that band 3 disappeared first, followed by band 2, and finally by band 1. Cross-reactions with histoplasmin were found in 8 patients who gave a line of identity with the M precipitin line of histoplasmin. It was found that the M line of histoplasmin was identical to *P. brasiliensis* band 3. Bands 1 and 2 were specific. An equivalence was proposed among band 1 from immunodiffusion tests (Restrepo and Moncada, 1974), precipitin arc A (Restrepo and Drouhet, 1970), and precipitation band E (Yarzabal et al., 1976), the latter two obtained by electrophoresis.

Using mixtures ("total antigen") of 3-month-old metabolic and somatic antigens of both the mycelial and the yeastlike phases, Yarzabal et al. (1973) demonstrated the existence of several proteic molecules, one of them migrating to the cathode at pH 8.2 in agar electrophoresis and agarose immunoelectrophoresis. This protein was absent in extracts prepared similarly from *B. dermatitidis, H. capsulatum, H. duboisii,* and *H. farciminosum.* Through immunoelectrophoretic analysis with rabbit hyperimmune sera, it was found that *P. brasiliensis* extracts contained 25 immunogens, 10 of them shared with *B. dermatitidis,* 9 with *H. farciminosum,* 7 with *H. capsulatum,* and 4 with *H. duboisii* (Yarzabal et al., 1973). These immunogens resolved into more than 60 when separated by bidimensional immunoelectrophoresis (Yarzabal, 1982). From the cathodic protein migrating at pH 8.2 in electrophoresis, two antigens seemed to be specific for *P. brasiliensis* (Yarzabal et al., 1976, 1977). Synthesized by both phases of the fungus, they appeared in cultures before the end of the log phase. Antigen E1 had an alkaline phosphatase activity and was responsible for the early appearance of precipitins in infected or immunized animals (Yarzabal et al., 1976). Although antigen E1 was initially considered as the specific antigen, it was soon eclipsed by a second arc (E2), which migrated very closely to E1 but which lacked the activity of alkaline phosphatase displayed by E1 (Yarzabal et al., 1977). E2 is released into the medium at 4 to 9 days of growth of *P. brasiliensis* in either phase, and is absent from species of *Aspergillus, Blastomyces, Candida, Coccidioides,* and *Histoplasma,* among others (Yarzabal, 1982), being considered therefore as the specific antigen. But since it was rather laborious to produce, its use for routine check-up was discouraged at that time.

Blumer et al. (1984) prepared a paracoccidioidin from the culture media of six different strains of *P. brasiliensis*, grown at 25°C as the mycelial phase. Media

Table 2 ELISA Titers for Cases of Paracoccidioidomycosis and Histoplasmosis Before and After Absorbing Sera with Several *H. capsulatum* Components (ABS-ELISA)

	\multicolumn{6}{c}{Serum titer}					
	ELISA	\multicolumn{5}{c}{ABS-ELISA[a]}				
		A	B	C	D	E
Paracoccidioido-mycosis						
1	5,120	5,120	5,120	5,120	5,120	5,120
2	5,120	5,120	5,120	5,120	5,120	5,120
3	1,280	1,280	1,280	1,280	1,280	1,280
4	320	320	320	320	320	320
Histoplasmosis						
5	320	160	80	160	80	<40
6	320	160	80	160	40	<40
7	80	40	40	40	40	<40

[a]Sera were absorbed with A, mycelial-phase culture filtrate diluted to 25 µg/ml; B, mycelial-phase culture filtrate diluted to 50 µg/ml; C, yeast-phase cell suspension, 6×10^6 cells per ml; D, yeast-phase cell suspension, 1×10^6 cells per ml; E, cell suspension (1×10^6 cells per ml) in culture filtrate (50 µg protein per ml).
Source: Mendes-Giannini et al., 1984.

used were potato glucose agar slants as the seed culture, and trypticase soy and Sabouraud broths as the growth media. The quantity of *P. brasiliensis* antigen produced and the optimal time for its production varied among the isolates tested and according to the medium use for growth. The sensitivity of this antigen was equivalent to that of Restrepo and Schneidau (1967), with a sensitivity of 90% in 114 proven cases of paracoccidioidomycosis and a specificity of 100% since no cross-reactions were observed with heterologous mycotic diseases. This antigens was useful for immunodiffusion tests, but not for complement fixation, and it was proven identical with the specific E antigen (Yarzabal et al., 1976) and with antigen 1 (Restrepo and Moncada, 1974).

The ELISA assay using yeast culture filtrates of *P. brasiliensis* proved specific if sera were previously absorbed with *C. albicans* whole cells (Camargo et al., 1984b). Only one of the control serum of candidiasis and one of histoplasmosis sera exhibited a weak ELISA titer for antibodies to *P. brasiliensis*, probably due to common antigenic determinants not eliminated by absorption with *C. albicans* cells. This approach was also attempted by Mendes-Giannini et al. (1984), who used *H. capsulatum* instead of *C. albicans* to absorb sera. To standardize conditions, three *P. brasiliensis* strains grown at 35°C in four different culture media (Sabouraud-dextrose broth, trypticase soy broth, proteose-peptone broth, and Casals medium) were studied. Among filtrates of *P. brasiliensis* cultures, those obtained in Casals medium showed the highest reactivity with paracoccidioidomycosis sera in ELISA. Although no false positive results were seen, cross-reactions were frequent with sera from patients with other mycoses (histoplasmosis, lobomycosis, cryptococcosis, candidiasis). In a search for better specificity, sera were absorbed with *H. capsulatum* antigens, using total culture filtrates of *P. brasiliensis* to sensitize the plates. The absorption-ELISA test showed a high sensitivity since it was positive with all 69 sera from cases of paracoccidioidomycosis. The test was also very specific, with negative results for all non-paracoccidioidomycosis sera studied, including those from patients with other mycoses (Table 2). Titration of antibodies to *P. brasiliensis* by erythroimmunoassay (EIA) (Camargo et al., 1984a) indicated that homologous and heterologous sera reacted with a culture filtrate antigen prepared according to Restrepo and Drouhet (1970). In this case, absorption of all sera with *C. albicans* cells decreased the reactions for both homologous and heterologous (candidiasis, aspergillosis) sera.

ORIGIN AND NATURE OF *P. brasiliensis* ANTIGENS

The classification of antigens according to their localization on extraction is useful for preparative purposes. However, it does not give information as to their origin or the mechanisms whereby a given antigen is exported outside the cell. The origin and nature of antigens from the yeast form of *P. brasiliensis* strain IVIC Pb73 were partially determined by serological and biochemical techniques (Hallak, J., and San-Blas, F., IX International Congress of ISHAM, Atlanta, GA, 1985). Cytoplasmic and membrane fractions obtained from protoplasts, as well as cell wall and extracellular fractions, were tested in the search for antigens. Antigenic activities were found in all fractions except cell walls. Immunodiffusion tests showed that the antigens gave reactions of identity among them and with Yarzabal's antigen (Fig. 5). These results and those obtained by affinity chromatography, gel filtration, electrophoresis and other techniques showed that two main antigens exist in the cytoplasmic fraction, one of which is glycosylated. The extracellular antigens were glycosylated and separable by electroimmunodiffusion and gel filtration. They differed in molecular weight. The smaller component seemed to originate by hydrolysis of the bigger one as a result of the action of fungal proteases. Those results suggested that these antigens were

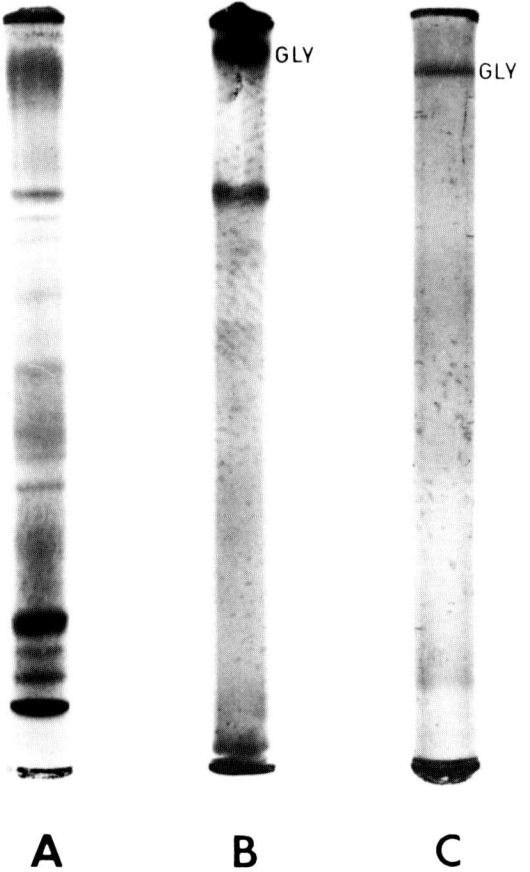

Figure 5 Disk electrophoresis of cytosol content released from protoplasts of the yeast phase of *P. brasiliensis* strain IVIC Pb73. (A) Cytosol proteins stained with Coomassie Blue. (B) Unstained cytosol proteins. The gel was washed with a patient's serum to visualize the antigen in the cytosol. Two main antigen–antibody complexes are clearly seen. (C) Cytosol protein colored with a specific glycoprotein stain.

derived from the nonglycosylated antigen found in the cytoplasmic fraction, which is glycosylated for transport outside the yeast cell and thereafter hydrylyzed by fungal proteases.

The glycosylated nature of exocellular antigens has also been reported by Puccia et al. (1986). Using strain B339, already reported by Restrepo and Drouhet (1970), these authors prepared crude antigen from supernatants taken from

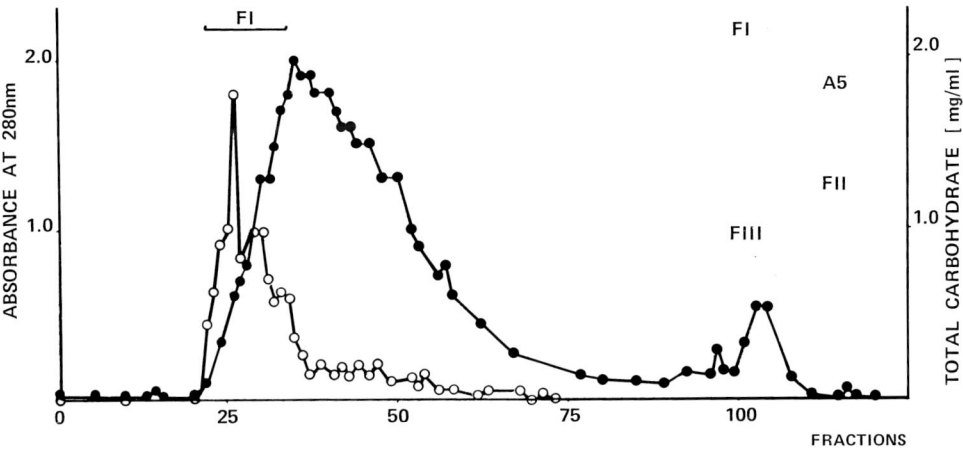

Figure 6 Elution of crude exocellular antigens from *P. brasiliensis* yeastlike phase through a Bio-Gel P30 column (47 × 2.6 cm). ●————●, Absorbance at 280 nm; o————o), total carbohydrate content. Immunodiffusion test (insert) shows the precipitation by a paracoccidioidomycosis serum (center well) of fraction I. (Puccia et al., 1986; reproduced by permission).

15- to 40-day-old cultures of *P. brasiliensis*, yeastlike phase. Components which account for the main antigenic activity were isolated by gel filtration on a BioGel P30 column, followed by affinity chromatography on a Sepharose-concanavalin A column, and elution (Fig. 6). Three main fractions were found: a 43,000 MW glycoprotein (gp43), a polydisperse high MW glycoconjugate, and a diffusely migrating 55,000 MW glycoprotein (gp55). Of these, gp43 behaved as a specific antigen and was further resolved by isoelectric focusing into three components of distinct pIs. This antigen had the same migration in immunoelectrophoresis as band E (Yarzabal et al., 1976) and was immunoprecipitated by antibodies specific for the E antigen (Puccia et al., 1986).

CONCLUDING REMARKS

The preparation of highly specific and reliable antigens from *P. brasiliensis* has become an important objective for investigators interested in the clinical and epidemiological aspects of paracoccidioidomycosis. Serological tests are well covered nowadays with the isolation of a specific antigen called precipitation band E (Yarzabal et al., 1976), immunodiffusion antigen 1 (Restrepo and Moncada, 1974), precipitin arc A (Restrepo and Drouhet, 1970), or glycoprotein gp

43 (Puccia et al., 1986), and whose identical nature has been postulated by these authors. Its excellent performance as a highly specific reagent makes it the antigen of choice for serological tests. Some limitations, though, exist. Because not every mycological laboratory has the facilities to prepare it, kits are to be provided by a few sources to all research and clinical laboratories dealing with paracoccidioidomycosis. This step is mandatory to assure the long desired standardization to a reference antigen with the additional benefit of making results from different laboratories comparable. Another limitation of these preparations is the presence of proteolytic enzymes which render the antigen useless with time. The addition of PMSF and other proteinase inhibitors has not resulted in the desired stability, suggesting that proteinases other than serine proteinases are present in these preparations. For this reason it is imperative to study further the nature of these antigens in order to avoid their inactivation.

A serious limitation of this antigen is its inability to induce intradermal reactions; therefore, the use of poorly specific paracoccidioidins is still recommended until a better reagent becomes available. These preparations have as their major handicap frequent cross-reactivity with other fungal antigenic extracts, particularly *H. capsulatum*, whose geographical area of distribution overlaps with that of *P. brasiliensis*. Again, the heterogeneity of such paracoccidioidins makes it difficult to compare results obtained with antigens from different sources. Furthermore, skin test reactivity depends on the competence of cells of the immune system which is adversely affected in patients with paracoccidioidomycosis. As a consequence, a negative skin test does not rule out the mycosis, while a positive test alone has no diagnostic value. Despite these limitations, skin tests are useful as complementary diagnostic tools, for follow-up studies and for epidemiological surveys. The design of more specific and sensitive paracoccidioidins is an important objective to pursue in order to achieve an early and precise diagnosis of this severe mycosis.

On February 1986 the 3rd International Meeting on Paracoccidioidomycosis was held in Colombia. A consensus was reached among the participants to create a committee to work on these lines. Under the coordination of Dr. Angela Restrepo and using Camargo et al. (1984c) antigen, members of the committee set up to prepare a reference antigen, which is currently being tested by several research teams throughout Latin America. Although it is still too early to have definitive results, this joint effort will certainly provide encouraging progress in the near future to move forward the research on paracoccidioidomycosis and its etiological agent, *P. brasiliensis*.

REFERENCES

Almeida, F. P., and Lacaz, C. S. (1941). Intradermo-reacao com paracoddidioidina no diagnostico do granuloma paracoccidioidico. II. A reacao de Monte-

negro no granuloma paracoccidioidico. *Fol. Clin. Biol.* (S. Paulo) *13*:177–182.
Almeida, F. P., Lacaz, C. S., and Cuhna, A. C. (1945). Intradermo-reacao para diagnostico da blastomicose sul-americana (granulomatose paracoccidioidica). *Arq. Bras. Med. 35*:267–272.
Azuma, I., Kanetsuna, F., Tanaka, Y., Yamamura, Y., and Carbonell, L. M. (1974). Chemical and immunological properties of galactomannans obtained from *Histoplasma duboisii, Histoplasma capsulatum, Paracoccidioides brasiliensis*, and *Blastomyces dermatitidis. Mycopathol. Mycol. Appl. 54*:111–125.
Blumer, S. O., Jalbert, M., and Kaufman, L. (1984). Rapid and reliable method for production of a specific *Paracoccidioides brasiliensis* immunodiffusion test antigen. *J. Clin. Microbiol. 19*:404–407.
Camargo, Z. P., Guesdon, J. L.,Drouhet, E., and Improvisi, L. (1984a). Titration of antibodies to *Paracoccidioides brasiliensis* by erythro-immunoassay. *Sabouraudia 22*:73–77.
Camargo, Z. P., Guesdon, J. L., Drouhet, E., and Improvisi, L. (1984b). Enzyme-linked immunosorbent assay in paracoccidioidomycosis. *Mycopathologia 88*: 31–37.
Camargo, Z. P., Guesdon, J. L., Drouhet, E., and Improvisi, L. (1984c). Magnetic enzyme-linked immunosorbent assay (MELISA) for determination of specific IgG in paracoccidioidomycosis. *Sabouraudia J. Med. Vet. Mycol. 22*:291–299.
Camargo, Z. P., Unterkircher, C., and Drouhet, E. (1986). Comparison between magnetic enzyme-linked immunosorbent assay (MELISA) and complement fixation test (CF) in the diagnosis of paracoccidioidomycosis. *J. Med. Vet. Mycol. 24*:77–79.
Cano, L. E., Brummer, E., Stevens, D. A., and Restrepo, A. (1986). An evaluation of the enzyme-linked immunoabsorbent assay (ELISA) for quantitation of antibodies to *Paracoccidioides brasiliensis. J. Med. Vet. Mycol. 24*:467–475.
Carbonell, L. M., and Rodriguez, J. (1968). Mycelial phase of *Paracoccidioides brasiliensis* and *Blastomyces dermatitidis*: and electron microscope study. *J. Bacteriol. 96*:533–534.
Carvalho, A. (1953). Sobre o emprego da paracoccidioidina na cicade do Rio de Janeiro: primeros resultados baseados no estudio de 475 individuos. *Rev. Bras. Tuberc. 21*:73–82.
Carvalho, A. (1958). Sobre o uso intradermico da paracoccidioidina (filtrado e suspensao) como meio auxiliar diagnostico na doenca de Lutz. *Rev. Tuberc. 2*:131–178.
Conti-Diaz, I. A. (1972). Skin tests with paracoccidioidin and their importance. In *Paracoccidioidomycosis, Proceedings of the First Panamerican Symposium*. Scientific Publication N. 254, PAHO, pp, 197–202.
Douat, N. E., and Dias, V. M. (1958). Intradermo-reacoes de paracoccidioidina, coccidioidina e histoplasmina: resultados de testes em 300 individuos. *Rev. Brasil Tuberc. 26*:663–668.

Fava-Netto, C. (1955). Estudos quantitativos sobre a fixacao do complemento na blastomicose sul-americana, com antigeno polissacaridico. *Arq. Cirurg. Clin. Exp. 18*:197–254.

Fava-Netto, C. (1961). Contribucao para o estudo imunologico da blastomicose de Lutz (blastomicose sul-americana). *Rev. Inst. A. Lutz* (Sao Paulo) *21*:99–194.

Fava-Netto, C. (1965). The immunology of South American blastomycosis. *Mycopathol. Mycol. Appl. 26*:349–358.

Fava-Netto, C., and Raphael, A. (1961). A reacao intradermica com polisacaride do *Paracoccidioides brasiliensis*, na blastomicose sul-americana. *Rev. Inst. Med. Trop. S. Paulo 3*:161–165.

Fava-Netto, C., Vegas, V. S., Sciannamea, I. M., and Guarnieri, D. B. (1969). Antigeno polissacaridico do *Paracoccidioides brasiliensis*. Estudo do tempo de cultivo do *P. brasiliensis*, necessario ao preparo do antigeno. *Rev. Inst. Med. Trop. S. Paulo 11*:177–181.

Franco, M. F., Fava-Netto, C., and Chamma, L. G. (1973). Reacao de imunofluorescencia indireta para o diagnostico sorologico da blastomicose sul-americana. Padronizacao da reacao e comparacao dos resultados com a reacao de fixacao do complemento. *Rev. Inst. Med. Trop. S. Paulo 15*:393–398.

Gomes, J. M., and Assumpcao, L. (1924). Em torno do genero *Coccidioides*. *Ann. Paul. Med. Cirurg. 15*:49–61.

Hounie, P., and Artagaveytia-Allende, R. C. (1957). Encuesta sobre la sensibilidad al agente de la blastomicosis sudamericana. *An. Fac. Med. Montev. 42*:27–32.

Inlow, D. I. (1979). Estudios sobre el dimorfismo de *Paracoccidioides brasiliensis*. M. Sc. Thesis, IVIC, Caracas, Venezuela.

Kanetsuna, F., Carbonell, L. M., Azuma, I., and Yamamura, Y. (1972). Biochemical studies on thermal dimorphism of *Paracoccidioides brasiliensis*. *J. Bacteriol. 110*:208–218.

Kanetsuna, F., Carbonell, L. M., Moreno, R. E., and Rodriguez, J. (1969). Cell wall composition of the yeast and mycelial forms of *Paracoccidioides brasiliensis*. *J. Bacteriol. 97*:1036–1041.

Lacaz, C. S. (1948). Blastomicose sul-americana: reacoes intradermicas com a paracoccidioidina, coccidioidina, e blastomicetina. *Rev. Hosp. Clin. Fac. Med. S. Paulo 3*:11–18.

Lacaz, C. S. (1949). Novos dados em relacao a blastomicose sul-americana e seu agente etiologico. *Rev. Med. Cir. S. Paulo. 9*:303–341.

Lacaz, C. S. (1951). Lesoes pulmonares na blastomicose sul-americana:inquerito preliminar realizado com a paracoccidioidina. *Hospital (Rio) 39*:405–422.

Lacaz, C. S., Passos-Filho, M. C. R., Fava-Netto, C., and Macarron, B. (1959). Contribucao para o estudo da "blastomicose-infeccao": inquerito com a paracoccidioidina; estudo sorologico e clinico-radiologico dos paracoccidioidina-positivos. *Rev. Inst. Med. Trop. S. Paulo 1*:245–259.

Lutz, A. (1908). Uma mycose pseudococcidica localizada na bocca e observada no Brazil. Contribucao ao conhecimento das hyphoblastomycoses americanas. *Brasil-med. 22*:121–124.

Mackinnon, J. E., Artagaveytia-Allende, R. C., and Arroyo, L. (1953). Sobre la especificidad de la intradermo reaccion con paracoccidioidina. *An. Fac. Med. Montev. 38*:363-382.
McGowan, K. L., and Buckley, H. R. (1985). Preparation and use of cytoplasmic antigens for serodiagnosis of paracoccidioidomycosis. *J. Clin. Microbiol. 22*: 39-43.
Mendes-Giannini, M. J. S., Camargo, M. E., Lacaz, C. S., and Ferreira, A. W. (1984). Immunoenzymatic absorption test for serodiagnosis of paracoccidioidomycosis. *J. Clin. Microbiol. 20*:103-108.
Moses, A. (1916). Fixacao do complemento na blastomicose. *Mem. Inst. Oswaldo Cruz 8*:68-70.
Negroni, R., Elias Costa, M. R., Bianchi, O., and Galimberti, R. (1976). Preparacion y estudio de un antigeno celular de *Paracoccidioides brasiliensis* util para pruebas cutaneas. *Sabouraudia 14*:265-273.
Negroni, R., and Negroni, P. (1968). Antigenos del *Paracoccidioides brasiliensis* para las reacciones serologicas. *Mycopathol. Mycol. Appl. 34*:285-288.
Nickerson, W. J., and Edwards, G. A. (1949). Studies on the physiological bases of morphogenesis in fungi. I. The respiratory metabolism of dimorphic pathogenic fungi. *J. Gen. Physiol. 33*:41-55.
Peat, S., Whelan, W. J., and Edwards, T. E. (1961). Polysaccharides of baker's yeast. Part IV. Mannan. *J. Chem. Soc.*, 29-34.
Puccia, R., Schenkman, S., Gorin, P. A. J., and Travassos, L. R. (1986). Exocellular components of *Paracoccidioides brasiliensis*: identification of a specific antigen. *Infect. Immun. 53*:199-206.
Restrepo-Moreno, A. (1970). Serological comparison of the two morphological phases of *Paracoccidioides brasiliensis*. *Infect. Immun. 2*:268-273.
Restrepo, A., Cano, L. E., and Ochoa, M. T. (1985). A yeast-derived antigen from *Paracoccidioides brasiliensis* useful for serological testing. *J. Med. Vet. Mycol. 23*:23-29.
Restrepo, A., Cano, L. E., and Tabares, A. M. (1983). A comparison of mycelial filtrate- and yeast lysate-paracoccidioidin in patients with paracoccidioidomycosis. *Mycopathologia 84*:49-54.
Restrepo, A., and Drouhet, E. (1970). Etude des anticorps precipitants dans la blastomycose sud-americaine par l'analyse immunoelectrophoretique des antigenes de *Paracoccidioides brasiliensis*. *Ann. Inst. Pasteur 119*:338-346.
Restrepo, A., and Moncada, L. H. (1974). Characterization of the precipitin bands detected in the immunodiffusion test for paracoccidioidomycosis. *Appl. Microbiol. 28*:138-144.
Restrepo, A., Robledo, M., Ospina, S., Restrepo, M., and Correa, A. (1968). Distribution of paracoccidioidin sensitivity in Colombia. *Amer. J. Trop. Med. 17*:25-37.
Restrepo, A., and Schneidau, Jr., J. D. (1967). Nature of the skin-reactive principle in culture filtrates prepared from *Paracoccidioides brasiliensis*. *J. Bacteriol. 93*:1741-1748.

San-Blas, F., and Centeno, S. (1977). Isolation and preliminary characterization of auxotrophic and morphological mutants of the yeastlike form of *Paracoccidioides brasiliensis*. *J. Bacteriol. 129*:138-144.

San-Blas, F., De Marco, G., and San-Blas, G. (1982). Isolation and partial characterization of a growth inhibitor of *Paracoccidioides brasiliensis*. *Sabouraudia 20*:159-168.

San-Blas, F., and San-Blas, G. (1985). *Paracoccidioides brasiliensis*. In *Fungal dimorphism*. Edited by P. Szaniszlo. Plenum Press Co., pp. 93-120.

San-Blas, G., and San-Blas, F. (1977). *Paracoccidioides brasiliensis*: cell wall structure and virulence. A review. *Mycopathologia 62*:77-86.

Yarzabal, L. (1982). Composicion antigenica de *Paracoccidioides brasiliensis*. In *Paracoccidioidomicose blastomicose sul-americana*. Edited by G. Del Negro, C. Da Silva Lacaz, and A. M. Fiorillo. Sarvier, Edusp, Sao Paulo, pp. 59-67.

Yarzabal, L. A., Andrieu, S., Bout, D., and Naquira, F. (1976). Isolation of a specific antigen with alkaline phosphatase activity from soluble extracts of *Paracoccidioides brasiliensis*. *Sabouraudia 14*:275-280.

Yarzabal, L. A., Biguet, J., Vaucelle, T., Andrieu, S., Torres, J. M., and Da Luz, S. (1973). Analisis immunoquimico de extracto soluble de *Paracoccidioides brasiliensis*. *Sabouraudia 11*:80-88.

Yarzabal, L. A., Bout, D., Naquira, F., Fruit, J., and Andrieu, S. (1977). Identification and purification of the specific antigen of *Paracoccidioides brasiliensis* responsible for immunoelectrophoretic band E. *Sabouraudia 15*:79-85.

8
Antigenic Structures of *Sporothrix schenckii*

LUIZ R. TRAVASSOS
Escola Paulista de Medicina, São Paulo, Brazil

INTRODUCTION

Comprehensive reviews on the biology of *Sporothrix schenckii* (Hektoen and Perkins, 1900) have appeared recently (Travassos and Lloyd, 1980; Travassos, 1985). The present review will focus mainly on the structural and immunological aspects of the major antigens of this fungus and how they can influence the infection process.

Being a dimorphic fungus, *S. schenckii* undergoes a phase transition from mycelium to yeast depending on environmental factors, including temperature, CO_2, other nutrients—among them certain amino acids and high glucose (Drouhet and Mariat, 1952; Romano, 1966; Mendonça, et al., 1976; Rodriguez-Del Valle et al., 1983)—as well as a low redox potential (Rippon, 1980). In a chemically defined medium, temperature (25°C) and low pH (4-5) are factors determining conversion of conidia into hyphae. Since the yeast phase is the infective form, attention has been drawn mainly to the antigenic characteristics of this cell type, although several reports have appeared on the ultrastructural alterations occurring during a differentiation process transforming hyphae into yeasts, conidia into hyphae, and conidia directly into yeasts (Lane and Garrison, 1970; Garrison et al., 1975; Travassos et al., 1977; Garrison, 1985).

Yeasts of *S. schenckii* can be demonstrated in tissue or biological specimens by reaction with fluorescent anti-*S. schenckii* rabbit antibodies (Kaplan and Ivens, 1960). Cross-reactivity has only been detected with nonpathogenic species of *Ceratocystis* (Harada et al., 1976), although more specialized structures such as perithecia and ascospores were unreactive. It seems that, based on DNA base composition and hybridization experiments, *S. schenckii* strains are

193

not anamorphs of different *Ceratocystis* species, but, rather, they represent a homogeneous species, which is the imperfect form of one as yet unidentified *Ceratocystis* species (Travassos, 1985).

Antigenic expression in *S. schenckii* has been used not only as markers of differentiation in this fungus (Mendonça, et al., 1976), but also as a diagnostic tool and as an indication of the surface changes that may determine evolution or regression of the infection (Travassos and Alviano, 1981). Surface polysaccharides of this human pathogen have had their fine structures studied in detail and in many cases completely characterized.

ISOLATION OF ANTIGENIC STRUCTURES

Crude purified antigens from *S. schenckii* have been prepared using cultures either predominantly in the yeast phase or containing various proportions of hyphae and conidia. Growth temperature—28-30°C or 35-37°C—and the culture medium—chemically defined, Sabouraud's, brain-heart infusion or a variety of other complex media—will determine the final morphology of the culture. As quoted before, morphogenetic factors such as CO_2 are important for yeast phase generation. Crude antigenic preparations from these cultures reflect their predominant cell morphology as well as the growth conditions, such as aeration and incubation temperature. Soluble as well as particulate antigens have been used in routine immunological tests. Most preparations resist lyophilization, heating at 60°C followed by cooling at 4°C (Rocha Posada, 1968), or even autoclaving. Crude antigenic preparations are apparently stable for years. Decreased reactivity is detected when molecular aggregation or particle clumping occurs and centrifugation becomes necessary.

From Whole Cells

Antigens can be extracted from cells of mixed morphology or from individual cell types. The second alternative is possible when the growth conditions permit formation of 100% cells in one phase or the other. Yeasts are readily recognized as fusiform and ovoid cells, often pleomorphic in liquid cultures, multiplying by single or multiple budding. Growth is abundant at 35°C. Cells can be readily centrifuged and washed 2 or 3 times before antigen extraction. In the case of the mycelium phase, various degrees of sporulation are obtained depending on the culture medium and incubation time. A conidialess mycelial liquid culture was obtained in one instance, helping to identify antigens exclusively present in hyphae (Mendonça et al., 1976). With a mycelial culture containing hypha-attached or free conidia, the usual way to separate both cell types is by filtration through gauze. The procedure can be repeated several times until pure conidial suspensions are obtained (Lloyd et al., 1978). Depending on the degree of sporulation, a pure hyphal preparation may not be possible to obtain because of

Antigenic Structures of *Sporothrix schenckii*

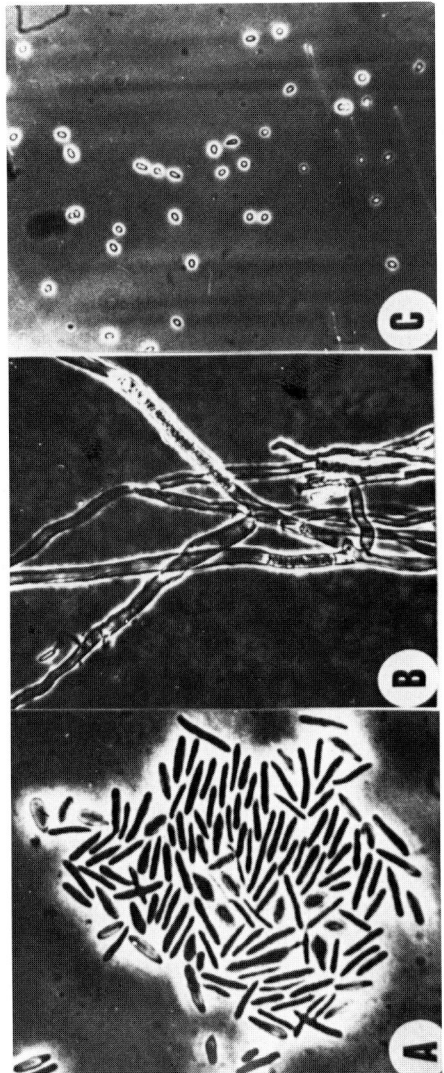

Figure 1 Isolated *Sporothrix schenckii* cell types. A, pure yeast culture obtained in BHI at 37°C (X 1723); B, isolated hyphae with some emerging preconidial structures (XX 282). C, pure suspension of conidia isolated by filtration of a mycelial culture (X 277).

the numerous emergences of preconidial structures. Figure 1 shows the isolated *S. schenckii* cell types.

Whatever the composition of the cell mass, this is collected by centrifugation, washed in water or saline, and extracted for antigens as follows.

Peptidorhamnomannans

Several extraction procedures have been used: (1) The fungus has been extracted by French press homogenization and a polysaccharide complex was recovered which was effective in eliciting a sporotrichin reaction. The preparation, however, was not specific since it cross-reacted with an *Aspergillus* antigen (Aoki et al., 1968); (2) Mycelium was defatted by repeated extraction with ethanol-ethyl ether (1:1); it was then extracted with 45% aqueous phenol (Noguchi, 1972) and centrifuged. The lyophilized water layer, after being dialized against water, was resolubilized and applied on a Sephadex G-100 column. The excluded fraction contained a crude preparation of a peptidopolysaccharide (Shimonaka et al., 1975); (3) Yeast cells have been autoclaved while still suspended in the culture medium or in water, saline, or buffer suspensions after centrifuged and washed. Culture media containing yeast extract or brain-heart infusion are normally dialized against water and the dialyzate inoculated with the fungus to avoid admixture with polysaccharides normally present in those materials. The autoclaved suspensions are centrifuged, and the supernatants, depending on the volume, are concentrated in vacuo and precipitated with 2-3 vols. of ethanol. Solubilization in water, dialysis and reprecipitation with ethanol are steps that can be repeated for further purification, especially if complex media are used. The crude antigen was fractionated by hexadecyltrimethylammonium bromide (Cetavlon) as described (Lloyd and Bitoon, 1971). *S. schenckii*-extracted cellular antigens are partly directly precipitated by Cetavlon. For the remaining part, precipitation is carried out at pH 8.5, with the antigen being pretreated with boric acid. Dissociation of the complex formed with Cetavlon is achieved with acetic acid in water, and the antigen is recovered by precipitation with ethanol or methanol. Precipitation with Cetavlon yields peptidopolysaccharides since no hydrolytic step is involved. Purification of peptidopolysaccharides on a diethylaminoethyl-Sephadex column provided a major product eluted with 0.3 M NaCl (Lloyd and Bitoon, 1971), which contained 33.5% rhamnose, 57% mannose, and 14.2% protein. Other preparations of peptidorhamnomannans from other strains using the same method (Travassos et al, 1977) gave Rha/Man ratios ranging from 1:1 to 1:1.6, depending on growth conditions.

Alkali-Extracted Rhamnomannans

Peptide-free rhamnomannans have been obtained by cell extraction with 2% KOH at 100°C (water bath with reflux) for 2 h. Alkaline extracts were neutralized with acetic acid to precipitate insoluble glucan and were then precipitated with ethanol. For large volumes a previous concentration step in vacuo at 45-50°C was necessary followed by centrifugation and ethanol precipitation of the supernatant fluid. The precipitate was thoroughly washed with ethanol on a glass filter and, as needed to increase dryness, also with methanol. The dry powder obtained represents the crude polysaccharide preparation. For rhamnomannan purification, a precipitation with Fehling solution was performed at 4°C. Insoluble copper complexes, which may be more or less abundant depending on the amount of the reactive polysaccharides, were centrifuged, washed on a glass filter with 2% KOH, then with ethanol, and the blue crystals dissociated by shaking with a cation-binding ion-exchange resin. The mixture was decanted and the yellowish fluid, mixed with the water wash of the resin, were precipitated with absolute ethanol, the precipitate washed with the same solvent or with methanol and dried over P_2O_5. Ratios of Rha/Man in these purified preparations ranged from 1:1 to 1:1.2 (Travassos et al., 1973). They contained less than 1% of total nitrogen and less than 0.4% phosphate.

Fehling-purified *S. schenckii* rhamnomannans may sometimes include neutral and acidic polysaccharides mainly due to the presence of variable amounts of β-D-glucuronic acid, a structural feature that is a constant characteristic of *Ceratocystis stenoceras* rhamnomannans. Neutral and acidic polymers have been separated in a DEAE-cellulose column eluted sequentially with water, 5% acetic acid, and 5% formic acid (Gorin et al., 1977). Acidic eluates were neutralized and precipitated with ethanol.

One should bear in mind that hot dilute alkaline extraction and even precipitation with Fehling reagent at 4°C are partly depolymerizing procedures so that the molecular weights of the purified rhamnomannans are significantly reduced at the final step of ethanol precipitation. That the fragments released can be relatively short is clearly shown by dialysis against water of the purified preparation: the yield of the ethanol-precipitated dialyzed polysaccharide can be significantly reduced on a weight basis as compared to the original polymer. Long chains are, however, still sufficient for a structural analysis that is representative of the undegraded polysaccharide. Low molecular weight fragments can still be recovered by gel filtration and individually analyzed.

From Culture Supernatants

Depending on the culture age, increasing amounts of surface antigenic materials are shed into the medium, particularly in shaken cultures. Cells that have lost most of their external coat are still viable (Travassos et al., 1977). Components released in the medium sometimes aggregate to form exogenous membranous materials, which may alter the host-fungus relationships in terms of immunological protection and resistance to infection (Travassos and Alviano, 1981). In a related fungus, *Ceratocystis ulmi*, formation of membranes (ceratoulmin) in the form of microtubules or vesicles is a factor leading to development of Dutch Elm disease (Takai, 1974).

Culture supernatants obtained by centrifugation of yeast or mycelial cultures are usually concentrated in vacuo, dialyzed against water, and centrifuged again to remove all insoluble material. The clear supernatant fluid is precipitated with 2-3 vols. ethanol and the precipitate washed with ethanol on a glass filter. When concentration of the culture medium does not precede ethanol precipitation and the polysaccharide molecules are more dispersed, addition of 1% Na acetate usually helps precipitation. Frequently, with dialyzed preparations or solutions that have already been precipitated once and the corresponding residues resolubilized in water, failure to precipitate with ethanol or methanol due to lack of ions can be overcome by adding a few drops of concentrated HCl. The same applies to resin-deionized solutions of polysaccharides (see isolation of galactomannans).

Rhamnomannan-Containing Antigens

Peptidopolysaccharides can be isolated by precipitation of the borate complexes with Cetavlon at pH 8.5 as mentioned before. Rhamnomannans derived from the peptidopolysaccharides or from the crude antigen are isolated after heating in dilute alkali, neutralization with acetic acid, and precipitation with ethanol as described (Travassos et al., 1973, 1974). Some contamination with galactose is sometimes observed in these preprations, but usually the proportion of this monosaccharide in the rhamnomannan antigens is not higher than 5%. Purity of the rhamnomannan is determined by assaying total nitrogen and phosphorus.

Galactomannans

With a few exceptions, *S. schenckii* galactomannans are not precipitable by Fehling reagent. This is probable due to the high degree of substitution of the α-linked mannopyranan main chain by side chains containing β-D-galactofuranosyl units. However, these polysaccharides are heterogeneous, and on separation by Cetavlon-borate precipitation at pH 8.0 and pH 9.5, the ratios of Man/Gal varied in *S. schenckii* polysaccharides from 1.6 to 1.2, respectively, and in the closely related species *C. stenoceras* from 1 to 0.56 (Mendonça-Previato et al., 1980). It is conceivable that a series of galactomannans with different galactose-to-mannose ratios are present. The little precipitation by Fehling reagent of molecules with less substitution by galactofuranosyl side chains may account for the small contamination found in some rhamnomannan preparations.

Galactomannans from *S. schenckii* and *C. stenoceras* were isolated from the same crude antigenic preparation of the culture supernatant. At the Fehling-precipitation step to isolate rhamnomannans, galactomannans, some amylose, and β-linked soluble glucan remain in the mother liquor and are not precipitated. washing to remove membranes and cytoplasmic contaminants contain only traces of galactose.

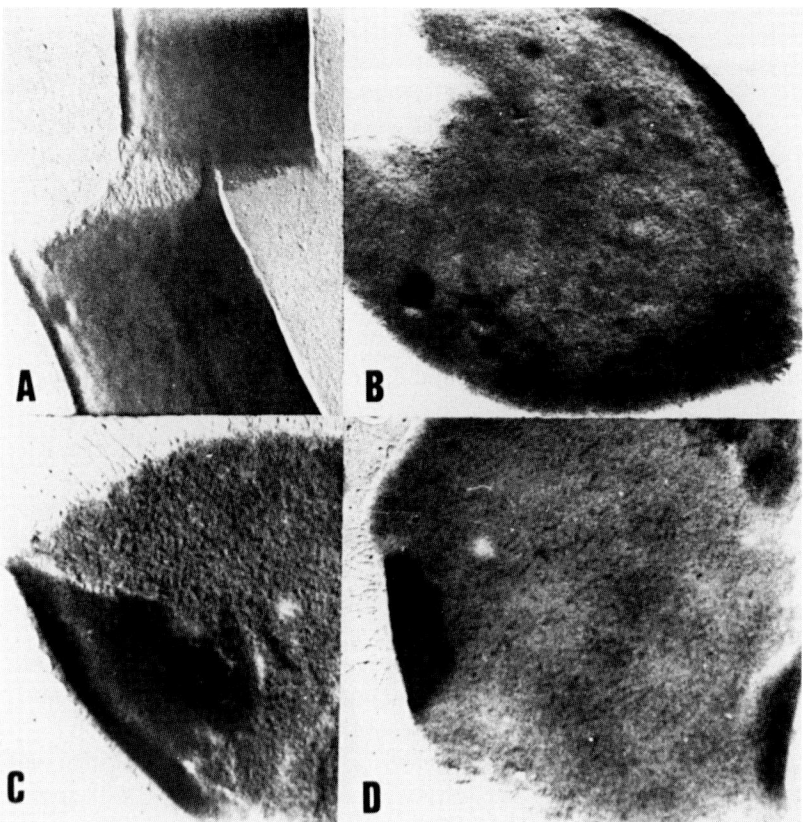

Figure 2 Isolated cell walls from different *S. schenckii* cell types. A, Cell wall from a fragmented hypha; B, wall from a conidium keeping the original pyriform shape; C and D, two wall preparations from yeast forms, one of them showing a bud-scar. (X 5500).

The supernatant fluid from the Fehling reaction mixture was then neutralized with acetic acid, dialyzed against distilled water, and deionized with an ion-exchange resin. The polysaccharide in the deionized solution was then concentrated in vacuo and precipitated with ethanol. The final preparation was enriched for galactomannans but contained some glucose (18-20%) and a small amount of rhamnose, possibly derived from a fraction of the rhamnomannans which was not precipitated by Fehling reagent.

Culture supernatant fluids are the best source for isolation of galactomannans due to their peripheral location on the cell wall and their tendency to be released into the medium. Cell walls isolated by mechanical disruption and exhaustive

From Isolated Cell Walls

Very few reports are available on the isolation of antigens from *S. schenckii* cell walls. Nielsen (1968) isolated defatted and detergent-extracted walls from *S. schenckii* cells prepared by mechanical disruption. Other treatments of the isolated walls included pronase and ribonuclease treatments. Tested intradermally, all treated and untreated cell wall preparations displayed the same reactivities as standard whole cell antigens.

β-Glucans, soluble and insoluble, have been isolated from *S. schenckii* cell walls (Previato et al., 1979a), but these polysaccharides are not recognized antigens of this fungus acting as immunopotentiating agents (see below).

The presence of rhamnomannan antigens, probably in a complexed form within the framework of the cell wall components, is inferred from the analysis of the total constituents of walls from the different *S. schenckii* cell types (Previato et al., 1979b). Both rhamnose and mannose were present in similar proportions in isolated walls from yeast forms, hyphae and conidia (Fig. 2).

Cell walls and membranes, which are obtained by disruption of *S. schenckii* yeast phase in a Ribi-Sorvall refrigerated fractionator, were also analyzed for presence of antigens (Nakamura, 1976). These were extracted with sodium deoxycholate (0.2%) and the extract precipitated with acetone (−20°C). The white precipitate was collected by centrifugation, dissolved in 0.5 M Tris-HCl buffer, and fractionated in a DEAE-Sephadex column. Two fractions were obtained, differing mostly in their nitrogen content. Surprisingly, rhamnose and galactose were the only sugars detected. According to the author, fraction I contained a peptidorhamnogalactan, which cross-reacted with a peptidorhamnomannan and produced delayed hypersensitivity skin tests in patients with sporotrichosis.

ANTIGENIC POLYSACCHARIDES

A detailed description of the fine structures of *S. schenckii* polysaccharides has appeared recently (Travassos, 1985). Attention has also been given to polysaccharides from the related genus *Ceratocystis*, especially *C. stenoceras*.

Identified Structures

Antigenic polysaccharides from *S. schenckii* were analyzed in a purified form by traditional chemical methods, such as methylation-fragmentation and several others, together with spectroscopic methods, the most useful of these being NMR spectroscopy. Among heteropolysaccharides, those of *S. schenckii* were the first to be thoroughly studed by ^{13}C NMR spectroscopy (Gorin, 1981).

Rhamnomannans

These polysaccharides are α-linked structures consisting of (1 → 6)-linked D-mannopyranosyl units in the main chain with almost all of them substituted at the O-3 positions by single-unit rhamnosyl or dirhamnosyl side chains. Generally, polymers isolated by alkali extraction do not contain α-D-mannopyranosyl nonreducing end-units (Travassos et al., 1973; 1974) and, consequently, are not precipitated by concanavalin A (Fig. 3). In contrast, peptido-polysaccharides isolated by the Cetavlon method react with concanavalin A (Travassos et al., 1977) and contain a porportion of terminal α-D-mannopyranosyl residues (Lloyd and Bitoon, 1971). Single units of α-L-rhamnopyranose (1 → 3)-linked to the mannopyranan main chain are typical of the rhamnomannans obtained at 37°C or isolated from yeasts or conidia (Travassos and Mendonça-Previato, 1978). This structure (I) is also found in rhamnomannan from *Ceratocystis*, particularly *C. stenoceras* and *C. ulmi*. Dirhamnosyl side chains—α-L-Rha$p$-(1 → 2)-α-L-Rha$p$-(1 → 3)-α-D-Man$p$—are mainly synthesized by hyphae at 25°C, and their proportion can represent 85% of all branches in the polysaccharide. This structure (II) is characteristic of *S. schenckii* mycelial rhamnomannans (Travassos, 1985) and differs from the acidic side chain of *C. stenoceras* (Gorin

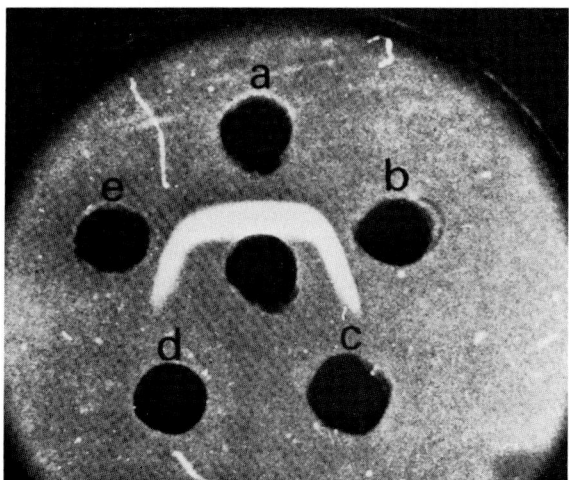

Figure 3 Double-diffusion pattern of concanavalin A with *S. schenckii* polysaccharides and peptidopolysaccharides. Center well, concanavalin A; a, *S. schenckii* galactomannan; b, *S. schenckii* 1099.18 peptidopolysaccharide; c, strain 1099.18 rhamnomannan; d, strain 1099.12 rhamnomannan; and e, strain 1099.12 peptido-polysaccharide. (From Travassos et al., 1977.)

et al., 1977), which has a β-D-glucopyranosyluronic acid unit wedged between the rhamnose units (III).

```
                                                    α-L-Rhap
                                                      1 ↓
                                                       4
                               α-L-Rhap              β-D-GlcpUA
                                 1 ↓                   1 ↓
                                  2                     2
       α-L-Rhap                α-L-Rhap              α-L-Rhap
         1 ↓                     1 ↓                   1 ↓
          3                       3                     3
  → 6)-α-D-Manp-(1 → 6)-  → 6)-α-D-Manp-(1 → 6)-  → 6)-α-D-Manp-(1 → 6)-

           I                      II                    III
```

Spectra and interpretations of the ^{13}C NMR signals obtained for the various rhamnomannan side chains as well as the complete assignments of all signals arising from a *S. schenckii* rhamnomannan obtained at 37°C have appeared in previous reviews (Travassos, 1985, and Travassos and Lloyd, 1980, respectively).

Another feature of the rhamnomannans from *S. schenckii* is the occurrence of 4-*O* and 2,4-di-*O*-substituted α-D-mannopyranose units. These units appear as a low-temperature (25°C) induced effect and are independent of the morphology of the culture. Smith degradation of *S. schenckii* rhamnomannans obtained at 25°C gave erythritol, 2-*O*-α-mannopyranosyl-D-erythritol, and *O*-α-D-mannopyranosyl-(1 → 4)-*O*-α-D-mannopyranosyl-(1 → 2)-D-erythritol (Travassos and Lloyd, 1980). Possible structures giving rise to these fragments might have α-D-mannopyranosyl nonreducing end-units, but since these units are not detectable by methylation analysis in the original polysaccharide, they are probably substituted by α-L-rhamnopyranosyl terminal units. The corresponding fragment containing rhamnose has not, however, been characterized, probably because of the lability of the rhamnose linkage. A probable structure is depicted below (IV).

```
          α-L-Rhap          α-L-Rhap
            1 ↓                1 ↓
             4                  4
          α-D-Manp          α-D-Manp
            1 ↓                1 ↓
             2                  2
    → 4)-α-D-Manp-(1 → 4)-α-D-Manp-(1 → 4)-α-D-Manp-

                         IV
```

Galactomannans

A complex structure for galactomannans of both *S. schenckii* and *C. stenoceras* was identified (Mendonça-Previato et al., 1980). The core of these polysaccharides consists of 2-*O*, 2,6-di-*O*, and perhaps also 2,3-di-*O*-substituted α-D-mannopyranosyl units. Three and four consecutive (1 → 2) linkages occur between the α-D-mannopyranosyl residues. Side chains are composed of β-D-galactofuranosyl units (removable by partial hydrolysis at pH 2) which are (1 → 6)-linked to each other and at least partly (1 → 2)-linked to the mannan core. Possible partial structures present in these polysaccharides are indicated (V).

```
     β-D-Gal f                    β-D-Gal f
        1↓                           1↓
        6                            6
     β-D-Gal f                    β-D-Gal f
        1↓                           1↓
        2                            2
→ 6)-α-D-Manp-(1 → 6)-           α-D-Manp
                                     1↓
                                     2
                                     6
                                6-α-D-Manp-(1→6)-

                    V
```

The importance of ^{13}C NMR spectroscopy in helping to determine the above structures is discussed elsewhere (Mendonça-Previato et al., 1980; Travassos, 1985).

Similarity with Other Polymers

The fine structures of antigenic polysaccharides, especially those from pathogenic fungi, have been determined in only a few instances. Most analyses are restricted to identification of the monosaccharide constituents of polysaccharides. Therefore, the structural comparison of antigens is usually made by determining their immunological cross-reactivity using reference antisera. Using rabbit antisera and human sporotrichosis sera against *S. schenckii* grown at 34-37°C and therefore containing antibodies directed mainly to the mannorhamnosyl epitope of the rhamnomannan, Ishizaki et al. (1978) compared the culture filtrate antigens of 100 strains of 75 species of the ***Sporothrix-Ceratocystis-Europhium-Graphium*** complex and one species of *Botrytis*. Cross-reactivity, as tested by capillary precipitin and double immunodiffusion assays, was high (80%) with species of *Ceratocystis, Europhium*, and several *Graphium*-form species. They all produced exoconidial forms and rhamnose- and mannose-containing polysaccharides. Cross-reactivities among antigens of *Sporothrix curviconia*, *S. inflata*, and *S. schenckii* var. *luriei* are assumed to be L-rhamnose-dependent (Ishizaki et al., 1979b). The antigenic relationship between *S. schenckii* and *Ceratocystis minor* (Polonelli and Morace, 1982) was shown by the exoantigen immunodiffusion technique of Kaufman and Standard (1978).

Cross-reactions between *S. schenckii* and several types of *Streptococcus pneumoniae* were reported. Cross-agglutination reactions were inhibited by purified pneumococcal S substance (Neill et al., 1955). Antigenic relationships were also described with certain strains of *Leuconostoc mesenteroides*, group B *Streptococcus* (Nakamura et al., 1977), and Klebsiella pneumoniae K47 (Ishizaki et al., 1979a). See Figure 4 for depiction of the structure of common cross-reacting epitopes.

Using a rabbit antiserum raised against an *S. schenckii* strain grown at 25°C, with a clear specificity for the dirhamnosyl side chains of the rhamnomannan (Travassos et al., 1978), we showed that a precipitation reaction occurred with the homologous antigen but that very poor reactivities were obtained with the *S. schenckii* rhamnomannan obtained at 37°C or with the rhamnomannans

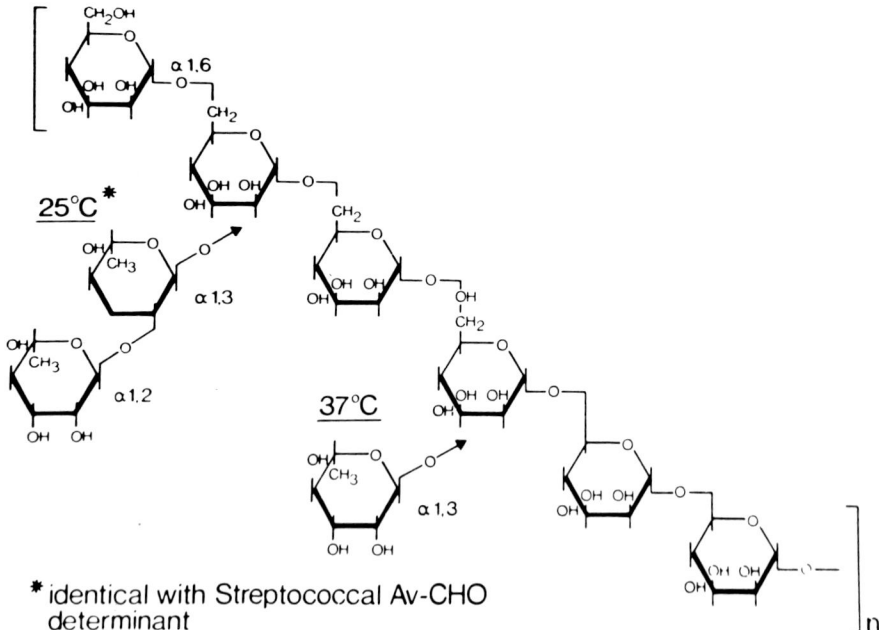

Figure 4 Depiction of an *S. schenckii* rhamnomannan structure showing the two main epitopes, one of them cross-reacting with streptococci.

from several *Ceratocystis* species. The specificity of the antibodies was confirmed by inhibition of the homologous precipitation reaction with *O*-α-L-rhamnopyranosyl-(1 → 2)-*O*-α-L-rhamnopyranosyl-(1 → 3)-D-mannopyranose (Lloyd and Travassos, 1975). As pointed out earlier, this structure is absent in *Ceratocystis* polysaccharides, being characteristic of *S. schenckii*.

Cross-reactions involving the galactomannan antigens are more universal since the reacting epitopes contain β-D-galactofuranosyl units, which are found in many other fungi and also in protozoa (Schnaidman et al., 1986). The first demonstration of this cross-reactivity was made when an anti-*Hormodendrum* serum was shown to strongly react with whole cells of *S. schenckii* (Travassos and Lloyd, 1980), and its specificity was directed against β-D-galactofuranosyl terminal units. It also reacted with galactomannans from *Aspergillus fumigatus, Exophiala werneckii*, and *Trichophyton rubrum* (Suzuki and Takeda, 1977).

Some cross-reactions with pathogenic fungi of a rabbit antiserum raised against yeast forms of *S. schenckii* were inhibited by serum absorption with *E. werneckii* whole cells or by adding galactose. Absorption with *Saccharomyces cerevisiae* or addition of mannose were ineffective (Ishizaki et al., 1981).

Use as Specific Diagnostic Reagents

Since rhamnomannans, the chief *S. schenckii* antigens, are specific of this pathogenic fungus, occurring only in nonpathogenic species, they should be the choice antigens for the diagnosis of extracutaneous sporotrichosis. Specific reactions for *S. schenckii* using agglutination and precipitin tests have indeed been reported (Seeliger, 1968; Karlin and Nielsen, 1970; Roberts and Larsh, 1971). The problem of using rhamnomannans precipitated by Fehling's reagent, and thus containing only traces of galactose, is the reduction of the molecular weight of the polymer. The resulting mixture of carbohydrate chains of different sizes in a nondialyzed preparation, which are still precipitable by ethanol at proportons of 3:1 to 5:1 (v/v), may not be a good antigen for routine precipitation in double immunodiffusion (ID) tests. It seems that peptidopolysaccharides that are constituents of the soluble sporotrichin preparations (Gonzalez-Ochoa and Figueroa, 1947) are the preferred antigens for diagnosis of sporotrichosis. The Cetavlon-precipitated peptido-polysaccharide should be highly specific for *S. schenckii*, but so far similar preparations have not been introduced in routine testing of sera. Soluble, crude antigens from culture filtrates have been used instead of purified antigens, and the results in ID tests are usually satisfactory (Karlin and Nielsen, 1970; Roberts and Larsh, 1971; McMillen, 1974). The technique using ID of specific exoantigens extracted directly from agar slant cultures with aqueous merthiolate (Kaufman and Standard, 1978), which permits an early identification of suspected cultures, has been employed for *S. schenckii* cultures, allowing identification of 10 isolates within 3 days (Polonelli and Morace, 1982).

Apart from ID tests, which are widely used, other serological tests for sporotrichosis including agglutination of the yeast-phase antigen at titers higher than 1:80 (Welsh and Dolan, 1973), and complement fixation (CF) tests (Jones et al., 1969) have been used. To minimize cross-reactions in the agglutination tests, yeast suspensions should be previously autoclaved (McMillen, 1974). It is possible that by using this procedure surface galactomannans are totally liberated from cells, thus reducing the cross-reactivity with other pathogenic fungi.

In contrast to the early negativity of serological tests in the cured cutaneous form of sporotrichosis, positive CF and ID reactions remain positive in the extracutaneous form for as long as 7 years after cure (McMillen, 1974).

Influence on Phagocytosis

The phagocytosis of yeast cells of two strains of *S. schenckii* was studied using thioglycollate-elicited mouse peritoneal macrophages (Oda et al., 1983). Phagocytosis was very much stimulated when yeasts were opsonized with concanavalin A. Reactivity with this lectin probably involved the peptidorhamnomannan and galactomannan components. Treatment with neuraminidase increased ingestion of unopsonized yeasts 7.7-fold. Two layers of components at the wall surface of *S. schenckii* yeasts could be observed that were rich in sialic acid (Fig. 5). Sialic acid residues could be located in the peptidopolysaccharide fraction—even on a lipid-extracted crude preparation—and on polar glycolipids (Alviano et al., 1982). A preparation of total sialoglycolipids from *S. schenckii* was hydrolyzed and the sialic acid units separated by thin layer chromatography. N-Glycoylneuraminic acid was the predominant derivative, although some N-acetylneuraminic acid was also detected. This was to our knowledge the first demonstration of sialic acids in pathogenic fungi. We recently extended these observations to *Fonsecaea pedrosoi*, in which conidia were found to contain also a mixture of N-glycolyl- and N-acetylneuraminic acids (Souza et al., 1986). As pointed out before, the relationship between surface components containing N-glycolylneuraminic acid and infection is unclear, however, one should remember that the N-glycolyl group is antigenic in man (Merrick et al., 1978). The disaccharide Neu-5-glycolyl-$(2 \rightarrow 3)$-β-Gal is the epitope reacting with serum sickness antibodies (Smith and Ginsburge, 1980). The only other glycolipid isolated from *S. schenckii* in a purified form was a monoglucosylceramide (Cardoso et al., 1987), but its biological role is unknown.

Mannose, rhamnose, and galactose, which are major constituents of *S. schenckii* surface antigens, reduced phagocytosis indexes to 40-50% (Oda et al., 1983). Equally effective were glucosamine, N-acetylglucosamine, and N-acetylneuraminic acid. Both a purified rhamnomannan and a partially purified galactomannan inhibited phagocytosis by about 60%. N-Acetylglucosamine inhibited binding of yeasts to macrophages at 4°C as well as phagocytosis. Since this is

Antigenic Structures of *Sporothrix schenckii*

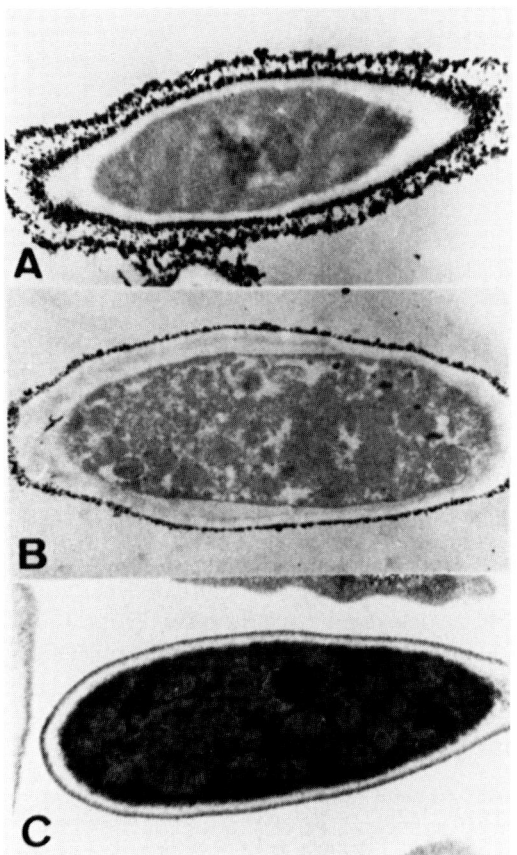

Figure 5 Staining with colloidal iron hydroxide of anionic groups at the cell surface of *S. schenckii* yeast form. A, Bilayer reactivity in a yeast before enzyme treatment; B, yeast showing residual reactivity with CIH after neuraminidase treatment; C, complete removal by neuraminidase of anionic groups reacting with CIH. (X 16,000) (From Alviano et al., 1982.)

not a prominent component of the cell wall, the inhibitor effect may be attributed to binding of N-acetylglucosamine to a receptor common to other sugars, similar to the (mannose/glucosamine)-reacting receptor identified in alveolar macrophages (Warr, 1980).

All interactions described above may have minor effects in advanced sporotrichotic infection when antibodies are already present, but may be quite relevant at the onset of infection when the unchallenged proliferation of yeast forms determine the initial course of the mycosis.

Immunopotentiating Polysaccharides

Linear insoluble glucans of *S. schenckii* at 10 mg/kg per day and following a standard therapy schedule (Tarnowski, 1975) caused regression of Sarcoma 180 implanted in mice. These linear polymers consist of mixed β-(1 → 4)-, β-(1 → 6)-, and β-(1 → 3)-linked D-glucopyranose units. Their antitumor activity follows similar effects observed with other fungal polysaccharides (Whistler et al., 1976). *S. schenckii* rhamnomannans were also tested for their activity to reduce growth of Sarcoma 180 in SW55 mice (Kubelka, 1980). A monorhamnosyl-rhamnomannan at 200 mg/kg in 10 doses inhibited the tumor growth in 62%. The antitumor effects of mono- or dirhamnosyl-rhamnomannans were comparable, after 40 days of tumor implantation, to those of *S. cerevisiae* glucan and scleroglucan. All effects depended on the doses and therapy schedule used. With mouse peritoneal macrophages, rhamnomannans increased the phagocytic indexes for normal and antibody-sensitized sheep erythrocytes, and this effect was more pronounced than that with *S. cerevisiae* mannan.

INDIRECT DETECTION OF ANTIGENS

Surface antigens of *S. schenckii* form a thick coat of surface constituents, especially in the yeast phase, which reacts with a variety of ligands including specific antibodies, lectins, and reagents for anionic groups.

Antibody-Dependent Reactions

Fluorescent antibodies have been used to stain both yeast and mycelial phases from culture or in clinical specimens (Kaplan and Ivens, 1960). Reactions are rather specific and helpful in the diagnosis of sporotrichosis.

The indirect immunofluorescence technique was used to demonstrate the specific reactivity of rabbit antibodies with certain cell types but not with others. The original sera were raised against whole cells of *S. schenckii* grown at 37°C, at 25°C, or in the yeast or mycelial phase, respectively. Each serum was cross-absorbed with the heterologous phase to remove common antibodies (Lloyd et al., 1978). Specific sera were indeed obtained; one of them (AY) reacted strongly with the yeast forms and conidia and very weakly with hyphae; the other serum (AH) reacted strongly with hyphae and weakly with the other cell types (Travassos, 1985). These results are in accordance with the chemical structures described for the different rhamnomannans of these cells.

Peroxidase-linked antibodies could localize fungal cells (e.g., *S. schenckii* and *Cryptococcus neoformans*) in tissue sections fixed with formaldehyde and embedded in paraffin without prior trypsinization (Russell et al, 1979). Peroxidase-labeled antibodies produced a staining differential comparable to that of immunofluorescence for detection of fungi in tissue and body fluids.

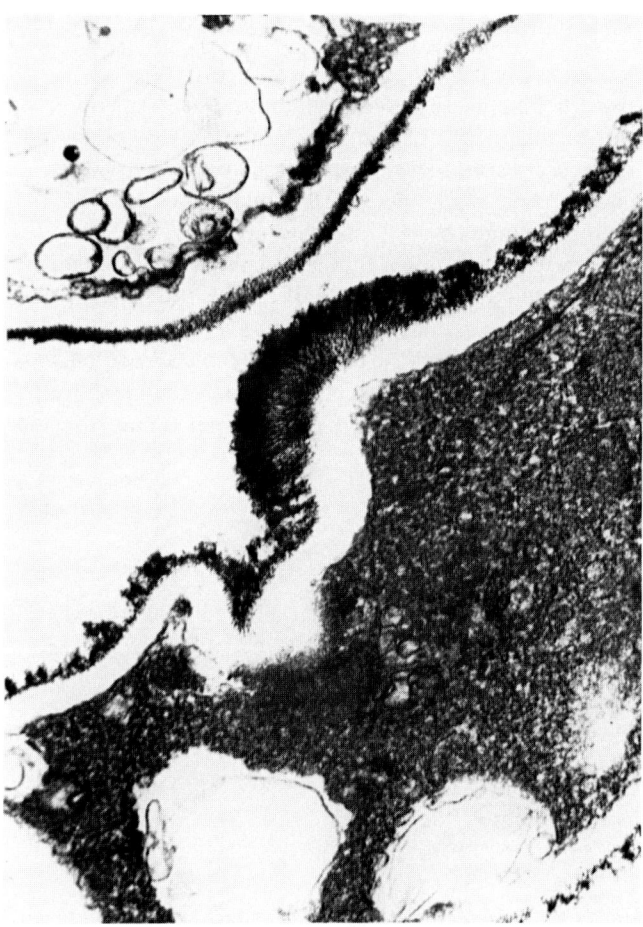

Figure 6 Microfibrillar material accumulating at the point of emergence of a yeast cell from a hypha. The electron opacity of the microfibrils has been enhanced by staining with acidified dialyzed iron. Glutaraldehyde-dialyzed iron-osmium tetroxide (X 46,376) (Courtesy of R. G. Garrison.)

Reactivity with Other Ligands

S. schenckii surface antigens heavily bind concanavalin A (Travassos et al., 1977). Yeast cells are strongly reactive so that mycelium-to-yeast transitions even at the early beginning of the differentiation step (primordia) can be located by reaction with fluorescent concanavalin A. The increased density of reaction with the lectin overlaps the thick microfibrillar material that is frequently shown in electron micrographs by R. G. Garrison (Fig. 6). The concanavalin A-stainable layer on hyphae is much thinner than that in yeasts; no transitional structure was detected in the conversion of conidia into hyphae (Fig. 7).

Among lectins, only concanavalin A and the lectin from *Limulus polyphemus* agglutinated yeast forms. All other purified 15 lectins tested were unreactive (Alviano et al., 1982). This is in agreement with the presence of D-mannose and sialic acid units on the cell surface as mentioned.

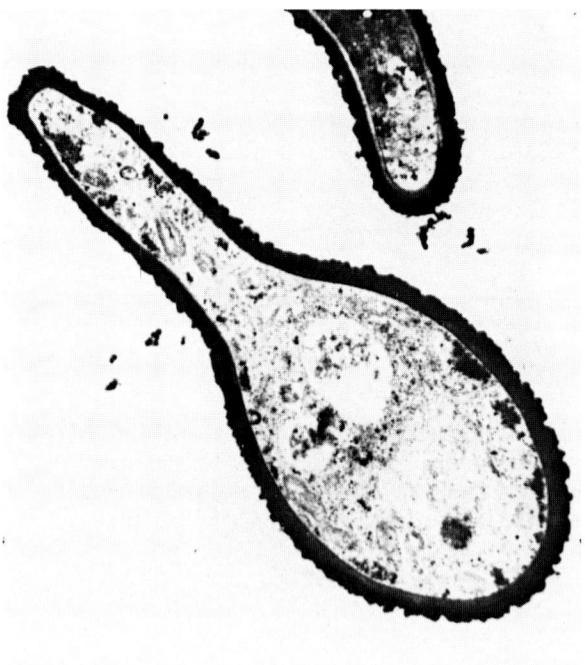

Figure 7 Homogeneous reaction with concanavalin A of the cell wall of a conidium with a germ tube showing no apparent signs of differentiation. (X 13,975) (From Travassos et al., 1977.)

Other surface markers expressed in *S. schenckii* together with the antigenic constituents are the numerous anionic groups. Microfibrillar structures are stained with acidified dialyzed iron (Garrison et al., 1975). Other reagents included cationized ferritin at neutral pH and colloidal iron hydroxyde at pH 1.8 (Benchimol et al., 1979). Using cationized ferritin, it was possible to visualize the detachment into the medium of the external layer of the yeast cell wall leaving behind a completely uncoated structure of glucan components among other constituents of the inner layers of the cell wall. The sloughing off of surface antigenic material can play an important role in the course of infection since the immunological defense mechanisms will be involved in interactions with organized extracellular membranous secretions rather than with the cells themselves, which can then proliferate unchallenged. As to the colloidal iron hydroxide (CIH) staining at pH 1.8, sialic acids are the best substrates due to their low pK in comparison with those of dicarboxylic amino acids: treatment of yeasts with neuraminidase completely abolished reactivity with CIH (Fig. 5; Alviano et al., 1982).

HYPERSENSITIVITY REACTIONS

Skin tests using soluble sporotrichin or particulate antigens have been used for the diagnosis of sporotrichosis or in epidemiological studies to detect previous immunosensitizing contact with this fungus (Gonzalez-Ochoa and Figueroa, 1947). Since antibodies can be elicited by inoculation of the sporotrichin, these tests should be limited to particular epidemiological studies since they can mask the immunological response to an ongoing sporotrichotic infection.

Specificity and Cross-Reactivities

After the immunosensitizing contact with *S. schenckii*, the cellular antigen elicits positive reactions for prolonged periods, but they are not necessarily related to ongoing disease (Mayorga et al., 1978). With the soluble sporotrichin, skin reactions can become negative 1 to 3 years after cure. The sensitivity is greater with yeast-phase antigens than with antigens from the mycelium phase. Carrado-Bravo (1975) testing 359 pulmonary patients by skin testing with both particulate (heat-killed yeasts) and soluble sporotrichin suggested the occurrence of cross-reactivity between *S. schenckii* and other fungal antigens. Using ethanol-precipitated culture filtrates of the yeast forms of *S. schenckii, C. stenoceras, C. ulmi, C. ips,* and *C. minor,* Ishizaki et al. (1976) found extensive cross-reactivities in delayed hypersensitivty tests in human patients, the best antigen being that of *C. stenoceras.*

Apart from these cross-reactivities which involve antigens from nonpathogenic fungi, the reactions with *S. schenckii* may be considered specific in terms of other pathogenic fungi. Histoplasmin, coccidiodin, yeasts of *Blastomyces* and *Candida* do not induce delayed-type reactions when tested in animals sensitized to *S. schenckii* (Nielsen, 1968).

Various degrees of cross-reactions were observed in cutaneous delayed hypersensitivity tests with antigens from *S. schenckii*, *S. schenkii* var. *luriei*, *S. curviconia*. *S. inflata*, and *S. ghanensis* (Ishizaki et al., 1980).

Natural Resistance and Cellular Immunity

Probably due to the many properties described for individual components of the cell surface of *S. schenckii* and their high immunogenicity, infection by this fungus is often halted before any noticeable symptoms appear, provided the host is not immunodepressed. In endemic areas, immunity develops in individuals who have had repeated contacts with *S. schenckii* (Gonzalez-Ochoa, 1965).

Cell-mediated immunity in patients with sporotrichosis was evaluated by lymphocyte transformation assays (Plouffe et al., 1979). Most patients had positive tests in vitro in response to yeast forms of *S. schenckii*. A mycelial antigen also induced lymphocyte transformation in a patient with sporotrichotic arthritis (Steele et al., 1976). However, healthy lymphocyte donors showing positive blastogenic indexes for *S. schenckii* antigen also showed high blastogenic responses to *Candida albicans*.

As to the time when cellular immunity becomes operative in halting growth of the fungus, Hachisuka and Sasai (1981) found that it was effective as a defense mechanism only after the eighth day of infection in cyclophosphamide-treated mice. Observations in humans and animals indicate that humoral immunity is also an important defense mechanism in early infection.

Localization of Epitopes

Immediate and delayed-type hypersensitivity reactions are both elicited by the peptidopolysaccharide complexes (Noguchi, 1972). Determinants of delayed-type hypersensitivity are most probably located in the peptide portion of the molecule. Lee and Lloyd (1975) determined that in the peptidogalactomannan from *E. werneckii*, the determinants of delayed-type hypersensitivity were peptide in nature, being stable even to rigorous acid conditions: two peptides which elicited skin test reactivity were isolated. With *S. schenckii*, a peptidopolysaccharide eliciting both immediate and delayed-type reactions was treated with 16 mM periodate at 4°C for 100 h and also with 0.01% papain at 37°C for 48 h. Periodate treatment did not inhibit the delayed-type reactions, whereas papain treatment diminished the delayed-type reactions, including both inhibition of peritoneal cell migration and skin tests (Shimonaka et al., 1975). Specific peptides eliciting delayed-type hypersensitivity have not so far been characterized.

COMPLEMENT ACTIVATION

Fungi are able to interact with the complement system, the main effect being leukocyte chemotaxis and increased phagocytosis due to recognition of bound C3 or its degradation products by macrophages. Activation of complement by fungal cells proceeds either by the classical or alternative pathway. Activation of C3 by *Cryptococcus neoformans* and the binding of cleavage fragments to the yeasts is mediated by the alternative pathway (Kozel and Pfrommer, 1986). C3b was not detected probably due to its rapid conversion to iC3b. In bacteria, in contrast to *C. neoformans*, there is little decay of C3b to iC3b. After functional enhancement by a lymphokine, the phagocyte receptor for iC3b (CR3) plays an important role in the endocytosis of the encapsulated yeast (Griffin, 1981). Other pathogenic fungi in which the alternative complement pathway was activated were *Histoplasma capsulatum* (Rattnoff et al., 1980), *Fonsecaea pedrosoi* (Torinuki et al., 1984), *Candida albicans* (Ray and Wuepper, 1976), *Malassezia ovalis* (*Pityrosporum ovale*) (Bellow et al., 1980), *Trychophyton rubrum* (Swan et al., 1983). The classical example that complex polysaccharides present in fungal cell walls have the capability to activate the alternative complement pathway is that of zymosan.

Activity of Whole *S. schenckii* Yeast Cells

Washed yeast cells grown in brain-heart infusion were used at a concentration of 10^9 cells/ml corresponding, for comparison, to similar wet weight of zymosan. These cells were able to consume all hemolytic complement activity in the presence of Mg-EGTA (Scott et al., 1986). Incubation of yeasts with nonimmune nonchelated serum resulted in deposition of C3 but not IgG on the yeast cell surface. Therefore, like most fungi (Fine, 1981), *S. schenckii* activates complement through the alternative pathway. Opsonization by C3 may be an important factor in the host defense mechanisms against *S. schenckii*.

Torinuki and Tagami (1985) used a boiled mycelium colony of *S. schenckii*, which was homogenized for complement activation of serum. In presence of *S. schenckii* C3 was converted from $\beta_1 C$ to $\beta_1 A$, and there was generation of C5a which was unaffected by addition of Mg-EGTA. The formation of C5a explains the chemotactic activity even in nonimmune serum.

Effect of Isolated Purified Fractions

Hamuro et al. (1978) did not find a clear correlation between the anticomplementary and tumor-regressing activity of 8 polysaccharides as tested by inhibition ratios against Sarcoma 180 transplanted in mice. On the other hand, these authors showed a contribution of both soluble and insoluble forms of pachyman to the initiation of the alternative complement pathway.

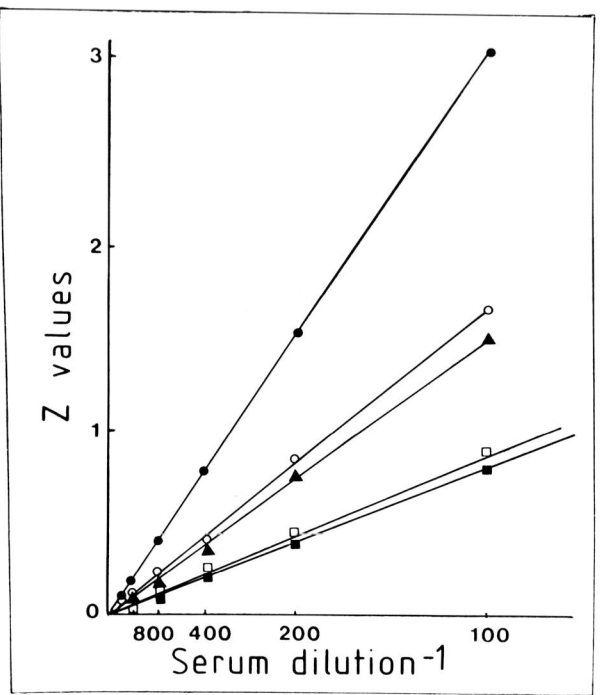

Figure 8 C3 consumption from normal human serum resulting from activation of ACP by fungal polysaccharides: (●), control with no polysaccharide added; (-○- and -□-), addition of peptidorhamnomannan and rhamnomannan, respectively, from *S. schenckii*; (▲ and ■), addition of peptidomannan and mannan from *S. cerevisiae*, respectively. Z, Number of hemolytic sites per cell ($Z = -\ln(1 - y)$, where y is the percentage of lysis of Cl, 4 bound sensitized sheep erythrocytes).

Czop and Austen (1985) came to the conclusion that at 2 mg/ml of serum diluted 1/2 in 8 mM EGTA/2 mM Mg, nonturbid preparations of mannan, laminarin, or pyrogen-free inulin failed to activate the alternative complement pathway. J. Aisengart (1987, to be published) came to a completely different conclusion using a different proportion of glycan-treated serum to rabbit erythrocytes. Under these conditions, soluble *S. cerevisiae* mannan and *S. schenckii* rhamnomannans were able to activate the alternative complement pathway (ACP); a double test was initially used: (1) inhibition of lysis of rabbit erythrocytes in presence of Mg-EGTA; and (2) cross-immune electrophoresis with the quantitative determination of the relative areas of C3 and C3b after incubation with the polysaccharide. In both tests, the isolated rhamnomannans (obtained at 37 or 25°C) showed a stronger anticomplementary activity than that of the corresponding peptidopolysaccharides. Figure 8 shows the quantitative determination of the hemolytic sites per cell (C1,4-bound sensitized sheep erythrocytes) after C3 consumption due to activation of the ACP by *S. cerevisiae* and *S. schenckii* (Gaither and Frank, 1979) polysaccharides.

CONCLUSIONS

Although several antigenic structures have been identified in *S. schenckii*, it is obvious that other structures are present in the cell wall which may be important as immunogens and as virulence factors. Lipids are only beginning to be studied in terms of fine structure determination. The reactivity of concanavalin A with peptidopolysaccharides could be due to a mannan core preceding the rhamnomannan long chains that would be unusually labile to alkali treatment, to agree with the methylation results or, rather, to the occurrence of numerous mannose-containing O-linked oligosaccharides. On β-elimination, Mendonça-Previato (personal communication) showed the presence of several O-linked substitutions in the peptide, but the corresponding saccharides liberated are still under study. About 50% of threonine and 63% of serine were lost on β-elimination of a yeast peptidopolysaccharide containing 17-18% protein. An analogous situation is that of *C. ulmi*: carbohydrate chains with an average length of 25 sugar residues are attached either to threonine or serine. A proportion of 70-85% of these amino acids were lost on mild alkaline treatment of the glycopeptide (Strobel et al., 1978).

The biosynthesis of the peptidorhamnomannans and the regulation of the rhamnosyl-transferases according to cell differentiation is another line open for investigation. An intriguing question concerns the structural binding of the sialic acid units, which are so conspicuous in the yeast phase, being constituents of two external layers of the cell wall. Other aspects that deserve attention are the peptide determinants of the delayed hypersensitivity reactions and their conservation in several peptidopolysaccharide molecules from different *Ceratocystis* species in addition to *S. schenckii*.

Finally, I would like to call the attention of the reader to the different behavior of rhamnomannans and peptidorhamnomannans towards complement activation. Apparently, the peptido-complexes are worse substrates for the C3-convertase activity than the isolated polysaccharides either by intrinsic structural hindrances or because they may be acceptors for inhibitory serum factors (e.g., H and I). This again is a very promising area of research involving *S. schenckii* antigens.

ACKNOWLEDGMENTS

I thank Drs. Margarita Silva-Hutner and Kenneth O. Lloyd for their advice and friendship during my early studies on *S. schenckii* while at Columbia University,

New York. The association with Dr. Philip A. J. Gorin has always been rewarding and stimulating. I also thank my former and present graduate students without whom most of the data herewith presented could not be obtained. While writing this review I benefited from grants and fellowship from the Brazilian National Research Council and several other agencies, including FINEP, FAPESP, and FIPEC (Banco do Brasil).

REFERENCES

Alviano, C. S., Pereira, M. E. A., Souza, W., Oda, L. M., and Travassos, L. R. (1982). Sialic acids are surface components of *Sporothrix schenckii* yeast forms. *FEMS Microbiol. Lett.* 15:223-227.

Aoki, Y., Nakayoshi, H., and Ono, M. (1968). Studies on the immunologically active substances of fungi. III. The chemical properties of skin test antigens from organisms of *Sporotrichum schenckii, Cryptococcus neoformans, Aspergillus fumigatus* and of serologically active compounds in several mycotic antigens. *Jpn. J. Allergy* 17:56-61.

Bellow, P. W., Rossenberg, E. W., and Jennings, B. R. (1980). Activation of the alternative pathway of complement by *Malassezia ovalis* (*Pityrosporum ovale*). *Mycopathologia* 70:187-191.

Benchimol, M., Souza, W., and Travassos, L. R. (1979). Distribution of anionic groups at the cell surface of different *Sporothrix schenckii* cell types. *Infect. Immun.* 24:912-919.

Cardoso, D. B. S., Angluster, J., Travassos, L. R., and Alviano, C. S. (1987). Isolation and characterization of a glucocerebroside (monoglucosylceramide) from *Sporothrix schenckii. FEMS Microbiol. Lett.* 43:279-282.

Carrada-Bravo, T. (1975). New observations on the epidemiology and pathogenesis of sporotrichosis. *Ann. Trop. Med. Parasitol.* 69:267-273.

Czop, J. K. and Austen, K. F. (1985). Properties of glycans that activate the human alternative complement pathway and interact with the human monocyte β-glucan receptor. *J. Immunol.* 135:3388-3393.

Drouhet, E. and Mariat, F. (1952). Etude des facteurs déterminant le développement de la phase levure de *Sporotrichum schenckii. Ann. Inst. Pasteur Paris* 83:506-514.

Fine, D. P. (1981). *Complement and infectious diseases.* CRC Press, Inc., Boca Raton, Fla.

Gaither, T. A. and Frank, M. M. (1979). Complement. In *Clinical Diagnosis and Management by Laboratory Methods.* Edited by J. B. Henry. W. B. Saunders Co., Philadelphia, PA., p. 1245.

Garrison, R. G. (1985). Cytological and ultrastructural aspects of dimorphism. In *Fungal Dimorphism, with Emphasis on Fungi Pathogenic for Humans.* Edited by P. J. Szaniszlo. Plenum Press, New York, pp. 15-47.

Garrison, R. G., Boyd, K. S., and Mariat, F. (1975). Ultrastructural studies of the mycelium-to-yeast transformation of *Sporothrix schenckii. J. Bacteriol.* 124:959-968.

Gonzalez-Ochoa, A. (1965). Contribuciones recientes al conocimiento de la esporotricosis. *Gac. Med. Mex. 95*:463-474.
Gonzalez-Ochoa, A. and Figueroa, E. S. (1947). Polisacaridos del *Sporotrichum schenckii*. Dados inmunologicos. Intra-dermoreaccion en el diagnostico de la esporotricosis. *Rev. Inst. Salubr. Enferm. Trop. Mexico City 8*:143-153.
Gorin, P. A. J. (1981). Carbon-13 nuclear magnetic resonance spectroscopy of polysaccharides. *Adv. Carbohydr. Chem. Biochem. 38*:13-104.
Gorin, P. A. J., Haskins, R. H., Travassos, L. R., and Mendonça-Previato, L. (1977). Further studies on the rhamnomannans and acidic rhamnomannans of *Sporothrix schenckii* and *Ceratocystis stenoceras*. *Carbohydr. Res. 55*:21-33.
Griffin, F. J., Jr. (1981). Roles of macrophage Fc and C3b receptors in phagocytosis of immunologically coated *Cryptococcus neoformans*. *Proc. Natl. Acad. Sci. 78*:3853-3857.
Hachisuka, H., and Sasai, Y. (1981). Development of experimental sporotrichosis in normal and modified animals. *Mycopathologia 76*:79-82.
Hamuro, J., Hadding, U., and Bitter-Suermann, D. (1978). Solid phase activation of alternative pathway of complement by β-1,3-glucans and its possible role for tumour regressing activity. *Immunology 34*:695-705.
Harada, T., Nishikawa, T., and Hatano, H. (1976). Antigenic similarity between *Ceratocystis* species and *Sporothrix schenckii* as observed by immunofluorescence. *Sabouraudia 14*:211-215.
Hektoen, L., and Perkins, C. F. (1900). Refractory subcutaneous abcesses caused by *Sporothrix schenckii*, a new pathogenic fungus. *J. Exp. Med. 5*: 77-89.
Ishizaki, H., Kurata, Y., Nakamura, Y., and Wheat, R. W. (1979a). Serological cross-reactivity between *Klebsiella pneumoniae* K47 and *Sporothrix* species. *Curr. Microbiol. 2*:355-356.
Ishizaki, H., Nakamura, Y., Kariya, H., Iwatsu, T., and Wheat, R. W. (1976). Delayed hypersensitivity cross-reactions between *Sporothrix schenckii* and *Ceratocystis* species in sporotrichotic patients. *J. Clin. Microbiol. 3*:545-547.
Ishizaki, H., Nakamura, Y., Kurata, Y., Iwatsu, T., and Wheat, R. W. (1980). Delayed hypersensitivity cross-reactions among *Sporothrix* species in sporotrichotic patients. *J. Dermatol. 7*:75-76.
Ishizaki, H., Nakamura, Y., and Wheat, R. W. (1979b). Comparative immunochemical studies on *Sporothrix* species. *J. Dermatol. 6*:317-320.
Ishizaki, H., Nakamura, Y., and Wheat, R. W. (1981). Serological cross-reactivity between *Sporothrix schenckii* and various unrelated fungi. *Mycopathologia 73*:65-68.
Ishizaki, H., Wheat, R. W., Kiel, D. P., and Conant, N. F. (1978). Serological cross-reactivity among *Sporothrix schenckii, Ceratocystis, Europhium*, and *Graphium* species. *Infect. Immun. 21*:585-593.
Jones, R. D., Sarosi, G. A., Parker, J. D., Weeks, R. J., and Tosh, F. E. (1969). The complement-fixation test in extracutaneous sporotrichosis. *Ann. Intern. Med. 71*:913-918.

Kaplan, W. (1970). The fluorescent antbody technique in the diagnosis of mycotic diseases. In *Proceedings of the Internaltion Symposium on Mycoses.* Pan American Health Organization, Washington, D.C., pp. 86-95.

Kaplan, W., and Ivens, M. S. (1960). Fluorescent antibody staining of *Sporotrichum schenckii* in cultures and clinical materials. *J. Invest. Dermatol. 35*: 151-159.

Karlin, J. V., and Nielsen, H. S., Jr. (1970). Serologic aspects of sporotrichosis. *J. Infect. Dis. 121*:316-327.

Kaufman, L., and Standard, P. G. (1978). Improved version of the exoantigen test for identification of *Coccidioides immitis* and *Histoplasma capsulatum* cultures. *J. Clin. Microbiol. 8*:42-45.

Kozel, T. R., and Pfrommer, G. S. (1986). Activation of the complement system by *Cryptococcus neoformans* leads to binding of iC3b to the yeast. *Infect. Immun. 52*:1-5.

Kubelka, C. F. (1980). Inibição de Sarcoma 180 e aumento de endocitose de macrógos peritoneais de camundongos por polissacaridios de fungos. Thesis, M.Sc.; Federal University of Rio de Janeiro, Brasil.

Lane, J. W., and Garrison, R. G. (1970). Electron microscopy of the yeast to mycelial phase conversion of *Sporotrichum schenckii*. *Can. J. Microbiol 16*: 747-749.

Lee, W.-L., and Lloyd, K. O. (1975). Immunological studies on a yeast peptidogalactomannan. Nature of antigenic determinants reacting with rabbit antisera and those involved in delayed hypersensitivity in guinea pigs. *Arch. Biochem. Biophys. 171*:624-630.

Lloyd, K. O., and Bitoon, M. A. (1971). Isolation and purification of a peptidorhamnomannan from the yeast form of *Sporothrix schenckii*. Structural and immunochemical studies. *J. Immunol. 7*:663-671.

Lloyd, K. O., Mendonça-Previato, L., and Travassos, L. R. (1978). Distribution of antigenic polysaccharides in different cell types of *Sporothrix schenckii* as studied by immunofluorescent staining with rabbit antisera. *Exp. Mycol. 2*:130-137.

Lloyd, K. O., and Travassos, L. R. (1975). Immunochemical studies on L-rhamno-D-mannans of *Sporothrix schenckii* and related fungi by use of rabbit and human antisera. *Carbohydr. Res. 40*:89-97.

Mayorga, R., Caceres, A., Toriello, C., Gutierrez, G., Alvarez, O., Ramirez, M. E., and Mariat, F. (1978). Etude d'une endemie sporotrichosique au Guatemala. *Sabouraudia 16*:185-198.

McMillen, S. (1974). The serology of sporotrichosis. In *The Diagnosis and Treatment of Fungal Infections.* Edited by H. M. Robinson, Jr. Charles C. Thomas, Publisher, Springfield, Ill., pp. 401-409.

Mendonça, L., Gorin, P. A. J., Lloyd, K. O., and Travassos, L. R. (1976). Polymorphism of *Sporothrix schenckii* surface polysaccharides as a function of morphological differentiation. *Biochemistry 15*:2423-2431.

Mendonça-Previato, L., Gorin, P. A. J., and Travassos, L. R. (1980). Galactose-containing polysaccharides from the human pathogen *Sporothrix schenckii* and *Ceratocystis stenoceras*. *Infect. Immun. 29*:934-939.

Merrick, J. M., Zadarlik, K., and Milgrom, F. (1978). Characterization of the Hanganutziu-Deicher (serum sickness) antigen as gangliosides containing *N*-glycolylneuraminic acid. *Intern. Arch. Allergy Appl. Immunol. 57*:477–480.

Nakamura, Y. (1976). Purification and isolation of a biologically active peptidorhamnogalactan from *Sporothrix schenckii*. *J. Dermatol. 3*:25-29.

Nakamura, Y., Ishizaki, H., and Wheat, R. W. (1977). Serological cross-reactivity between group B *Streptococcus* and *Sporothrix schenckii*, *Ceratocystis* species and *Graphium* species. *Infect. Immun. 16*:547-549.

Neill, J. M., Castillo, C. G., and Pinks, A. H. (1955). Serological relationships between fungi and bacteria. I. Cross reactions of *Sporotrichum schenckii* with pneumococci. *J. Immunol. 74*:120-125.

Nielsen, H. S., Jr. (1968). Biological properties of skin test antigens of yeast form *Sporotrichum schenckii*. *J. Infect. Dis. 118*:173-180.

Noguchi, T. (1972). Immunochemical studies on *Sporotrichum schenckii*. I. Extraction and chemical composition of the immunologically active substances. *Acta Sch. Med. Univ. Gifu 20*:335-343.

Oda, L. M., Kubelka, C.F., Alviano, C. S., and Travassos, L. R. (1983). Ingestion of yeast forms of *Sporothrix schenckii* by mouse peritoneal macrophages. *Infect. Immun. 39*:497-504.

Plouffe, J. F., Jr., Silva, J., Jr., Fekety, R., Reinhalter, E., and Browne, R. (1979). Cell-mediated immune responses in sporotrichosis. *J. Infect. Dis. 139*:152-157.

Polonelli, L., and Morace, G. (1982). Exoantigen studies of *Sporothrix schenckii*, *Ceratocystis minor*, and *Graphium penicillioides* cultures. *J. Clin. Microbiol. 15*:362-365.

Previato, J. O., Gorin, P. A. J., Haskins, R. H., and Travassos, L. R. (1979a). Soluble and insoluble glucans from different cell types of the human pathogen *Sporothrix schenckii*. *Exp. Mycol. 3*:92-105.

Previato, J. O., Gorin, P. A. J., and Travassos, L. R. (1979b). Cell wall composition in different cell types of the dimorphic species *Sporothrix schenckii*. *Exp. Mycol. 3*:83-91.

Ratnoff, W. D., Pepple, J. M., and Winkelstein, J. A. (1980). Activation of the alternative complement pathway by *Histoplasma capsulatum*. *Infect. Immun. 30*:147-149.

Ray, T., and Wuepper, K. D. (1976). Activation of the alternative (properdin) pathway of complement by *Candida albicans* and related species. *J. Invest. Dermatol. 67*:700-703.

Rippon, J. W. (1980). Dimorphism in pathogenic fungi. *CRC Crit. Rev. Microbiol. 8*:49-97.

Roberts, G. D., and Larsh, H. W. (1971). The serologic diagnosis of extracutaneous sporotrichosis. *Am. J. Clin. Pathol. 56*:597-600.

Rocha-Posada, H. (1968). Prueba cutanea con esporotricina. Su sensibilidad y especificidad. *Mycopathol. Mycol. Appl. 36*:42-54.

Rodriguez-Del Valle, N., Rosario, M., and Torres-Blasini, G. (1983). Effects of pH, temperature, aeration and carbon source on the development of the mycelial or yeast forms of *Sporothrix schenckii* from conidia. *Mycopathologia* *82*:83-88.

Romano, A. H. (1966). Dimorphism. In *The Fungi*, Vol. II. Edited by G. C. Ainsworth and A. S. Sussman. Academic Press, New York, pp. 181-209.

Russell, B., Beckett, J. H., and Jacobs, P. H. (1979). Immunoperoxidase localization of *Sporothrix schenckii* and *Cryptococcus neoformans*. *Arch. Dermatol. 115*:433-435.

Schnaidman, B.B., Yoshida, N., Gorin, P. A. J., and Travassos, L. R. (1986). Cross-reactive polysaccharides from *Trypanosoma cruzi* and fungi (especially *Dactylium dendroides*). *J. Protozool. 33*:186-191.

Scott, E. N., Muchmore, H. G., and Fine, D. P. (1986). Activation of the alternative complement pathway by *Sporothrix schenckii*. *Infect. Immun. 51*: 6-9.

Seeliger, H. P. R. (1968). Serology as an aid to taxonomy. In *The Fungi*, vol. III. Edited by G. C. Ainsworth and A. S. Sussman. Academic Press, New York, pp. 597-624.

Shimonaka, H., Noguchi, T., Kawai, K., Hasegawa, I., Nasawa, I., and Ito, Y. (1975). Immunochemical studies on the human pathogen *Sporothrix schenckii*: effects of chemical and enzymatic modification of the antigenic compounds upon immediate and delayed reactions. *Infect. Immun. 11*:1187-1194.

Smith, D. F., and Ginsburg, V. (1980). Antibodies against sialyloligosaccharides coupled to protein. *J. Biol. Chem. 255*:55-59.

Souza, E. T., Silva-Filho, F. C., Souza, W., Alviano, C. S., Angluster, J., and Travassos, L. R. (1986). Identification of sialic acids on the cell surface of hyphae and conidia of the human pathogen *Fonsecaea pedrosoi*. *J. Med. Vet. Mycol. 24*:145-153.

Steele, R. W., Cannady, P. B., Jr., Moore, W. L., Jr., and Gentry, L. O. (1976). Skin test and blastogenic responses to *Sporotrichum schenckii*. *J. Clin. Invest. 57*:156-160.

Strobel, G., Alfen, N., Hapner, K. D., McNeil, M., and Albersheim, P. (1978). Some phytotoxic glycopeptides from *Ceratocystis ulmi*, the Dutch elm disease pathogen. *Biochim. Biophys. Acta 538*:60-75.

Suzuki, S., and Takeda, N. (1977). Immunochemical studies on the galactomannans isolated from mycelia and culture broths of three *Hormodendrum* strains. *Infect. Immun. 17*:483-490.

Swan, J. W., Dahl, M. V., Coppo, P. A., and Hammerschmidt, D. E. (1983). Complement activation by *Trichophyton rubrum*. *J. Invest. Dermatol. 80*: 156-158.

Takai, S. (1974). Pathogenicity and ceratoulmin production in *Ceratocystis ulmi*. *Nature* (London) *252*:124-126.

Tarnowski, G. S. (1975). Approaches to the immunotherapy of experimental tumours. In *Host Defense Against Cancer and its Potentiation*. Edited by D. Mizuna. University Park Press, Baltimore, pp. 389-396.

Torinuki, W., Okohchi, K., Takematsu, H., and Tagami, H. (1984). Activation of the alternative complement pathway by *Fonsecaea pedrosoi. J. Invest. Dermatol. 83*:308-310.

Torinuki, W., and Tagami, H. (1985). Complement activation by *Sporothrix schenckii. Arch. Dermatol. Res. 277*:332-333.

Travassos, L. R. (1985). *Sporothrix schenckii*. In *Fungal Dimorphism, with Emphasis on Fungi Pathogenic for Humans*. Edited by P. J. Szaniszlo. Plenum Press, New York, pp. 121-163.

Travassos, L. R., and Alviano, C. S. (1981). Cell surface structures of *Sporothrix schenckii* in relation to pathogenicity. *An. Acad. Bras. Ciencias 53*:847.

Travassos, L. R., and Lloyd, K. O. (1980). *Sporothrix schenckii* and related species of *Ceratocystis. Microbiol. Revs. 44*:683-721.

Travassos, L. R., Gorin, P. A. J., and Lloyd, K. O. (1973). Comparison of the rhamnomannans from the human pathogen *Sporothrix schenckii* with those from the *Ceratocystis* species. *Infect. Immun. 8*:685-693.

Travassos, L. R., Gorin, P. A. J., and Lloyd, K. O. (1974). Discrimination between *Sporothrix schenckii* and *Ceratocystis stenoceras* rhamnomannans by proton and carbon-13 magnetic resonance spectroscopy. *Infect. Immun. 9*: 674-680.

Travassos, L. R., and Mendonça-Previato, L. (1978). Synthesis of monorhamnosyl L-rhamno-D-mannans by conidia of *Sporothrix schenckii. Infect. Immun. 19*:1-4.

Travassos, L. R., Mendonça-Previato, L., and Gorin, P. A. J. (1978). Heterogeneity of the rhamnomannans from one strain of the human pathogen *Sporothrix schenckii* determined by ^{13}C nuclear magnetic resonance spectroscopy. *Infect. Immun. 19*:1107-1109.

Travassos, L. R., Souza, W., Mendonça-Previato, L., and Lloyd, K. O. (1977). Location and biochemical nature of surface components reacting with concanavalin A in different cell types of *Sporothrix schenckii. Exp. Mycol. 1*: 293-305.

Warr, G. A. (1980). A macrophage receptor for (mannose/glucosamine) glycoproteins of potential importance in phagocytic activity. *Biochem. Biophys. Res. Commun. 93*:737-745.

Welsh, R. D., and Dolan, C. T. (1973). *Sporothrix* whole yeast agglutination test: low-titer reactions of sera of subjects not known to have sporotrichosis. *Am. J. Clin. Pathol. 59*:82-85.

Whistler, R. L., Bushway, A. A., Singh, P. P., Nakahara, W., and Tokuzen, R. (1976). Non-cytotoxic, antitumor polysaccharides. *Adv. Carbohydr. Chem. 32*:235-275.

Part II
HOST RESPONSE TO INVASIVE FUNGI
MECHANISMS OF RESISTANCE

9
The Role of Complement in Host Resistance to Systemic Fungal Infection

RICHARD A. CALDERONE
Georgetown University, Washington, D.C.

LISA LINEHAN
University of Vermont Medical Center, Burlington, Vermont

INTRODUCTION

The complement system is thought to play an important role in resistance to a variety of medically important fungi (Reiss, 1986). In general, when coated with complement component C3 and C3 fragments, fungi such as *Cryptococcus neoformans* are readily phagocytized and often killed by such phagocytic cells (Kozel et al., 1984). Of the two major arms of the complement system, the classical and alternative pathways, the alterntive pathway seems to be of greater significance in regard to the functions described above. While the role of complement in protection against fungi will be discussed below, most of the literature on complement and fungi involves studies of *Candida* spp. and *Cryptococcus neoformans*. Thus, emphasis will be placed on these organisms. However, when possible, other fungi will also be mentioned.

In more recent investigations, fungal receptors for complement components have been characterized. This is especially true with receptors for the alternative pathway components including C3 conversion products such as iC3b and C3d. Thus, attention will be given to these recent developments in fungal-host interactions.

COMPLEMENT-CANDIDA INTERACTIONS

Anti-*Candida* antibodies, when complexed with whole fungi or fungal products, may activate the classical complement pathway. An example of this is the mannan-antibody complexes present in the sera of some patients with disseminated candidiasis. Conversely, the alternative complement pathway may be activated directly by the components of fungi in the absence of antibody. The alternative pathway appears to be of primary importance in host defenses against *Candida* infections. This observation is based upon at least two experimental approaches including selective differential inhibition of each pathway by chelating agents and the use of complement-deficient animals. Thus, guinea pig serum that has been depleted of the fourth component of complement, and therefore does not have a functional classical pathway, still produces a normal level of phagocytosis of *Candida albicans*. This observation was made by Gelfand et al. (1978), who performed studies using guinea pigs with congenic defects in C4 as well as animals treated with cobra venom factor (CVF). A defect in the production of C4 would preclude any contribution of the classical pathway, while CVF directly depletes C3 to C9. Normal NIH guinea pigs and C4-deficient guinea pigs, the latter group also treated with cobra venom factor to deplete all complement components, were challenged intravenously with *C. albicans*. At 72 h postinfection, all the normal animals were alive compared to 30% of those devoid of complement. This experiment was repeated using normal, C4-deficient, or CVF-treated animals, and autopsies were performed to detect the number of recovered organisms in various organs. Previous studies had identified the kidney as the target organ infected by *C. albicans*. Neither the normal nor the C4-deficient animal showed any increase of *Candida* counts in the kidneys compared to unchallenged animals. Those animals lacking their alternative pathway, however, showed a marked increase (172-fold) in colony-forming units (cfu) infection by 72 h (Gelfand et al., 1978).

Further studies by this group showed that 24 h after i.v. injection of 2×10^6 *C. albicans*, normal and C4-depleted guinea pigs were found to have 0.2 *Candida* cfu/ml blood, while CVF-treated mice had 6.3 cfu/ml. These observations indicate an absolute requirement for the alternate complement pathway. In some cases, the intact classical pathway may function to provide more efficient conversion of C3 and to optimize the kinetics of opsonization, but clearly activation of the alternative pathway is essential.

The direct participation of complement in ingestion and intracellular killing of *Candida* was demonstrated by Yamamura and Valdimarsson (1977), who investigated the two processes separately by using a dual assay. These studies were performed in the presence of untreated, heat-inactivated, C2-, C3-, or C6-deficient sera at concentrations of 2.5 and 5.0%. The various complement-deficient sera all had normal opsonizing activity for *C. albicans*, whereas intracellular killing was not observed unless C3 was present. Even at high concentrations of C3-deficient serum, killing was not observed until 25 μg/ml of C3 was added. C6-deficient serum had a strong phagocytic candidacidal activity at 2.5 and 5.0%. Serum deficient in C3 induced minimal phagocytic candidacidal activity at 2.5%, but a 5% concentration of such serum supported a normal level of killing. This was taken to be additional proof that *Candida* is capable of directly activating the alternative pathway, as it is well established that a higher serum concentration is required for classical pathway activation by C3 convertase (Yamamura and Valdimarsson, 1977). Pereira and Hosking (1984), however, found that depletion of complement (or antibody) did not abrogate the killing of *C. albicans* but reduced it. Their investigations suggested that *C. albicans* when opsonized in fresh normal pooled human serum were phagocytized and killed more efficiently than those opsonized with either antibody or complement.

In response to results obtained in vivo showing that mice with complement-deficient sera are less resistant to experimental candidiasis than mice with intact complement systems, Morelli and Rosenberg (1971) examined the role of C5 in phagocytosis. Leukocytes from normal and C5-deficient mice were incubated with either complement positive sera (C^+), complement negative sera (C^-), or hyperimmune sera (Ab^+) and used in a phagocytosis assay. They found that the highest degree of phagocytosis occurred when leukocytes were incubated with $C^+ Ab^+$ serum, followed by C^+ serum, $Ab^+ C^-$ serum, and C^- serum. The difference in phagocytosis between $C^+ Ab^+$ serum and the C^+ serum was not a large one; however, these experiments provide evidence that opsonic components such as complement may be most important in controlling *Candida* infections in this model.

Studies by Ray and Wuepper (1976) have shown that serum chelated with Mg-EGTA (which eliminates the elements of the classical pathway of complement but leaves those of the alternative pathway intact) would support phagocytosis of yeast, but serum chelated with EDTA (which eliminates the elements of both classical and alternative) showed no opsonic activity. Again, this study establishes the primary role of the alternative pathway in protection against candidiasis.

Additional proof for the prominent role of complement in the phagocytosis of *C. albicans* is provided by the observation that phagocytosis by murine neutrophils was greatly reduced by heat inactivation at 56°C (Morrison and Cutler, 1981) although other factors may also be required for phagocytosis by macrophages. Other investigators have reported similar effects with heat-treated sera (Solomokin et al., 1978; Ferrante and Thong, 1979).

COMPLEMENT AND CUTANEOUS CANDIDIASIS

C. albicans is a frequent cause of acute infections of the skin and mucous membranes. For this reason numerous cutaneous animal models have been developed to define host-parasite interactions which occur during this infection (Ray, 1985). In rodent models cutaneous infections were established with cotton patches applied over inoculated areas of the skin (Ray and Wuepper,

1972). The overall rate of infection was found to depend on the number of organisms seeded and, as in human infections, epidermal and subcorneal microabscesses containing polymorphonuclear leukocytes (PMN) were observed. A rapid, directed migration of PMNs from cutaneous blood vessels was observed at sites of *Candida* hyphal invasion to the malpighian layer of the skin, which was thought to indicate the in vivo activation of the complement alternative pathway by the yeast. This concept was supported by other studies in which serum complement was depleted by the administration of cobra venom factor (CVF). Thus, animals depleted of C3 and C5 by treatment of CVF or congenic C5-deficient mice failed to mount a neutrophilic inflammatory response, indicating that chemotactic factors derived from C3 and C5 were essential to this biological response (Ray and Wuepper, 1978).

A guinea pig model of cutaneous candidiasis has also been established by Van Cutsem and Thienpont (1971) and more recently by Sohnle, Frank, and Kirkpatrick (1976). In the latter instance both occlusive and nonocclusive methods were used to establish an infection. In occlusive infections, neutrophil-rich, subcorneal infiltrates were observed at 24 h along with the appearance of PAS-positive mycelia. The deposition of the third component of complement, C3, in the stratum corneum as well as on the invading mycelial elements was observed. Immunoglobulin and C4 were not observed in the lesions. These authors also found that C4-deficient guinea pigs still developed lesions with deposits of C3, much like normal animals, supporting the notion that activation of the alternative pathway of complement is critical to the influx of PMN (Sohnle and Kirkpatrick, 1976).

COMPLEMENT RECEPTORS OF *C. ALBICANS*

Complement activation by *Candida* appears to be a function of the cell wall. In order to biochemically identify the complement-activating constituents, Ray et al. (1979) prepared various extracts of *C. albicans*. A cell wall polysaccharide (mannose 82%, glucose 11%, fucose 7%) from an ethylene glycol extract of yeast phase organisms was found to activate the alternative complement pathway, induce neutrophil chemotaxis, and react antigenically with antisera to *C. albicans*. The extract produced precipitin lines in double immunodiffusion studies with serum from patients with invasive candidiasis and rabbit antisera to both mycelial and yeast forms of *C. albicans* but not with normal serum of either humans or rabbit origin. Thus, mannan-containing polysaccharides of *C. albicans* directly activate the alternate pathway and trigger PMN influx. Cell wall mannoprotein (which is 90% mannan and 10% protein) has been identified as the immunodominant antigen as measured by agglutination and precipitation reactions (Wilton and Lehner, 1980).

The structural configuration of mannan, which is composed especially of α-linkages, and its polymeric nature conforms to other know polysaccharide activators of the alternative pathway. However, yeast glucan particles have a capacity to activate the alternative complement pathway comparable to that of zymosan particles, suggesting that the active constituents of zymosan are its β-(1,3) and/or β-(1,6) glucans rather than its mannan component (Czop and Austen, 1985). The water-soluble glucans of *Streptococcus mutans*, which contain both α-(1,3) and α-(1,6) glucosidic bonds, are also effective alternative pathway activators, as are certain pneumococcal capsular polysaccharides (Winkelstein et al., 1976).

Thus, a cell wall polysaccharide may elicit leukocyte migration in the presence of serum via complement activation and explains the observations of Escobar et al. (1975) regarding mannan-induced chemotaxis as well as earlier studies by Denning and Davies (1973). However, *C. albicans* culture filtrates also induced a neutrophil chemotaxis which was apparently independent of serum (Cutler, 1977).

As stated above, complement component C3 plays an important role in the complement cascade. C3 fragments are deposited on the *Candida* cell surface as a result of activation of the alternative pathway. As such, these C3 fragments can serve as opsonic ligands for phagocytosis. The identity of the fragments and mechanism of C3 binding to the *C. albicans* cell wall has been described recently by Kozel et al. (1987). In this study, ^{125}I-labeled C3 was incubated with yeast cells in 20% serum. Binding of C3 to the *Candida* cell surface was found to occur through both ester and amide linkages and was absent in the presence of EDTA (indicating the primary role of the alternative pathway for activation of C3). Using specific monoclonal antibodies to complement components, C3b was found to be the primary fragment bound to the yeast cell wall. In this same study *Cryptococcus neoformans* was found to differ from *C. albicans* in that C3 molecules were bound by an ester linkage almost exclusively. Also, iC3b was the complement fragment most commonly observed on the cell surface of *C. neoformans*. These studies demonstrate clearly that the activation and binding of C3 on the cell surface of these pathogenic yeasts is significantly different (Kozel and Pfrommer, 1986). Since the surface composition of each organism is different (heteropolysaccharide capsule of *C. neoformans* vs. the mannoprotein complex of the outer cell wall of *Candida*), it is not surprising that the mechanism of complement activation also differs. A summary of the activation and binding of C3 fragments by *C. neoformans* and *C. albicans* is presented in Table 1.

When complement is activated by either the classic or the alternative pathway, molecular fragments (C3b, C3d, iC3b, C3d,g) are generated which are recognized by specific receptors on a variety of mammalian cells. The binding of these components to their cellular receptors is thought to initiate a number of

Table 1 Activation and Binding of C3 Fragments to *C. albicans* and *C. neoformans*

	C. albicans	*C. neoformans*
Binding site	Cell wall	Capsule
C3 molecules/yeast	$2\text{-}4 \times 10^5$	$5\text{-}10 \times 10^6$
Inhibition by EDTA	Marked	Marked
Inhibition by Mg-EGTA	30-40%	30-40%
Release by NH_2OH	50%	90-100%
C3 fragment(s)	C3b	C3b and iC3b
Activation kinetics	Rapid	Relatively slow

Source: Kozel, Brown, and Pfrommer, 1987.

responses such as phagocytosis by neutrophils and macrophages or other specific immune functions (Berger et al., 1981-1982).

The best studied cellular receptors for complement components are those involved in enhancement of phagocytosis of bacteria, yeast, antibody-sensitized erythrocytes, and other particles. It has been shown that receptors on phagocytic cells for fragments of C3 as well as Fc receptors are vital to phagocytosis. Four receptors have been identified thus far, and they are designated as CR1, CR2, CR3, and CR4 (CR = Complement Receptor). The first discovery of cellular complement receptors was made by Nelson (1953), who showed that phagocytosis of bacteria sensitized by antibody and complement by neutrophils was more efficient in the presence of human erythrocytes. Bacteria were shown to bind to erythrocytes because of the interaction of organism-bound C3b with the C3b receptor on erythrocytes, a phenomenon known as immune adherence. All primate erythrocytes possess immune adherence receptors. Specifically, receptors for C3b have been shown to exist on monocytes, lymphocytes, eosinophils, and neutrophils (Berger et al., 1981-1982). The isolated C3b receptor, which is a glycoprotein called CR1, is approximately 1.2×10^6 daltons, with a single 225,000 dalton band visible on SDS-PAGE. In addition, CR1 has been found on mast cells, basophils and renal glomerular cells. Under experimental conditions, C3b-coated particles can be shown to adhere to other cell types such as fibroblasts, endothelial cells, macrophages, and certain lymphocyte tumor lines (Brown et al., 1984).

C3b is cleaved by a serum serine protease, Factor I (C3b inactivator), a reaction in which a cofactor (Factor H) is necessary. The product of this cleavage is a 3-chain molecule, iC3b. There are receptors for iC3b on a number of cell types, including erythrocytes, neutrophils, lymphocytes, monocytes, and mast cells. The binding of iC3b-bearing particles to leukocytes possessing this receptor (CR3) is not inhibited by antibodies to CR1 or to the C3d receptor (CR2) (Vik and Fearon, 1985).

C3d is the product of iC3b cleavage by trypsin, plasmin, elastase, or cathepsin; the receptor for C3d is called CR2. This receptor has been demonstrated on B-lymphocytes and lymphoblastoid cell lines as well as adherent monocytes. The receptor is a 145 kD glycoprotein with an affinity for other C3 ligands. Finally, CR4 recognizes the C3dg/C3d ligand and is found on platelets and neutrophils (Vik and Fearon, 1985).

COMPLEMENT RECEPTORS OF *C. Albicans* FOR iC3b AND C3d

Initially, Heidenreich and Dierich (1985) and then Edwards et al. (1986) found that *C. albicans* was able to mimic the human complement receptors by binding erythrocytes coated with specific human C3 fragments. The binding was specific for C3 fragments because erythrocytes coated with hemolysin alone and hemolysin plus complement components 1, 4 and 2 did not bind to the microorganism. Of the C3 conversion products, iC3b and C3d-coated erythrocytes bound most strongly to *Candida*, whereas C3b-coated cells bound specifically but only weakly. The binding of iC3b and C3d occurred with the pseudohyphal forms of the organism rather than to the yeast although the work of Heidenreich and Dierich indicated some binding to yeast forms. Strains of *Candida* that vary in their pathogenicity were examined for the presence of these receptors. *S. cerevisiae*, a nonpathogen with a cell wall resembling that of *Candida*, was also examined. Among *Candida* species, *C. albicans* bound the complement-coated erythrocytes the most strongly. Attempts were made to block adherence of complement-coated erythrocytes with monoclonal and polyclonal antibodies to human complement receptors to determine whether there was a structural similarity between the *Candida* and the mammalian cell CR (Edwards et al., 1986). If the complement receptors on mammalian cells have a similar configuration to the CR of *Candida*, then blocking of receptor activity of *Candida* with antibodies to the human CR may be possible. All blocking studies with anti-human CR1 and CR3 antibodies were negative. However, blocking of iC3b and C3d binding to *Candida* occurred with the anti-CR2 antibodies, but only when a very high concentration of antibody was used and the erythrocytes were coated with very small amounts of the C3 degradation products.

In contrast to the blocking results obtained with the use of anti-CR2 antibodies, a strong binding of the anti-human CR3 Mol was shown in immunofluorescence assays compared with the lack of binding of a purified mouse IgM preparation of no known specificity. The binding was restricted to the pseudohyphal forms of the organism. This suggests a close similarity between the molecular structure on the surface of the yeast with a human CR3.

Gilmore et al. (1988) also demonstrated by fluorescence microscopy that other anti-CR3 monoclonals, OKM-1 and M1/170, bound to *C. albicans*. The demonstration of the binding of these two anti-human CR3 monoclonals to the *C. albicans* surface and the lack of blocking of adherence of iC3b-coated erythrocytes suggests that these fluorescing epitopes may not be within the binding sites or that they are cross-reacting epitopes. Although C3d binds strongly to *C. albicans* and both monoclonal and polyclonal anti-CR2 blocked binding of C3d to *C. albicans*, no staining by immunofluorescence was found with the monoclonal anti-human CR2. This may be due to weak affinity between the antibody and the *Candida* cell surface. Increased receptor expression was observed following conversion of yeast to pseudohyphae as well in yeast cells grown in the presence of 50 mM D-glucose. These observations were made using quantitative flow cytometry of monoclonal-treated cells or by direct binding of iC3b. Finally, increased receptor expression was correlated with a 30% decrease in neutrophil phagocytosis. Thus, the biological significance of such receptor-like binding activity on *C. albicans* may relate to a mechanism through which the organism evades normal defense processes.

The first description of iC3b and C3d receptors in *Candida* was made by Heidenreich and Dierich (1985) when yeast cells contaminating a lymphoblastoid culture reacted with complement-coated sheep erythrocytes in the same way as lymphoblasts. To identify the complement component which was responsible for adherence, they studied complement intermediates (erythrocytes coated with purified C4, C2, C3b, iC3b, and C3d) and their ability to bind to *C. albicans*. Only a few adhering erythrocytes, if any, were seen with all intermediates except iC3b and C3d. They also noted that adherence was achieved only with intact organisms since fragmented material obtained by mechanical grinding or by sonication of yeast cells had a limited inhibitor effect. Neither mannan from baker's yeast nor the monosaccharides D-galactose, L-mannose, and N-acetyl glucosamine showed a reproducible inhibitory effect on the adherence of iC3b and C3d intermediates. However, preincubation of erythrocytes with D-mannose or D-glucose resulted in an inhibition of the adherence of these two intermediates to *C. albicans* by about 30% compared to the control. This suggested a lectinlike interaction between iC3b or C3d and the mannoproteins on the surface of the cell wall of *C. albicans*. When other *Candida* species were examined, only *C. stellatoideae* (which is closely related to *C. albicans*) was able to bind iC3b and C3d. Here, as suggested in the studies of Edwards, et al. (1986), the presence of receptors on pathogenic *Candida* could imply a role in pathogenesis. *Candida* cells may be opsonized via C3 and form complement bridges with other *Candida* cells that have free binding sites for iC3b and C3d, thus making the cells within the clumps less susceptible to phaocytosis. If opsonins such as iC3b are bound to specific receptors,

Table 2 Comparative Analysis of CR2/CR3 Receptors from Mammalian Cells and *C. albicans*

	Mammalian cells		*C. albicans*	
	CR2	CR3	CR2	CR3
Molecular weight	145 kD	200 kD	60,70 kD[a]	N.D.[b]
Oligosaccharide	Yes	Yes	Yes	N.D.
Antibody reactivity of:				
HB5	+		+ (weakly)[c]	
anti-GP140	+		+ (weakly)	
OKM-1		+		+
M1/170		+		+
Mo1		+		+[d]

[a]*Source*: Calderone et al., 1988.
[b]N.D. = not determined.
[c]Each of the antibodies listed blocked rosetting of EAC3d by pseudohyphae but at high concentrations of antibody and at low (C3d or iC3b) ligand concentrations. *Source*: Edwards et al., 1986
[d]Reactivity determined by immunofluorescence. *Source*: Gilmore et al. and Hostetter, 1988.

iC3b-mediated phagocytosis would be impaired. Also, receptor secreted by *Candida* could compete with phagocyte receptor for C3 fragments, thus reducing phagocytosis. The role of the CR2 of human B cells has been studied (Berger, Gaither, and Frank, 1981-1982); it is only recently, however, that a receptor for C3d has been described in *C. albicans*. This is intriguing, as the cell wall of this fungus differs dramatically from the lipoprotein bilayer of mammalian cells.

Preliminary characterization of the C3d receptor of *C. albicans* has been completed (Calderone et al. 1988). Whole pseudohyphal extracts were fractionated by DEAE-column chromatography. A fraction was recovered, which, when incubated with EAC3d, blocked rosetting when pseudohyphae were added subsequently. This fraction was subjected to affinity chromatography using C3d-Thiol-Sepharose. The material which was eluted from this column was analyzed by SDS-PAGE, and by Western blotting with an anti-*Candida* monoclonal antibody (CA-A), which had been shown to block rosetting of EAC3d when preincubated with pseudohyphae. The C3d receptor of *C. albicans* was resolved in two bands of 62 and 70 kD, respectively. The C3d receptor of *C. albicans* appeared to be a mannoprotein since a fraction from whole cell extracts, obtained by Con-A affinity chromatography, blocked rosetting of EAC3d-pseudohyphae. In this respect, the *Candida* CR2 is much like the mammalian receptor since both are glycosylated. Rosetting activity of Con A-treated pseudohyphae was not inhibited; however, the oligosaccharide component is probably not involved in binding to C3d but could serve to stabilize the receptor as has been shown for the CR2 of mammalian cells.

A comparison of the *Candida* and mammalian cell CR2 and CR3 is presented in Table 2. Monoclonal antibody to the CR2 of mammalian cells (as well as a polyclonal, anti-GP-140) blocks rosetting of EAC3d by pseudohyphae (at high concentrations) but does not react in immunofluorescence assays. Conversely, monoclonals reactive with mammalian CR3 (OKM-1, Mol, M1/170) gave a positive fluorescence reaction with *Candida* pseudohyphae but did not block rosetting. Those results would suggest that these anti-CR3 monoclonals react at a site on the *Candida* receptor different from the binding site for iC3b.

Cryptococcus neoforman's-COMPLEMENT INTERACTIONS

As with *Candida* spp., complement components play an important role in resistance to infection by *C. neoformans* (Kozel et al., 1984). C3 fragments or IgG have been shown to bind to the cell surface capsule of *C. neoformans* while promoting phagocytosis by human neutrophils (Diamond et al., 1973; Diamond et al., 1974; Rhodes et al., 1980; Griffin, 1981; Davies et al., 1982). Thus, guinea pigs depleted of the C3 to C9 complement components by CVF showed a diminished clearance of cryptococci when compared with untreated controls (Diamond et al., 1973).

Studies of cryptococcosis in inbred mouse strains have resulted in the identification of three genetic loci which influence susceptibility: Hc, nu, and xid (Rhodes Wicker, and Urba, 1980; Marquis et al., 1985). Mutations at each of these loci are thought to affect the immune system. Other studies by Marquis et al. (1985) have established that the beige mutation in mice is associated with increased susceptibility to cryptococcosis and that other autosomal genes are likely to be involved in the genetic control of susceptibility. Unlike *C. albicans*, binding of C3 to *C. neoformans* most commonly occurs through ester bonds and results in the deposition of iC3b fragments on the capsular surface of the yeast (Kozel and Pfrommer, 1986; Kozel et al., 1987) (see Table 1).

COMPLEMENT INTERACTIONS WITH OTHER HUMAN PATHOGENIC FUNGI

In Table 3, a list of human pathogenic fungi which activate the alternative pathway is presented along with information as to whether or not phagocytosis of complement-coated fungi is augmented. In most cases activation of the

Table 3 The Alternative Complement-Fixation Pathway and Human Pathogenic Fungi

Organism	Activation of pathway	Augmentation of phagocytosis and or killing	References
Torulopsis glabrata	+	+	(Ferrante and Thong, 1979)
Paracoccidioides brasiliensis	+	+	(McEwen et al., 1987)
Pityrosporum orbiculare	+/−	+	(Sohnle and Collins-Lech, 1983)
Aspergillus fumigatus	+/−	+/−	(Lehrer and Jan, 1970; Waldorf and Diamond, 1985)
Rhizopus oryzae	+/−	N.D.	(Waldorf and Diamond, 1985)
Coccidioides immitis	+[a]	N.D.	(Galgiani, et al., 1978; Galgiani et al., 1980)
Coccidioides immitis	N.D.	+[b]	(Frey and Drutz, 1986)
Blastomyces dermatitidis	+	+	(Drutz and Frey, 1985)
Blastomyces dermatitidis	N.D.	+	(Sixbey et al., 1979)
Histoplasma capsulatum	+	N.D.	(Ratnoff et al., 1980)
Sporothrix schenckii	+	N.D.	(Scott et al., 1986; Cunningham et al., 1979)
Fosecaea pedrosi	+	N.D.	(Torinuki et al., 1984)

[a]Activation of both classical and alternate pathways observed by Galgiani et al., 1980.
[b]Immune serum significantly higher than nonimmune in promoting pharocytosis and killing of specific growth forms (Frey and Drutz, 1986).

alternative pathway by many of the human pathogenic fungi has been observed (see Table 3). The interaction of *Aspergillus* and *Rhizopus* species with the complement system deserves further discussion. As described by Waldorf and Diamond (1985), freshly grown conidia of *A. fumigatus* induced a significant increase in the absolute number of neutrophils recovered from lung lavages following intranasal inoculation of mice while fresh spores of *R. oryzae* did not induce such a response. However, when *R. oryzae* spores were incubated to induce a pregermination swelling, neutrophil migration was observed. In vitro chemotaxis assays using neutrophils were also performed with both fungi in this study. Chemotaxis was only observed with both fungi at very high spore/conidia concentrations in serum when freshly prepared spores/conidia were used. In contrast, preswollen spores of both organisms when incubated in serum induced chemotaxis to a much greater extent than fresh spores. The authors believed that spores and conidia activate the complement system which in turn releases chemotactic complement components since anti-C5 and anti-C3 treatment of activated serum neutralized neutrophil migration. Thus, as indicated in Table 3, these fungi may/may not activate complement so that activation is dependent upon the growth form of the fungus.

The inability of freshly obtained spores of *R. oryzae* to activate the alternative pathway may in part be related to the mechanism of complement inactivation as described by Washburn, Hammer, and Bennett (1986). These investigators have described a heat-sensitive component from culture filtrates of *A. fumigatus* which inhibited complement-mediated phagocytosis and killing by monocytes of *A. fumigatus, A. niger*, and *Cryptococcus neoformans*. The latter two species did not produce such an inhibitor. The inhibitor was specific for the alternative pathway and the mechanism of this inhibition did not involve chelation of calcium or magnesium or proteolysis of complement or a direct interference with monocytes. This inhibitor has a molecular weight of 10,000 and was thought to act (at least with *C. neoformans*) by preventing the binding of complement components to the yeast cell surface.

While cell-mediated immunity (CMI) is thought to be associated with resistance to coccidioidomycosis, recent studies point to the significance of human neutrophils in the eradication of *Coccidioides immitis*, especially during the early phase of disease when CMI has not yet been established (Galgiani et al., 1978; 1980; 1984; Frey and Drutz, 1986). It has been demonstrated that both the mycelial and spherule phases of *C. immitis* were able to activate complement resulting in complement-mediated chemotaxis of neutrophils (Galgiani et al., 1978, 1980). PMNs were able to inhibit the incorporation of chitin precursors into mycelial forms of *C. immitis*, an effect which was facilitated by a heat-labile factor from nonimmune serum (Galgiani et al., 1984). More recently, Frey and Drutz (1986) were able to show that the extent of phagocytosis and killing of *C. immitis* by neutrophils was strain-dependent and variable, being conditioned on the growth form of the organism. Immune serum significantly increased phagocytosis (over nonimmune) but did not promote a corresponding increase in killing. Thus, both nonimmune and immune sera may be necessary for phagocytic cell function. In comparison to the studies presented above, however, Beaman and Holmberg (1980) demonstrated little killing (13% by 4 h) of *C. immitis* by leukocytes from rhesus macaques even though *C. albicans* and *Listeria monocytogenes* cells were readily killed.

Resistance to *Blastomyces dermatitidis* probably follows a pattern similar to many other systemically invading fungi, but the details of resistance mechansims are not as thoroughly understood. Although CMI is most likely important, recent studies by Drutz and Frey (1985) point to the role of nonimmune phagocytic cells in protection. These investigators found that conidia of *B. dermatitidis* were readily phagocytized by human phagocytes (PMN). Serum was required for optimal phagocytosis and of the serum components, complement appeared to be essential since heat-inactivation as well as the removal of divalent cations lessened the serum effect. Probably by virtue of their larger size, yeast cells were not phagocytized (97%, extracellular) and few cells (approximately 20% vs. 60% with conidia) were killed. A similar observation has been made by Brummer and Stevens (1982). They reported that the replication of *B. dermatitidis* yeast cells was inhibited by human monocytes and macrophages while PMN actually stimulated their replication. Marginal amounts of killing of yeast cells (29%) by human granulocytes have been noted by Sixbey et al. (1979) when either 10% normal or immune human serum was added to their in vitro system.

Activation of the alternate pathway has been observed with several pathogenic fungi, although the specific role of complement in resistance is uncertain (52-55, see Table 3). Cunningham et al. (1979) have shown that *Sporothrix schenckii* yeast cells can be phagocytized and killed by human neutrophils in the presence of 10% unheated serum.

COMPLEMENT AND DISEASE PROGRESSION

While complement contributes directly to host defense mechanisms, there is some indication that the severity of disease is correlated with the levels of complement-fixing (CF) antibody in patient sera. Thus, in coccidioidomycosis, disease progression and remission are correlated with an increase or decrease, respectively, in the concentration of complement-fixing antibody (Smith et al., 1956). A hyporeactivity of cell-mediated responses to antigens of *C. immitis* has been observed to coincide with levels of circulating immune complexes (Cox et al., 1982). Although the concentration of immune complexes did not correlate directly with CF-antibody titers, immune complexes were quantitated by a

Cl_q-binding assay, and Cl_q was detected in purified immune complexes. Thus, activation of the classical complement pathway by antigen-antibody complexes may result in defective immune responses. Such a sequence of events may actually contribute to disease progression. More recent studies by Cox and Pope (1982) suggest that immunoglobulin G, either monomeric or complexed with antigen, mediated suppression. A similar type of suppressive mechanism may also be operative in histoplasmosis (Cox, 1979) or contribute to the pathogenesis of allergic bronchopulmonary aspergillosis (Geha, 1977).

CONCLUSION

Many of the fungi that cause the systemic mycoses are able to fix complement either through the classical or alternative pathways. Of the two pathways, the alternative pathway seems to play a primary role in resistance to infection by promoting phagocytosis. In *Candida*, activation of the alternative pathway is associated with polysaccharide-rich fractions of cell wall. More recently, the presence of surface receptors for C3 conversion products, iC3b and C3d, have been described for *C. albicans*. As these receptors are also found on mammalian cells, their role in virulence of the organism is not known but is inferred by their presence on *C. albicans* and not nonpathogenic *Candida* Sp. With coccidioidomycosis and other systemic mycoses, disease progression is associated with increasing levels of complement-fixing antibody.

ACKNOWLEDGMENTS

The author wishes to thank Ms. Stephanie Coleman for typing this manuscript. Some of the data presented was supported by NIH grant HLB-21370.

REFERENCES

Beaman, L., and Holmberg, C. A. (1980). Interaction of nonhuman primate peripheral blood leukocytes and *Coccidioides immitis in vitro*. *Infect. Immun. 29*: 1200-1201.

Berger, M., Gaither, T. A., and Frank, M. M. (1981-1982). Complement receptors. *Clin. Immunol. Rev. 1*: 471-545.

Brown, E. J., Joiner, K. A., and Frank, M. M. (1984). Complement. In *Fundamental Immunolongy*. Raven Press, New York.

Brummer, E., and Stevens, D. (1982). Opposite effects of human monocytes, macrophages and polymorphonuclear neutrophils on replication of *Blastomyces dermatitidis in vitro*. *Infect. Immun. 36*: 297-303.

Calderone, R. A., Linehan, L., Wadsworth, E., and Sandberg, A. (1988). Identification of C3d receptors on *Candida albicans*. *Infect. Immun. 56*: 252-258.

Cox, R. A. (1979). Immunologic studies of patients with histoplasmosis. *Am. Rev. Respir. Dis. 120*: 143-149.
Cox, R. A., and Pope, R. M. (1982). Serum-mediated suppression of lymphocyte transformation response in coccidioidomycosis. *Infect. Immun. 55*: 1058-1062.
Cox, R., Pope, R. M., and Stevens, D. A. (1982). Immune complexes in coccidioidomycosis. Correlation with disease involvement. *Am. Rev. Respir. Dis. 126*: 439-443.
Cunningham, K. M., Bulmer, G. S., and Rhoades, E. R. (1979). Phagocytosis and intracellular fate of *Sporothrix schenckii*. *J. Infect. Dis. 140*: 815-817.
Cutler, J. E. (1977). Chemotactic factor produced by *Candida albicans*. *Infect. Immun. 18*: 568-573.
Czop, J. K., and Austen, K. F. (1985). Properties of glycans that activate the human alternative complement pathway and interact with the human monocyte β-glucan receptor. *J. Immunol. 135*: 3388-3393.
Davies, S. F., Clifford, D. P., Hoidal, J. R., and Repine, J. E. (1982). Opsonic requirements for the uptake of *Cryptococcus neoformans* by human polymorphonuclear leukocytes and monocytes. *J. Infect. Dis. 145*: 870-874.
Denning, T. J. V., and Davies. R. R. (1973). *Candida albicans* and the chemotaxis of polymorphonuclear neutrophils. *Sabouraudia 11*: 210-215.
Diamond, R. D., May, J. E., Kane, M. A., Frank, M. M., and Bennett, J. E. (1974). The role of the classical and alternative complement pathways in host defenses against *Cryptococcus neoformans*. *J. Immunol. 112*: 2260-2270.
Drutz, D. J., and Frey, C. L. (1985). Intracellular and extracellular defenses of human phagocytes against *Blastomyces dermatitidis* conidia and yeast. *J. Lab. Clin. Med. 105*: 737-750.
Edwards, J. E., Jr., Gaither, T. A., O'Shea, J. J., Rotrosen, D., Lawley, T. L., Wright, S. A., Frank, M. M., and Green, I. (1986). Expression of specific binding sites on *Candida* with functional and antigenic characteristics of human complement receptors. *J. Immunol. 137*: 3577-3583.
Escobar, M. R., Hamilton, P. B., and Weeks, B. A. (1975). Neutrophil chemotaxis by cell wall mannans of *Candida albicans*. *J. Reticuloendothel. Soc. 18*-28A.
Ferrante, A., and Thong, Y. H. (1979). Requirement of heat-labile osponins for maximal phagocytosis of *Candida albicans*. *Sabouraudia 17*: 293-297.
Frey, C. L., and Drutz, D. J. (1986). Influence of fungal surface components on the interaction of *Coccidioides immitis* with polymorphonuclear leukocytes. *J. Infect. Dis. 153*: 933-943.
Galgiani, J. N., Issenberg, R. A., and Stevens, D. A. (1978). Chemotaxigenic activity of extracts from the mycelial and spherules phases of *Coccidioides immitis* for human polymorphonuclear leukocytes. *J. Infect. Dis. 21*: 862-865.
Galgiani, J. N., Yam, P., Petz, L. D., Williams, P. L., and Stevens, D. A. (1980). Complement activation by *Coccidioides immitis*: in vitro and clinical studies. *Infect. Immun. 28*: 944-949.

Galgiani, J. N., Payne, C. M., and Jones, J. F. (1984). Human PMN-leukocyte inhibition of incorporation of chitin precursors into mycelial of *Coccidioides immitis*. *J. Infect. Dis. 149*: 404-412.

Geha, R. S. (1977). Circulating immune complexes and activation of the complement sequence in acute bronchopulmonary aspergillosis. *J. All. Clin. Immun. 60*: 357-359.

Gelfand, J. A., Hurley, D. L., Fauci, A. S., and Frank, M. M. (1978). Role of complement in host defense against experimental disseminated candidiasis. *J. Infect. Dis. 138*: 9-16.

Gilmore, B. J., Retsinas, E. M., Lorenz, J. S., and Hostetter, M. K. (1988). An iC3b receptor on *Candida albicans*. *J. Infect. Dis. 157*: 38-46.

Griffin, F. M. (1981). Roles of macrophage Fc and C3b receptors in phagocytosis of immunologically coated *Cryptococcus neoformans*. *Proc. Natl. Acad. Sci. USA 78*: 3853-3857.

Heidenreich, F., and Dierich, M. P. (1985). *Candida albicans* and *Candida stellatoidea*, in contrast to other *Candida* species, bind iC3b and C3d but not C3b. *Infect. Immun. 50*: 598-600.

Kozel, T. R., Highson, B., and Stratton, C. J. (1984). Localization on encapsulated *Cryptococcus neoformans* of serum components opsonic for phagocytosis by macrophage and neutrophils. *Infect. Immun. 43*: 574-579.

Kozel, T. R., and Pfrommer, G. S. T. (1986). Activation of the complement system by *Cryptococcus neoformans* leads to binding of iC3b to the yeast. *Infect. Immun. 52*: 1-5.

Kozel, T. R., Brown, R. R., and Pfrommer, G. S. T. (1987). Activation and binding of C3 by *Candida albicans*. *Infect. Immun. 55*: 1890-1894.

Lehrer, R. I., and Jan, R. G. (1970). Interaction of *Aspergillus fumigatus* spores with human leukocytes and serum. *Infect. Immun. 1*: 345-350.

Marquis, G., Montplaisir, S., Pelletier, M., Mousseau, S., and Auger, P. (1985a). Genetic resistance to murine cryptococcosis: increased susceptibility in the CBA/N XID mutant strain of mice. *Infect. Immun. 47*: 282-287.

Marquis, G., Montplaisir, S., Pelletier, M., Mousseau, S., and Auger, P. (1985b). Genetic resistance to murine cryptococcosis: the beige mutation (Chediak-Higashi Syndrome) in mice. *Infect. Immun. 47*: 288-293.

McEwen, J. G., Brummer, E., Stevens, D. A., and Restrepo, A. (1987). Effect of murine polymorphonuclear leukocytes on the yeast form of *Paracoccidioides brasiliensis*. *Amer. J. Trop. Med. Hyg. 36*: 603-608.

Morelli, R., and Rosenberg, L. (1971). The role of complement in the phagocytosis of *Candida albicans* by mouse peripheral blood leukocytes. *J. Immunol. 107*: 476-481.

Morrison, R. P., and Cutler, J. E. (1981). In vitro studies of the interaction of murine phagocytic cells with *Candida albicans*. *J. Reticuloendothel. Soc. 29*: 23-34.

Nelson, R. A. (1953). The immune adherence phenomenon: an immunologically specific reaction between microorganisms and erythrocytes leading to enhanced phagocytosis. *Science 118*: 733-737.

Pereira, H. A., and Hosking, C. S. (1984). The role of complement and antibody in opsonization and intracellular killing of *Candida albicans*. *Clin. Exp. Immunol. 57*: 307-314.

Ratnoff, W. D., Pepple, J. M., and Winkelstein, J. A. (1980). Activation of the alternate complement pathway by *Histoplasma capsulatum*. *Infect. Immun. 30*: 147-149.

Ray, T. L., and Wuepper, K. D. (1972). Experimental cutaneous candidiasis in rodents. *J. Invest. Derm. 66*: 29-33.

Ray, T. L., and Wuepper, K. D. (1976). Activation of the alternative (properdin) pathway of complement of *Candida albicans* and related species. *J. Invest. Derm. 67*: 700-703.

Ray, T. L., and Wuepper, K. D. (1978). Experimental cutaneous candidiasis in rodents II. Role of stratum corneum barrier and serum complement as a mediator of a protective inflammatory response. *Arch. Dermatol. 1149: 539-*543.

Ray, T. L., Hanson, A., Ray, L. F., and Wuepper, K. D. (1979). Purification of a mannan from *Candida albicans* which activates serum complement. *J. Invest Derm. 73*: 269-274.

Ray, T. L. (1985). Animal models of experimental *Candida* infections of the skin. In *Models in Dermatology*. Edited by Marbach and Lowe. S. Karger, Basel.

Reiss, E. (1986). *Molecular Immunology of Mycotic and Actinomycotic Infections*. Elsevier, New York.

Rhodes, J. C., Wicker, L. S., and Urba, W. J. (1980). Genetic control of susceptibility to *Cryptococcus neoformans* in mice. *Infect. Immun. 29*: 494-499.

Scott, E. N., Muchmore, H. G., and Fine, D. P. (1986). Activation of the alternate pathway of *Sporothrix schenckii*. *Infect. Immun. 51*: 6-9.

Sixbey, J. W., Fields, B. T., Sun, C. N., Clark, R. A., and Nolan, C. M. (1979). Interactions between human granulocytes and *Blastomyces dermatitidis*. *Infect. Immun. 23*: 41-44.

Smith, C. E., Saito, M. T., Simons, S. A. (1956). Pattern of 39,500 serologic tests in coccidioidomycosis. *J. Am. Med. Assoc. 160*: 546-552.

Sohnle, P. G., and Kirkpatrick, C. H. (1976). Deposition of complement in the lesions of experimental cutaneous canididasis in guinea pigs. *J. Cutan. Pathol. 3*: 232-238.

Sohnle, P. G., Frank, M. M., and Kirkpatrick, C. H. (1976). Deposition of complement components to the cutaneous lesions of chronic mucocutaneous candidiasis. *Clin. Immunol. Immunopathol. 5*: 340-350.

Sohnle, P. G., Frank, M. M., and Kirkpatrick, C. H. (1976). Mechanisms involved in the elimination of organisms from experimental cutaneous *Candida albicans* infections in guinea pigs. *J. Immunol. 117*: 523-530.

Sohnle, P. G., and Collins-Lech, C. (1983). Activation of complement by *Pityrosporium orbiculare*. *J. Invest. Derm. 80*: 93-97.

Solomokin, J. S., Mills, E. L., Giebink, G. S., Nelson, R. D., Simmons, R. L., and Quie, P. Q. (1978). Phagocytosis of *Candida albicans* by human leukocytes: opsonic requirements. *J. Infect. Dis. 137*: 30-37.

Torinuki, W., Okohchi, K., Takematsu, H., and Tagami, H. (1984). Activation of the alternative complement pathway by *Fonsecaea pedrosi*. *J. Invest. Dermatol. 83*: 308-310.

Van Cutsem, J., and Thienpont, D. (1971). Experimental cutaneous *Candida albicans* in guinea pigs. *Sabouraudia 9*: 17-20.

Vik, D. P., and Fearon, D. T. (1985). Neutrophils express a receptor for iC3b and C3d that is distinct from CR1, CR2 and CR3. *J. Immunol. 134*: 2571-2579.

Waldorf, A. R., and Diamond, R. D. (1985). Neutrophil chemotactic responses induced by fresh and swollen *Rhizopus oryzae* spores and *Aspergillus fumigatus* conidia. *Infect. Immun. 48*: 458-463.

Washburn, R. G., Hammer, C. H., and Bennett, J. E. (1986). Inhibition of complement by culture supernatants of *Aspergillus fumigatus*. *J. Infect. Dis. 00*: 944-951.

Wilton, J. M. A., and Lehner, T. (1980). Immunology of candidiasis. *Comprehen. Immunol. 8*: 525-559.

Winkelstein, J. A., Bocchini, J. A., and Schiffman, G. (1976). The role of capsular polysaccharide in the activation of the alternative pathway by the pneumococcus. *J. Immunol. 116*: 367-370.

Yamamura, M., and Valdimarsson, V. (1977). Participation of C3 in intracellular killing of *Candida albicans*. *Scand. J. Immunol. 6*: 591-596.

10
Pulmonary Defense Mechanisms Against Opportunistic Fungal Pathogens

ALAYN R. WALDORF
University of California, Berkeley, Berkeley, California

INTRODUCTION

The occurrence of opportunistic mycoses is rising rapidly. The increase in leukemia, lymphoma, organ transplantation, cytotoxic chemotherapy, acquired immunodeficiency syndrome (AIDS), or other immunosuppressed patients contributes to the increased importance of these infections. The use of more potent (and more effective) cytotoxic agents, wide use of glucocorticoids and broad spectrum antibiotics, and use of other supportive measures to sustain life during intense cytotoxic therapy are also important in increasing the occurrence of opportunistic mycoses. However, the nature and the degree of the immune effect which predisposes to a particular opportunistic fungal infection depends on the underlying disease and, often, the treatment given. The number of fungi with the potential for causing pulmonary infections is quite large (those organisms that grow at 37°C and with infectious particles small enough to reach the lung), but the number of fungi that have been shown to cause disease is actually quite limited. These organisms usually infect only patients predisposed by a distinct underlying disease or defense mechanism defect, and it is thus possible to utilize this to determine which aspect of the host's defense are of importance in controlling a particular organism.

Aspergillosis and mucormycosis encompass a broad range of disease states following the inhalation of the ubiquitous conidia or spores (Lehrer et al., 1980). There are several unique features of aspergillosis and mucormycosis, unlike the traditional pathogenic dimorphic fungi, that make their study particularly interesting. These organisms are not dimorphic, like the true pathogenic dimorphic fungi, as they grow both in the environment and within the host as hyphal forms. However, the host must contend with several forms of the organism to successfully eliminate them. A resting, nonmetabolically active spore or conidium enters the host, it swells and germinates (in the susceptible host) and invades tissue as a hyphal organism. Each form displays different antigenic and surface features and elicits different host responses. Finally, if germination of the spore or conidium occurs, the host must compete with a rapidly growing organism that is too large to be ingested by a single cell and so must be handled by extracellular defense mechanisms.

Candida pneumonia is a rare event. *Candida* are part of the human endogenous flora found on mucosal surfaces, in the stool, and sometimes on the skin. Small numbers are normally present, but the numbers increase with alterations in the normal flora by antibiotics or alteration in the immune competence of the individual. In candidiasis, *Candida* are dimorphic fungi with pseudohyphal and yeast forms occurring in invasive lesions, so that the host again must eliminate several very different forms of the organism. If *Candida* blastoconidia form pseudohyphae, the host must also compete with an organism that is often too large to be ingested by a single cell and so must be handled by extracellular defense mechanisms.

Cryptococcosis is an invasive disease caused by *Cryptococcus neoformans*. Unlike most other deep mycoses, isolated pulmonary disease accounts for at most about 10% of patients with cryptococcosis. However, occasional patients, probably because of impaired host defenses, have overwhelming diffuse pneumonitis. A primary pulmonary infection follows inhalation of the organism with subsequent localization. Unlike the other fungi mentioned above, *C. neoformans* grows in tissue exclusively in the yeast form.

Other organisms have also been reported that cause opportunistic pulmonary infections including *Pseudoallescheria boydii* and members of certain species of *Penicillium, Hansenula, Geotrichum*, and others (Winston et al., 1977; Blazer et al., 1984). However, because little is known about host responses in these infections, inclusion of these organisms awaits further research.

The bronchoalveolar macrophage is the resident mononuclear phagocyte of the lung and functions as the primary defense against inhaled particulate matter (Hocking and Golde, 1979). The alveolar macrophage is unique among mononuclear phagocytes in several ways. Their residence at an air-tissue interface results in direct exposure to inhaled microorganisms and environmental toxins, and the aerobic environment of the alveolus has led these cells to develop special metabolic adaptations (Hocking and Golde, 1979). However, the lung is not an isolated organ, and other defense mechanisms also play a role in host defenses against opportunistic pulmonary fungal infections. These include the polymorphonuclear cell (neutrophils), circulating monocyte, circulating antibodies, complement, and T lymphocytes. Since these latter cells and immune mechanisms

CLINICAL ASPECTS OF OPPORTUNISTIC PULMONARY FUNGAL INFECTIONS

Aspergillosis

The aspergilli are among the most ubiquitous saprophytic fungi in the environment. Phialoconidia (conidia) from these organisms are easily isolated from decaying vegetation, soil, and air worldwide (Rippon, 1982). The term aspergillosis encompasses a broad range of disease states whose etiologic agents are members of the genus *Aspergillus*. Although there are about 600 known species of *Aspergillus*, most cases of human disease are caused by *A. fumigatus, A. flavus,* and *A. niger*, although *A. oryzae, A. glaucus,* and *A. nidulans* have also been implicated (Rinaldi, 1983).

The term aspergillosis covers a broad spectrum of manifestations in human disease, which fall into four main categories: allergic aspergillosis, aspergilloma, invasive aspergillosis, and chronic necrotizing pulmonary aspergillosis (Greenberger, 1986). It is important to remember, however, that these represent a spectrum of disease states and that there is some overlap between groups. Moreover, a particular manifestation of the disease is related primarily to the immunological status, the environmental exposure, and the genetic makeup of the host, rather than to the organism that causes the infection.

Like other types of aspergillosis, invasive aspergillosis is presumably acquired by inhalation of spores into the respiratory tract. Thus, most cases of human disease originate in the lungs or sinuses. The term invasive aspergillosis is used generally to imply histopathologically demonstrated invasion into tissue. Aspergillosis in the lung arises by endobronchial proliferation of hyphae followed by invasion across bronchi into pulmonary arterioles and lung parenchyma, with subsequent ischemic necrosis (Orr et al., 1978). The propensity of *Aspergillus* hyphae to invade blood vessels can also result in disseminated lesions with thrombosis, infarction, and hemorrhage of the organ involved.

A. fumigatus and *A. flavus* are responsible for most invasive disease, and undoubtedly multiple factors, both host- and species-related, account for their greater involvement in both animal and human disease; i.e., concentration in the atmosphere, conidial size and shape, thermotolerance, and immune competence of the host. Invasive aspergillosis most frequently occurs in the severely immunocompromised patient. Cases of invasive aspergillosis have been reported in apparently immunocompetent patients, those only mildly to moderately immunocompromised, or in individuals with no known predisposing factors (D'Silva et al., 1982, Karam and Griffin, 1986). Risk factors predisposing patients to invasive aspergillosis include corticosteroid therapy (Young et al., 1970; Meyer et al., 1973; Gustafson et al., 1983), cytotoxic chemotherapy (Young et al., 1970; Meyer et al., 1973), transplantation, especially during immunosuppressive therapy (Burton et al., 1972; Kyriakideo et al., 1976; Gustafson, et al., 1983), and chronic granulomatous disease, a qualitative disorder of neutrophil function (Cohen et al., 1981). Neutropenia remains the single most important predisposing factor.

Mucormycosis

The agents of mucormycosis (rhinocerebral mucormycosis, phycomycosis, or zygomycosis) are classified in the order Mucorales, containing at least 12 pathogenic species (Lehrer et al., 1980), of which *Rhizopus oryzae* (*R. arrhizus*), *Rhizomucor pusillus*, and *Absidia corymbifera* are the most common. The Mucorales are among the most ubiquitous saprophytic fungi in the environment, and sporangiospores (spores) from these organisms are easily isolated from decaying vegetation, soil, and air worldwide. Mucormycosis is typically characterized by tissue invasion and a predilection of the organism for blood vessel invasion with hemorrhage, necrosis, and infarction. In the diverse clinical manifestations of mucormycosis, most can be categorized as rhinocerebral, pulmonary, widely disseminated or cutaneous, though other miscellaneous forms also occur rarely (Meyer et al., 1972). Like aspergillosis, the form of mucormycosis that occurs is often related to the predisposing factor(s) or underlying disease state of the host in addition to the route of inoculation.

Pulmonary infections are characterized by vascular thrombosis, infarction, and necrosis, again following inhalation and germination of the spores. The infection usually progresses rapidly, but chronic pulmonary lesions have been reported (Gale and Leitsch, 1972). Complications can include severe hemoptysis, erosion of the right mainstem bronchus with performation of its superior branch, bronchopleural fistulae, granulomatous mediastinitis, and embolic dissemination to the brain or gastrointestinal tract.

The epidemiology of mucormycosis, like that of other opportunistic infections, reflects decreased resistance of the compromised host. Rhinocerebral mucormycosis is most often seen in a setting of uncontrolled diabetes mellitus, especially ketoacidosis (Pollock et al., 1975; Pillsbury and Fisher, 1977; Hale, 1979). Although patients with leukemia or renal transplant recipients have developed rhinocerebral mucormycosis, some of them had coexisting diabetes or had been treated with steroids (Meyer et al., 1972; Morduchowicz et al., 1986). In most instances pulmonary mucormycosis is associated with leukemia, lymphoma, severe neutropenia, massive corticosteroid therapy, and renal transplant recipients (Stahel et al., 1982; Morduchowicz et al., 1986). Other predisposing factors include diabetes mellitus, renal failure, and burns. There also are rare reports of pulmonary mucormycosis occurring in an individual with no known predisposing factors (Meyer and Armstrong, 1973).

Pulmonary Candidiasis

Candidiasis is a term used to describe a primary or secondary infection caused by an organism belonging to the genus *Candida*. These are usually *Candida albicans, Candida tropicallis*, or rarely *Candida glabrata, Candida kruseii*, and others. *Candida* species are widespread in nature. *Candida* are dimorphic fungi and grow both in the yeast (blastoconidium) form and as hyphal (pseudohyphal) organisms. Candidiasis can be manifested as a superficial skin problem, a chronic infection of nails, a disease of the mucocutaneous tissues, or a frequently fatal systemic disease that may involve the lungs, heart, gastrointestinal tract, and other organs (Ahearn, 1978). Few organisms present such a diversity of clinical pictures as seen with the various candidiases. All of the tissue and organ systems are subject to invasion, and the pathology evoked is as variable as the clinical syndromes.

Systemic candidal infection is a relatively rare condition except as the terminal event of a debilitating illness or as the result of chronic insult and continued seeding of *Candida* into the body. Systemic candidiasis is a frequently fatal disease that may involve one or several of the following: the lungs, heart, gastrointestinal tract, urinary tract, or other organs. A combination of overgrowth of *Candida* in the gastrointestinal tract and cytotoxic injury to gastrointestinal mucosa permits invasion and spread to the lungs. Alternatively, overgrowth of *Candida* on the skin can provide the source of infection and an intravenous catheter the portal of entry. Abdominal surgery, especially with bowel anastomosis, is a risk factor for invasive candidiasis. Finally, infection of the lungs may occur due to aspirated organisms from the oropharynx with secondary seeding of distant tissue, although this is not the usual route of infection, and chest reoentgenogram usually show diffuse nodular lesions suggesting hematogenous spread of *Candida*

Pulmonary Cryptococcosis

C. neoformans has been widely identified in nature, particularly in pigeon droppings and excreta of other birds (Powell et al., 1972). The infectious particle, or propagule, has not as yet been identified. Cryptococci were thought to grow in nature as yeast and to infect man in that form. When cryptococcal yeasts are mixed with dry sterilized soil and allowed to dessicate, the size of the yeast cells progressively shrinks over time, and the organisms become small enough to be aspirated into small respiratory airways of mice (Smith et al., 1964). More recently, Kwon-Chung (1975, 1976) has identified two sexual forms of *C. neoformans* that can be mated to produce a basidiomycete, *Filobasidiella neoformans*.

The basidiospores have no capsule, yet within 24 h after intranasal inoculation into mice they revert to typical encapsulated cryptococci and are lethal to mice. However, since *F. neoformans* has not been found in nature, the pathogenic role of this sexual form of *C. neoformans* is uncertain, although the basidiospore is an attractive candidate for the infectious particle.

Cryptococcosis is an invasive disease caused by *C. neoformans*. Cryptococci are found throughout the world. Although aerosolization with inhalation of the organism is suspected as the route of infection, rarely is a positive exposure history confirmed. Pulmonary cryptococcosis was once thought to be a lethal rarity. However, the increased use of immunosuppressive agents for a variety of medical problems has increased the frequency of pulmonary cryptococcosis. Pulmonary disease may be the first sign leading to the detection of disseminated disease, or patients may present with meningitis and subsequently be found to have pulmonary disease. The organism has a predilection for the central nervous system, and meningitis is the most common clinical form of presentation of the disease.

Like many of the opportunistic pulmonary fungal infections, the broad spectrum of pulmonary cryptococcosis cannot be overemphasized. The pulmonary involvement may be subacute or chronic, it may be walled off with little clinical consequence, or it may result in multiorgan involvement. Patients with underlying lung diseases such as chronic bronchitis, chronic obstructive pulmonary disease, pulmonary fibrosis, asthma, and tuberculosis are often only colonized with *C. neoformans*. The patient with altered host defenses is usually at the other end of the spectrum with disseminated cryptococcosis.

Although no clearly definable immunologic deficit has been identified in over 50% of cryptococcosis patients, the immunocompromised patient has a marked increased susceptibility, and, more importantly, these patients have an increased risk of dissemination of cryptococcosis. Among the more common underlying diseases associated with cryptococcosis are reticuloendothelial malignancies, organ transplantation with immune suppression, AIDS, Cushing's syndrome, and treatment with glucocorticoids for a variety of indications.

PHAGOCYTIC CELLS IN PULMONARY DEFENSES

The function of the phagocytic system as a part of the innate defense system is difficult to separate from the phagocytosis that is mediated by acquired immune mechanisms. Since products of lymphocytes (i.e., lymphokines, antibodies) can influence any or all of the fixed and circulating phagocytes, an assessment of the innate ability of a host to phagocytize and kill an organism is difficult. The situation with the opportunistic fungi is particularly difficult since even "normal" humans and animals are commonly exposed to these organisms and their antigens. It is the rare individual that has had no previous contact with the ubiquitous aspergilli and mucorales. Similarly, exposure to *Candida* or *Cryptococcus* results from environmental contact from the commensal relationship of *Candida* or other yeasts expressing similar antigens, which colonize the gastrointestinal tract. The wide variation and disagreement in the results of studies that investigate innate host responses may be partially explained by variations in previous exposure of individuals to the organisms and, therefore, an unequal contribution to the immune status at the time of the experiment.

Aspergillus and Rhizopus

Alveolar Macrophage Defense Against *Aspergillus* Conidia

Experimental studies have been undertaken to develop animal models that simulate human aspergillosis. The importance of the alveolar macrophage in defense against *Aspergillus* was established early (Merkow et al., 1971). In animal models, germination of *Aspergillus* conidia plays an important role in the early pathogenesis of the disease which leads to lethal infection. The alveolar macrophage is the first phagocytic cell to encounter inhaled conidia and its primary role in defense appears to be prevention of conidial germination. In normal animals, following intranasal inoculation of *Aspergillus* conidia, alveolar macrophages form an efficient early defense system against resting conidia. Within 15 minutes of inoculation conidia are encircled by phagocytic membranes, and lysosomes degranulate within two hours (Merkow et al., 1971). There is rapid killing of *Aspergillus* conidia by alveolar macrophages both in vitro and in vivo so that there is no mortality in normal animals (Schaffner, et al., 1982, 1983; Waldorf et al., 1984a; Kurup, 1984).

The administration of various immunosuppressants to mice lowers their resistance to experimental aspergillosis, though the degree varies with each of the different treatments. In contrast to the normal animals, animals treated with cortisone are highly susceptible to *Aspergillus* infections. A single or multiple dose of cortisone produces 100% mortality within 10 to 18 days of infection with *A. fumigatus*. Silica, presumed to be toxic to macrophages, is also effective, resulting in an 80% mortality (Monga, 1983). Increased susceptibility of mice to *A. fumigatus* infection after treatment with gold sodium thiomalate, which is known to interfere with macrophages and monocytes, also has been shown (Williams et al., 1981). Additionally, adoptive protection against lethal *Aspergillus* infection is produced by transferring macrophages from immunized mice (Williams et al., 1981). Thus, macrophages seem to be clearly involved in natural immunity to *A. fumigatus*.

Cortisone significantly impairs the ability of macrophages to prevent germination of conidia in vitro and in vivo (Merkow et al., 1968; Merkow et al., 1971; Schaffner et al., 1982; Schaffner et al., 1983). Following intranasal inoculation of *Aspergillus* conidia in cortisone-treated mice, alveolar macrophages phagocytize conidia. However, unlike the normal macrophages, lysosomal granules of the cortisone-treated macrophages do not fuse with the phagosomes containing the conidia, and conidia begin to germinate within 2-4 h of inoculation (Merkow et al., 1971). Once hyphae develop, the fungus readily pierces the phagocyte in vitro and in vivo (Schaffner et al., 1982).

Cortisone appears to act on the mononuclear phagocyte directly rather than through T lymphocytes or other systems that modify macrophage function (Schaffner et al., 1982). It has been suggested that the indirect stabilization of alveolar macrophage membranes by corticosteroids prevents the interaction of lysosomes with phagocytized conidia, thus interfering with the intracellular destruction of conidia, allowing them to germinate (Merkow et al., 1968).

Other studies in in vitro systems indicate that direct damage by glucocorticoids to the antimicrobial activity of alveolar macrophages are critical to the control of *Aspergillus* (Schaffner, 1985). When alveolar macrophages are exposed to therapeutic ranges of glucocorticoids for 24-36 h, antifungal activity of the phagocytes was fully suppressed. The suppression of the antifungal activity appears to be mediated by high-affinity glucocorticoid receptors on the macrophages. Since most receptor-mediated glucocorticoid effects are related to a modulation of protein synthesis, such as the reduction of the synthesis of neutral proteases, the authors speculate that the effect of corticosteroids on the antifungal activity of macrophages similarly reflects the suppression of the synthesis of proteins of the antimicorbial armature of the phagocyte (Schaffner, 1985). This study suggests that *Aspergillus* conidia are not killed by macrophage oxidative mechanisms since glucocorticoids, at therapeutic levels, did not affect macrophage activation. Gamma-interferon also had no effect on the activity of corticosteroid-treated macrophages against *Aspergillus* conidia, as it seems to enhance oxidative killing only.

When *Aspergillus* conidia are incubated in broth they become metabolically active and swollen. In vitro killing of swollen *Aspergillus* conidia by rabbit alveolar macrophages occurs earlier and is of greater magnitude, compared to that of resting *Aspergillus* conidia (Levitz et al., 1986). These results suggest that before killing of the conidia can occur, the conidia must become metabolically active, perhaps increasing the permeability of the conidial wall and allowing better activity of the macrophage microbicidal products.

The mechanism(s) by which alveolar macrophages kill conidia is still being investigated. Alveolar macrophages have the capacity to kill *Aspergillus* conidia by both oxidative and nonoxidative mechanisms, as the conidia are killed at equal rates under aerobic and anaerobic conditions (Schaffner et al., 1983). Moreover, macrophages lacking the ability to generate reactive oxygen intermediates, because of prolonged culture, kill conidia as effectively as macrophages (in culture for only 2 days) with high oxidative capacity.

Myeloperoxidase, hydrogen peroxide, and a halide play an important role in the antimicrobial activity of neutrophils and monocytes. However, tissue macrophages, including alveolar macrophages, do not contain myeloperoxidase, and so these cells must use other mechanisms for killing spores and conidia. An antimicrobial system composed of ferrous ion, hydrogen peroxide, and iodide appears to have potent anticonidial activity and is one possible alternative mechanism for oxidative antifungal activity (Levitz and Diamond, 1984). The ferrous ion system, in cell free conditions, is comparable in activity to the myeloperoxidase system of neutrophils in its ability to kill *A. fumigatus* conidia and *Rhizopus* spores (Levitz and Diamond, 1984). When other halides replace the iodide there is no inhibition of *A. fumigatus* conidial germination. Ferrous ion appears to react with hydrogen peroxide to generate hydroxyl radical, which then react with iodide to form the toxic product(s) of the ferrous ion system. (Kelbanoff, 1982). Since alveolar macrophages kill *Aspergillus* conidia, but lack myeloperoxidase, a myeloperoxidase-independent fungicidal mechanism must be operative. The in vivo significance of the ferrous ion system, if any, is yet to be determined.

Alveolar Macrophage Defense Against *Rhizopus* Spores

Diabetes mellitus, especially with ketoacidosis, has been the most commonly recognized underlying disease associated with mucormycosis (Meyer and Armstrong, 1973). Animal models of mucormycosis in diabetic mice and rabbits have been used to determine the influence of diabetes in predisposition to mucormycosis, as well as host defense mechanisms against the mucorales in general (Reinhardt et al., 1970, 1981; Waldorf and Diamond, 1984; Waldorf et al., 1984b; Levitz et al., 1986). The models of mucormycosis provide an opportunity for evaluation of the relative importance of different cell populations in the pathogenesis of mucormycosis. Diabetic animals have rapidly fatal infection with *R. oryzae, Rhizomucor pusillus, R. arrhizus, R. microsporus, Rhizopodiformis* (Reinhardt et al., 1981), or *Absidia corymbifer* (Sheldon and Bauer, 1959). Intranasal or intrasinus inoculation of *R. oryzae* into streptozotocin-induced diabetic mice results in mucormycotic infections with pathology closely resembling pulmonary infections in humans: extensive tissue necrosis in the vicinity of hyphae and the penetration of hyphae into vascular structures (Merkow et al., 1968; Waldorf and Diamond, 1984). Unlike other models of mucormycosis (Bauer et al., 1955; Sheldon and Bauer, 1959; Sheldon and Bauer, 1960; Waldorf et al., 1982), these diabetes models lead to induction of specific enhanced susceptibility to infection by *Rhizopus*, resembling the predisposition to mucormycosis which occurs in uncontrolled human diabetic patients. Thus, in these diabetic animals susceptibility to aspergillosis does not occur.

Rhizopus spore germination occurs in tissues of dibetic or cortisone-treated mice following intranasal inoculation, but not in normal animals (Merkow et al.,

1968; Lundborg and Holma, 1972; Smith, 1976; Waldorf et al., 1982; Waldorf et al., 1982; Waldorf and Diamond, 1984). Normal alveolar macrophages are able to phagocytize *Rhizopus* spores in vitro and in vivo and form an efficient defense against spores by inhibiting spore germination (Merkow et al., 1968; Lundborg and Holma, 1972). In contrast, there is a significant decrease in inhibition of spore germination by diabetic bronchoalveolar macrophages both in vivo and in vitro (Waldorf et al., 1984b). The decrease is not attributable to defects in phagosome-lysosome fusion of diabetic alveolar macrophages.

Animal data suggest that, although prevented from germinating, *Rhizopus* spores are not easily killed within normal murine lungs as are *A. fumigatus* conidia (Waldorf et al., 1984a). *Rhizopus* spores remain viable in lung tissue of normal mice for as long as 10 days after intranasal inoculation, without germinating, whereas *A. fumigatus* conidia are removed within 2 days of inoculation. Normal alveolar macrophages participate in defense against *Rhizopus* spores by inhibiting germination in vivo and immediately after lavage. However, with subsequent in vitro incubation in media which favors germination, spores lavaged from normal diabetic and cortisone-treated mice germinate at comparable levels (Waldorf and Diamond, 1984; Waldorf et al., 1984a).

Data from an in vivo macrophage killing assay demonstrates the contributory effects of various conditions that predispose to aspergillosis and mucormycosis, including uncontrolled diabetes or cortisone treatment (Waldorf et al., 1984a). Alveolar macrophages from normal, diabetic, and cortisone-treated mice are unable to kill *Rhizopus* spores in vivo. These studies indicate that inhibition of germination in vivo need not imply sporicidal activity, and that inhibition of germination and killing of fungal spores appear to constitute separate mechanisms of host protection (Waldorf et al., 1984a).

The mechanism(s) by which alveolar macrophages kill *Rhizopus* spores is still being investigated. Hydrogen peroxide, in cell-free assays, significantly inhibits *Rhizopus* spore germination (Waldorf and Pelosi, 1989). The results suggest that hydrogen peroxide or its byproducts are an important oxidative mechanism by which alveolar macrophages mediate inhibition of spore germination. Since diabetic alveolar macrophages are unable to inhibit spore germination and produce significantly less hydrogen peroxide in response to opsonized *Rhizopus* spores, a decreased production of hydrogen peroxide in diabetic alveolar macrophages would constitute an important defect in host defense against mucormycosis (Waldorf and Pelosi, 1989). In addition to defects in oxidative metabolism in diabetic alveolar macrophages, histochemical studies of these cells also indicate that alveolar macrophages from diabetic animals had decreased acid phosphatase and lysozyme activity and contain fewer mature resident macrophages and either newly recruited or less mature cells than alveolar macrophages obtained from normal animals (Waldorf and Pelosi, 1989).

Circulating Monocyte Defense Against *Aspergillus* Conidia and *Rhizopus* Spores

In vitro studies on the capacity of macrophages to kill spores or inhibit germination have produced conflicting results. This may be due, in part, to differences in the anticonidial activity of circulating monocytes and macrophages from different anatomical sites and from different animals, as well as the use of different assays for measuring inhibition of germination or killing. For example, peritoneal macrophages from mice and rabbits cannot inhibit germination of *Aspergillus* conidia and are less active than alveolar cells which can inhibit germination (Schaffner et al., 1983; Williams et al., 1981). Moreover, blood monocytes kill *Aspergillus* conidia at rates similar to those of alveolar macrophages. Since peritoneal and alveolar macrophages are derived from blood monocytes (van Furth et al., 1980), it is possible that local factors present in the lung (but not in the peritoneum) promote development of strong anticonidial activity.

Like the alveolar macrophage, the mechanism(s) by which monocytes kill or inhibit germination of conidia and spores is not clear. The combination of myeloperoxidase, hydrogen peroxide, and a halide appears to play an important role in the anticonidial activity of human blood monocytes (Levitz and Diamond, 1984). Damage to *Aspergillus* conidia and *Rhizopus* spores by the meyloperoxidase system can be inhibited by azide and catalase, confirming the requirement for hydrogen peroxide.

Peripheral blood monocytes can also kill *Aspergillus* conidia by mechanisms other than oxidative systems. Monocytes from patients with chronic granulomatous disease, which are unable to generate comparable amounts of reactive oxygen intermediates, are as sensitive to the *Aspergillus* conidia killing inhibition effects of corticosteroids as were normal cells (Schaffner, 1985). These results suggest that the effects of corticosteroids on monocytes, like alveolar macrophages, reflects the suppression of synthesis of neutral proteases and does not affect monocyte activation.

Alveolar Macrophage and Circulating Monocyte Defense Against *Aspergillus* and *Rhizopus* Hyphae

Invasive aspergillosis and mucormycosis are characterized by tissue invasion by hyphae that are too large to be ingested completely by phagocytic cells. Therefore, any antifungal activity by alveolar macrophages must be carried out extracellularly. Normal alveolar macrophages can attach to and damage *Rhizopus* hyphae in the absence of complete ingestion even though the hyphae are too large to be ingested completely (Waldorf et al., 1984b). Normal human peripheral blood monocytes also can damage and apparently kill *Rhizopus* and *Aspergillus* hyphae by oxidative microbicidal systems (Diamond et al., 1978, 1982, 1983) Normal human monocytes attach to and damage *Rhizopus* hyphae, as evidenced by a reduction in radiolabeled uracil uptake by hyphae and by light microscopy (Diamond et al., 1982). The mean amount and range of damage to hyphae by human peripheral blood monocytes is consistent with that observed with normal alveolar macrophages (Diamond et al., 1978a, 1982). With all cell types, attachment of hyphae occurs in the absence of serum. Autologous serum does not affect results, regardless of whether the sera were fresh, heated at 56°C, or contained antibodies to *Rhizopus*.

Results of inhibitors of leukocyte function and normal monocytes support the importance of the myeloperoxidase-peroxide-halide system in the damage of *Aspergillus* and *Rhizopus* hyphae (Diamond et al., 1982). In vivo, alternate mechanisms to the myeloperoxidase system are likely to operate in mediating hyphal damage particularly because granule-associated myeloperoxidase is lost during differentiation of monocytes into macrophages. This suggests that oxidative microbicidal products independent of the myeloperoxidase system may be important (Lehrer, 1978). These oxidative products may include superoxide anion, hydrogen peroxide (Drath and Karnovsky, 1975; Babior, 1976; Lehrer, 1978), singlet oxygen, and hydroxyl radical (Drath and Karnovsky, 1975; Babior, 1978; Harrison et al., 1978; Held and Hurst, 1978). Oxidative microbicidal mechanisms seem more critical in monocyte-mediated damage to *Rhizopus* hyphae than to *Aspergillus* hyphae.

At least under some conditions, there exists the potential for activity of nonoxidative antihyphal mechanisms (Diamond et al., 1983). Potentially fungicidal cationic proteins may be present in monocytes. Such cationic proteins might have microbial effects by themselves. Even if present in suboptimal levels for fungicidal activity, cationic proteins can also interact synergistically with oxidative mechanisms (Diamond et al., 1983).

In contrast with *Aspergillus*, nonoxidative mechanisms, including granule-associated cationic proteins, do not appear to be involved in damage of *Rhizopus* hyphae by monocytes (Diamond et al., 1982). Separated monocyte lysosomal granules also do not affect *Rhizopus* hyphal viability unless a hydrogen peroxide source and halides are added to supplement granule-associated myeloperoxidase.

Polymorphonuclear Leukocyte Defense Against *Aspergillus* and *Rhizopus*

The cause of the significant increase in opportunistic fungal infections observed in the last several years is unknown but is believed to be the result of an increased susceptibility secondary to more aggressive cytotoxic therapy (Stahel et al., 1982). In addition to the association of invasive aspergillosis and pulmonary mucormycosis with neutropenia and chronic granulomatous disease, data from experimental animal models suggest an important role for the neutrophil in host defense against these organisms.

The roles of myelosuppression and neutropenia in host defense become evident only after *Aspergillus* conidia escape from the reticuloendothelial system and start mycelial growth (Schaffner et al., 1982). Similarly, if alveolar macrophages are altered by the induction of diabetes, they can no longer prevent *Rhizopus* spore germination, and the importance of the neutrophil in defense becomes apparent (Waldorf et al., 1984a, 1984b). Once *Aspergillus* conidia or *Rhizopus* spores are induced to swell prior to initiation of germination, an impressive ability of neutrophils to damage these organisms is evident.

In vitro data appears to confirm the evidence obtained from animal models. Human neutrophils can damage and kill hyphae of *A. fumigatus* and *R. oryzae* (Diamond and Clark, 1982; Schaffner et al., 1986). Of importance in damaging hyphae by human neutrophils are the oxidative microbicidal mechanisms (Cohen et al., 1981).

In addition to oxidative mechanisms, neutrophils and other phagocytes can kill fungi by nonoxidative mechanisms, by the action of microbicidal substances released from their granules during phagocytosis (Selsted et al., 1985; Spitznagel and Shafer, 1985; Levitz et al., 1986). Antihyphal mechanisms of intact leukocytes do not seem to be identical for *Aspergillus* and *Rhizopus* hyphae. Moreover, interactions of several potential oxidative and nonoxidative antihyphal mechanisms, in vivo, may define the host's ability to limit fungal infections. These differences may be related to the availability of alternative, nonoxidative mechanisms active against *R. oryzae* but not *A. fumigatus* in host neutrophils. Differing susceptibility of *Aspergillus* and *Rhizopus* hyphae to granule-associated cationic proteins, acting alone or together with other constituents (such as elastase or lysozyme), might explain such a process. When concentrations of oxidative or nonoxidative substances are limiting or suboptimal, interactions of mechanisms may have important implications in activity of defense mechanisms against opportunistic mycoses in the intact host.

Candida

Alveolar Macrophage Defense Against *Candida* Blastoconidia

Candida infections in the lung are uncommon except as terminal complications (Nugent and Onofrio, 1983). The difference in the relative frequency of candidiasis in the gastrointestinal tract and lower respiratory tract in immunocompromised patients may reflect differences in innate tissue resistances. Animal models of candidiasis do suggest that host defenses of the lung are important in controlling pulmonary candidiasis. Normal mice survive intratracheal inoculation of *C. albicans* in doses which kill 85% of mice infected intravenously. Mouse alveolar macrophages ingest *Candida* blastoconidia in vivo, and rabbit alveolar macrophages are able to inhibit 71-93% of the macromolecular synthesis of *Candida* (Peterson and Calderone, 1978a) and phagocytize and reduce viable numbers in vitro (Zeligs et al., 1984).

In a recent report, the interactions of mature cells of the monocyte-macrophage system with *Candida* blastoconidia were analyzed. Mature macrophage populations behave heterogenously in their ability to inhibit the growth of *Candida* (Decker et al., 1986). At the highest effector:target ratios tested (5:1), Kupffer cells, splenic macrophages, and alveolar macrophages had comparable candidastatic activity and inhibited *Candida* significantly above that exhibited by peritoneal macrophages. The heterogeneity in the ability of these macrophage populations to inhibit growth of *Candida* was not overcome when the macrophages were stimulated with lymphokine, nor was it observed in the case of other targets (tumor cells).

When the various mature macrophage populations were evaluated for their ability to produce reactive oxygen intermediates (as measured by chemiluminescence), the macrophage population with the highest activity was from the peritoneal cavity. This included both in vitro activated and nonactivated macrophages (Decker et al., 1986). Alveolar macrophages emitted peak chemiluminescence of approximately one-fourth that of peritoneal macrophages. Thus, the ability of certain macrophage populations to inhibit *Candida* appears not to be linked to their production of reactive oxygen intermediates following phagocytosis of the yeast.

The mechanism(s) of candidacidal activity in different macrophage populations appears to vary. *Candida* inhibition assays using the various macrophage populations and performed in the presence of superoxide dismutase or catalase indicated that only peritoneal macrophages are sensitive to the activity of these enzymes (Decker et al., 1986). The authors conclude that oxygen-dependent mechanisms (superoxide) do contribute to the inhibition of *Candida* yeast cell growth, but only by macrophage populations which produce large quantities of reactive oxygen species, i.e., resident peritoneal macrophages. Moreover, these same macrophage populations are less efficient in *Candida* inhibition than other macrophage populations. In contrast, the highly efficient *Candida* fungistasis exerted by alveolar macrophages appears to be performed by oxygen-independent mechanisms, requires up to 8 hours, and can be increased by macrophage activation (Decker et al., 1986). Differences in the phagocytosis of *Candida* were not involved, since all macrophages, activated or not, behaved similarly.

The ability of immature and mature cells of the monocyte-macrophage system to inhibit *Candida* blastoconidia growth has also been studied (Zeligs et al., 1984; Decker et al., 1986). Bone marrow-derived macrophage precursors displayed strong candidastatic potential. However, peritoneal-derived macrophage precursors had reduced abilities to inhibit *Candida* growth and were similar to the results observed with mature peritoneal macrophages (Decker et al., 1986). In contrast to the results observed with peritoneal macrophages and their precursors, alveolar macrophages from neonatal mice exerted poor fungistatic activity, but mature alveolar macrophages had strong candidacidal activity. These results suggest that candidastatic activity is a feature of a definite maturation stage and is not found in all cells of the monocyte-macrophage lineage. Thus, it appears that alveolar macrophages (as well as liver and splenic macrophages) have an increased candidastatic activity over other macrophage populations, which may contribute to the infrequency of pulmonary candidiasis.

The relative infrequency of candidiasis in the lower respiratory tract in immunocompromised patients may also reflect differences in the recruitment of systemic host defense factors to the lung. It has been suggested that neutrophils play a significant role in clearance of *Candida* from the lung, as there is at least a 6-h delay before the number of viable intrapulmonary *Candida* begin to fall, during which time neutrophils appear in the bronchoalveolar spaces (Nugent and Onofrio, 1983). Since neither cortisone nor nitrogen mustard treatment (inhibitors of alveolar macrophages) significantly reduced the neutrophil influx into the bronchoalveolar spaces and yet delayed the clearance of *C. albicans* blastoconidia from the lungs, it would appear that alveolar macrophages play a decisive role in *Candida* clearance (Nugent and Onofrio, 1983). Moreover, if nonoxidative mechanisms are involved in alveolar macrophage defense and require up to 8 h, as suggested by Decker et al. (1986), the arrival of neutrophils at this time may be coincidental.

The mechanism by which alveolar macrophages kill *Candida* blastoconidia are still being investigated. Lysosome-rich materials isolated from mouse macrophages are capable of causing a reduction in both *C. albicans*-specific amino acid uptake and the number of viable blastoconidia cells, suggesting that alveolar macrophages can damage *Candida* blastoconidia by nonoxidative mechanisms (Peterson and Calderone, 1978a, 1978b). Rabbit neutrophils and alveolar macrophages have been shown to contain cysteine-rich, low molecular weight peptides with candidacidal activity (Selsted et al., 1983, 1985). *C. albicans* blastoconidia are susceptible to several of these purified cationic proteins. These cationic proteins may represent an important oxygen-independent antifungal defense mechanism of alveolar macrophages.

Candidacidal activity of resident or activated macrophages is inhibited effectively by superoxide dismutase, which removes superoxide anion, and less effectively but significantly by catalase, which removes hydrogen peroxide (Lehrer, 1975; Sasada and Johnson, 1980). Inhibition is also achieved with benzoate, which is believed to scavenge hydroxyl radical, and with azide, which is thought to remove singlet oxygen. Mannitol, a second scavenger of hydroxyl radical, inhibited candidacidal activity only slightly.

An antimicrobial system composed of ferrous ion, hydrogen peroxide, and iodide appears to have potent antiyeast activity and is one possible mechanism of oxidate antifungal activity (Levitz and Diamond, 1984). The ferrous ion system (as discussed above), in cell free conditions, is comparable in activity to the myeloperoxidase system of neutrophils in its ability to kill *C. albicans* blastoconidia (Levitz and Diamond, 1984). Since alveolar macrophages damage *C. albicans* blastoconidia, but lack myeloperoxidase, a myeloperoxidase-independent fungicidal mechanism must be operative. The in vivo significance of the ferrous ion system, if any, is yet to be determined.

Circulating Monocytes Defense Against *Candida* Blastoconidia

Like the alveolar macrophage, the mechanism(s) by which monocytes kill or inhibit germination of blastoconidia is not clear. The combination of myeloperoxidase, hydrogen peroxide and a halide appears to play an important role in the anticonidial activity of human blood monocytes (Levitz and Diamond, 1984). The role of oxidative metabolism in the killing of *Candida* blastoconidia by monocytes was confirmed using scavengers of toxic oxygen metabolites.

Alveolar Macrophage and Circulating Monocyte Defense Against *Candida* Hyphae

Like invasive aspergillosis and mucormycosis, candidiasis is characterized by tissue invasion by hyphae that are too large to be ingested completely by phagocytic cells. Therefore, any antifungal activity of macrophages and monocytes must be carried out extracellularly. Normal human blood monocytes attach to and damage *C. albicans* hyphae, inducing morphological alterations and a reduction in radiolabeled uracil uptake by hyphae (Diamond and Haudenschild, 1981). Autologous serum does not affect results, regardless of whether the sera were fresh, heated at 56°C, or contained antibodies to *Candida*. Functional studies with inhibitors of leukocyte function and with normal monocytes indicate the importance of the myeloperoxidase-peroxide-halide system in the damaging of *C. albicans* hyphae. Singlet oxygen may also be involved in damage to *Candida* while hydroxyl radicals, by themselves, are not fungicidal. It has been speculated that a loss in oxidative antifungal mechanisms seen in myeloperoxidate-dificient monocyles may be compensated for by an increase in other mechanisms including certain enzymes, such as alkaline phosphatase (Larrocha et al., 1982).

In vivo, alternative mechanisms to the myeloperoxidase system, which may operate to induce hyphal damage, are particularly relevant because granule-associated myeloperoxidase is lost during differentiation of monocytes into macrophages. This suggests that oxidative microbicidal products independent of the myeloperoxidase system may be important (Lehrer, 1978). These oxidative products may include superoxide anion, hydrogen peroxide (Drath and

Karnovsky, 1975; Babior, 1978; Lehrer, 1978), singlet oxygen, and hydroxyl radicals (Drath and Karnovsky, 1975; Babior, 1978; Harrison et al., 1978; Held and Hurst, 1978). Singlet oxygen may be important in damaging *Candida* hyphae.

Although normal human monocytes can damage hyphal forms of *C. albicans*, monocyte-mediated damage can be inhibited by hydrocortisone, both in the presence and in the absence of serum (Diamond, 1983). Inhibition of damage occurs at concentrations of hydrocortisone readily achievable in vivo by pharmacological doses. Thus, it is possible that pharmacologic doses of corticosteroids might produce serum hormone concentrations that could interfere with monocyte activity against *Candida* hyphae. These findings may, in part, explain the increased incidence of candidiasis in cortisone-treated patients. However, the mechanism(s) of inhibition of fungal damage by steroids is not known.

Polymorphonuclear Leukocyte Defense Against *Candida*

The association of invasive pulmonary candidiasis with data from experimental neutropenia animal models suggests an important role for the neutrophil in host defense against *Candida*. Normal mice exposed to the inhalation of viable *C. albicans* blastoconidia are resistant to lethal infection (Chow et al., 1982; Nugent and Onofrio, 1983; Nugent and Onofrio, 1984). In contrast, an increased susceptibility to pulmonary candidiasis develops when animals are suppressed by subcutaneous inoculations of cortisone, lymphoid leukemia, x-irradiation, or cytotoxic drugs.

In vitro data appears to confirm the evidence obtained from animal models (Lehrer and Cline, 1969a; Lehrer, 1970; Taschdijian et al., 1971; Ishikawa et al., 1972a, 1972b; Lehrer et al., 1975; Leijh et al., 1977; Richardson and Smith, 1983; Bjerknes, 1984; Cockayne and Odds, 1984). However, studies aimed at quantitative determinations of the candidacidal activities of phagocytes have shown considerable variations. The neutrophil appears to be the most efficient of the leukocyte populations tested in their candidacidal activity, although blastoconidia were less effectively killed than *Candida* hyphae (Lehrer and Cline, 1969a; Cockayne and Odds, 1984). The increased susceptibility to killing of hyphae remains unexplained but may be affected by the different composition and structure of the cell walls in blastoconidia and *Candida* hyphaes. The low percentage of *Candida* blastoconidia killed following ingestion by neutrophils may be related to the finding of unsealed phagolysosomes, which would be expected to have diminished anticandidal activity (Cech and Lehrer, 1984). Incomplete sealing of the phagolysosomes occurs and may allow leakage of reactants, such as myeloperoxidase and hydrogen peroxide to the cells exterior.

The differences in the magnitude of the respiratory burst to varying fungal stimuli, as well as the differences in susceptibility to neutrophil oxidants, most likely relate to differences in the structure and chemistry of the fungal conidial wall. Unopsonized blastoconidia of *C. albicans* are phagocytized by murine neutrophils, in vitro, and following phagocytosis there is an increased oxidative metabolism but suppression of hydrogen peroxide production (Danley and Hilger, 1981).

Studies on the mechanism of killing blastoconidia by phagocytes have shown that several candidacidal substances in neutrophils occur. Hydrogen peroxide with a reactive halide and myeloperoxidase is lethal for *Candida* blastoconidia (Lehrer and Cline, 1969b; Lehrer, 1972; Klebanoff, 1980). Moreover, patients with CGD and, perhaps, myeloperoxidase deficiencies are more susceptible to systemic candidiasis. Neutrophils from myeloperoxidase-deficient patients phagocytize *Candida*, but their ability to kill the ingested *Candida* is reduced (Lehrer, 1970; Lehrer and Cline, 1969b). Neutrophils from CGD patients are unable to damage *Candida* blastoconidia.

Alternative fungicidal mechanisms exist in neutrophils. Human myeloperoxidase-deficient neutrophils kill *C. parapsilosis* and *C. pseudotropicallis* by a mechanism completely independent of myeloperoxidase, iodination, and hydrogen peroxide (Lehrer, 1972). It has been shown that *C. albicans* blastoconidia are susceptible to several cationic peptides purified from rabbit granulocytes (Lehrer et al., 1985; Selsted et al., 1985). It is possible that these cationic peptides are an important component of the antifungal defense mechanism of neutrophils.

Following germ tube formation of blastoconidia, *C. albicans* characteristically form pseudohyphae which are too large to be ingested by phagocytic cells. Human neutrophils can damage and kill *C. albicans* hyphae (Diamond and Krzesicki, 1978; Diamond et al., 1978b, 1980; Diamond and Clark, 1982; Cockayne and Odds, 1984). Even in the absence of serum, neutrophils attached to and spread over the surfaces of hyphae which were too large to be completely phagocytized and induced morphologic changes and metabolic impairments in the hyphae as shown by alterations in the uptake of radioisotopes. Moreover, where *C. albicans* germ tubes and hyphae are completely phagocytized, significantly more hyphal forms are killed than blastoconidia (Cockayne and Odds, 1984). Of central importance in damaging hyphae by neutrophils are the oxidative microbicidal mechanisms. The major proteins of the nonoxidative antifungal mechanisms of neutrophils appear to correspond to granule-associated cationic proteins (Diamond and Clark, 1982).

Cryptococcus

Alveolar Macrophage Defense Against *Cryptococcus*

Experimental murine models have demonstrated that there is a strong host tissue response to *C. neoformans* and suggest that tissue macrophages play an important role in host defense. A granulomatous response against *C. neoformans* infection occurs in organs that are rich in macrophages and lymphocytes such as the lung (Shadomy and Lurie, 1971; Miyaji and Nishimura, 1981). In contrast, in organs poor in macrophages, such as the brain, the histopathological feature is a gelatinous cyst containing numerous yeast cells and a scarcity of host immune cells. Ultrastructural studies on the interaction between macrophages and cells of encapsulated cryptococci show long, cytoplasmic processes from the macrophages penetrating the cryptococcal capsule (Papadimitriou et al., 1978). The macrophage causes degradative changes in the phagocytized yeast, including disruption of the yeast cytoplasm, thinning of the capsule, and occasional fragmentation of the cell wall. The yeast cell wall is the most resistant structure to the microbicidal activity of macrophages.

The ability of alveolar macrophages to kill *C. neoformans* remains in doubt. Guinea pig alveolar macrophages have been shown to ingest nonencapsulated yeasts but are unable to kill the phagocytized cells during several hours of exposure (Bulmer and Tacker, 1975). Alveolar macrophages from sensitized mice are also unable to kill engulfed cryptococci (Karaoui et al., 1977a, 1977b). In contrast to the results obtained with alveolar macrophages, peritoneal macrophages inhibit the intracellular replication of phagocytized yeasts (Mitchel and Friedman, 1972). The apparent inability of alveolar macrophages to kill engulfed *C. neoformans* compared to other tissue macrophages and neutrophils may be attributed to the lack of peroxidase-mediated killing mechanisms or to a reduced macrophage oxidative metabolism.

The capsule of *C. neoformans* appears to play an important role in allowing the organism to evade host defenses. The capsular polysaccharide of *C. neoformans* inhibits macrophage attachment to and phagocytosis of the yeast as nonencapsulated yeasts are more readily phagocytized than encapsulated cryptococci (Bulmer and Sans, 1968). This inhibition of macrophage phagocytosis is related to the presence of the capsular polysaccharide at the yeast surface rather than for some indirect effect of the polysaccharide on serum components necessaary for phagocytosis (Kozel et al., 1984). Since opsonizing IgG plays an important role in the attachment phase of *C. neoformans* to macrophages (Kozel and McGraw, 1979; Kozel and Folette, 1981), interference with phagocytosis by capsular material is a major problem in effective host response to cryptococcosis. In in vitro studies a lymphokine which enhances macrophage complement receptor function appears, in part, to overcome the antiphagocytic effect of the capsular material (Griffin, 1981). However, its role in vivo is unclear.

Macrophages play a critical role in the induction of an immune response. The participation of the macrophage in altered immune states induced by crytococcosis has been investigated. The interaction between lymphocytes and macrophages may be critical in primary immunity against cryptococcosis, since phagocytosis alone by alveolar macrophages is not sufficient to prevent infection by the yeast (Karaouri et al., 1977b). In vitro studies have suggested that a T cell lymphokine can suppress macrophage phagocytosis (Morgan et al., 1983). The magnitude of the suppression is not great, suggesting that it is attributable to selective suppression of a particular subset of macrophages.

Polymorphonuclear Leukocyte and Circulating Monocyte Defense Against *Cryptococcus*

A critical element of defense against *C. neoformans* appears to be the neturophil. After experimental respiratory infection of animals, cryptococci can be identified in the lung alveoli and alveolar septae. Organisms are rapdily destroyed by an influx of neutrophils, with surviving yeasts generating large capsules. Human neutrophils and monocytes require serum for maximal phagocytosis of poorly and well-encapsulated *C. neoformans* yeast cells in vitro. Cryptococci that are phagocytized by human neutrophils are killed by mechanisms dependent on the generation of hydrogen peroxide (Diamond et al., 1972) and by nonoxidative mechanisms (Miller and Kohn, 1983). However, as capsule size increases, ingestion and killing diminishes.

SUMMARY

Though of critical importance, nonimmune host defense mechanisms against aspergillosis and mucormycosis are not completely understood. Prevention of these infections presumably requires control of either spore germination and/or hyphal growth by the host. The data suggest that the host provides an important barrier to infection by control of spore or conidia germination, the critical step involving conversion of the fungus to its tissue-invasive form. The mechanisms of host defense against *A. fumigatus* are not strictly dependent on inhibition of conidia germination. Rather, pulmonary defense against *Aspergillus* appears to depend to a greater degree on early killing of fungal conidia by alveolar macrophages. In contrast, prevention of mucormycosis appears to require inhibition of fungal spore germination by the bronchoalveolar macrophage, thereby preventing conversion of the fungus to its hyphal form, although resident bronchoalveolar macrophages are unable to kill *R. oryzae* spores. Thus, host pulmonary defenses to *Rhizopus* and *Aspergillus* vary, even in normal animals.

The tissue-invasive hyphal forms of the fungi which cause appergillosis and mucormycosis are too large to be ingested by phagocytic cells. Although macrophages and monocytes can damage hyphae, the bulk of this role appears to fall upon the neutrophil. However, antihyphal mechanisms of neutrophils may not necessarily be identical for all types of hyphae. Moreover, interactions of several potential oxidative and nonoxidative antihyphal mechanisms may define the host's ability to limit fungal infections. In individuals where concentrations of oxidative or nonoxidative substances are limiting or suboptimal, interactions of mechanisms may be required for antihyphal activity, and studies of these interactions are important to gain better knowledge of the defense mechanisms against opportunistic mycoses in the intact host. In summary, at least two distinct lines of defense against *Aspergillus* and *Rhizopus* are known in the normal host. Alveolar macrophages kill *Aspergillus* conidia and prevent germination of *Rhizopus* spores. The neutrophil damages the hyphal form of *Aspergillus* and Rhizopus. Thus, neutrophils and monocytes or macrophages act as distinct components of host defenses against aspergillosis and mucormycosis, and they may also operate by different mechanisms.

Both invasion of tissue and environmental contact with aspergilli and the Mucorales induce the production of antibodies to these organisms. However, any definite role of antibodies or B cell-dependent immunity in effective host defense against initial invasion is still in doubt. It appears that the natural resistance of the host to *Aspergillus* depends primarily upon nonimmune factors. However, if these factors are breached, it remains possible that humoral immunity may play a role in host defense by limiting hyphal growth, perhaps in conjunction with phagocytic cells. The importance of cell-mediated immunity in resistance to aspergillosis and mucormycosis is not yet understood. It appears that the initial susceptibility to lethal infection is substantially T cell independent. The protection observed in older animals and following immunization to *Aspergillus*, and to repeat challenge with organisms, does however, indicate a role for the T cell, presumably in conjunction with macrophages.

Adequate defense against pulmonary candidiasis presumably requires control of blastoconidia replication, prevention of conidia germination, and/or killing of *Candida* hyphae. Early host defense appears to be dependent upon killing of blastoconidia by the alveolar macrophage. Although monocytes have been shown to damage and kill hyphae, the bulk of this role again falls upon the neutrophil. Interactions of several potential oxidative and nonoxidative antihyphal mechanisms may define the ability of the host to limit *Candida* infections. In individuals where concentrations of oxidative or nonoxidative substances are limiting or suboptimal, interactions of mechanisms may be required for antihyphal activity. Although they may be of critical importance, immune host defense mechanisms against *Candida* are not completely understood. Studies indicate that cell-mediated immunity plays a role in resistance to candidiasis. Nevertheless, it is obvious that there is more than one single mechanism for resistance to these opportunistic infections, reflected by the spectrum of underlying diseases and conditions that predispose to candidiasis.

Cryptococcus neoformans is inhaled into the lungs and causes pulmonary infection that may or may not be controlled in the lungs. A number of years ago, host-*Cryptococcus* interactions were thought to be clearly understood. Cell-mediated immunity was considered the only important defense against cryptococcosis. However, numerous studies have indicated this theory is too simplistic. Initial nonspecific mechanisms, depending on complement, neutrophils, and perhaps alveolar macrophages, may restrict dissemination from the lung. Although the alveolar macrophage appears to play a minimal role in innate defense against cryptococcal yeasts, it does play a role as an accessory cell in immunity. In contrast to the alveolar macrophage, neutrophils can ingest and kill *C. neoformans* apparently by oxidative and nonoxidative mechanisms. In addition, there is a complex pattern of interactions between host and yeast, which involves multiple arms of immune defense, and multiple virulence factors of the organism. Host-fungal interactions are clearly complex, and much remains to be done in elucidating this relationship.

REFERENCES

Ahearn, D. G. (1978). Medically important yeasts. *Ann. Rev. Microbiol. 32*: 59-68.

Babior, B. M. (1978). Oxygen-dependent microbial killing by phagocytes. *N. Engl. J. Med. 298*: 659-666.

Bauer, H., Flanagan, J. F., and Sheldon, W. H. (1955). Experimental cerebral mucormycosis in alloxan-diabetic rabbits. *Yale J. Biol. Med. 28*: 29-36.

Bjerknes, R. (1984). Flowcytometric assay for combined measurement of phagocytosis and intracellular killing of *Candida albicans*. *J. Immunol. Meth. 72*: 229-241.

Blazer, B. R., Hurd, D. D., and Snover, D. C. (1984). Invasive *Fusarium* infections in bone marrow transplant recipients. *Am. J. Med. 77*: 645-651.

Boxer, L. A. Ghazally, I., Allen, J. M., and Baehner, R. L. (1976). Oxidative metabolic responses of rabbit pulmonary alveolar macrophages. *Blood 53*: 486-491.

Bulmer, G. S., and Sans, M. D. (1968). *Cryptococcus neoformans*. III Inhibition of phagocytosis. *J. Bacteriol. 95*: 5-8.

Bulmer, G. S., and Tacker, J. R. (1975). Phagocytosis of *Cryptococcus neoformans* by alveolar macrophages. *Infect. Immun. 11*: 73-79.

Burton, J. R., Zachery, J. B., Bessin, R., Rathbun, H. K., Greenough, W. B., III, Sterioff, S., Wright, J. R., Slavin, R. E., and Williams, G. M. (1972). Aspergillosis in four renal transplant recipients: diagnosis and effective treatment with amphotercin B. *Ann. Intern. Med. 77*: 383-387.

Cech, P., and Lehrer, R. I. (1984). Heterogeneity of human neutrophil phagolysosomes: function and consequences for candidacidal activity. *Blood 64*: 147-151.

Chow, H. S., Sarpel, S. C., and Epstein, R. B. (1982). Experimental candidiasis in neutropenic dogs: tissue burden of infection and granulocyte transfusion effects. *Blood 59*:328-333.

Cockayne, A., and Odds, F. C. (1984). Interactions of *Candida albicans* yeast cells, germ tubes, and hyphae with human polymorphonuclear leukocytes in vitro. *J. Gen. Microbiol. 130*: 465-471.

Cohen, M. S., Isturiz, R. E., Malech, H. L., Root, R. K., Wilfert, C. M., Gutman, L., and Buckley, R. H. (1981). Fungal infection in chronic granulomatous disease: the importance of the phagocyte in defense against fungi. *Am. J. Med. 71*: 59-69.

Danley, D. L., and Hilger, A. E. (1981). Stimulation of oxidative metabolism in murine polymorphonuclear leukocytes by unopsonized fungal cells: evidence for a mannose-specific mechanism. *J. Immunol. 127*: 551-556.

Decker, T., Lohmann-Matthes, M-L., and Baccarini, M. (1986). Heterogenous activity of immature and mature cells of the murine monocyte-macrophage lineage derived from different anatomical districts against yeast-phase *Candida albicans*. *Infect. Immun. 54*: 477-486.

Diamond, R. D. (1983). Inhibition of monocyte-mediated damage to fungal hyphae by steroid hormones. *J. Infect. Dis. 147*: 160.

Diamond, R. D., and Clark, R. A. (1982). Damage to *Aspergillus fumigatus* and *Rhizopus oryzae* hyphae by oxidative and nonoxidative microbicidal products of human neutrophils in vitro. *Infect. Immun. 38*: 487-495.

Diamond, R. D., Clark, R. A., and Haudenschild, C. C. (1980). Damage to *Candida albicans* hyphae and pseudohyphae by the meyloperoxidase system and oxidative products of neutrophil metabolism in vitro. *J. Clin. Invest. 66*: 908-917.

Diamond, R. D., and Haudenschild, C. C. (1981). Monocyte mediated serum-independent damage to hyphal and pseudohyphal forms of *Candida albicans* in vitro. *J. Clin. Invest. 67*: 173-182.

Diamond, R. D., Haudenschild, C. C., and Erickson III, N. F. (1982). Monocyte-mediated damage to *Rhizopus oryzae* in vitro. *Infect. Immun. 38*: 292-297.

Diamond, R. D., Huber, E., and Haudenschild, C. C. (1983). Mechanisms of destruction of *Aspergillus fumigatus* hyphae mediated by human monocytes. *J. Infect. Dis. 147*: 474-483.

Diamond, D.R., and Krzesicki, R. (1978). Mechanisms of attachment of neutrophils to *Candida albicans* pseudohyphae in the absence of serum, and of subsequent damage to pseudohyphae by microbicidal processes of neutrophils in vitro. *J. Clin. Invest. 61*: 360-369.

Diamond, R. D., Krzesicki, R., Epstein, B., and Jao, W. (1978a). Damage to hyphal forms of fungi by human leukocytes in vitr. *Am. J. Pathol. 91*: 313-323.

Diamond, R. D., Krzesicki, R., and Jao, W. (1978b). Damage to pseudohyphal forms of *Candida albicans* by neutrophils in the absence of serum in vitro. *J. Clin. Invest. 61*: 349-359.

Diamond, R.D., Root, R. K., and Bennett, J. E, (1972). Factors influencing killing of *Cryptococcus neoformans* by human leukocytes in vitro. *J. Infect. Dis. 125*: 267-376.

Drath, D. B., and Karnovsky, M. L. (1975). Superoxide production by phagocytic leukocytes. *J. Exp. Med. 141*: 257-265.

D'Silva, H., Burke, J. F., and Cho, S. Y. (1984). Disseminated aspergillosis in a presumably immunocompetent host. *JAMA 248*: 1495-1499.

Gale, A. M., and Leitsch, W. P. (1972). Solitary pulmonary nodule due to phycomycosis

Greenberger, P.A. (1986). Aspergillosis–Clinical aspects. *Zbl. Bakt. Hyg. [A]. 261*: 487-495.

Griffin, F. M., Jr. (1981). Roles of macrophage Fc and C3b receptors in phagocytosis of immunologically coated *Cryptococcus neoformans*. *Proc. Natl. Acad. Sci. 78*: 3835-3857.

Gustafson, T. L., Schaffner, W., Lavely, G. B., Stratton, C. W., Johnson, H. K., and Hutcheson, R. H., Jr. (1983). Invasive aspergillosis in renal transplant recipients: correlation with corticosteroid therapy. *J. Infect. Dis. 148*: 230-238.

Hale, L.M. (1979). Orbital phycomycosis, *South. Med. J. 63*: 886-890.

Harrison, J. E., Watson, B. D., and Schultz, J. (1978). Myeloperoxidase and singlet oxygen: a reappraisal. *FEBS Lett. 92*: 327-339.

Held, A. M., and Hurst, J. K. (1978). Ambiguity associated with use of singlet oxygen trapping agents in myeloperoxidase catalyzed oxidations. *Biochem. Biophys. Res. Commun. 81*: 878-890.

Hocking, W. G., and Golde, D. W. (1979). The pulmonary-alveolar macrophage. *N. Engl. J. Med. 301*: 580-587.

Ishikawa, T., Yu, M. C., and Arbesman, C. E. (1972a). Electron microscopic demonstration of phagocytosis of *Candida albicans* by human eosinophilic leukocytes. *J. Allergy Clin. Immunol. 50*: 183-195.

Ishikawa, R. I., Dalton, A. C., and Arbesman, C. E. (1972b). Phagocytosis of *Candida albicans* by eosinophilic leukocytes. *J. Allergy Clin. Immunol. 49*: 311-322.

Karam, G. H., and Griffin, F. M. (1986). Invasive pulmonary aspergillosis in nonimmunocompromised, nonneutropenic hosts. *Rev. Infect. Dis. 8*: 357-363.

Karaoui, R. M., Hall, N. K., and Larsh, H. W. (1977a). Role of macrophages in immunity and pathogenesis of experimental cryptococcosis induced by the airborne route. Part I: Pathogenesis and acquired immunity of *Cryptococcus neoformans. Mykosen 20*: 380-388.

Karaoui, R. M., Hall, N. K., and Larsh, H. W. (1977b). Role of macrophages in immunity and pathogenesis of experimental cryptococcosis induced by the airborne route. Part II: Phagocytosis and intracellular fate of *Cryptococcus neoformans. Mykosen 20*: 409-422.

Klebanoff, S. J. (1980). Oxygen metabolism and the toxic properties of phagocytes. *Ann. Intern. Med. 93*: 480-492.

Kelbanoff, S. J. (1982). The iron-H_2O_2-iodide cytotoxic system. *J. Exp. Med. 156*: 1262-1269.

Kozel, T. R., and Folette, J. F. (1981). Opsonization of encapsulated *Cryptococcus neoformans* specific anticapsular antibody. *Infect. Immun. 31*: 978-984.

Kozel, T. R., Highison, B., and Stratton, C. J. (1984). Localization on encapsulated *Cryptococcus neoformans* of serum components opsonic for phagocytosis by macrophage and neutrophils. *Infect. Immun. 43*: 574-579.

Kozel, T. R., and McGraw, T. G. (1979). Opsonization of *Cryptococcus neoformans* by human immunoglobulin G: Role of IgG in phagocytosis by macrophages. *Infect. Immun. 25*: 255-261.

Kurup, V. P. (1984). Interaction of *A. fumigatus* spores with pulmonary alveolar macrophages of rabbits. *Immunobiology 166*: 53-61.

Kwon-Chung, K. J. (1975). *Filobasidiella* the perfect state of *Cryptococcus neoformans*. *Mycologia 67*: 1197-1200.

Kwon-Chung, K. J. (1976). A new species of *Filobasidiella* the sexual state of *Cryptococcus neoformans* B and C serotypes. *Mycologia 68*: 942-946.

Kyriakideo, G. K., Zinneman, H. H., Hall, W. H., Arora, V. R., DeWolf, W. C., and Miller, J. (1976). Immunologic monitoring and aspergillosis in renal transplant patients. *Am. J. Surg. 131*: 246-254.

Larrocha, C., Fernandez De Castro, M., Fontan, G., Viloria, A., Fernandez-Chacon, J., and Jimenez, C. (1984). Hereditary myeloperoxidase deficiency: study of 12 cases. *Scan. J. Haematol. 29*: 389-397.

Lehrer, R. I. (1970). Measurement of candidacidal activity of specific leukocyte types in mixed cell populations. I. Normal, myeloperoxidase deficient, and chronic granulomatous disease neutrophils. *Infect. Immun. 2*: 41-47.

Lehrer, R. I. (1972). Functional aspects of a second mechansim of candidacidal activity of human neutrophils. *J. Clin. Invest. 41*: 2566-2572.

Lehrer, R. I. (1975). The fungicidal mechanisms of human monocytes. I. Evidence for myeloperoxidase-linked and myeloperoxidase-independent candidiacidal mechanisms. *J. Clin. Invest. 55*: 338-346.

Lehrer, R. I. (1978). Metabolism and microbicidal function. *Ann. Intern. Med. 88*: 79-91.

Lehrer, R. I., and Cline, M. J. (1969a). Interaction of *Candida albicans* with human leukocytes and serum. *J. Bacteriol. 98*: 998-1011.

Lehrer, R. I., and Cline, M. J. (1969b). Leukocyte myeloperoxidase deficiency and disseminated candidiasis: the role of myeloperoxidase in resistance to *Candida* infection. *J. Clin. Invest. 48*: 1478-1488.

Lehrer, R. I., Howard, D. H., Sypherd, P. S., Edwards, J. E., Segal, J. P., and Winston, D. J. (1980). Mucormycosis. *Ann. Intern. Med. 93*: 93-108.

Lehrer, R. I., Ladra, K. M., and Hake, R. B. (1975). Nonoxidative fungicidal mechanisms of mammalian granulocytes: demonstration of components with candidacidal activity in human, rabbit, and guinea pig leukocytes. *Infect. Immun. 11*: 1226-1234.

Lehrer, R. I., Szklarek, D., Ganz, T., and Selsted, M. E. (1985). Correlation of binding of rabbit granulocyte peptides to *Candida albicans* with candidacidal activity. *Infect. Immun. 49*: 207-211.

Leijh, P. C. J., Van den Barselarr, M. T., and van Furth, R. (1977). Kinetics of phagocytosis and intracellular killing of *Candida albicans* by human granulocytes and monocytes. *Infect. Immun. 17*: 313-320.

Levitz, S. M., and Diamond, R. D. (1984). Killing of *Aspergillus fumigatus spores and Candida albicans* yeast phase by the iron-hydrogen peroxide-iodide cytotoxic system: comparison with the myeloperoxidase-hydrogen peroxide-halide system. *Infect. Immun. 43*: 1100-1002.

Levitz, S. M., Selsted, M. E., Ganz, T., Lehrer, R. I., and Diamond, R. D. (1986). In vitro killing of spores and hyphae of *Aspergillus fumigatus* and *Rhizopus oryzae* by rabbit neutrophil cationic peptides and bronchoalveolar macrophages. *J. Infect. Dis. 154*: 483-489.

Lundborg, M., and Holma, B. (1972). In vitro phagocytosis of fungal spores by rabbit lung macrophages. *Sabouraudia 10*: 152-156.

Merkow, L. P., Epstein, S. M. Sidransky, H., Verney, E., and Pando, M. (1971). The pathogenesis of experimental pulmonary aspergillosis. An ultrastructural study of alveolar macrophages after phagocytosis of *Aspergillus flavus* spores in vivo. *Am. J. Pathol. 62*: 57-74.

Merkow, L. L., Prado, M., Epstein, S. M., Verney, E., and Sidransky, H. (1968). Lysosomal stability during phagocytosis of *Aspergillus flavus* spores by alveolar macrophages of cortisone-treated mice. *Science 160*: 79-81.

Meyer, R. D., and Armstrong, D. (1973). Mucormycosis-changing status. *CRC Crit. Rev. Clin. Lab. Sci. 4*: 421-451.

Meyer, R. D., Rosen, P., and Armstrong, D. (1972). Phycomycosis complicating leukemia and lymphoma. *Ann. Intern. Med. 77*: 872-879.

Meyer, R. D., Young, L. S., Armstrong, D., and Yu, B. (1973). Aspergillosis complicating neoplastic disease. *Am. J. Med. 54*: 6-9.

Miller, G. P. G., and Kohn, S. (1983). Antibody dependent leukocyte killing of *Cryptococcus neoformans*. *J. Immunol. 131*: 1455-1459.

Mitchel, T. G., and Friedman, L. (1972). In vitro phagocytosis and intracellular fate of variously encapsulated strains of *Cryptococcus neoformans*. *Infect. Immun. 5*: 491-498.

Miyaji, M., and Nishimura, K. (1981). Studies on organ specificity in experimental murine cryptococcosis. *Mycopathologia 76*: 145-154.

Monga, D. P. (1983). Studies on experimental aspergillosis in immunodeficient mice. *Zbl. Bakt. Hyg. I Abt. Orig. A 254*: 552-562.

Morduchowicz, G., Shmueli, D., Shapira, Z., Cohen, S. L., Yussim, A., Block, C. S., Rosenfeld, J. R., and Pitlik, S. D. (1986). Rhinocerebral mucormycosis in renal transplant recipients: report of three cases and review of the literature. *Rev. Infect. Dis. 8*: 441-446.

Morgan, M. A., Blackstock, R. A., Bulmer, G. S., and Hall, N. K. (1983). Modification of macrophage phagocytosis in murine cryptococcosis. *Infect. Immun. 40*: 493-500.

Nugent, K. M., and Onofrio, J. M. (1983). Pulmonary tissue resistance to *Candi-*

da albicans in normal and in immunosuppressed mice. *Am. Rev. Respir. Dis. 128*: 909-914.

Nugent, K. M., and Onofrio, J. M. (1984). Pulmonary clearance of *Candida albicans* in neutropenic mice. *J. Infect. Dis. 149*: 27-35.

Orr, D. P., Myerowitz, R. L., and Dubois, P. J. (1978). Patho-radiologic correlation of invasive pulmonary aspergillosis in the compromised host. *Cancer 41*: 2028-2032.

Papadimitriou, J. M., Robertson, T. A., Kletter, Y., Aronson, M., and Walters, M.N-I. (1978). An ultrastructural examination of the interaction between macrophages and *Cryptococcus neoformans. J. Pathol. 124*: 103-109.

Peterston, E., and Calderone, R. (1978a). Growth inhibition of *Candida albicans* by rabbit alveolar macrophages. *Infect. Immun. 15*: 910-920.

Peterson, E., and Calderone, R. (1978b). Inhibition of specific amino acid uptake in *Candida albicans* by lysosomal extracts from rabbit alveolar macrophages. *Infect. Immun. 21*: 506-517.

Pillsbury, H. C., and Fisher, N. D. (1977). Rhinocerebral mucormycosis. *Arch. Otolaryngol. 103*: 600-607.

Pollock, R. A., Pratt, R. C., Shulman, J. A., and Turner, J. S. (1975). Nasal mucormycosis: early detection and treatment without radical surgery or amphotercin B. *South. Med. J. 68*: 1279-1285.

Powell, K. E., Dahl, B. A., Weeks, R. J., and Tosh, F. E. (1972). Airborne *Cryptococcus neoformans*: Particles from pigeon excreta compatible with alveolar deposition. *J. Infect. Dis. 125*: 412-415.

Reinhardt, D. J., Daplan, W., and Ajello, L. (1970). Experimental cerebral zygomycosis in alloxan-diabetic rabbits. 1. Relationship of temperature tolerance of selected zygomycetes to pathogenicity. *Infect. Immun. 2*: 404-413.

Reinhardt, D. J., Licata, I., Kaplan, W., Ajello, L., Chandler, F. W., and Ellis, J. J. (1981). Experimental cerebral zygomycosis in alloxan-diabetic rabbits: variation in virulence among zygomycetes. *Sabouraudia 19*: 245-255.

Richardson, M. D., and Smith, H. (1983). Ultrastructural features of phagocytosis and intracellular killing of *Candida albicans* by mouse polymorphonuclear phagocyte monolayers. *Mycopathologia 83*: 97-102.

Rinaldi, M. G. (1983). Invasive aspergillosis. *Rev. Infect. Dis. 5*: 1061-1077.

Rippon, J. W. (1982). Aspergillosis. In *Medical Mycology*. W. B. Saunders Co., Philadelphia.

Sasada, M., and Johnson, R. B., Jr. (1980). Macrophage microbicidal activity. Correlation between phagocytosis-associated oxidative metabolism and the killing of *Candida* by macrophages. *J. Exp. Med. 152*: 85-98.

Schaffner, A. (1985). Therapeutic concentrations of glucocorticoids suppress the antimicrobial activity of human macrophages without impairing their responsiveness to gamma interferon. *J. Clin. Invest. 76*: 1755-1764.

Schaffner, A., Davis, C. E., Schaffner, T., Market, M., Douglas, H., and Braude, A. I. (1986). In vitro susceptibility of fungi to killing by neutrophil granulocytes discriminates between primary pathogeneicity and opportunism. *J. Clin. Invest. 78*: 511-524.

Schaffner, A., Douglas, H., and Braude, A. I. (1982). Selective protection against conidia by mononuclear and against mycelia by polymorphonuclear phagocytes in resistance to *Aspergillus*. Observations of these two lines of defense in vivo and in vitro with human and mouse phagocyte. *J. Clin. Invest. 69*: 617-631.

Schaffner, A., Douglas, H., Braude, A. I., and Davis, C. E. (1983). Killing of *Aspergillus* spores depends on the anatomical source of the macrophage. *Infect. Immun. 42*: 1109-1115.

Selsted, M. E., Brown, D. M., DeLange, R. J., and Lehrer, R. I. (1983). Primary structures of MCP-1 and MCP-2, natural peptide antibiotics of rabbit lung macrophages. *J. Biol. Chem. 258*: 14485-14489.

Selsted, M. F., Szklarek, D., Ganz, T., and Lehrer, R. I. (1985). Activity of rabbit leukocyte peptides against *Candida albicans*. *Infect. Immun. 49*: 202-206.

Shadomy, H. J., and Lurie, H. I. (1971). Histopathological observations in experimental cryptococcosis caused by a hypha-producing strain of *Cryptococcus neoformans* (Coward strain) in mice. *Sabouraudia 9*: 6-9.

Sheldon, W. H., and Bauer, H. (1959). The development of the acute inflammatory response to experimental cutaneous mucormycosis in normal and diabetic rabbits. *J. Exp. Med. 110*: 845-852.

Sheldon, W. H., and Bauer, W. (1960). Tissue mast cells and acute inflammation in experimental cutaneous mucormycosis of normal 48/80-treated and diabetic rats. *J. Exp. Med. 112*: 1069-1093.

Smith, J. M. (1976). In vivo development of spores of *Absidia ramosa*. *Sabouraudia 14*: 11-15.

Smith, C. D., Ritter, R., Larsh, H. W., and Furculow, M. L. (1964). Infections of white mice with airborne *Cryptococcus neoformans*. *J. Bacteriol. 87*: 1363-1368.

Spitznagel, J. K., and Shafer, W. M. (1985). Neutrophil killing of bacteria by oxygen-independent mechanisms: a historical summary. *Rev. Infect. Dis. 7*: 398-403.

Stahel, R. A., Vogt, P., Schuler, G., Ruttner, J. R., Frick, P., and Oelz, O. (1982). Systemic fungal infections in hematological malignancies; a growing problem. *J. Infect. 5*: 269-275.

Taschdjian, C., Toni, E. F., Jsu, K. C., Seelig, M. S., Cuestra, M. B., and Koznin, P. J. (1971). Immunofluorescence studies of *Candida* in human reticuloendothelial phagocytes: implications for immunogenesis and pathogenesis of systemic candidiasis. *Am. J. Clin. Pathol. 56*: 50-59.

Waldimarsson, H., Higgs, J., Wells, R., Yamamura, M., Hobbs, J., and Holt, P. (1973). Immune abnormalities associated with chronic mucocutaneous candidiasis. *Cell. Immunol. 6*: 348-357.

van Furth, R., Diesselhoff-denDulk, M. M. C., Raeburn, J. A., vanZweith, T. L., Crofton, R. W., and Blusse van Oud Alblas, A. (1980). Characteristics, origin and kinetics of human and murine mononuclear phagocytes. In *Mononuclear Phagocytes*. Edited by R. vanFurth. Nijhoff, The Hague.

Waldorf, A. R., and Diamond, R. D. (1984). Cerebral mucormycosis in diabetic mice after intrasinus challenge. *Infect. Immun. 44*: 194-195.

Waldorf, A. R., Halde, C., and Vedros, N. A. (1982). Murine model of pulmonary mucormycosis in cortisone-treated mice. *Sabouraudia 20*: 217-224.

Waldorf, A. R., Halde, C., and Vedros, N. A. (1983). Passive immunization in murine mucormycosis. *Mycopathologia 83*: 149-155.

Waldorf, A. R., Levitz, S. M., and Diamond, R. D. (1984a). In vivo bronchoalveolar macrophage defense against *Rhizopus oryzae* and *Aspergillus fumigatus*. *J. Infect. Dis. 150*: 752-760.

Waldorf, A.R., and Pelosi, C.A. (1989). Defects of bronchoalveolar macrophages and deficient defenses against fungi in diabetic mice. (in press).

Waldorf, A. R., Ruderman, N., and Diamond, R. D. (1984b). Specific suceptibility to mucormycosis in murine diabetes and bronchoalveolar macrophage defense against *Rhizopus*. *J. Clin. Invest. 74*: 150-160.

Williams, D. M., Weiner, M. H., and Drutz, D. J. (1981). Immunologic studies of disseminated infection with *A. fumigatus* in the nude mouse. *J. Infect. Dis. 143*: 726-739.

Winston, D. J., Jordan, M. C., and Rhodes, J. (1977). *Alescheria boydii* infections in the immunosuppressed host. *Am. J. Med. 63*: 830-835.

Young, R. C., Bennett, J. E., Vogel, C. L., Carbone, P. P., and DeVita, V. T. (1970). Aspergillosis: The spectrum of the disease in 98 patients. *Medicine 49*: 147-155.

Zeligs, B. J., Nerurkar, L. S., and Bellanti, J. A. (1984). Chemotactic and candidacidal responses of rabbit alveolar macrophages during development and the modulating roles of surfactant in these responses. *Infect. Immun. 44*: 379-385.

11
The Role of Neutrophils and Macrophages in Host Resistance to Systemic Fungal Infections

ELMER BRUMMER
Institute for Medical Research, San Jose and Stanford University School of Medicine, Palo Alto, California

INTRODUCTION

Blastomycosis, coccidioidomycosis, histoplasmosis, and paracoccidioidomycosis are true systemic mycoses caused by thermally dimorphic fungi *Blastomyces dermatitidis, Coccidioides immitis, Histoplasma capsulatum*, and *Paracoccidioides brasiliensis*, respectively. In nature these fungi live in the saprophytic phase growing in soil as mycelia and under certain conditions forming conidia or arthroconidia. Usually infection is established by inhalation of mycelial fragments and/or conidia, which at body temperature transform into the parasitic form, i.e., budding yeasts, or in the case of *C. immitis* endosporulating spherules.

Initially pulmonary host defenses, especially the alveolar macrophages, have to deal with conidia, arthroconidia, and mycelial fragments. Several hours later an inflammatory response recruits polymorphonuclear neutrophils (PMN) and monocytes. However, during this time conversion to the more resistant parasitic form takes place. In early primary infections the battle is waged between the parasitic forms of the fungi and nonspecific cellular host defenses. Within weeks specific immune responses, both humoral and cellular, can be measured. Whether the infection is resolved or progresses to chronic pulmonary and disseminated disease depends upon numerous factors. Hosts that have resolved these fungal infections present a picture quite different from the one described above upon reinfection. In general, they resist inocula which would cause serious infections in the naive host.

The role of PMN and macrophages in nonimmune resistance against the infective form of pathogenic thermally dimorphic fungi continues to be defined.

Our understanding of how the host defenses resist the parasitic forms of these fungi is clearly more advanced. For example, the crucial role of lymphokines in the activation of macrophage fungicidal activity has now been established. These and other recent developments in our knowledge of how PMN and macrophages function in resistance to primary exposure and infection are the subject of this chapter.

BLASTOMYCOSIS

PMN and Macrophages in Human Blastomycosis

The involvement of PMN, tissue Mϕ, and giant cells was described in histological studies by Baker (1942) in chronic pulmonary and disseminated blastomycosis. Although some lesions in systemic cases resembled lesions seen in cases of tuberculosis, other lesions were interpreted as being primarily pyogenic with abscesses containing mainly PMN along with giant cells. Granuloma formation as reported by Baker (1942) indicated that monocytes and tissue macrophages can play an important role in containing this infection. Experimental evidence from in vitro studies of monocyte and monocyte-derived macrophage interaction with *B. dermatitidis* supports this view. For example, replication of the yeast form of *B. dermatitidis* was dramatically inhibited (85%) by monocyte-derived macrophages in 3-day cocultures (Brummer and Stevens, 1982). In this high macrophage to yeast ratio, macrophages migrated and formed large clumps around *B. dermatitidis*. Monocytes behaved in a similar manner, but were less effective in inhibiting replication, e.g., 45% in 3-day cocultures. In contrast to another report (Drutz and Frey, 1985), a reduction in colony forming units (CFU) below inoculum size was not observed. Similar results have been reported by Bradsher et al. (1987) using a system where clumping of macrophages was prevented and phagocytosis promoted. Intracellular replication of *B. dermatitidis* was inhibited more by monocyte-derived macrophages from patients who had recovered from blastomycosis than by those from normal healthy donors. Furthermore, alveolar macrophages from both cured patients and healthy skin test negative donors restricted the replication of *B. dermatitidis* in vitro (Bradsher et al., 1985). Although alveolar macrophages treated with supernatants from antigen-stimulated sensitized lymphocytes showed enhanced inhibition of replication, killing was not measurable.

Despite their prominence in lesions, PMN do not appear to be particularly effective in eradicating *B. dermatitidis*. Support for this view has accumulated from results of in vitro experiments where PMN were challenged with yeast-form *B. dermatitidis*. Peripheral blood PMN, which killed *Candida albicans* blastoconidia failed to kill isolates of the larger yeast *B. dermatitidis* as determined by the reduction of inoculum CFU (Brummer and Stevens, 1982). Shaffner et al. (1986) reported similar results using PMN and yeast-form *B. dermatitidis* in a limiting dilution assay. Conversely, there are two reports that PMN have modest, but significant ability to kill (20%) yeast-form *B. dermatitidis* (Sixbey et al., 1979; Drutz and Frey, 1985). These studies used an assay where suspensions of PMN and *B. dermatitidis* were mixed in effector to target cell ratios of 1:1 or 10:1. Sixbey et al. (1979) used a pour plate method with incubation at room temperature (mycelial growth) to assess reduction of inoculum CFU, whereas Drutz and Frey (1985) used a more subjective method to determine viability, namely the methylene blue dye exclusion technique. In any event, PMN were not very effective against the yeast form of *B. dermatitidis*. Whether inflammatory PMN are more effective and play a protective role, as suggested by studies using murine inflammatory PMN (Brummer et al., 1986), has not been determined at this time.

Resistance to Initial Infection

The role of PMN and macrophages in initial resistance to natural infection with mycelial fragments or conidia of *B. dermatitidis* has received little attention. Early studies by Denton and DiSalvo (1968) demonstrated that conidia laden soil was infective for mice. A model of pulmonary infection using conidia has been described only recently (Williams and Moser, 1987). Thirty-seven percent of an inoculum 10^4 conidia instilled intratracheally were recovered from the lungs as CFU shortly after infection. Conversion of conidia to the yeast form was not documented by histology, however, lethal pulmonary and disseminated blastomycosis developed with a 100% mortality rate by 210 days. A 10-fold higher inoculum achieved a 100% mortality rate by 68 days. These findings suggest that alveolar macrophages plus subsequent inflammatory PMN and monocytes are not very effective in killing conidia or preventing their conversion to the yeast form. No data have been published on the killing of conidia in vitro by murine alveolar or peritoneal macrophages, monocytes, or PMN.

On the other hand, Drutz and Frey (1985) have reported that human PMN readily phagocytose (90.5 ± 51.5%) and killed (52.2 ± 5.7%) conidia from 4 different isolates of *B. dermatitidis*. Human monocytes and monocyte-derived macrophages were also reported to kill conidia, 34.1 ± 3.2% and 89.0 ± 5.2%, respectively. Susceptibility of conidia from these isolates to killing by PMN may reflect their low virulence in mice because 5×10^4 conidia given intranasally did not produce serious disease (Drutz et al., 1985), whereas 10^4 conidia of FW isolate produced 100% fatal blastomycosis after intratracheal inoculation (Williams and Moser, 1987). While the susceptibility of conidia to killing by PMN remains controversial, mycelial elements of *B. dermititidis* appear to be readily killed by human PMN as demonstrated by Schaffner et al. (1986): yeast were allowed to form mycelia by incubation at room temperature and then reacted with PMN at 37°C for 2 h, resulting in 40-60% sterilization of the cultures. Parallel cultures of yeast-form *B. dermatitidis* could not be sterilized by PMN. It would be interesting to know if conidia are phagocytosed and killed by alveolar macrophages. To our knowledge such investigations are yet to be done.

Taken together, conidia from virulent isolates of *B. dermatitidis* appear to evade murine pulmonary host defenses long enough to convert to the parasitic phase (Williams and Moser, 1987), and this resistance yeast-form cause serious disease (Schaffner et al., 1986).

Murine Blastomycosis Studies

Murine models of blastomycosis have proven to be very useful in defining the roles of PMN and macrophages in resistance to *B. dermatitidis*. Inflammatory PMN elicited intraperitoneally with antigen in immunized mice were (Brummer et al., 1984) efficient in killing yeast-form *B. dermatitidis* (Morozumi et al., 1982). It is postulated that such PMN play a major role in resistance of the immunized host against reinfection. Even inflammatory PMN elicited by certain nonspecific irritants, e.g., caseinate but not thioglycollate, were very effective in killing *B. dermatitidis* (93 ± 3%) compared to peripheral blood PMN (38 ± 4%) (Brummer et al., 1986). The killing of *B. dermatitidis* by PMN involve an enhanced oxidative burst as measured by luminol-dependent chemiluminescence (Brummer et al., 1985a). Products of the hydrogen peroxide-myeloperoxidase-halide system were shown to kill *B. dermatitidis* in vitro (Sugar et al., 1983) and were implicated in PMN killing of *B. dermatitidis* (Brummer et al., 1985a). On the other hand, the role of PMN in resistance of normal mice to primary infectin with *B. dermatitidis* is less clear. After pulmonary infection of mice with yeast-form *B. dermatitidis*, an influx of PMN was documented by Sugar and Fields (1985) using lung lavage and histology. Fatal blastomycosis occurred despite an increase in inflammatory PMN in the lungs from 1 to 9 days after infection and PMN constituting 80% of the lavaged cells. Apparently modest fungicidal activity of normal murine inflammatory PMN cannot handle a pulmonary infection caused by a few hundred yeast-form *B. dermatitidis*.

Resting murine alveolar and peritoneal macrophages are not able to play much of a defensive role against yeast-form *B. dermatitidis*. Neither is able to reduce inoculum CFU in vitro; however, resident peritoneal macrophages have been shown to inhibit replication (Brummer et al., 1980; Brummer et al., 1981; McDaniels and Cozad, 1983). On the other hand, peritoneal macrophages when activated in vivo by treatment of mice with *Mycobacterium bovis* (BCG) of Concanavalin A (Con A) intraperitoneally are able to kill *B. dermatitidis* in vitro (Brummer et al., 1983). Peritoneal or alveolar macrophages can be activated in vitro by recombinant gamma-interferon or supernatants from Con A-stimulated lymphocyte cultures (Brummer et al., 1985; Brummer and Stevens, 1987).

These findings clearly indicate that activated macrophages play a crucial role in resistance to infection with *B. dermatitidis*. Taken together, these findings show that resistance to *B. dermatitidis* develops in conjunction with sensitization of T lymphocytes (Brummer et al., 1982), followed by lymphokine production and activation of macrophages.

Summary

In summary, PMN, alveolar macrophages, monocytes, and tissue macrophages play a containment role in the early stages of blastomycosis, which buys time for the development of immunological responses. Once cell-mediated immune responses come into play, more effective fungicidal mechanisms develop such as activation of macrophages (Brummer et al., 1985a) and granuloma formation (Miyaji and Nishimura, 1983).

COCCIDIOIDOMYCOSIS

Cellular Resistance in Naive Host

In experimental coccidioidomycosis early cellular events after pulmonary infection with arthorconidia of *C. immitis* have been described by Kong et al. (1964) and Kong and Levine (1967). In naive mice pulmonary defenses were ineffective against as few as 7 arthroconidia, and 230 CFU of *C. immitis* were recovered after 5 days. Both spherules and arthroconidia were found in macrophages within the first 2 days after infection. Maximal inflammatory responses, which consisted of polymorphonuclear neutrophils (PMN), lymphocytes, and macrophages, occurred 8 days after infection. In vitro studies have supported in vivo observations about ineffective pulmonary defenses in that alveolar macrophages (Beaman and Holmberg, 1980a) and PMN (Beaman and Holmberg, 1980b) lacked significant fungicidal activity for arthroconidia in vitro. Limited fungicidal activity of human PMN against arthroconidia has been reported by Drutz and Huppert (1983). Killing ranged from 3.1 ± 2.8 to $29.6 \pm 6.0\%$ and was influenced by the hyphal outer wall layer (HOWL) (Frey and Drutz, 1986). Considering the ineptitude of alveolar macrophages and PMN in pulmonary defenses against the infective form of *C. immitis* cited above, it is remarkable that only 60% of coccidiocidal infections in humans are asymptomatic and that the 40% symptomatic infections range from mild influenzalike complaints to pneumonia with only a small percentage (5%) developing serious disease (Pappagianis, 1980). This paradox has been considered by Galgiani et al. (1984). They postulated that inflammatory PMN in the initial site of infection exert an inhibitory effect on chitin synthesis and retard growth of *C. immitis* thus buying time for development of cellular immune responses.

Development of a delayed-type hypersensitivity state to *C. immitis* antigens by natural subclinical exposure to this pathogen is responsible for a protective cellular immune response against reinfection. For example, in the presence of autologous lymphocytes, monocytes from skin test positive individuals were able to kill 63 ± 7% of arthroconidia, whereas monocytes or lymphocytes alone were ineffective. Monocytes and lymphocytes from skin test negative individuals failed to kill arthroconidia (Beaman and Pappagianis, 1985). Presumably, the ability of sensitized lymphocytes to respond to *C. immitis* antigens by proliferation and lymphokine production activates monocytes for killing of arthroconidia (Cox et al., 1977). Killing of arthroconidia by activated monocytes was shown to be associated with enhanced phagosome-lysosome fusion (Beaman and Papagianis, 1985). It is likely that human alveolar macrophages can be activated in the same manner that monocytes are activated to kill arthroconidia. Such pulmonary cellular immune responses to arthroconidia in the sensitized host have been demonstrated in animal models, which will be described in the next section.

Studies of Macrophages and PMN in Murine Models

Development of resistance in mice to *C. immitis* by immunization with killed vaccines was first reported by Levine et al. (1960), and with viable *C. immitis* by Pappagianis et al. (1961). This model was subsequently used by Kong et al. (1964), who observed that following single and especially repeated immunization vaccinated mice showed pulmonary resistance against arthroconidia in terms of multiplication of *C. immitis* in the lungs, e.g., 160 CFU at day 0, 63,000 CFU at day 13, and ultimate survival. The importance of T lymphocytes in the acquired resistance *C. immitis* was demonstrated by Beaman et al. (1977). Later, the role of sensitized lymphocytes in macrophage-induced killing of *C. immitis* in vitro was delineated by Beaman et al. (1981). Only lymphocytes from immunized mice were able to collaborate with macrophages in the killing of *C. immitis* arthroconidia or endospores. The interaction between antigen-sensitized lymphocytes and macrophages, which resulted in activation of macrophages, was shown to involve lymphokines (Beaman et al., 1983). More recently it was reported that recombinant gamma-interferon (γ IFN) was able to activate murine alveolar or peritoneal macrophages to kill either arthroconidia or endospores of *C. immitis* (Beaman, 1987). Taken together, it is abundantly clear that activation of host macrophages is central to resistance against either reinfection by arthroconidia or control of endosporulating spherules and elimination of endospores.

The role of PMN in acquired resistance to fungal infections has received scant attention until recently (Brummer et al., 1985c). PMN elicited intraperitoneally in immunized mice by homologous antigen were shown to have a greater fungicidal activity than peripheral blood PMN (Brummer et al., 1986). Such activated

PMN were unable to kill arthroconidia in vitro; however, they were able to significantly kill endospores (77.5 ± 7%) of *C. immitis* (Brummer et al., 1985c). these findings suggest a definite role for the inflammatory reaction and activated PMN in resistance to the parasitic form of *C. immitis*, particularly in the immunized host.

PARACOCCIDIOIDOMYCOSIS

Introduction

Paracoccidioidomycosis, caused by the dimorphic fungus *Paracoccidioides brasiliensis*, is the major systemic mycosis in Latin America. Histological studies of tissue lesions in human paracoccidioidomycosis of the chronic type provided clear evidence for cell-mediated immunity and macrophage involvement in the disease process (Fava-Netto, 1965). This was manifested by granuloma formation, necrosis, and fibrosis. Although immunological studies have been made and the importance of cellular versus humoral immune responses documented in patients (Fava-Netto, 1965; Mendes et al., 1971; Mussatti et al., 1976; Mok and Greer, 1977; Restropo et al., 1978), there is no detailed information available about the fungicidal activity of the alveolar macrophages, the infiltrating monocytes, or the inflammatory PMN.

Human PMN vs. *P. brasiliensis*

In contrast to macrophages, the fungicidal activity of human peripheral blood PMN has been studied. Some reports indicate that PMN may be able to kill yeast-form *P. brasiliensis* in vitro (Restropo and Velez, 1975; Goihman-Yahr et al., 1981), whereas others found that PMN failed to sterilize cocultures containing as few as 12 CFU of *P. brasiliensis* (Schaffner et al., 1986). Using a different assay system, Brummer et al. (1988c) reported that PMN from healthy donors failed to significantly kill *P. brasiliensis* isolates, whereas *Candida albicans* was readily killed in the same system. These results are surprising because PMN make oxidative burst responses to *P. brasiliensis*, and thus fungus has been shown to be sensitive to the products of the hydrogen peroxide-myeloperoxidase-halide system (McEwen et al., 1984). However, it would be more appropriate to test inflammatory PMN in order to verify a role of PMN in resistance to *P. brasiliensis* rather than peripheral blood PMN, inflammatory PMN at the site of infection, or facsimiles thereof. Inflammatory PMN, as will be detailed below in a murine system, acquire enhanced microbicidal capacity and kill some isolates of *P. brasiliensis* (McEwen et al., 1984).

Studies in Murine Models

The sequence of events in primary pulmonary *P. brasiliensis* infection has been described in a murine model by Bedoya et al. (1986). Histological studies performed at time intervals after nonlethal yeast infection showed that the initial host response was an infiltration of PMN to the acinopulmonary lumen and interalveolar septum. PMN were seen in contact with *P. brasiliensis*. Lymphocytes and macrophages were clearly visible in pulmonary lesions 6-7 days after infection, and macrophages with ingested yeast cells were also observed. After 5 weeks of infection, the acute process had declined and lesions acquired a granulomatous configuration. In chronic disease of mice the histological picture was similar to that seen in human chronic paracoccidioidomycosis (Defaveri et al., 1982; Brummer et al., 1984). Although it is clear that PMN are involved in primary and chronic paracoccidioidomycosis, their possible role in resistance is yet to be defined.

Recently, methods have been developed for the production of *P. brasiliensis* conidia in sufficient quantities to do in vivo experiments (Restrepo et al., 1986). In mice infected with conidia, histological studies showed that by 18 h after infection, conidia converted to budding yeasts in the lungs (McEwen et al., 1987). The accumulation of PMN and macrophages in this natural type of infection paralleled that described above for infection with yeast cells. However, the presence of macrophages and PMN in *P. brasiliensis* lesions does not define the role of inflammatory cells in resistance to the infection.

In Vitro Studies of Murine PMN and Macrophages

Studies in vitro show that resident peritoneal macrophages ingest yeast cells of *P. brasiliensis* (Calich et al., 1979) but do not kill isolates of *P. brasiliensis* (Brummer et al., 1988b). However, when macrophages are activated by an overnight coculture with lymph node cells plus concanavalin A (Con A) or by exposure to supernatants from Con A-stimulated spleen cell cultures, or 1,000 to 10,000 γIFN units per ml, they killed 33-35% of isolates of *P. brasiliensis* (Brummer et al., 1988b). Since macrophage activation is likely to take place in vivo as the result of the interaction of sensitized lymphocytes with *P. brasiliensis*, activated macrophages would be expected to play a crucial role in resistance to and clearance of *P. brasiliensis*. Additional studies showed that killing of *P. brasiliensis* by activated macrophages could not be inhibited by superoxide dismutase, catalase, or azide. This indicated that activated macrophages killed *P. brasiliensis* by nonoxidative mechanisms (Brummer et al., 1988b).

The fungicidal activity of alveolar macrophages for *P. brasiliensis* is weakly potent (18 ± 4%) but readily demonstrable (Brummer et al., 1988c). Like peritoneal macrophages, alveolar macrophages can be activated in vitro by treatment with lymph node cells plus Con A, Con A spleen cell supernatants, or γIFN. Killing was greatest by alveolar macrophages treated with Con A supernatants (72 ± 4%). Alveolar macrophages treated with lymph node cells plus Con A or IFN killed *P. brasiliensis* 45 ± 2% and 51 ± 1%, respectively (Brummer et al., 1988c). Alveolar macrophages could also be activated in vivo by treatment of mice with a single dose of IFN (4×10^5 units) intraperitoneally. Alveolar macrophages lavaged from the lungs of mice 24 h after IFN treatment had enhanced capacity to kill *P. brasiliensis*, e.g., 47 ± 7% vs. 25 ± 4% (Brummer et al., 1988c). These findings suggest a therapeutic use of γIFN in treatment of paracoccidioidomycosis in conjunction with ketoconazole, the current drug of choice, which is only fungistatic.

HISTOPLASMOSIS

Introduction

Histoplasmosis is a systemic mycosis caused by the thermally dimorphic fungus *Histoplasma capsulatum* (DeMonbreun, 1934). *H. capsulatum* is found worldwide (Ajello, 1971) and is readily isolated from soil contaminated with bird dung or bat guano (Seidlberg and Ajello, 1952; DiSalvo and Ajello, 1969). In endemic areas only a few individuals develop chronic progressive disease. This suggests that normal host defenses are effective against ordinary levels of exposure. On the other hand, outbreaks of acute disease are seen with exposure to heavily contaminated soil as in the Mason City, Iowa epidemic of 1962 (D'Alessio et al., 1965) or following cave explorations (Sacks et al., 1986).

Role of PMN and Macrophages in Primary Infection

The involvement of PMN and macrophges in primary resolving pulmonary histoplasmosis has been described as PMN bronchopneumonia. It is followed by nodular "tubercle" formation within a few days with subsequent caseation and calcification (Kipkie and Howell, 1951; Schwarz, 1958). Human PMN do not appear to be effective in killing *H. capsulatum* (Schaffner et al., 1986), but they probably play a role in "walling-off" the infection by their accumulation. Human resident alveolar macrophages, judging from animal studies, readily ingest *H. capsulatum*, but do not kill the parasitic yeast-form, which multiplies inside macrophages as an intracellular parasite (Howard, 1965). Although ineffective in the killing *H. capsulatum*, normal human PMN and alveolar macrophages play a role in limiting the infection to discrete loci. The subsequent development of a cellular immune response with granuloma formation provides an ultimate mean of resistance to moderate infection with *H. capsulatum* (Schwarz, 1958). Immunologically activated macrophages probably play a crucial role in the resistance and clearance process described above. However, the fungicidal effects of human alveolar macrophages, monocytes, or monocyte-derived macrophages, resident or activated, are not known at this time.

Role of Macrophages from Immunized Mice in Resistance

Murine models of histoplasmosis have been used extensively to study the immunological responses to infection and the effect of these responses on the ability of macrophages to control the replication of *H. capsulatum*. In vitro studies demonstrated that lymphocytes from immunized mice collaborated with peritoneal macrophages from normal mice to restrict intracellular replication of ingested yeast-form *H. capsulatum* (Howard et al., 1971; Howard, 1973). Recent studies have established that lymphokines produced by stimulated T lymphocytes, particularly γIFN, activate macrophages to be fungistatic for ingested *H. capsulatum* (Wu-Hsieh et al., 1984; Wu-Hsieh and Howard, 1987). In similar in vitro studies peritoneal macrophages, activated with lymphokines, were shown to be fungicidal for *H. capsulatum* as determined by reduction of inoculum colony forming units (Patino et al., 1987). Interestingly, in these studies the susceptibility of beige mice compared to C57BL/6 mice correlated with failure of macrophages from beige mice to be activated by lymphokines to kill *H. capsulatum*. This finding supports the position that activated macrophages are the crucial effector cells in resistance to *H. capsulatum*.

Role of Alveolar Macrophages in Resistance

Although peritoneal macrophages can deal effectively with yeast-form *H. capsulatum*, little is known about the role of alveolar macrophages against *H. capsulatum* in the pulmonary compartment, which is the initial site of infection. Likewise, little is known about the interaction of the natural infective unit, the microconidia, with alveolar macrophages. A study by Kimberlin et al. (1981) found that 35-37% of alveolar macrophages from normal or vaccinated mice ingested microconidia, although only the alveolar macrophages from vaccinated mice restricted intracellular replication. A greater proportion (66%) of the alveolar macrophages from BCG-treated mice ingested conidia. Alveolar macrophages from BCG-treated mice were the only ones to significantly kill ingested microconidia; e.g., by 72 h only 10% of the macrophages contained recognizable fungal particles. The ability and rapidity of ingested microconidia to convert to the yeast phase is an important factor in the pathogenesis of histoplasmosis. Microconidia in normal alveolar macrophages began to convert to yeast cells by 12-18 h. Yeast cells, unlike microconidia, were resistant to the microbicidal activity of alveolar macrophages from BCG-treated mice, e.g., they are not killed, nor was their replication restricted (Kimberlin et al., 1981). Taken together alveolar macrophages in vaccinated histoplasmin skin test positive mice play an important role in resistance to the natural infective unit, the microconidia, by restricting conversion and replication, whereas resident alveolar macrophages from normal mice provide little or no resistance. Whether a similar condition exists in histoplasmin skin test positive humans is not known at this time and awaits future investigations.

Experimental Infections with Microconidia

A model of pulmonary infection with aerosolized microconidia was devised by Schlitzer et al. (1981) using guinea pigs. Relatively small inocula (1-3,000 CFU) resulted in an acute type of infection with dissemination to the lymph nodes. It took pulmonary defenses considerable time to clear *H. capsulatum* from bronchoalveolar washes (2 weeks) and from the lungs (3 weeks). Such experiments give some idea about the level of exposure to natural infective units which can be handled by alveolar macrophages and immune responses. Similar pulmonary infection experiments in mice with yeast-form *H. capsulatum*, e.g., 16,000 CFU in the lungs at 8 h, showed dissemination to the spleen in one week. However, by 4 weeks *H. capsulatum* was cleared from lungs and spleen (Baughman et al., 1986). Again these results indicate the infectious inoculum that normal pulmonary defenses can handle.

Role of PMN in Resistance

Few studies have been done on the interaction of animal PMN with microconidia or yeast cells of *H. capsulatum*. The report by Howard (1973) on killing of *H. capsulatum* yeast cells by guinea pig peripheral blood PMN stands alone. Killing was assessed by stainability in these studies; however, the validity of using vital exclusion dye staining has been questioned by Schaffner et al. (1986) and Kwon-Chung and Tewari (1987). Consequently the role of PMN in resistance to initial infection with microconidia and primary acute histoplasmosis in animals is controversial at this time.

REFERENCES

Ajello, L. (1971). Coccidioidomycosis and histoplasmosis: a review of their epidemiology and geographical distribution. *Mycopathol. Mycol Appl. 45*: 221-230.

Baker, R. D. (1942). Tissue reactions in human blastomycosis: an analysis of tissue from twenty three cases. *Am. J. Pathol. 18*: 479-495.

Baughman, R. P., Kim, C. K., Vinegar, A., Hendricks, D. E., Schmidt, D. J., and Bullock, W. E. (1986). The pathogenesis of experimental pulmonary histoplasmosis: correlative studies of histopathology, bronchoalveolar lavage and respiratory function. *Am. Rev. Respir. Dis. 134*: 771-776.

Beaman, L., Pappagianis, D., and Benjamini, E. (1977). Significance of T cells in resistance to experimental murine coccidioidomycosis. *Infect. Immun. 17*: 580-585.

Beaman, L., and C. A. Holmberg (1980a). In vitro response of alveolar macrophages to infection with *Cocciodioides immitis. Infect. Immun. 28*: 594-600.

Beaman, L., and Holmberg, C. A. (1980b). Interaction of nonhuman primate peripheral blood leukocytes and *Coccidioides immitis* in vitro. *Infect. Immun. 29*: 1200-1201.

Beaman, L., Benjamini, E., and Pappagianis, D. (1981). Role of lymphocytes in macrophage-induced killing of *Coccidioides immitis* in vitro. *Infect. Immun. 34*: 347-353.

Beaman, L., Benjamini, E., and Pappagianis, D. (1983). Activation of macrophages by lymphokines: enhancement of phagosome-lysosome fusion and killing of *Coccidioides immitis. Infect. Immun. 39*: 1201-1207.

Beaman, L., and Pappagianis, D. (1985). Fate of *Coccidioides immitis* arthroconidia in human peripheral blood monocytes cultures in vitro. In *Coccidioidomycosis: Proceedings of the 4th International Conference*. Edited by Einstein, H. E., and Catanzaro, A. Pub. Nat. Found. Infect. Dis., Washington, D.C., pp. 170-180.

Beaman, L. (1987). Fungicidal activation of murine alveolar macrophages by recombinant gamma-interferon. *Infect. Immun. 55*: 2951-2955.

Bedoya, V., McEwen, J., Tabares, A. M., Uribe, F., and Restrepo, A. (1986). The pathogenesis of paracoccidioidomycosis: a histopathological study of the experimental murine infection. *Mycopathologia 94*: 133-144.

Bradsher, R. W., Ulmer, W. C., Marmer, D. J., Townsend, J. W., and Fields, R. F. (1985). Intracellular growth and phagocytosis of *Blastomyces dermatitidis* by monocyte-derived macrophages from previously infected and normal subjects. *J. Infect. Dis. 151*: 57-64.

Bradsher, R. W., Balk, R. A., and Jacobs, R. F. (1987). Growth inhibition of *Blastomyces dermatitidis* in alveolar and peripheral macrophages from patients with blastomycosis. *Am. Rev. Respir. Dis. 135*: 412-417.

Brummer, E., Morozumi, P. A., and Stevens, D. A. (1980). Macrophages and fungi: effect of macrophage induction, activation by different stimuli, and soluble factors on *Blastomyces* in vitro. *RES: J. Reticuloendothel. Soc. 28*: 507-518.

Brummer, E., Morozumi, P. A., and Stevens, D. A. (1981). Virulence of fungi: correlation of virulence of *Blastomyces dermatitidis* in vivo with escape from macrophage inhibition of replication in vitro. *Infect. Immun. 32*: 864-871.

Brummer, E., and Stevens, D. A. (1982). Opposite effects of human monocytes, macrophages and polymorphonuclear neutrophils on replication of *Blastomyces dermatitidis* in vitro. *Infect. Immun. 36*: 297-303.

Brummer, E., Morozumi, P. A., Vo, P. T., and Stevens, D. A. (1982). Protection against pulmonary blastomycosis: adoptive transfer of T lymphocytes, but not serum, from resistant mice. *Cell. Immunol. 73*: 349-359.

Brummer, E., and Stevens, D. A. (1983). Enhancing effect of murine polymorphonuclear neutrophils (PMN) on the multiplication of *Blastomyces dermatitidis* in vitro and in vivo. *Clin. Exp. Immunol. 54*: 587-594.

Brummer, E., Sugar, A. M., and Stevens, D. A. (1983). Activation of peritoneal activity against *Blastomyces dermatitidis* and effect of specific antibody and macrophages by Concanavalin A or *Mycobacterium bovis* (BCG) for fungicidal complement. *Infect. Immun. 39*: 817-822.

Brummer, E., Sugar, A. M., and Stevens, D. A. (1984). Immunological activation of polymorphonuclear neutrophils for fungal killing: studies with murine cells and *Blastomyces dermatitidis* in vitro. *J. Leukocyte Biol. 36*:505-520.

Brummer, E., Restrepo, A., Stevens, D. A., Azzi, R., Gomez, A., Hoyos, G. L., McEwen, J. G., Cano, L. E., and de Bedout, C. (1984). Murine model of paracoccidicidomycosis. Production of fatal acute pulmonary or chronic pulmonary and disseminated disease: immunological and pathological observations. *J. Exp. Path. 1*: 241-255.

Brummer, E., Sugar, A. M., and Stevens, D. A. (1985a). Enhanced oxidative burst in immunologically activated but not elicited polymorphonuclear leukocytes correlates with fungicidal activity. *Infect. Immun. 49*: 396-401.

Brummer, E., Morrison, C., and Stevens, D. A. (1985b). Recombinant and natural gamma-interferon activation of macrophages in vitro: different dose requirements for induction of killing activity against phaogcytizable and nonphagocytizable fungi. *Infect. Immun. 49*: 724-730.

Brummer, E., Beaman, L., and Stevens, D. A. (1985c). Killing of endospores, but not arthroconidia, of *Coccidioides immitis* by immunologically activated polymorphonuclear neutrophils. In *Coccidioidomycosis: Proceedings of the 4th International Conference*. Edited by Einstein, H. E., and Cantazaro, A. Pub. Nat. Found. Infect. Dis., Washington, D.C., pp. 201-213.

Brummer, E., McEwen, J. G., and Stevens, D. A. (1986). Fungicidal activity of murine inflammatory polymorphonuclear neutrophils: comparison with murine peripheral blood PMN. *Clin. Exp. Immunol. 66*: 681-690.

Brummer, E., and Stevens, D. A. (1987). Activation of pulmonary macrophages for fungicidal activity by gamma-interferon or lymphokines. *Clin. Exp. Immunol. 70*: 520-528.

Brummer, E., McEwen, J., Hanson, L., Restrepo, A., and Stevens, D. A. (1988a). Fungicidal activity of human and murine polymorphonuclear neutrophils for *Paracoccidioides brasiliensis*: susceptibility of different isolates. Abstract, X Intranational Congress, Int. Soc. Human & Animal Mycology (ISHAM), Barcelona, Spain.

Brummer, E., Hanson, L., and Stevens, D. A. (1988b). Activation of macrophages for killing of *Paracoccidioides brasiliensis*: evidence for nonoxidative mechansims. *Int. J. Immunopharmac. 10*:945-952.

Brummer, E., Hanson, L., Restrepo, A., and Stevens, D. A. (1988c). In vivo and in vitro activation of alveolar macrophages by gamma-interferon for enhanced killing of *Paracoccidioides brasiliensis* or *Blastomyces dermatitidis*. *J. Immunol. 140*:2786-2789.

Calich, V. L., Kipnis, T. L., Mariano, M., Fava-Netto, C., and Dias de Silva, W. (1979). The activation of the complement system by *Paracoccidioides brasiliensis* in vitro: its opsonic effect and possible significance for an in vitro model of infection. *Clin. Immunol. Immunopathol. 12*: 20-30.

Cox, R. A., Brummer, E., and Lecara, G. (1977). In vitro lymphocyte responses of cocciodioidin skin test positive and negative persons to coccidioidin, spherulin and Coccidioides cell wall antigen. *Infect. Immun. 15*: 751-755.

D'Alessio, D. J., Heeren, R. H., Hendricks, S. L., Ogilvie, P., and Furcolow, M. L. (1965). A starling roost as the source of urban epidemic histoplasmosis in an area of low incidence. *Am. Rev. Respir. Dis. 92*: 725-731.

Defaveri, J., Rezkallah-Iwassson, and de Franco, M. F. (1982). Experimental pulmonary paracoccidioidomycosis in mice: morphology and correlation of lesions with humoral and cellular immune response. *Mycopathologia 77*: 3-11.

DeMonbreun, W. A. (1934). The cultivation and cultural characteristics of Darling's *Histoplasma capsulatum. Am. J. Trop. Med. 14*: 93-125.

Denton, J. R., and DiSalvo, A. F. (1968). Respiratory infection of laboratory animals with conidia of *Blastomyces dermatitidis. Mycopathol. Mycol. Appl. 36*: 129-136.

DiSalvo, A. F., and Ajello, L. (1969). Isolation of *Histoplasma capsulatum* from Arizona bats. *Am. J. Epidemiol. 89*: 606-614.

Drutz, D. J., and Huppert, M. (1983). Coccidioidomycosis: factors affecting host-parasite interactions. *J. Infect. Dis. 147*: 372-390.

Drutz, D. J., and Frey, C. L. (1985). Intracellular and extracellular defenses of human phagocytes against *Blastomyces dermatitidis* conidia and yeasts. *J. Lab. Clin. Med. 105*: 737-750.

Drutz, D. J., DeMarsh, P. L., and Frey, C. L. (1985). Divergent patterns of murine pulmonary blastomycosis induced by conidia vs. yeast. *Abst. Ann. Meet. Am. Soc. Microbiol.*, Abstr. No. F4, p. 365.

Fava-Netto, C. (1965). The immunology of South American blastomycosis. *Mycopathologia 26*: 349-358.

Frey, C. L., and Drutz, D. J. (1986). Influence of fungal surface components on the interaction of *Cocciodioides immitis* with polymorphonuclear neutrophils. *J. Infect. Dis. 153*: 933-943.

Furcolow, M. L. (1958). Recent studies on the epidemiology of histoplasmosis. *Ann. N.Y. Acad. Sci. 72*: 127-163.

Galgiani, J. N., Payne, C. M., and Jones, J. F. (1984). Human polymorphonuclear leukocyte inhibition of incorporation of chitin precursors into mycelia of *Cocciodioides immitis. J. Infect. Dis. 149*: 404-412.

Goihman-Yahr, M., Essenfeld,-Yahr, E., Albornoz, M. C., Yarzabel, D., de Gomez, M. H., San Martin, B., Ocanto, A., Gil, F., and Convit, J. (1981). Defect of in vitro digestive ability of polymorphonuclear leukocytes in paracoccidioidomycosis. *Infect. Immun. 28*: 557-566.

Howard, D. H. (1965). Intracellular growth of *Histoplasma capsulatum. J. Bacteriol. 89*: 518-523.

Howard, D. H., Otto, V., and Gupta, R. K. (1971). Lymphocyte mediated cellular immunity in histoplasmosis. *Infect. Immun. 4*: 605-610.

Howard, D. H. (1973). Fate of *Histoplasma capsulatum* in guinea pig polymorphonuclear leukocytes. *Infect. Immun. 8*: 412-419.

Howard, D. H. (1973). Further studies on the inhibition of *Histoplasma capsulatum* within macrophages from immunized animals. *Infect. Immun. 8*: 577-581.

Kinberlin, C. L., Hariri, A. R., Hempel, H. O., and Goodman, N. L. (1981). Interactions between *Histoplasma capsulatum* and macrophages from normal and treated mice: comparison of the mycelial and yeast phases in alveolar and peritoneal macrophages. *Infect. Immun. 34*: 6-10.

Kipkie, G. F., and Howell, A. (1951). Histopathology of experimental histoplasmosis. *Arch. Pathol. 51*: 312-318.

Kong, Y. M., Levine, H. B., Madin, S. H., and Smith, C. E. (1964). Fungal multiplication and histopathologic changes in vaccinated mice infected with *Cocciodioides immitis*. *J. Immunol. 92*: 779-790.

Kong, Y. M., and Levine, H. B. (1967). Experimentally induced immunity in the mycoses. *Bact. Rev. 31*: 35-53.

Kwon-Chung, K. J., and Tewari, R. P. (1987). Determination of viability of *Histoplasma capsulatum* yeast cells grown in vitro: comparison between dye and colony count method. *J. Med. Vet. Mycol. 25*: 107-114.

Levine, H. S., Cobb, J. M., and Smith, C. E. (1960). Immunity to coccidioidomycosis induced in mice by purified spherule, arthrospore and mycelial vaccines. *Trans. N.Y. Acad. Sci. 22*: 436-449.

McDaniels, L. S., and Cozad, G. C. (1983). Immunomodulation by *Blastomyces dermatitidis*: functional activity of murine peritoneal macrophages. *Infect. Immun. 40*: 733-740.

McEwen, J., Sugar, A. M., Brummer, E., Restrepo, A., and Stevens, D. A. (1984). Toxic effect of products of oxidative metabolism on *Paracoccidioides brasiliensis*. *J. Med. Microbiol. 18*: 423-428.

McEwen, J. G., Bedoya, V., Patino, M. M., Salazar, and Restrepo, A. (1987). Experimental murine paracoccidioidomycosis induced by the inhalation of conidia. *J. Med. Vet. Mycol. 25*: 165-175.

Mendes, N. F., Mussatti, C. C., Leao, R. C., Mendes, E., and Naspits, C. K. (1971). Lymphocyte culture and skin allograft survival in patients with South American blastomycosis. *J. Allergy 48*: 40-45.

Miyaji, M., and Nishimura, K. (1983). Granuloma formation and killing functions of granuloma in congenitally athymic nude mice infected with *Blastomyces dermatitidis* and *Paracoccidioides brasiliensis*. *Mycopathology 82*: 129-141.

Mok, P. W., and Greer, D. (1977). Cell mediated immune responses in patients with paracoccidioidomycosis. *Clin. Exp. Immunol. 28*: 89-98.

Morozumi, P. A., Brummer, E., and Stevens, D. A. (1982). Protection against pulmonary blastomycosis: correlation with cellular and humoral immunity in mice after subcutaneous nonlethal infection. *Infect. Immun. 37*: 670-678.

Mussatti, C. C., Rezkallah, M. T., Mendes, E., and Mendes, N. F. (1976). In vivo and in vitro evaluation of cell mediated immunity in patients with paracoccidioidomycosis. *Cell. Immunol. 24*: 365-378.

Patino, M. M., Williams, D., Ahrens, J., and Graybill, J. R. (1987). Experimental histoplasmosis in the beige mouse. *J. Leuk. Biol. 41*: 228-235.

Pappagianis, D., Levine, H. B., Smith, C. E., Berman, A. J., and Kobayashi, G. A. (1961). Immunization of mice with viable *Coccidioides immitis*. *86*: 28-34.

Pappagianis, D. (1980). Epidemiology of coccidioidomycosis. In *Coccidioidomycosis*. Edited by Stevens, D. A. Plenum, New York, pp. 63-85.

Restrepo, A., Greer, D., and Vasconcellos, M. (1973). Paracoccidioidomycosis: a review. *Rev. Med. Vet. Mycol. 8*: 97-123.

Restrepo, A., and Velez, H. (1975). Efectos de la facocitosis in vitro sobre *Paracoccidioides brasiliensis*. *Sabouraudia 13*: 10-21.

Restrepo, A., Restrepo, M., Restrepo, F., Aristizabal, L. H., Mocada, L. H., and Velez, H. (1978). Immune responses in paracoccidioidomycosis and controlled study of 16 patients before and after treatment. *Sabouraudia 16*: 151-163.

Restrepo, A., Salazar, M. E., Cano, L. E., and Patino, M. M. (1986). A technique to collect and dislodge conidia produced by *Paracoccidioides brasiliensis* mycelial form. *J. Med. Vet. Mycol. 24*: 245-248.

Sacks, J. J., Ajello, L., and Crockett, L. K. (1986). An outbreak and review of cave-associated *Histoplasma capsulatum*. *J. Med. Vet. Mycol. 24*: 313-327.

Schaffner, A., Davis, C. E., Schaffner, T., Market, M., Douglas, H., and Braudet, A. I. (1986). In vitro susceptibility of fungi to killing by neutrophil granulocytes discriminates between primary pathogeneicity and opportunism. *J. Clin. Invest. 78*: 511-524.

Schlitzer, R. L., Chandler, F. W., and Larsh, H. W. (1981). Primary acute histoplasmosis in guinea pigs exposed to aerosolized *Histoplasma capsulatum*. *Infect. Immun. 33*: 575-582.

Schwarz, J. (1958). The pathogenesis of histoplasmosis. *Trans. N.Y. Acad. Sci. 20*: 541-548.

Sixbey, J. W., Fields, B. T., Sun, C. N., Clark, R. A., and Nolan, C. M. (1979). Interaction between human granulocytes and *Blastomyces dermatitidis*. *Infect. Immun. 23*: 41-44.

Sugar, A. M., Chahal, R. S., Brummer, E., and Stevens, D. A. (1983). Susceptibility of *Blastomyces dermatitidis* strains to products of oxidative metabolism. *Infect. Immun. 41*: 908-912.

Sugar, A. M., and Fields, K. G. (1985). Characteristics of the pulmonary cellular immune response to two strains of *Blastomyces dermatitidis* in the mouse. *Am. Rev. Respir. Dis. 132*: 1319-1323.

Williams, J. E., and Moser, A. A. (1987). Chronic murine pulmonary blastomycosis induced by intratracheally inoculated *Blastomyces dermatitidis* conidia. *Am. Rev. Respir. Dis. 135*: 17-25.

Wu-Hsieh, B., Zlotnik, A., and Howard, D. H. (1984). T-cell hybridoma produced lymphokine that activate macrophages to suppress intracellular growth of *Histoplasma capsulatum*. *Infect. Immun. 43*: 380-385.

Wu-Hsieh, B., and Howard, D. H. (1987). Inhibition of intracellular growth of *Histoplasma capsulatum* by recombinant murine gamma-interferon. *Infect. Immun. 55*: 1014-1016.

Zeidlberg, L. D., and Ajello, L. (1952). Isolation of *Histoplasma capsulatum* from soil. *Am. J. Public Health 42*: 930-935.

Part III
HOST RESPONSE TO INVASIVE FUNGI
IMMUNOREGULATORY ASPECTS

12
Immunomodulation in Response to *Candida*

JUDITH E. DOMER and RONALD E. GARNER
Tulane University School of Medicine, New Orleans, Louisiana

INTRODUCTION

Candidiasis is the classic protean disease, in that many different clinical forms, ranging from cutaneous to mucocutaneous to systemic, of varying severity and of varying consequence to the patient are regularly encountered. Several species of *Candida* are capable of causing disease, but *C. albicans* and *C. tropicalis* greatly outnumber the other species in incidence. While some of the *Candida* species of lesser pathogenicity have been isolated from skin, *C. albicans* and *C. tropicalis* are rarely isolated from skin, and instead they reside predominantly in the gastrointestinal tract (Cohen et al., 1969; Wingard et al., 1979).

Since the candidal organisms that cause disease are all members of the normal flora and enjoy a commensal relationship with the normal host, when candidiasis occurs it is an opportunistic disease precipitated when one of a number of potential compromising conditions creates an imbalance in the normal defenses. The clinical form of candidiasis observed often correlates with the compromising condition. The first line of defense against systemic candidiasis, for example, is the polymorphonuclear leukocyte (PMN), and when individuals become granulocytopenic, usually as a result of cytoreductive drug therapy for an underlying disease, they become particularly susceptible to candidemia (Rifkind et al., 1967; Myerwitz et al., 1977). If the candidemia is uncorrected, systemic disease may result involving any of the organs of the body, but the kidney is involved in about 80% of such patients, with the other organs having less involvement (Parker et al., 1976). On the other hand, when granulocyte numbers are normal or near normal, and cellular immunity appears to be compromised, as in some patients with chronic mucocutaneous candidiasis (CMC) or acquired immunodeficiency disease syndrome (AIDS), the form of candidiasis observed is cutaneous and/or mucocutaneous (Valdimarsson et al., 1973; Kirkpatrick and Smith, 1974; Klein et al., 1984).

When serious candidal disease occurs, it usually involves large numbers of organisms growing in the internal organs or on the cutaneous and mucocutaneous surfaces. As a consequence of this overgrowth of *Candida*, candidal antigens, presumably as a result of metabolic activity of the living cells or degradation of dead cells, are introduced into the circulation and may be detected by serological methods (Weiner and Yount, 1976; Kerkering, et al., 1979; Lehmann and Reiss, 1980; Araj et al., 1982). The precise nature of the antigen that circulates is not clear and, in fact, may vary, in that in some instances cytoplasmic antigens are detected (Araj et al., 1982), and in other instances cell wall antigens, i.e., antigens containing mannan, have been detected (Weiner and Yount, 1976; Kerkering et al., 1979; Lehmann and Reiss, 1980).

The observation that candidal antigens, in particular cell wall antigens, may circulate during disease is of considerable interest in view of the fact that the yeast cell wall, or components of the yeast cell wall, have been considered as agents capable of influencing immune reactions for many years (von Dungern, 1900; Pillemer and Ecker, 1941; Riggi and Di Luzio, 1961). The early work was done with the nonpathogenic yeast, *Saccharomyces cerevisiae*, but in more recent years immunomodulatory activities have been ascribed to components of *C. albicans*, the subject of this review, as well as to components of *Cryptococcus neoformans* (Breen, et al., 1982; Murphy and Moorhead, 1982; Blackstock, et al., 1987), *Paracoccidioides brasiliensis* (Silva and Fazioli, 1985), and nonpathogenic mycelial fungi such as *Peziza vesiculosa* (Ohno et al., 1985).

Published studies having as their topic immunoregulation in candidasis or in response to viable or nonviable preparations of *Candida* can be organized into two broad categories, namely (1) immunosuppression as observed during human clinical disease, and (2) immunomodulation, including both enhancement and suppression, as observed under laboratory conditions using experimental systems. Each topic will be presented separately below.

CLINICAL STUDIES SUGGESTING IMMUNOSUPPRESSIVE PHENOMENA

Introduction

The idea that specific suppressive mechanisms might be functional in individuals with chronic candidal infections took root in the late 1960s when two types of evidence surfaced. Both types of evidence were noted in patients with a relatively rare but nevertheless important clinical syndrome, chronic mucocutaneous candidiasis (CMC). First, CMC patients were described who had defects in their ability to mount delayed hypersensitivity responses (Chilgren et al., 1967; Buckley et al., 1968; Kirkpatrick et al., 1970), and second, there were the first reports of the presence of inhibitors of various immunologic functions in the sera of patients with CMC (Canales et al., 1969; Paterson et al., 1971). It soon became apparent, however, that CMC patients as a group possessed highly variable defects in their immune systems (Valdimarsson et al., 1973), that some, but not all, had demonstrable regulatory serum factors (Lehner et al., 1972; Verhaegen et al., 1976; Fischer et al., 1978), and that at times cell-mediated immune function was intact, but phagocytic cell function was suppressed (Bortolussi et al., 1981; Fischer et al., 1982). In fact, when analyzing the dat data published for patients in whom suppressive phenomena could be demonstrated, e.g., patients with CMC (Canales et al., 1969; Lehner et al., 1972; Patterson et al., 1971; Twomey et al., 1975; Takeya et al., 1976; Djawari et al., 1977; Mobacken et al., 1977; Fischer et al., 1978; Aronson et al., 1979; Sacchi et al., 1979; Cates et al., 1980; Gatenby et al., 1980; Walker and Urbaniak, 1980; Bortolussi et al., 1981; Ruiz-Arguelles et al., 1983; Barnaba et al., 1985), disseminated candidiasis (Laforce et al., 1975), pseudomembranous or chronic esophageal candidiasis (Verhaegen et al., 1976; Dutta and Al-Ibrahim, 1978; Lee et al., 1986), or recurrent candidal vaginitis (Verhaegen et al., 1976; Witkin et al., 1983, 1986), it is clear that there appears to be no single mechanism responsible for the suppressive phenomena observed.

While there appears to be no single factor responsible for the suppression observed in the various forms of candidiasis, the inhibitors described fall into several broadly based categories, namely, soluble serum or plasma factors (Canales et al., 1969; Lehner et al., 1972; Paterson et al., 1971; Laforce et al., 1975; Twomey et al., 1975; Verhaegen et al., 1976; Mobacken et al., 1977; Fischer et al., 1978; Van Der Meer et al., 1978; Aronson et al., 1979; Cates, et al., 1980; Gatenby et al., 1980; Walker and Urbaniak, 1980; Kennedy et al., 1981; Witkin et al., 1983; Lee et al., 1986), basic defects in phagocytic function at either the level of the macrophage or the polymorphonuclear leukocyte (PMN) (Djawari et al., 1977; Sacchi et al., 1979; Bortolussi et al., 1981; Witkin et al., 1986) and changes in the type or ratio of specific types of lymphocytes (Ruiz-Arguelles et al., 1983; Witkin et al., 1983; Barnaba et al., 1985). It is obvious from the number of publications involved that serum/plasma factors have received the most attention.

Serum/Plasma Factors

The involvement of serum or plasma factors in immunosuppressive phenomena has been demonstrated in in vitro systems, either with the PMN (predominantly) or mononuclear (MN) phagocyte, or in assays for lymphoproliferative activity in response to mitogens and/or antigens. Some of the studies involving pri-

Table 1 Serum-Dependent Cellular Immune Defects in Patients with Candidiasis

No. pts.	Skin test	Lymphocyte stimulation		Inhibition of normal lymphocytes	Characteristics of inhibitor	Reference
		Mitogens	Antigens			
1	complete anergy, 8 antigens	normal PHA	*Candida* neg, mumps neg	Yes	circulating component of *Candida*	Canales et al., 1969
1	anergic to *Candida*	normal PHA	*Candida* neg	Yes	disappeared after anti-*Candida* therapy	Patterson et al., 1971
1	complete anergy, *Candida*, mumps, SK-SD	normal PHA, PWM	*Candida* neg mumps neg	Yes	disappeared after plasmaphoresis: cold-labile; no correlation with anti-*Candida* antibody; may have interfered with macrophage function	Twomey et al., 1975
1	complete anergy, 6 antigens	normal PHA	*Candida* neg, mumps pos, SK-SD pos		persisted after clinical remission	Aronson et al., 1979

1	anergic to *Candida* and mumps, SK-SD pos	PHA↓ in pooled AB serum greatly↓ in autologous serum	*Candida* neg, mumps pos, SK-SD pos		disappeared after thymoma disappeared: heat-stable, 56°C, 1 hr; stable to freezing-thawing; mol. wt. 50–65,000 daltons	Gatenby et al., 1980
6/23	anergic to *Candida*, PPD pos, SK-SD pos	normal PHA, Con A, PWM	*Candida* neg, PPD pos, SK-SD pos	Yes	disappeared after anti-*Candida* therapy: nontoxic; thermostable; nonprecipitable with ammonium sulfate; absorbed out with anti-*Candida* antibodies	Fischer et al., 1978
1	complete anergy	Con A		Yes	low mol. wt. (<10,000 daltons); heat-labile; removed by anti-*Candida* antibodies; nonprecipitable with ammonium sulfate	Lee et al., 1986

Abbreviations: pos, positive; neg, negative; PHA, phytohemagglutinin; PWM, pokeweed mitogen; SK-SD, steptokinase-streptodornase; mol. wt., molecular weight; Con A, concanavalin A; PPD, protein purified derivative.

marily lymphoproliferative activity are summarized in Table 1. In all of the studies presented in Table 1, serum or plasma factors were involved in the suppression observed. In most instances, mitogen responses of the lymphocytes of patients were normal, proliferative responses to *Candida* antigens did not occur, proliferative lymphocyte or skin test responses to heterologous antigens such as mumps, streptokinase-streptodornase, or purified protein derivative were variable, and the addition of serum or plasma from the patient to cultures of normal lymphocytes resulted in suppression of normal responses. On balance, *Candida*-specific suppressor mechanisms were noted more frequently than pansuppressor mechanisms.

While the preceding generalizations can be made about the phenomenon itself, no such generalizations are possible with respect to the nature of the inhibitor. In some cases it disappeared after anti-*Candida* therapy (Paterson et al., 1971; Fischer et al., 1978), in other cases it persisted after clinical remission (Aronson et al., 1971); in some cases it was heat- and cold-stable and estimated to have a molecular weight of 50-65,000 daltons (Gatenby et al., 1980), and in other cases it was heat-labile with a low molecular weight, less than 10,000 daltons (Lee et al., 1986). In one case it was suggested to be the product of a thymoma, since it disappeared after successful treatment of the thymoma (Gatenby et al., 1978). In several instances it could be removed from the serum or plasma by anti-*Candida* antibodies (Fischer et al., 1978; Lee et al., 1986), suggesting, of course, that the factor was a component of *Candida*. However, the inhibitors described by Fischer et al. (1978) and Lee et al. (1986) were probably different components, since in one instance the factor was heat-stable (Fischer et al., 1978) and in the other case it was heat-labile (Lee et al., 1986).

Serum-dependent abnormalities have been observed at the level of the phagocytic cell as well (Laforce et al., 1975; Verhaegen et al., 1976; Van der Meer et al., 1978; Cates et al., 1980; Kennedy et al., 1981; Walker and Urbaniak, 1980; Witkin, 1986). In all but one instance, those abnormalities were associated with PMN activity. The predominant defect noted was a reduced ability to kill *Candida* after it had been ingested by the PMN (Laforce et al., 1975; Walker and Urbaniak, 1980; Kennedy et al., 1981) or mononuclear (MN) cells (Van der Meer et al., 1978). More rarely the defect described involved the phagocytic event itself (Verhaegen et al., 1976) or chemotaxis for the PMN (Cates et al., 1980). The inhibitor of chemotaxis was described as being heat-stable, reversible, nondialyzable, and precipitable with 40% ammonium sulfate from both patient and normal plasma (Cates et al., 1980), while the inhibitor of phagocytosis was heat-labile (Verhaegen et al., 1976). In two of the studies where candidacidal activity was suppressed, it was suggested that the inhibiting factor was *Candida*-specific immunoglobulin G (IgG) (Laforce et al., 1975; Walker and Urbaniak, 1980). Immunoglobulin was also implicated in the inhibition of chemotaxis (Cates et al., 1980), since the inhibitor was precipitable with 40% saturated ammonium chloride and migrated with IgG on immunoelectrophoresis. Further evidence for the involvement of IgG, in this instance *Candida*-specific IgG, in suppression of immune phenomena was suggested by the studies of Witkin (1986) as well. He added anti-*Candida* IgG to human lymphocyte cultures and determined that the lymphoproliferative response to *Candida* antigen was suppressed in a dose-dependent fashion. Since preincubation of lymphocytes with anti-*Candida* IgG did not suppress their response to *Candida* antigens, but addition of the IgG at the same time, or after, the addition of the antigen did suppress the response, the author felt that the uptake and processing of *Candida* by the macrophage, and the recognition by lymphocytes of *Candida* antigen on the macrophage surface, were inhibited by the antibody. While the data clearly demonstrate inhibition by antibody, caution must be exercised in extrapolating the in vitro data to the human situation, in that high-titered rabbit antibody was used with human cells. The use of heterologous antibody could alter the interaction of cells, especially when the incubation periods are 6 days, in that the interaction of rabbit Fc with human Fc receptors may or may not occur with some IgG subsets. If the interaction does occur, the immunomodulatory effect may be different from that observed with human IgG.

Serum-Independent Phagocytic Cell Defects

The serum-independent phenomena that have been described have included basic defects in PMN phagocytic (Djawari et al., 1977), candidacidal (Djawari et al., 1977; Sacchi et al., 1979), or bactericidal (Mobacken et al., 1977) activities and the inability of MN to respond with an oxidative burst when phagocytizing *Candida* (Bortolussi et al., 1981). Witkin et al. (1986) implicated the macrophage in a *Candida*-induced suppressive phenomenon as well. Their work implied that macrophage production of prostaglandins was increased in the presence of *Candida* and that the elevated levels of prostaglandins blocked the lymphocyte proliferative response.

Alterations in Lymphocyte Subsets

Alterations within lymphocyte subsets, including increases in suppressor cell activity (Witkin et al., 1983; Barnaba et al., 1985) and a decrease in helper cell numbers (Ruiz-Arguelles et al., 1983) have been described in patients with CMC or recurrent vaginitis. While on the surface these data would appear to be contradictory, i.e., increase of suppressor cells in one instance and decrease in helper cells without a concomitant increase in suppressor cells in the other, CMC does not always occur on the same clinical background, and the underlying disease may have a great deal to do with the type of deficiency observed. For example, the two patients with *Candida*-specific suppressor cells described by Barnaba et al. (1985) were individuals with no known predisposing factors, such as endocrinopathies or malignant disease of the thymus. On the other hand, the patient with decreased helper cell function but normal suppressor cell function (Ruiz-Arguelles et al., 1983) had a malignant thymoma as the background on which the CMC occurred.

The suppressor cells described by Witkin et al. (1983) were in patients with recurrent vaginitis, and they were induced by preincubation of the patients' lymphocytes with *Candida* antigen followed by the addition of the preincubated lymphocytes to suspensions of normal lymphocytes. Presumably preincubation resulted in proliferation of suppressor cells to a level sufficient to effect suppression of normal lymphocytes when the two populations were mixed. Regrettably, no data were provided for attempts to mix patient and control lymphocytes directly to demonstrate suppression or to remove cells with the suppressor phenotype from the suspension to determine if the suppressor activity was abrogated. Borkowsky and Valentine (1979) had reported a similar finding, i.e., the development of suppressor cells in populations of peripheral blood leukocytes (PBL) incubated for 5-7 days with *Candida* antigen, but those investigators were using "normal" PBL, not PBL from patients with candidiasis. Later, Piccolella et al. (1981a) reported data with normal human PBL that corroborated the observations of Borkowsky and Valentine (1979). The observations of the latter two sets of investigators confuse the issue with respect to the interpretation of the data gathered by Witkin et al. (1983), because Witkin et al. (1983) included lymphocytes from normal individuals in their study and, in contrast to the observations with patient lymphocytes, no suppressor cells were generated in the cultures of normal lymphocytes.

Finally, an interesting observation was made by Melbye et al. (1985) who examined 79 Zairians with unspecified clinical backgrounds, 64 males and 15 females, and found that oral carriage of *C. albicans* had a positive correlation with a decreased helper:suppressor T lymphocyte ratio. The issue in question, of course, is which of these factors has influenced the other, if, indeed, the two phenomena are related. Does carriage of *C. albicans* increase because the normal ratio of suppressor cells to helper cells is altered in favor of the suppressor cell, interfering with the normal mechanisms that control the level of colonization with *Candida*, or does some undefined predisposing condition allow for the overgrowth of *Candida*, the fungus then being responsible for orchestration of immune dysfunction?

IMMUNOMODULATION IN EXPERIMENTAL SYSTEMS

Introduction

In view of the fact that zymosan and glucan, both extracted from *S. cerevisiae*, had been shown many years ago to have the capacity to modulate complement activity (von Dungern, 1900) and macrophage activiation (Riggi and Di Luzio, 1961), it is somewhat surprising that investigations of the immunomodulatory potential of *C. albicans*, a fungus containing similar components in its cell wall, did not begin to appear in the literature until the mid-1970s. When such studies did begin to appear, it was apparent that three lines of investigation were proceeding, namely, studies of the influence of polysaccharides derived from *Candida* on human PBL or PMN in in vitro culture systems, and studies in in vivo models in which *Candida* or its components were examined for their modulatory effects on acquired immune responses or for their antitumor effects. Depending upon the assay system and candidal preparations used, both stimulatory and suppressive effects have been noted. These are detailed and references cited in the next two sections.

In Vitro Studies with Human Cells

In a rather extensive series of experiments done in the same laboratory, the effects of a polysaccharide containing glucose, mannose, and 3.1% protein, designated MPPS, have been studied on human PBL cultured in vitro (Piccolella et al., 1980, 1981a, 1981b; Lombardi et al., 1984, 1985a; Wirz et al., 1984). Initially, these investigators described the effect of MPPS on PBL as being "mitogenic" (Piccolella et al., 1980, 1981b), but in view of the fact that most humans have an underlying sensitivity to candidal antigens, it is difficult to envision the responses they observed as being truly mitogenic. They had suggested they were mitogenic because they occurred in the absence of monocytes. The proliferative responses were more likely *Candida*-specific, however, as suggested by these same investigators in a later study (Lombardi et al., 1984), as well as by Ausiello et al. (1986). In addition, some confusion arises as to the cellular requirements of the proliferative response. In their early work, Piccolella et al. (1980) had suggested that monocytes were not involved in the response of PBL to MPPS, but in a later study they suggested that lymphocyte proliferation in response to MPPS was monocyte-dependent (Lombardi et al., 1984).

Ausiello et al. (1986) used a preparation of the *Candida* cell extracted in a manner similar to that described for MPPS, and compared the proliferative response of human PBL to that extract as well as to mannan, insoluble glucan, a mannan-rich preparation removed from cell walls with hot alkali, and inactivated whole cells. The only fractions that did not stimulate the PBL were the glucan and the mannan-rich extract. Interestingly, they did find that the mannan, while it was a minimally stimulatory agent for PBL, and a mannan-rich preparation that was never stimulatory for PBL, could both inhibit the proliferative response to intact *C. albicans* if the PBL were preincubated with the extracts prior to the addition of the intact yeasts. Since mannan has been shown by others to be immunosuppressive by virtue of its ability to interfere with monocyte processing of candidal antigen (Fischer et al, 1982), it is possible that the immunosuppressive phenomenon described by Ausiello et al. (1986) was the result of inhibition of normal monocyte function.

The results of several recent studies, in fact, suggest that during short incubation periods, MPPS activates macrophages to release interleukin-1 (IL-1) (Lombardi et al., 1984), which then promotes increased proliferation of T cells, but that with prolonged incubation, i.e., 6 days, nonspecific inhibitors are produced (Lombardi et al., 1985a, 1985b) which block the production of IL-1 by macrophages. Further, the production if interleukin-2 (IL-2) may be suppressed when MPPS-stimulated cells are incubated with the nonspecific inhibitor. The number of IL-1-responsive cells appears to remain the same, however, because the addition of exogenous IL-2 reversed the effect. What is clear from the studies is that the in vitro response of PBL to specific antigens—the same phenomena occurred with purified protein derivative (Lombardi et al., 1984)—is a modulated event which progresses through a series of stages involving initially stimulatory events and eventually inhibitory events, all of which are orchestrated by the production of soluble factors, and that mannan, if presented to monocytes in the appropriate form, suppresses the development of the stimulatory phase of the response.

The most definitive experiments with mannan and its suppressive effects on human cells have been done by Fischer et al. (1982) with MN, Nelson et al. (1984) with PBL, and by Wright et al. (1981, 1983, 1984) and Nelson et al. (1984) with PMN. Fischer et al. (1978) had noted in an earlier study that polysaccharide from *Candida* appeared to be involved in the depression of cellular immune phenomena. Subsequently, they determined that mannan interfered with *Candida* antigen presentation by adherent cells to autologous T lymphocytes. No suppressor cells were induced, nor was the mannan cytotoxic, but by comparing the uptake of radiolabeled mannan and pneumococcal polysaccharide, they observed that only the response of the MN to mannan was defective. Pneumococcal polysaccharide was ingested and degraded effectively. The cellular target for the inhibition of lymphoproliferative activity observed by Nelson et al. (1984) was not identified, but the phenomenon observed by them would appear to be different than that observed by Fischer et al. (1982). The inhibition observed by Nelson et al. (1984) in response to *S. cerevisiae* mannan was not limited to *Candida*-specific responses; bacterial and viral responses were inhibited as well. They felt, however, that residual copper in their mannan preparation contributed to the broadly-based suppression observed, because mannan which had passed through a copper-chelating resin was only inhibitory to lymphoproliferative responses initiated by *Candida* extract.

Interference by mannan with PMN function appears to be intimately associated with the interaction of mannan with myeloperoxidase (MPO) (Wright et al., 1981, 1983, 1984). The binding of MPO to *Candida* is required for killing, and soluble mannan antagonized that binding in a dose-dependent fashion (Wright et al., 1983). Furthermore, the inclusion of mannan in the incubation mixture with PMN and serum-opsonized zymosan resulted in inhibition of the respiratory burst and release of MPO. The release of other enzymes such as acid phosphatase, beta-glucuronidase, and lysozyme, was not inhibited (Wright et al., 1981), nor was the phagocytic event (Wright et al., 1983). Supporting evidence for the importance of MPO in the protective response has been provided by Wright and Nelson (1985) as well. They observed that mice inoculated intravenously (i.v.) with viable *C. albicans* could be significantly protected from death if treated with exogenous human MPO. For example, survival in the MPO-treated mice over the 60-day observation period was 80%, whereas only 25% of the controls survived over the same period. Further, the protection was abrogated if the MPO was mixed with mannan prior to inoculation into the infected animals. The interaction of mannan with MPO appears to occur through phosphate groups of the mannan outer chain (Wright et al., 1984). While the interference by mannan with PMN function may appear to have little to do with immunoregulation in candidiasis, there is evidence in the literature supporting the concept that oxygen metabolites interfere with normal lymphoproliferative responses (Metzger et al., 1980; Aune and Pierece, 1981; Deshazo et al., 1981; Nishida et al., 1981).

In Vitro Studies in Animal Models

Modulation of Acquired Immune Responses

Viable and nonviable intact *C. albicans*, as well as cell walls, and cytoplasmic or cell wall extracts, have been used to attempt to modulate various immune responses in animal models. Depending upon the system and modulator, suppression of immune responses, enhancement, or both, have been noted. Some of these responses have been summarized for presentation in Table 2. The primary assay systems involved for the detection of modulation were lymphocyte stimulation (LS), an in vitro correlate of cellular immunity, and plaque-forming cell assays (PFC), a measurement of the production of antigen-specific antibody-forming cells in the spleen or lymph nodes of animals. In many instances the responses measured were to mitogens or non-*Candida* antigens.

In general, i.v. inoculation of viable or nonviable *C. albicans* into conventional animals resulted in the suppression of LS responses to both mitogens and *Candida* antigen (Rogers and Balish, 1978a, 1978b; Skerl et al., 1980; Rivas and Rogers, 1983), whereas intraperitoneal (i.p.) inoculation of germ-free rats or conventional mice resulted in the observation of enhancement of LS or PFC phenomena (Rogers and Balish, 1978b; Cutler and Lloyd, 1982). The modulation observed was sometimes dose dependent (Skerl et al., 1980; Cutler and

Table 2 Evidence for Immunomodulation of Cellular or Humoral Responses by *C. albicans* or its Components In Vivo in Animal Models or In Vitro Using Animal Cells

Agent	Route/ animal	Lymphocyte stimulation Suppress/Enhance	Mitogens/Antigens	Humoral response (PFC)	DH	Reference
viable *C. albicans*	i.v./ mice	+ / − transient	Con A, PHA/PPD			Rogers and Balish, 1978a
	i.v./ germ-free rats	− / +	Con A, PHA/CA			Rogers and Balish, 1978b
	i.v./ conventional rats	+ / −	Con A, PHA/CA			Rogers and Balish, 1978b
	i.m./ mice	+ / −		SRBC,LPS unaffected		Vardinon and Segal, 1979
	i.v./ rabbits endocarditis	+ / −	CA			Skerl et al., 1980
	i.p./ mice	− / +		SRBC		Cutler and Lloyd, 1982

nonviable C. albicans	i.p./mice	− / +		SRBC/HE-CGG Cutler and Lloyd, 1982
	i.v./mice	+ / − transient	Con A, PHA LPS	Rivas and Rogers, 1983
	murine splenocytes in vitro	+ / −	PHA	Cuff et al., 1986
cell walls, polysaccharide	i.p./mice	− / +		SRBC Cutler and Lloyd, 1982
glycoprotein	i.v./mice	+ / −	GP	Carrow and Domer, 1985
mannan	i.p., i.v./mice	+ / +		SRBC/SSS-III Domer et al., 1986
glycoprotein	i.p., i.v./mice	− / +		SRBC/SSS-III Domer et al., 1987

Abbreviations: i.v., intravenous; i.m., intramuscular; i.p., intraperitoneal; Con A, concanavalin A; PHA, phytohemagglutinin; PPD, purified protein derivative; CA, C. albicans; LPS, lipopolysaccharide; SRBC, sheep erythrocytes; HE-CGG, horse-erythrocytes-chicken gamma globulin; GP, cell wall glycoprotein; SSS-III, pneumococcal polysaccharide Type III.

Lloyd, 1982) and in some instances transient (Rogers and Balish, 1978a; Rivas and Rogers, 1983), suggesting cyclic regulatory phenomena.

Modulation observed with subcellular components of *Candida* has again been variable, although investigators have relied more heavily on PFC data to prove their point than on manifestations of cell-mediated immunity. An exception to this generalization is the work of Carrow and Domer (1985), who showed that the i.v. inoculation of a cell wall-derived glycoprotein (GP) prior to immunization with viable *Candida* suppressed the inductive phase of cellular immunity demonstrable with the same preparation. The suppression was evident in vitro in an LS assay, as well as in vivo in the detection of delayed hypersensitivity. An important point with regard to their study was that the suppression was antigen specific. Mitogen responses, and in vitro and in vivo responses to other *Candida* antigens were not suppressed. The same extract, i.e., GP, an ethylenediamine extract, however, as well as a similar but not identical glycoprotein extracted with cold dilute NaOH, when inoculated into mice under the appropriate conditions with non-*Candida* antigens, either sheep erythrocytes (SRBC) or pneumococcal polysaccharide Type III, elicited only enhancement of the PFC response to both antigens (Domer et al., 1987). Cutler and Lloyd (1982) noted enhancement of the PFC response to SRBC as well when mice were immunized by the i.p. inoculation of a mixture of SRBC and cell wall polysaccharide removed with a phenol-water extraction procedure.

As mannan is believed to circulate during serious forms of candidiasis (Weiner and Yount, 1976; Kerkering et al., 1979; Lehmann and Reiss, 1980), it is of particular interest to attempt to determine the effect of mannan on immune responses. Domer et al. (1986), in fact, have published a study in which they investigated the effect of mannan on the PFC response to two different antigens, the T-helper cell-independent antigen, SSS-III, and the T-helper cell dependent antigen, SRBC. As a result of that investigation, it became clear that mannan extracted by the traditional method of Peat et al. (1961) was a heterogeneous mixture which contained components capable of either suppressing or enhancing antibody responses to non-*Candida* antigens. These components could be separated on the basis of size and charge into components having either no effect, or being capable of either stimulating or suppressing the PFC response. In general, the components of smaller molecular size were suppressive, while those which had lesser overall charge were enhancing. The modulatory effect observed did not seem to correlate with residual copper remaining from the extraction procedure, as had been suggested by Nelson et al. (1984), or with phosphorus or protein content. Furthermore, the mannan extracted contained mannose as the only sugar residue; there was no glucose, thus no glucan, present to stimulate the reticuloendothelial system (DiLuzio, 1983).

Determination of the mechanisms responsible for the observations of enhancement or suppression have been examined in only a few instances with varying degrees of success. Rivas and Rogers (1983) and Cuff et al. (1986) have evidence that the suppressor cell they generated in response to formalin-killed *C. albicans* was a B lymphocyte. It was not depleted by anti-Thy 1 or anti-Mac-1 treatment, and it was not plastic adherent. Moreover, it was susceptible to anti-Lyt 2.1, and could be separated from a mixture of cells by adherence to dishes coated with anti-mouse immunoglobulin F(ab')$_2$ fragments. Although Domer et al. (1986, 1987) have not yet determined the cell responsible for enhancement or suppression observed in their system, several possibilities have been ruled out, namely, neither of the two cell wall glycoproteins used by them, nor mannan (or its fractions) were mitogenic for normal mouse splenocytes, and the wall glycoproteins did not appear to stimulate the production of B-cell growth factors or IL-2. To the contrary, others have reported that the mannan from *S. cerevisiae* is mitogenic for normal mouse splenocytes (Mikami et al., 1982).

A plausible but unproven explanation for the modulatory effects noted with mannan or the wall glycoproteins is that they exert their effect by influencing regulatory T cells. Antibody responses to the two antigens used, SRBC and SSS-III, are known to be regulated by T lymphocytes (Markham et al., 1977a, 1977b; Yamaguchi and Kishimoto, 1978; Baker et al., 1982; Taylor et al., 1983a, 1983b). In addition, it is possible that one or all of the cell wall components activates macrophages. Both of the wall glycoproteins used by Domer et al. (1987) contained glucan, and glucan is known to be a stimulant of the reticuloendothelial system (Wooles and DiLuzio, 1963, 1964; DiLuzio, 1983). The extracts that are enhancing may stimulate the production of IL-1 by macrophages, which may in turn stimulate lymphocyte proliferation and enhanced antibody production. Although our initial unpublished observations were negative with regard to enhanced IL-1 production in the presence of wall glycoproteins, it is possible that the macrophage did, in fact, produce more IL-1, but it was not released under the in vitro culture conditions employed. Enhanced IL-1 production has been demonstrated in cultures of human cells which are incubated with a *Candida* polysaccharide (Lombardi et al., 1984), but that is presumed to be a *Candida*-specific response, not a nonspecific enhancement phenomenon.

Antitumor Effects

The antitumor effects of *C. albicans*, intact or fractionated, were not reported until the late 1970s (Bistoni et al., 1979; Weinberg and Hibbs, 1979), well after the antitumor effects of the cell wall components of *S. cerevisiae* (Diller et al., 1964; Suzuki et al., 1969) and other fungi (Chihara et al., 1969) had been noted. The initial studies were done with heat- (Weinberg and Hibbs, 1979) or merthiolate-inactivated (Bistoni et al., 1979) intact cells. Weinberg and Hibbs (1979) inoculated both *Candida* and tumor, a line-10 hepatoma cell, intradermally into guinea pigs, while Bistoni et al. (1979) inoculated the yeasts and tumor both i.p. A regimen involving the i.p. inoculation of *Candida* on days −14 and +1 with respect to inoculation of the tumor seemed most efficacious. Weinberg and Hibbs (1979) also looked at the tumoricidal capacity of murine peritoneal exudate cells three days after the i.p. inoculation of killed *Candida* and noted the presence of a highly tumoricidal population of macrophages. In addition to activated macrophages, however, the inoculation of inactivated *C. albicans* into the peritoneal cavity also induces the appearance of an expanded population of natural killer (NK) cells (Marconi et al., 1985a, 1985b). Some of the antitumor effects noted in systems where the tumor and the fungus are both inoculated i.p. could, therefore, be the result of both heightened NK and macrophage activity. The NK cells induced lysed only NK targets and had other properties consistent with NK cells.

The component(s) of *C. albicans* responsible for antitumor effect was in one case reported to have been glucan—not soluble mannan or glucan-protein complexes (Cassone et al., 1981, 1982)—in another case an extracellular polysaccharide consisting of about 75% mannose and 25% glucose (Saltarelli and Coppola, 1980), and in a third situation cell wall but not "particulate glucan" (Cassone et al., 1983). In each of these instances, the tumor against which the treatment was directed was different, but it is not clear why there should be such a discrepancy in protective effects. In view of the fact that different antitumor mechanisms may function against different tumors, e.g., NK cells versus activated macrophages, perhaps the differences in sensitivity to the various wall components reflects the differences in cell types stimulated by the fungal cell walls. Another possibility for the discrepancies observed between different systems is the possibility that different laboratories extract cell wall components in slightly different ways, and thus end up with products that are slightly different, e.g., in molecular size or proportion of specific intramolecular linkages. Matsumoto et al. (1980), for example, showed that mannan could be an effective antitumor agent against a mouse-implanted sarcoma 180 and Ehrlick-carcinoma solid tumor if the mannan had a minimum molecular weight. If, however, the molecular weight was cut by two-thirds, it reduced the antitumor activity by one-half.

SUMMARY

Candidiasis may either precede or follow severe modulations in the immune system of the host. The focus of this review has been to survey the data and current interpretations for potential factors responsible for these events of immunomodulation. The mere fact that *Candida* infections persist is evidence of some underlying abnormality, often associated with, but not exclusively restricted to, the cell-mediated immune system. In some instances, however, the cause and effect relationship is not clear, i.e., did infection with *Candida* initiate the immunosuppression, or did the underlying condition result in immunosuppression allowing for *Candida* to initiate disease? It is possible, however, that candidal infections may begin during minor immunosuppressive events, e.g., stress, pregnancy, or selected other primary infections, but then persist beyond these events because of an intrinsic or innate immunomodulatory defect. Under such circumstances, the initial imbalance of immune function should be corrected by normal homeostatic mechanisms, unless persistent colonization with *Candida* perpetuates the imbalance through the production or release of immunomodulatory factors.

One important target for research in this area, then, is the identification and purification of immunomodulatory factors produced or released during disease. To date, only preliminary data are available showing that the immunoregulatory potential of *Candida* resides in various candidal extracts, especially in the cell wall. Although the relevance of the data gathered in the experimental models might initially appear questionable, the fact that mannan, or molecules containing mannan, are known to circulate during disease (Weiner and Yount, 1976; Kerkering et al., 1979; Lehmann and Reiss, 1980) lends credence to the hypothesis.

A second important target for future research is the identification of the cellular target within the immune system that responds to the *Candida*-derived immunomodulators. The success of these studies may well depend upon the degree of purification of the responsible factors. In fact, much of the variability observed to date in modulatory events may result from the heterogeneity of the modulator, including the possibility that antagonistic or synergistic interactions of the individual components occur. The variability observed in certain clinical settings could result from basic flaws in the normal immunoregulatory pathways in the host also, and if a link could be established between the basic flaws, the candidal extracts, and the target cell of the candidal extracts, it may be possible to manipulate the system through immunotherapy. Finally, the characterization of the candidal substances may provide yet another clinical tool for use as an immunomodulator in such disorders as cancer, inheritable immunodeficiencies, and AIDS.

REFERENCES

Araj, G. F., Hopfer, R. L., Chesnut, S., Fainstein, V., and Body, G. P., Sr. (1982). Diagnostic value of the enzyme-linked immunosorbent assay for detection of *Candida albicans* cytoplasmic antigen in sera of cancer patients. *J. Clin. Microbiol.* 16: 46-52.

Aronson, I. K., Rieger, C. H. L., Soltani, K., Tkalcevic, V., Chan, W. C., Lorincz, A. L., and Matz, G. (1979). Late onset chronic mucocutaneous candidiasis with lymphoma and specific serum inhibitory factor. *Cancer 43*: 101-108.

Aune, T. M., and Pierce, C. W. (1981). Conversion of soluble immune response suppressor to macrophage-derived suppressor factor by peroxide. *Proc. Natl. Acad. Sci. U.S.A. 78*: 5099-5103.

Ausiello, C. M., Spagnoli, G. C., Boccanera, M., Casalinuovo, I., Malavasi, F., Casciani, C. U., and Cassone, A. (1986). Proliferation of human peripheral blood mononuclear cells induced by *Candida albicans* and its cell wall fractions. *J. Med. Microbiol. 22*: 195-202.

Baker, P. J., Amsbaugh, D. F., Stashak, P. W., Caldes, G., and Prescott, B. (1982). Direct evidence for the involvement of T suppressor cells in the expression of low-dose paralysis to type III pneumococcal polysaccharide. *J. Immunol. 128*: 1059-1062.

Barnaba, V., Zaccari, C., Levrero, M., and Balsano, F. (1985). Suppressor T cells role in the unresponsiveness to *Candida albicans* in chronic mucocutaneous candidiasis. *Boll. Ist. Sieroter. Milan 64*: 126-130.

Bistoni, F., Marconi, P., Pitzurra, M., Frati, L., Spreafico, F., Goldin, A., and Bonmassar, E. (1979). Combined effects of BCG or *Candida albicans* (CA) with antitumor agents against a virus-induced lymphoma in mice. *Europ. J. Cancer 15*: 1305-1314.

Blackstock, R., McCormack, J. M., and Hall, N. K. (1987). Induction of a macrophage-suppressive lymphokine by soluble cryptococcal antigens and its association with models of immunologic tolerance. *Infect. Immun. 55*: 233-239.

Borkowsky, W., and Valentine, F. T. (1979). The proliferative response of human lymphocytes to antigen is suppressed preferentially by lymphocytes precultured with the same antigen. *J. Immunol. 122*: 1867-1873.

Bortolussi, R., Faulkner, G., Lee, S. H. S., and Ozere, R. (1981). Phagocytosis of *Candida albicans* in chronic mucocutaneous candidiasis. *Pediatr. Res. 15*: 1287-1292.

Breen, J. F., Lee, I. C., Vogel, F. R., and Friedman, H. (1982). Cryptococcal capsular polysaccharide-induced modulation of murine immune responses. *Infect. Immun. 36*: 47-51.

Buckley, R. H., Lucas, Z. J., Hattler, B. G., Jr., Zmijewski, C. M., and Amos, D. B. (1968). Defective cellular immunity associated with chronic mucocutaneous moniliasis and recurrent staphylococcal botryomycosis: immunologic reconstitution by allogenic bone marrow. *Clin. Exp. Immunol. 3*: 152-169.

Canales, L., Middlemas, R. O., III, Louro, J. M., and South, M. A. (1969). Immunological observations in chronic mucocutaneous candidiasis. *Lancet 2*: 567-571.

Carrow, E. W., and Domer, J. E. (1985). Immunoregulation in experimental murine candidiasis: specific suppression induced by *Candida albicans* cell wall glycoprotein. *Infect. Immun. 49*: 172-181.

Cassone, A., Bistoni, F., Cenci, E., Pesce, C. D., Tissi, L., and Marconi, P. (1982). Immunopotentiation of anticancer chemotherapy by *Candida albicans*, other yeasts and insoluble glucan in an experimental lymphoma model. *Sabouraudia 20*: 115-125.

Cassone, A., Marconi, P., Bistoni, F., Mattia, E., Sbaraglia, G., Garaci, E., and Bonmassar, E. (1981). Immunoadjuvant effects of *Candida albicans* and its cell wall fractions in a mouse lymphoma model. *Cancer Immunol. Immunother. 10*: 181-189.

Cassone, A., Scaringi, L., Pesce, C. D., Titti, F., Bistoni, F., Marconi, P., and Rossi, G. B. (1983). Suppression of Friend leukemia cell-induced tumors by cellular preparations of *Candida albicans*. *Microbiologica 6*: 207-220.

Cates, K. L., Grady, P. G., Shapira, E., and Davis, A. T. (1980). Cell-directed inhibition of polymorphonuclear leukocyte chemotaxis in a patient with mucocutaneous candidiasis. *J. Allergy Clin. Immunol. 65*: 431-435.

Chihara, G., Maeda, Y., Hamuro, J., Sasaki, T., and Fukuoka, F. (1969). Inhibition of mouse sarcoma 180 by polysaccharides from *Lentinus edodes* (Berk.) Sing. *Nature 222*: 687-688.

Chilgren, R. A., Meuwissen, H. J., Quie, P. G., and Hong, R. (1967). Chronic mucocutaneous candidiasis, deficiency of delayed hypersensitivity, and selective local antibody defect. *Lancet 2*: 688-693.

Cohen, R., Roth, F. J., Delgado, E., Ahearn, D. G., and Kalser, M. H. (1969). Fungal flora of the normal human small and large intestine. *N. Engl. J. Med. 280*: 638-641.

Cuff, C. F., Rogers, C. M., Lamb, B. J., and Rogers, T. J. (1986). Induction of suppressor cells in vitro by *Candida albicans*. *Cell. Immunol. 100*: 47-56.

Cutler, J. E., and Lloyd, R. K. (1982). Enhanced antibody responses induced by *Candida albicans* in mice. *Infect. Immun. 38*: 1102-1108.

Deshazo, R. D., Ewel, C., Londono, S., Metzger, Z., Hoffeld, J. T., and Oppenheim, J. J. (1981). Evidence for the involvement of monocyte-derived toxic oxygen metabolites in the lymphocyte dysfunction of Hodgkin's disease. *Clin. Exp. Immunol. 46*: 313-320.

Diller, I. C., Fisher, M. E., and Gable, D. (1964). Effect of glucan on sarcoma 37. *Proc. Soc. Exp. Biol. Med. 117*: 107-110.

DiLuzio, N. R. (1983). Immunopharmacology of glucan: a broad spectrum enhancer of host defense mechanisms. *Trends Pharmaceut. Sci. 4*: 344-347.

Djawari, D., Bischoff, T., and Hornstein, O. P. (1978). Impairment of chemotactic activity of microphages in chronic mucocutaneous candidosis. *Arch. Dermatol. Res. 262*: 247-253.

Djawari, D., Hornstein, O. P., Gross, J., and Meinhof, W. (1977). Defect of phagocytosis and intracellular killing of *Candida albicans* by granulocytes in patients with familiar and non-familiar chronic mucocutaneous candidosis. *Arch. Dermatol. Res. 260*: 159-161.

Domer, J. E., Elkins, K., Ennist, D., Stashak, P. W., Garner, R. E., and Baker, P. J. (1987). Immunoenhancement of non-*Candida* antibody responses by *Candida albicans* cell wall glycoprotein. *Infect. Immun. 55*: 2619-2624.

Domer, J. E., Stashak, P. W., Elkins, K., Prescott, B., Caldes, G., and Baker, P. J. (1986). Separation of immunomodulatory effects of mannan from *Candida albicans* into stimulatory and suppressive components. *Cell. Immunol. 101*: 403-414.

Dutta, S. K., and Al-Ibrahim, M. S. (1978). Immunological studies in acute pseudomembranous esophageal candidiasis. *Gastroenterology 75*: 292-296.

Fischer, A., Ballet, J. J., and Griscelli, C. (1978). Specific inhibition of in vitro *Candida*-induced lymphocyte proliferation by polysaccharidic antigens present in serum of patients with chronic mucocutaneous candidiasis. *J. Clin. Invest. 62*: 1005-1013.

Fischer, A., Pichat, L., Audinot, M., and Griscelli, C. (1982). Defective handling of mannan by monocytes in patients with chronic mucocutaneous candidiasis resulting in a specific cellular unresponsiveness. *Clin. Exp. Immunol. 47*: 653-660.

Gatenby, P., Basten, A., and Adams, E. (1980). Thymoma and late onset mucocutaneous candidiasis associated with a plasma inhibitor of cell mediated immune function. *J. Clin. Lab. Immunol. 3*: 209-216.

Kennedy, C. T. C., Valdimarsson, H., and Hay, R. J. (1981). Chronic mucocutaneous candidiasis with a serum-dependent neutrophil defect: response to ketoconazole. *J. Royal Soc. Med. 74*: 158-161.

Kerkering, T. M., Espinel-Ingroff, A., and Shadomy, S. (1979). Detection of *Candida* antigenemia by counterimmunoelectrophoresis in patients with invasive candidiasis. *J. Infect. Dis. 140*: 659-664.

Kirkpatrick, C. H., Chandler, J. W., and Schimke, R. N. (1970). Chronic mucocutaneous moniliasis with impaired delayed hypersensitivity. *Clin. Exp. Immunol. 6*: 375-385.

Kirkpatrick, C. H., and Smith, T. K. (1974). Chronic mucocutaneous candidiasis: immunologic and antibiotic therapy. *Ann. Intern. Med. 80*: 310-320.

Klein, R. S., Harris, C. A., Small, C. B., Moll, B., Lesser, M., and Friedland, G. H. (1984). Oral candidiasis in high-risk patients as the initial manifestation of the acquired immunodeficiency syndrome. *N. Engl. J. Med. 311*: 354-358.

Laforce, F. M., Mills, D. M., Iverson, K., Cousins, R., and Everett, E. D. (1975). Inhibition of leukocyte candidacidal activity by serum from patients with disseminated candidiasis. *J. Lab. Clin. Med. 86*: 657-666.

Lee, W. M., Holley, H. P., Jr., Stewart, J., and Galbraith, G. M. P. (1986). Case report: refractory esophageal candidiasis associated with a low-molecular weight plasma inhibitor of T-lymphocyte function. *Am. J. Med. Sci. 292*: 47-52.

Lehmann, P. F., and Reiss, E. (1980). Detection of *Candida albicans* mannan by immunodiffusion, counterimmunoelectrophoresis, and enzyme-linked immunoassay. *Mycopathologia 70*: 83-88.

Lehner, T., Wilton, J. M. A., and Ivanyi, L. (1972). Immunodeficiencies in chronic muco-cutaneous candidosis. *Immunology 22*: 775-786.

Lombardi, G., Piccolella, E., Vismara, D., Colizzi, V., and Asherson, G. L. (1984). *Candida albicans* polysaccharide extract (MPPS) and PPD stimulate the production of interleukin-1 and lymphocyte proliferation. *Clin. Exp. Immunol. 58*: 581-586.

Lombardi, G., Piccolella, E., Vismara, D., Colizzi, V., and Zembala, M. (1985a). Monocyte subsets in the production of inhibitory factor by *Candida albicans*-activated human T cells. *Immunology 56*: 373-376.

Lombardi, G., Vismara, D., Piccolella, E., Colizzi, V., and Asherson, G. L. (1985b). A non-specific inhibitor produced by *Candida albicans* activated T cells impairs cell proliferation by inhibiting interleukin-1 production. *Clin. Exp. Immunol. 60*: 303-310.

Marconi, P., Scaringi, L., Cassone, A., and Tissi, L. (1985a). Effect of inactivated *albicans* on natural killer (NK) activity and blastogenesis in mice. In *Genetic and Phenotypic Markers of Tumors*. Edited by Aronson, S. A., Frati, L., and Verna, R. Plenum Publishing Corporation, New York, pp. 145-151.

Marconi, P., Scaringi, L., Tissi, L., Boccanera, M., Bistoni, F., Bonmassar, E., and Cassone, A. (1985b). Induction of natural killer cell activity by inactivated *Candida albicans* in mice. *Infect. Immun. 50*: 297-303.

Markham, R. B., Reed, N. D., Stashak, P. W., Prescott, B., Amsbaugh, D. F., and Baker, P. J. (1977a). Effect of concanavalin A on lymphocyte interactions involved in the antibody response to type III pneumococcal polysaccharide. II. Ability of suppressor T cells to act on both B cells and amplifier T cells to limit the magnitude of the antibody response. *J. Immunol. 119*: 1163-1168.

Markham, R. B., Stashak, P. W., Prescott, B., Amsbaugh, D. F., and Baker, P. J. (1977b). Effect of concanavalin A on lymphocyte interactions involved in the antibody response to type III pneumococcal polysaccharide. I. Comparison of the suppression induced by Con A and low dose paralysis. *J. Immunol. 118*: 952-956.

Matsumoto, T., Takanohashi, M., Okubo, Y., Suzuki, M., and Suzuki, S. (1980). Growth-inhibitory activity of the D-mannan of *Saccharomyces cerevisiae* X2180-1A-5 mutant strain against mouse-implanted sarcoma 180 and Ehrlick-carcinoma solid tumor. *Carbohydr. Res. 83*: 363-370.

Melbye, M., Schonheyder, H., Kestens, L., Stenderup, A., Gigase, P. L., Ebbesen, P., and Biggar, R. J. (1985). Carriage of oral *Candida albicans* associated with high number of circulating suppressor T lymphocytes. *J. Infect. Dis. 152*: 1356-1357.

Metzger, A., Hoffeld, J. T., and Oppenheim, J. J. (1980). Macrophage-mediated suppression. I. Evidence for participation of both hydrogen peroxide and prostaglandins in suppression of murine lymphocyte proliferation. *J. Immunol. 124*: 983-988.

Mikami, T., Nagase, T., Matsumoto, T., Suzuki, M., Suzuki, S., and Kumano, N. (1982). Mitogenic effect of the mannans from *Saccharomyces cerevisiae* on mouse spleen lymphocytes. *Microbiol. Immunol. 26*: 913-922.

Mobacken, H., Lindholm, L., and Olling, S. (1977). Deficient neutrophil function in a patient with chronic mucocutaneous candidiasis, thymoma and myasthenia gravis. *Acta Dermatovener. 57*: 335-339.

Murphy, J. W., and Moorhead, J. W. (1982). Regulation of cell-mediated immunity in cryptococcosis. I. Induction of specific afferent T suppressor cells by cryptococcal antigen. *J. Immunol. 128*: 276-283.

Myerwitz, R. L., Pazin, G. J., and Allen, C. M. (1977). Disseminated candidiasis—changes in incidence, underlying diseases, and pathology. *Am. J. Clin. Pathol. 68*: 29-38.

Nelson, R. D., Herron, M. J., McCormack, R. T., and Gehrz, R. Z. (1984). Two mechanisms of inhibition of human lymphocyte proliferation by soluble yeast mannan polysaccharide. *Infect. Immun. 43*: 1041-1046.

Nishida, Y., Tanimoto, K., and Akaoka, I. (1981). Effect of free radicals on lymphocyte response to mitogens and rosette formation. *Clin. Immunol. Immunopathol. 19*: 319-324.

Ohno, N., Mimura, H., Suzuki, I., and Yadomae, T. (1985). Antitumor activity and structural characterization of polysaccharide fractions extracted with cold alkali from a fungus, *Peziza vesiculosa*. *Chem. Pharm. Bull. 33*: 2564-2568.

Parker, J. C., Jr., McCloskey, J. J., and Knauer, K. A. (1976). Pathobiologic features of human candidiasis. A common deep mycosis of the brain, heart and kidney in the altered host. *Am. J. Clin. Pathol. 65*: 991-1000.

Paterson, P. Y., Semo, R., Blumenschein, G., and Swelstad, J. (1971). Mucocutaneous candidiasis, anergy and a plasma inhibitor of cellular immunity: reversal after amphotericin B therapy. *Clin. Exp. Immunol. 9*: 595-602.

Peat, S., Whelan, W. J., and Edwards, T. E. (1961). Polysaccharide of baker's yeast. Part IV. Mannan. *J. Chem. Soc. (London) 1*: 29-34.

Piccolella, E., Lombardi, G., and Morelli, R. (1980). Human lymphocyte-activating properties of a purified polysaccharide from *Candida albicans*: B and T cell cooperation in the mitogenic response. *J. Immunol. 125*: 2082-2088.

Piccolella, E., Lombardi, G., and Morelli, R. (1981a). Generation of suppressor cells in the response of human lymphocytes to a polysaccharide from *Candida albicans*. *J. Immunol. 126*: 2151-2159.

Piccolella, E., Lombardi, G., and Morelli, R. (1981b). Mitogenic response of human peripheral blood lymphocytes to a purified *C. albicans* polysaccharide fraction: lack of helper activities is responsible for the in vitro unresponsiveness to a second antigenic challenge. *J. Immunol. 126*: 2156-2160.

Piccolella, E., Vismara, D., Lombardi, G., Guerritore, D., Piantelli, M., and Ranelletti, F. O. (1985). Effect of glucocorticoids on the development of suppressive activity in human lymphocyte response to a polysaccharide purified from *Candida albicans*. *J. Immunol. 134*: 1166-1171.

Pillemer, L., and Ecker, E. E. (1941). Anticomplementary factor in fresh yeast. *J. Biol. Chem. 137*: 139-142.

Rifkind, D., Marchioro, T. L., Schneck, S. A., and Hill, R. B., Jr. (1967). Systemic fungal infections complicating renal transplantation and immunosuppressive therapy. *Am. J. Med. 43*: 28-38.

Riggi, S. J., and DiLuzio, N. R. (1961). Identification of a RE-stimulating agent in zymosan. *Am. J. Physiol. 200*: 297-300.

Rivas, V., and Rogers, T. J. (1983). Studies on the cellular nature of *Candida albicans*-induced suppression. *J. Immunol. 130*: 376-379.

Rogers, T. J., and Balish, E. (1978a). Suppression of lymphocyte blastogenesis by *Candida albicans*. *Clin. Immunol. Immunopathol. 10*: 298-305.

Rogers, T. J., and Balish, E. (1978b). Effect of systemic candidiasis on blastogenesis of lymphocytes from germfree and conventional rats. *Infect. Immun. 20*: 142-150.

Ruiz-Arguelles, A., Jett, J. R., and Ritts, R. E., Jr. (1983). Impaired generation of helper T cells in a patient with chronic mucocutaneous candidiasis and malignant thymoma. *J. Clin. Lab. Immunol. 10*: 165-169.

Sacchi, F., Cisternino, M. A., Clivio, A., Ferrari, F. A., Forunato, A., Nespoli, L., Beretta, A., and Siccardi, A. G. (1979). Disorders of candidacidal activity and other neutrophil functions in three cases of chronic candidiasis. *Helv. Paediat. Acta 34*: 607-616.

Saltarelli, C. G., and Coppola, C. P. (1980). Inhibitory effects of *Candida albicans* extracellular polysaccharides on mouse sarcoma 180. *J. Surg. Oncol. 15*: 99-106.

Silva, C. L., and Fazioli, R. A. (1985). A *Paracoccidioides brasiliensis* polysaccharide having granuloma inducing, toxic and macrophage-stimulating activity. *J. Gen. Microbiol. 131*: 1497-1501.

Skerl, K. G., Scheld, W. M., Alliegro, G. M., and Calderone, R. A. (1980). Lymphocyte blastogenesis during experimental endocarditis caused by *Candida albicans*. *J. Reticuloendothel. Soc. 28*: 495-506.

Suzuki, S., Suzuki, M., Hatsukaiwa, H., Sunayama, H., Suzuki, T., Uchiyama, M., Fukuoka, F., Nakanishi, M., and Akiya, S. (1969). Anti-tumor activity of polysaccharides. III. Growth-inhibitory activity of purified mannan and glucan fractions from baker's yeast against sarcoma-180 solid tumor. *Gann 60*: 273-277.

Takeya, K., Nomoto, K., Matsumoto, T., Miyake, T., and Himeno, K. (1976). Chronic mucocutaneous candidiasis accompanied by enhanced antibody production. *Clin. Exp. Immunol. 25*: 497-500.

Taylor, C. E., Amsbaugh, D. F., Stashak, P. W., Caldes, G., Prescott, B., and Baker, P. J. (1983a). Cell surface antigens and other characteristics of T cells regulating the antibody response to type III pneumococcal polysaccharide. *J. Immunol. 130*: 19-23.

Taylor, C. E., Stashak, P. W., Caldes, G., Prescott, B., Chused, T. E., Brooks, A., and Baker, P. J. (1983b). Activation of antigen-specific suppressor T cells by B cells from mice immunized with type III pneumococcal polysaccharide. *J. Exp. Med. 158*: 703-717.

Twomey, J., Waddell, C. C., Krantz, S., O'Reilly, R., L'Esperance, P., and Good, R. A. (1975). Chronic mucocutaneous candidiasis with macrophage dysfunction, a plasma inhibitor, and co-existent aplastic anemia. *J. Lab. Clin. Med. 85*: 968-977.

Valdimarsson, H., Higgs, J. M., Wells, R. S., Yamamura, M., Hobbs, J. R., and Holt, P. J. L. (1973). Immune abnormalities associated with chronic mucocutaneous candidiasis. *Cell. Immunol. 6*: 348-361.

Van Der Meer, J. W. M., Leijh, P. C. J., Van Den Barselarr, M., and Van Furth, R. (1978). Functions of phagocytic cells in chronic mucocutaneous candidiasis. *Br. Med. J. 1*: 147-148.

Vardinon, N., and Segal, E. (1979). Suppressive action of *Candida albicans* on the immune response in mice. *Exp. Cell Biol. 47*: 275-280.

Verhaegen, H., DeCock, W., and DeCree, J. (1976). In vitro phagocytosis of *Candida albicans* by peripheral polymorphonuclear neutrophils of patients with recurrent infections. Case reports of serum-dependent abnormalities. *Biomedicine 24*: 164-170.

von Dungern, F. (1900). Beiträge zur Immunitätslehre. *Munch. med. Woch. 47*: 677-680.

Walker, S. M., and Urbaniak, S. J. (1980). A serum-dependent defect of neutrophil function in chronic mucocutaneous candidiasis. *J. Clin. Pathol. 33*: 370-372.

Weinberg, J. B., and Hibbs, J. B., Jr. (1979). Enhanced macrophage tumoricidal activity and tumor suppression or regression caused by heat-killed *Candida albicans*. *J. Nat. Cancer Inst. 63*: 1273-1278.

Weiner, M. H., and Yount, W. J. (1976). Mannan antigenemia in the diagnosis of invasive *Candida* infections. *J. Clin. Invest. 58*: 1045-1053.

Wingard, J. R., Merz, W. G., and Saral, R. (1979). *Candida tropicalis*. A major pathogen in immunocompromised patients. *Ann. Intern. Med. 91*: 539-543.

Wirz, M., Lombardi, G., Pugliese, O., Morelli, R., and Piccolella, E. (1984). A purified polysaccharide isolated from *Candida albicans* induces antibody response in vitro by human peripheral blood lymphocytes and discriminates between sera from normal and *Candida albicans*-infected individuals. *Clin. Immunol. Immunopathol. 33*: 199-209.

Witkin, S. S. (1986). Inhibition of *Candida*-induced lymphocyte proliferation by antibody to *Candida albicans*. *Obstet. Gynecol. 68*: 696-699.

Witkin, S. S., Hirsch, J., and Ledger, W. J. (1986). A macrophage defect in women with recurrent *Candida* vaginitis and its reversal in vitro by prostaglandin inhibitors. *Am. J. Obstet. Gynecol. 155*: 790-795.

Witkin, S. S., Yu, R., and Ledger, W. J. (1983). Inhibition of *Candida albicans*-induced lymphocyte proliferation by lymphocytes and sera from women with recurrent vaginitis. *Am. J. Obstet. Gynecol. 147*: 809-811.

Wooles, W. R., and DiLuzio, N. R. (1963). Reticuloendothelial function and the immune response. *Science 142*: 1078-1080.

Wooles, W. R., and DiLuzio, N. R. (1964). The phagocytic and proliferative response of the reticuloendothelial system following glucan administration. *J. Reticuloendothel. Soc. 1*: 160-169.

Wright, C. D., Bowie, J. U., Gray, G. R., and Nelson, R. D. (1983). Candidacidal activity of myeloperoxidase: mechanisms of inhibitory influence of soluble cell wall mannan. *Infect. Immun. 42*: 76-80.

Wright, C. D., Bowie, J. U., and Nelson, R. D. (1984). Influence of yeast mannan in release of myeloperoxidase by human neutrophils: determination of structural features of mannan required for formation of myeloperoxidase-mannan-neutrophil complexes. *Infect. Immun. 43*: 467-471.

Wright, C. D., Herron, M. J., Gray, G. R., Holmes, B., and Nelson, R. D. (1981). Influence of yeast mannan on human neutrophil functions: inhibition of release of myeloperoxidase related to carbohydrate-binding property of the enzyme. *Infect. Immun. 32*: 731-738.

Wright, C. D., and Nelson, R. D. (1985). Candidacidal activity of myeloperoxidase: therapeutic influence of the enzyme in vivo. *Infect. Immun. 47*: 363-365.

Yamaguchi, K., and Kishimoto, S. (1979). Distinction between suppressors of the delayed-type hypersensitivity and the humoral response to sheep erythrocytes. *Immunology 35*: 721-731.

13
Immunoregulation in Cryptococcosis

JUNEANN W. MURPHY
University of Oklahoma, Norman, Oklahoma

INTRODUCTION

Cryptococcosis is a mycotic disease that is acquired by inhalation of the desiccated blastospores or possibly the basidiospores of *Cryptococcus neoformans* or *Filobasidiella neoformans*, respectively (Gordon, 1975; Staib, 1981; Zimmer et al., 1981). Upon inhalation, the organisms may be trapped and reside in the upper respiratory tract for a time before getting into the lungs, or the cryptococci may enter directly into the alveolar spaces (Gordon, 1975; Randhawa and Paliwal, 1977; Cohen, 1982). In either case, pulmonary cryptococcosis is generally subclinical and self-limiting (Campbell, 1966; Hatcher et al., 1971; Duperval et al., 1977; Emmons et al, 1977; Kerkering et al., 1981; Rippon, 1982; Feigin, 1983). If symptoms occur, they may range from very mild to severe pneumonialike (Emmons et al., 1977; Kerkering et al., 1981; Rippon, 1982; Feigin, 1983). In about 10% of the individuals who have primary pulmonary cryptococcosis, the cryptococci get into the blood stream and dessiminate to other tissues (Gordon, 1975). Since the organism has a predilection for the central nervous system, the majority of cryptococcal disease is diagnosed as meningeal cryptococcosis (Littman and Zimmerman, 1956; Emmons et al., 1977; Rippon, 1983). Once the cryptococci disseminate from the primary lung foci, patients usually have detectable levels of cryptococcal antigen in their serum and/or spinal fluid (Neill et al., 1951; Bloomfield et al., 1963; Gordon and Vedder, 1966; Bindschadler and Bennett, 1968; Bennett and Bailey, 1971; Goodman et al., 1971; Diamond and Bennett, 1974; Wold et al., 1980; Young et al., 1980; Kauffman et al., 1981; Eng et al., 1983). It is well documented that increasing cryptococcal antigen titers are indicative of a poor prognosis, and conversely, decreasing antigen titers signify a good prognosis (Gordon and Vedder, 1966; Diamond and Bennett, 1974).

C. neoformans, an encapsulated, yeastlike organism, is an ubiquitous microbe; therefore, humans are frequently exposed yet the incidence of cryptococcosis in normal individuals is relatively low (Emmons et al., 1977; Rippon, 1982). In contrast, individuals with depressed cell-mediated immune function due to underlying malignancy, chemotherapy, or infectious disease such as acquired immune deficiency syndrome (AIDS) are very susceptible to infection with *C. neoformans* (Lewis and Rabinovich, 1972; Emmons et al., 1977; Rippon, 1982; Perfect et al., 1983; Kovacs et al., 1985). Thus, it appears that normal host defense mechanisms are relatively effective against *C. neoformans*.

Most likely, when an individual is exposed to a low infective dose of *C. neoformans*, the majority of the organisms entering the lungs are eliminated by natural cellular resistance mechanisms mediated by polymorphonuclear leukocytes (PMNL), natural killer (NK) cells, and, to a lesser extent, alveolar macrophages. However, when the initial infective dose of cryptococci is large, then the natural cellular defenses usually are not capable of completely clearing the organisms from the lungs, and the host must rely on cell-mediated immune mechanisms to effect the clearance of the organisms. In the majority of immunologically competent individuals infected with *C. neoformans*, eradication of the organism results from a combination of natural and immune resistance mechanisms.

The humoral immune response to *C. neoformans* has been studied in both humans (Gordon and Vedder, 1966; Walter and Jones, 1968; Diamond and Bennett, 1974; Reiss et al., 1984) and animal models (Cauley and Murphy, 1979; Scott et al., 1981). Anticryptococcal antibodies appear early in the disease in patients with nonmeningeal cryptococcosis and occasionally after successful therapy (Gordon and Vedder, 1966; Reiss et al., 1984). Although anticryptococcal antibodies are not detectable in most cryptococcosis patients, when antibodies are detected, rising titers signal a favorable outcome (Gordon and Vedder, 1966; Diamond and Bennett, 1974). These observations may seem to suggest that anticryptococcal antibodies play a role in protection, but this is a controversial issue because animal studies have not consistently supported this concept (Goren, 1967; Monga et al., 1979; Graybill et al., 1981; Perfect et al., 1981; Dromer et al., 1987). The fact that demonstration of anticryptococcal antibodies during cryptococcosis is highly variable, together with the fact that Henderson et al. (1982, 1986) could not stimulate the production of anticryptococcal antibodies in humans cured of cryptococcosis to the same level as produced in control subjects, suggests that anticryptococcal antibody production is controlled either by the genetic makeup of the host or by immunoregulatory mechanisms or both. Animal studies tend to support the concept that the humoral immune response to cryptococcal antigen is easily down-regulated, although this does not eliminate the possibility of the genotype of the host being important as well.

Unlike the humoral immune response, the cell-mediated immune (CMI) response against cryptococci is clearly an important protective mechanism (Abrahams and Gilleran, 1960; Abrahams, 1966; Adamson and Cozad, 1969; Diamond, 1977; Graybill and Drutz, 1978; Cauley and Murphy, 1979; Fromtling et al., 1979a; Graybill et al., 1979; Lim and Murphy, 1980; Lim et al., 1980; Fung and Murphy, 1982). In humans, cell-mediated immunity has been assessed by measuring in vivo and in vitro responses to cryptococci or cryptococcal antigens (Salvin and Smith, 1961; Bennett et al., 1965; Atkinson and Bennett, 1968; Muchmore et al., 1969; Diamond and Bennett, 1973; Graybill and Alford, 1974; Schimpff and Bennett, 1975; Miller and Puck, 1984). Individuals who have recovered from cryptococcosis have positive CMI responses to cryptococcal antigen (Atkinson and Bennett, 1968; Diamond and Bennett, 1973; Graybill and Alford, 1974; Schimpff and Bennett, 1975; Miller and Puck, 1984); however, frequently those responses are lower than responses in normal control subjects who have been sensitized to the organism (Diamond and Bennett, 1973; Schimpff and Bennett, 1975). These observations, as with the humoral immune response, suggest immunoregulatory mechanisms are affecting the anticryptococcal CMI response.

In many respects, cryptococcosis is an excellent model for studying immunoregulation in an infectious disease. First, immunoregulation of the immune response to *C. neoformans* or cryptococcal antigens is a prominent feature in cryptococcosis (Diamond and Bennett, 1973; Diamond, 1977; Lim et al., 1980; Henderson et al., 1982; Henderson et al., 1986). Early investigators demonstrated that large doses of cryptococcal vaccines, i.e., *C. neoformans* cells or soluble polysaccharide antigen, resulted in reduced to absent protective immune responses, which indicated to them that high dose immunological paralysis was being induced (Gadebusch, 1958; Abrahams and Gilleran, 1960). Second, cryptococcosis is a chronic disease, so immune responses can be monitored over a reasonable time span. Third, experimental cryptococcosis in certain inbred mouse strains bears resemblance to human cryptococcosis and a large bank of knowledge and immunological reagents are available to facilitate studies on host-*C. neoformans* interactions in the mouse model. Fourth, it has been established that T-cell function is essential to effective host resistance against *C. neoformans* (Abrahams and Gilleran, 1960; Abrahams, 1966; Adamson and Cozad, 1969; Diamond, 1977; Graybill and Drutz, 1978; Cauley and Murphy, 1979; Fromtling et al., 1979a; Graybill et al., 1979; Lim et al., 1980; Fung and Murphy, 1982), so investigations can be focused immediately on the mechanisms which contribute to up- and down-regulation of T-cell functions. Fifth, during infection, cryptococcal antigen appears in the bloodstream in measurable quantities, thereby providing an almost continuous, quantifiable stimulus of the immune response. Finally, determinations of the numbers of viable cryptococci in infected tissues are relatively easy to perform, so it is possible to study clearance of the organism under different immunoregulating conditions. In this review, immunoregulation of both the humoral and cell-mediated anticryptococcal immune responses will be examined. Despite the fact that to some the term *immunoregulation* implies *specificity*, in this review findings concerned with nonspecific as well as specific modulation of the immune responses will be considered.

REGULATION OF THE HUMORAL IMMUNE RESPONSE

As indicated above, detection of anticryptococcal antibodies in humans during and following cryptococcosis has been erratic (Gordon and Vedder, 1966; Walter and Jones, 1968; Diamond and Bennett, 1974; Reiss et al., 1984). Similarly in experimental animals, attempts to stimulate antibodies by injecting viable or killed cryptococci or cryptococcal antigens have provided variable results (Gadebusch, 1958; Cozad et al., 1963; Bennett and Haseclever, 1965; Goren, 1967; Goren and Middlebrook, 1967; Kozel and Cazin, 1972; Murphy and Cozad, 1972; Kozel and Cazin, 1974; Kozel et al., 1977; Breen et al., 1982; Dromer et al., 1987), and when anticryptococcal antibodies were detected, titers were generally low (Kozel and Cazin, 1972; Murphy and Cozad, 1972; Kozel and Cazin, 1974; Kozel et al., 1977). Antibodies produced in response to *C. neoformans* in humans and in experimental animals are directed mainly at the major polysaccharide constituent of the cryptococcal capsule, i.e., glucuronoxylomannan (GXM) and a mannoprotein component (Reiss et al., 1984). The polysaccharide and glycoprotein nature of the cryptococcal antigens may contribute to the ability of these antigens to induce immunoregulation early in the response, which results in the characteristically low humoral immune response. For many years, these low levels of anticryptococcal antibodies were considered to indicate that *C. neoformans* was a weak antigen. More recent data suggest, however, that antibodies can be stimulated with cryptococcal antigen(s), but care must be given to the route of injection, the amount of antigen injected, and the nature of the antigen (Goren and Middlebrook, 1967; Murphy and Cozad, 1972; Kozel and Cazin, 1972; Kozel and Cazin, 1974; Kozel et al., 1977; Breen et al., 1982; Dromer et al., 1987). A feature that is clearly evident from the combined results of human and animal studies is that the anticryptococcal humoral immune response is easily down-regulated (Cozad et al., 1963; Murphy and Cozad, 1972; Kozel et al., 1977; Lim and Murphy, 1980, Breen et al., 1982; Reiss et al., 1984; Henderson et al., 1986).

The characteristics of the anticryptococcal humoral immune response are reminiscent of those of the antibody response to the type II T-independent antigen, pneumococcal polysaccharide SIII. For instance, cryptococcal cells or capsular polysaccharide antigen preparations: (1) induce antibody responses in T-cell-deficient, nu/nu mice (Cauley and Murphy, 1979) and in mice treated with antilymphocyte or antithymocyte serum (Breen et al., 1982) suggesting that cryptococcal antigens are T-independent antigens; (2) stimulate predominantly an IgM response (Kozel and Cazin, 1972; Murphy and Cozad, 1972; Reiss et al., 1984), which is characteristic of T-independent antigens; and (3) induce immunological unresponsiveness (Murphy and Cozad, 1972; Kozel et al., 1977; Breen et al., 1982).

Augmentation of the humoral immune response to cryptococcal antigen has been achieved only in experimental animal models by the careful selection of the appropriate dose and route of administering the antigen (Dromer et al., 1987), by mixing the cryptococcal cells or soluble antigen(s) with adjuvants (Murphy and Cozad, 1972) and by coupling the capsular polysaccharide to sheep red blood cells (Eckert and Kozel, 1987) or to protein carriers such as bovine gamma globulin (Goren, 1967) or methylated bovine serum albumin (Kozel and Cazin, 1974). Increasing levels of anticryptococcal antibodies are detected only rarely in human cryptococcosis; however, when titers are increasing, it is a favorable sign for the patient (Gordon and Vedder, 1966; Diamond and Bennett, 1974).

Suppression of the anticryptococcal antibody response rather than augmentation is the usual finding in human and animal studies (Cozad et al., 1963; Bennett and Hasenclever, 1965; Goren and Middlebrook, 1967; Bindschadler and Bennett, 1968; Kozel and Cazin, 1972; Murphy and Cozad, 1972; Diamond and Bennett, 1974; Kozel and Cazin, 1974; Kozel et al., 1977; Scott et al., 1981; Breen et al., 1982; Henderson et al., 1982; Reiss et al., 1984; Henderson et al., 1986). Several early reports noted that high doses of cryptococcal cells or large doses of capsular polysaccharide reduced the protective responses in mice challenged with viable cryptococci (Gadebusch, 1958; Abrahams and Gilleran, 1960; Bennett and Hasenclever, 1965). At the time those early studies were done, the investigators considered only the humoral immune response in their interpretations, and thus they thought paralysis of the humoral immune response was the cause of the reduced protection (Abrahams and Gilleran, 1960). Although now we know that the reduced protection may not have resulted from paralysis of the humoral immune response but rather from depression of the CMI response, the prediction that the humoral immune response was paralyzed or suppressed was valid. This was demonstrated by several subsequent investigations in which animal models were used to establish that high doses of cryptococcal polysaccharide indeed did induce suppression of the anticryptococcal antibody response (Cozad et al., 1963; Murphy and Cozad, 1972; Kozel et al., 1977; Breen et al., 1982). In 1963, Cozad et al. (1963) noted that rabbits pretreated with high doses of cryptococcal capsular polysaccharide responded to a subsequent immunization with lower levels of anticryptococcal antibodies than did the control rabbits, which were immunized but not pretreated with cryptococcal antigen. Using a murine model, Murphy and Cozad (1972) confirmed that high levels of cryptococcal capsular polysaccharide could affect adversely the antibody response. They showed that an intraperitoneal injection of 500, 50, or 5 µg of cryptococcal polysaccharide initially stimulated increases in the numbers of anticryptococcal antibody-forming cells in the spleens (Fig. 1), which peaked at 4 days after antigen injection and then declined to background levels by 14 days (Fig. 1). The two higher concentrations of antigen stimulated the greatest primary responses; however, those responses were only slightly higher than the background numbers of antibody-forming cells. In contrast to the primary response, when the same mice were given an immunizing challenge dose of cryptococcal antigen (0.13 µg of polysaccharide in incomplete Freunds adjuvant) on day 14 after initial antigen treatment, the numbers of antibody-forming cells in the spleens of mice previously treated with 50 or 500 µg of polysaccharide were depressed as compared to animals pretreated with saline or 0.5 µg of cryptococcal polysaccharide. The depression in numbers of antibody forming cells was dose dependent, with the greatest degree of depression being in the group pretreated with 500 µg of polysaccharide (Murphy and Cozad, 1972). This study demonstrated that the depressed antibody titers in mice given large doses of cryptococcal antigen were not due solely to neutralization of the antibodies by the antigen, but rather were due in part to the lack of antibody forming cells. Later, Kozel et al. (1977) confirmed tht 100 to 800 µg of *C. neoformans*-soluble polysaccharide reduced the ability of mice to respond with the normal level of antibody to a challenge dose of cryptococcal antigen. In addition, Kozel and coinvestigators (1977) demonstrated that the animals treated with high doses of cryptococcal polysaccharide had reduced responsiveness to a challenge immunization through 8 weeks but had recovered from the "paralyzing" effect of the polysaccharide by 12 weeks. Breen et al. (1982) reported that mice pretreated with 1 µg of cryptococcal capsular polysaccharide and then given antilymphocyte or antithymocyte serum at the time of challenge immunization had sixfold higher numbers of antibody-forming cells in their spleens than did mice given only the cryptococcal polysaccharide and challenge immunization. Their observations suggest that suppression of the anticryptococcal antibody response is mediated by T-suppressor cells. None of these studies addressed the question of the specificity of suppression of the anticryptococcal humoral immune response; however, Breen et al. (1982) did note in their discussion that they had observed modulation of antibody responses to other antigens such as sheep red blood cells after treating mice with cryptococcal polysaccharide.

There is also evidence that immunological unresponsiveness is induced in humans during active cryptococcosis and that the suppression of the anticrypto-

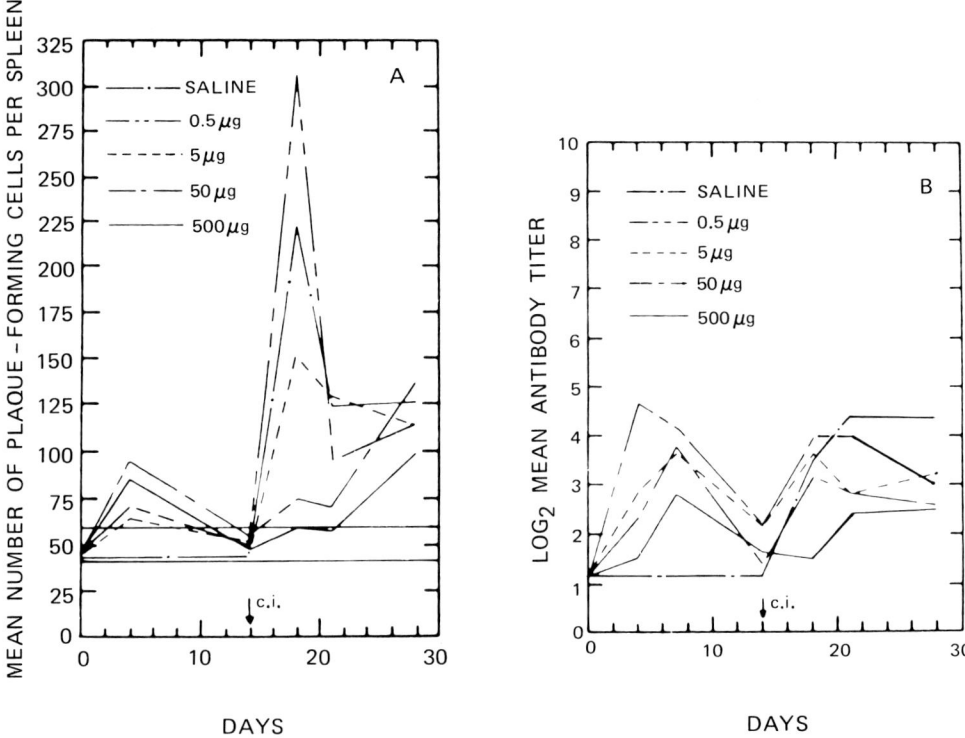

Figure 1 (A) Mean numbers of *C. neoformans*-specific plaque-forming cells in CBA/J mouse spleens after treatment with various amounts of cryptococcal polysaccharide on day 0. (B) Mean $\log_2$ agglutinin titers of sera from mice pretreated with the various amounts of cryptococcal polysaccharide on day 0. Arrow indicates challenge immunization with 0.125 µg of polysaccharide in incomplete Freund adjuvant. Reprinted with permission from *Infection and Immunity*. (From Murphy and Cozad, 1972.)

coccal antibody response is specific and of long duration (Henderson et al., 1982). Henderson et al. (1982) immunized 8 cured cryptococcosis patients and 10 normal control subjects with cryptococcal polysaccharide and, for comparison, type III pneumococcal polysaccharide and found that the cured patients responded with significantly lower antibody titers to cryptococcal polysaccharide than did the control subjects; however, there was no difference in the antibody responses of the two groups to pneumococcal polysaccharide. Based on the findings in the mouse model, it could be proposed that the inability to detect anticryptococcal antibodies after immunization of the patients is the result of reduced numbers of antibody-forming cells rather than neutralization of the antibody. This concept is supported by data published by Henderson et al. (1986) showing that anticryptococcal IgM or IgG could not be detected in supernatants from 9 of the 10 peripheral blood mononuclear cell samples collected 7 days after immunization of the cured cryptococcosis patients with cryptococcal polysaccharide. In contrast to the patients' cells, cells taken from the normal subjects 7 days after immunization secreted significant amounts of anticryptococcal immunoglobulins (Henderson et al., 1986). Thus, there is evidence from both animal and human studies that the antigenemia occurring during systemic cryptococcosis can induce suppression of the anticryptococcal antibody response. Moreover, the data of Henderson et al. (1982, 1986) suggest that suppression of the anticryptococcal antibody response is specific, especially with respect to the long-term suppression observed in humans.

REGULATION OF THE CELL-MEDIATED IMMUNE RESPONSE

As mentioned previously, cell-mediated immunity is an important protective immune mechanism, if not the most important defensive measure in cryptococcosis (Abrahams and Gilleran, 1960; Abrahams, 1966; Adamson and Cozad, 1969; Diamond, 1977; Graybill and Drutz, 1978; Fromtling et al., 1979a; Graybill et al., 1979; Lim and Murphy, 1980; Lim et al., 1980; Fung and Murphy, 1982). This may account for the fact that in recent years, most immunological studies on cryptococcosis have focused on the CMI response and its modulation. Despite this focus, our knowlege of factors which up-regulate the anticryptococcal CMI response is limited; however, a reasonably good understanding of the mechanisms responsible for suppression of the CMI response is beginning to emerge.

Development of the Anticryptococcal CMI Response

The anticryptococcal CMI response has been assessed in humans and animals by measuring various correlates of CMI such as the in vivo correlate delayed-type hypersensitivity (DTH) (Salvin and Smith, 1961; Bennett et al., 1965; Atkinson and Bennett, 1968; Muchmore et al., 1969), and the in vitro correlates lymphocyte transformation (LT) (Diamond and Bennett, 1973; Graybill and Alford, 1974; Schimpff and Bennett, 1975; Miller and Puck, 1984), interleukin-2 (IL-2) production by lymphocytes in response to *C. neoformans* (Miller and Puck, 1984), IL-2 receptor expression (Miller and Lewis, 1987), and production of macrophage migration inhibition factor (MIF) (Schimpff and Bennett, 1975).

Skin testing of humans with cryptococcal antigen preparations has been reported by several investigators (Salvin and Smith, 1961; Bennett et al., 1965; Atkinson and Bennett, 1968; Muchmore et al., 1969; Diamond and Bennett, 1973). Using an antigen prepared according to the procedure of Salvin and Smith (1961), Bennett et al. (1965) reported positive skin tests in 68% of the 22 normal subjects tested, in 44% of the 9 patients with active cryptococcosis, and in 88% of the 16 patients with treated, inactive cryptococcosis. With a similar antigen preparation, Muchmore et al. (1969) found that 32% of the healthy individuals from an area of high incidence of cryptococcosis had positive skin reactions. Slightly different results were obtained when a 1/100 dilution of a urea extract of *C. neoformans* was used as a skin test antigen (Atkinson and Bennett, 1968). None of the 21 control subjects reacted to the diluted urea extract antigen; however, 44% of the 16 laboratory workers who had been exposed previously to cryptococci were reactive (Atkinson and Bennett, 1968). When cryptococcosis patients were skin tested with the urea extract antigen, 50% of those with active disease and 59% of those with inactive, treated disease had positive skin test reactions when patients with generalized anergy were excluded (Atkinson and Bennett, 1968). These data are somewhat difficult to interpret and thus difficult to relate to data obtained with other cryptococcal antigen preparations, because the urea extract antigen also induced positive responses in patients with other mycoses (Atkinson and Bennett, 1968). For example, 50% of the blastomycosis patients, 22% of the histoplasmosis patients, and 25% of the coccidioidomycosis patients reacted positively to the urea extract antigen (Atkinson and Bennett, 1968). The cross-reactivity of the urea extract antigen was confirmed in guinea pigs sensitized to other mycotic agents (Atkinson and Bennett, 1968).

Anticryptococcal CMI responses in humans also have been assessed by measuring proliferation of peripheral blood lymphocytes after stimulation with either heat-killed cryptococci or soluble cryptococcal antigen. Diamond and Bennett (1973) found that lymphocytes from subjects who were skin test positive to the urea extract antigen proliferated significantly more in response to heat-killed *C. neoformans* cells than did lymphocytes from skin test negative subjects. When LT responses of cryptococcosis patients were compared to responses of skin test positive control subjects, the patients' lymphocytes were not as responsive to the heat-killed cryptococci as were the lymphocytes from the controls (Diamond and Bennett, 1973; Schimpff and Bennett, 1975). Graybill and Alford (1974) reported that an antigen prepared using a modification of the urea extract procedure (Atkinson and Bennett, 1968) induced positive lymphocyte transformation reactions in cells from half of their cryptococcosis patients. In this study (Graybill and Alford, 1974), lymphocytes from only 5 of the 24 healthy subjects gave positive responses to the soluble antigen, whereas lymphocytes from 18 of the 24 responded positively to heat-killed cryptococcal cells. One meningeal cryptococcosis patient followed by Graybill and Alford (1974) acquired a strong LT response to the soluble cryptococcal antigen during the course of treatment of the disease. The preceding studies established that lymphocytes from cryptococcosis patients generally do not display as strong a proliferative response to cryptococci or cryptococcal antigen as do cells from skin test positive control subjects or from cured cryptococcosis patients (Schimpff and Bennett, 1975). The depressed proliferative responses of the cells from cryptococcosis patients as compared to responses of cells from skin test positive control subjects appeared to be specific for cryptococcal antigens, because the mean proliferative responses to SK-SD (streptokinase and streptodornase) were similar for the two groups of individuals (Diamond and Bennett, 1973).

In comparing LT responses of normal subjects with those from cryptococcosis patients, Miller and Puck (1984) found the kinetics of the proliferative responses to be quite different between the two groups. In most cases, LT with patients' cells peaked after 5 to 7 days in culture; whereas the responses of cells from normal control subjects peaked later, usually about 9 days. When the cultured cells from normal control subjects were subcultured and restimulated with *C. neoformans* cells, the peak proliferative response occurred at 4-5 days, which was much earlier than the peak response of the primary culture (Miller and Puck, 1984). By day 2 in culture, the lymphocytes from cryptococcosis patients had produced significant amounts of the lymphokine IL-2. In contrast, cells from normal subjects in primary culture under the same conditions as patients' cells did not respond to stimulation with cryptococcal cells with measurable levels of IL-2 (Miller and Puck, 1984).

The data available from human studies are not sufficient to gain a satisfactory understanding of the developmental profiles of the anticryptococcal CMI response. In addition to the lack of data, there are several other factors that make interpretation of human studies difficult. For instance, the immunological status of the cryptococcosis patients prior to acquiring the disease seldom, if ever, is known, so a weak or absent CMI response in cryptococcosis patients could be due to an inherent immunological deficiency, an insufficient stimulation of the immune response, or a disease-related, acquired suppression. Moreover, with human subjects, it is frequently difficult to follow the immunological responses to *C. neoformans* from the onset of disease to the termination.

In contrast, it is much easier to acquire interpretable information concerning the developmental profile of the CMI response during cryptococcosis by studying animal models. With experimental animals, immunocompetent hosts can be followed throughout the disease process, and the various immunological functions, the numbers of organisms in specific tissues, and the serum cryptococcal antigen titers can be monitored as the disease progresses. Using this approach, it is possible to define the normal host responses and relate those responses to the level of infection. This type of approach has been taken with murine cryptococcosis as the experimental model. When mice having normal immune function were infected intranasally (i.n.) with an intermediate dose (10^3 cells) of *C. neoformans*, CMI responses, as measured by DTH reactivity, became positive by 3 weeks after infection and continued to rise until they peaked at about 6-7 weeks (Fig. 2A) (Lim et al., 1980). As the DTH reactivity of the animals increased, the cryptococcal cell burden declined (Fig. 2B) (Lim et al., 1980). The cryptococcal antigen levels in the sera (Fig. 2C) peaked at approximately the same time the DTH response peaked, and then the DTH response began to diminish slightly (Lim et al., 1980). When mice were given a larger infecting dose of *C. neoformans* by the same route (Murphy, J. W., unpublished data) or a similar infecting dose by the intraperitoneal route (Lim et al., 1980), the numbers of cryptococci in the tissues increased faster and the DTH response remained lower than in mice infected with smaller numbers of cryptococci by the i.n. route. Results from these animal studies suggest that when the numbers of cryptococci are low, the CMI response develops and hastens a reduction in the numbers of organisms, but as the fungal load increases and the level of antigen in the circulation rises, there is suppression of the CMI response. If the original organism burden is high, then the CMI response is down-regulated before it has sufficient time to develop to a maximum. This results in uncontrolled proliferation of the cryptococci, and, if unchecked, the host eventually dies of fulminating disease.

Anticryptococcal cell-mediated immunity has been induced experimentally in animals by immunizing with viable (Bennett et al., 1965; Murphy et al., 1975; Dykstra and Friedman, 1978; Hay and Reiss, 1978; Murphy and Pahlavan, 1979; Lim et al., 1980; Jones et al., 1981; Moser et al., 1982; Domer et al., 1983) or nonviable (Atkinson and Bennett, 1968; Bennett, 1981; Fung and Murphy, 1982; Murphy et al., 1988) virulent isolates of *C. neoformans*, avirulent mutants (Fromtling et al., 1979a, 1979b; Robinson et al., 1982), or soluble cryptococcal antigen preparations (Graybill and Taylor, 1978; Graybill et al., 1982; Murphy and Moorhead, 1982; Murphy et al., 1983; Hall et al., 1984; Murphy and Mosley, 1985; Mosley et al., 1986; Murphy, 1986a; Khakpour and Murphy, 1987; Fidel and Murphy, 1988; Murphy and Cox, 1988; Murphy et al., 1988) with or without adjuvant. An immunization which induces the anticryptococcal CMI response stimulates immature T cells that recognize cryptococcal antigen(s) to proliferate and differentiate into mature sensitized T cells (Fig. 3). Upon restimulation with cryptococcal antigen, the sensitized T cells will produce lymphokines that are responsible for initiating the DTH reaction. Such anticryptococcal sensitized T cells have been shown to be capable of transferring DTH reactivity and protection to naive syngeneic mice (Lim and Murphy, 1980). The

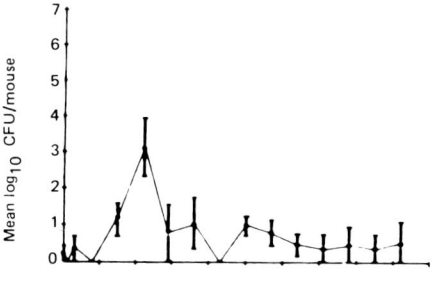

A

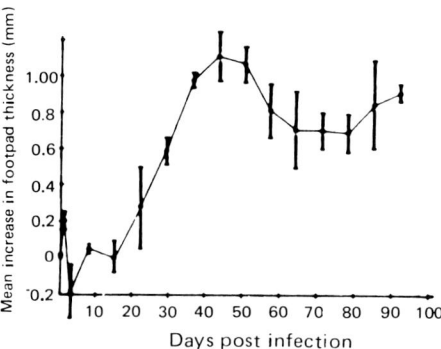

B

Figure 2 (A) Mean numbers of cryptococcal colony-forming units cultured from five CBA/J mice at various times after infecting i.n. with 10^3 viable *C. neoformans* cells on day 0. (B) Mean DTH responses to cryptococcal culture filtrate (CneF) antigen of the same mice as described in A. (C) Mean titers of cryptococcal polysaccharide antigen in sera from mice described in A (solid circle). Reprinted with permission from *Infection and Immunity*. (From Lim, Murphy, and Cauley, 1980.)

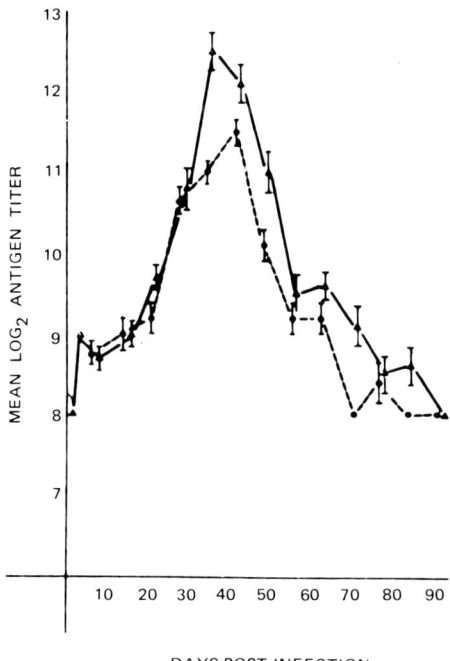

C

Figure 2 (Continued)

cells that transfer DTH reactivity are referred to as T_{DH} cells, and the anticryptococcal T_{DH} cells, like T_{DH} cells in other systems, are L3T4$^+$ (CD4), Ia$^-$ T cells (Khakpour and Murphy, 1987). More recently, evidence has been provided indicating that two other T-cell populations are induced upon immunization with cryptococcal cells or antigen (Fig. 3) (Khakpour and Murphy, 1987; Fidel and Murphy, in press). One of these cell populations is responsible for the amplified DTH response that is seen after a secondary stimulation with cryptococcal antigen (Fig. 3, Tamp) (Fidel and Murphy, in press). Potentially, this newly defined cell which augments the DTH response could be of great benefit to the host; however, considerably more information concerning this cell and its function(s) must be acquired before predictions on the value of the cell can be made. The third cell population induced by immunization with cryptococcal antigen is a third-order suppressor T cell, referred to as a Ts3 cell (Fig. 3) (Khakpour and Murphy, 1987), and it will be discussed below in the section on suppression of the CMI response.

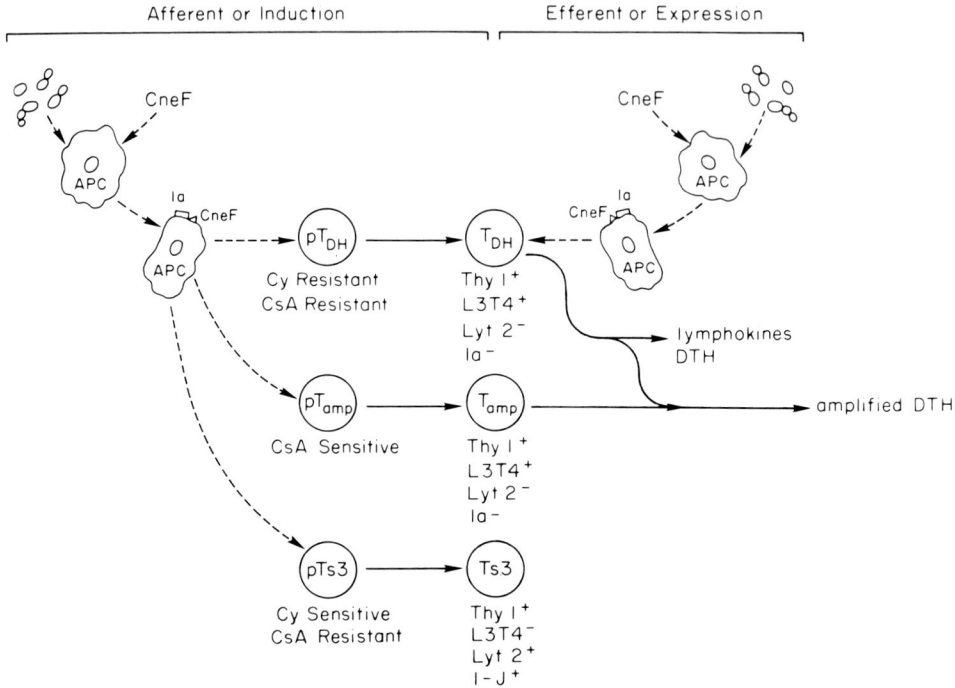

Figure 3 Illustration of induction and expression phases of various effector cells after immunizing with either cryptococcal cells or antigen. T_{DH} cells are responsible for the anticryptococcal DTH response. Tamp cells are responsible for the amplified DTH response observed after a secondary immunizing dose of cryptococcal antigen. Ts3 cells are third-order suppressor cells. The p before the cell type indicates precursor cells. Dotted lines designate hypothetical pathways; solid lines indicate the pathway has been demonstrated experimentally. Characteristics of each cell are listed below the cell.

Several different antigen preparations have been used for elicitation of the DTH response or for stimulation of LT or MIF (Salvin and Smith, 1961; Atkinson and Bennett, 1968; Murphy et al., 1974; Hay and Reiss, 1978; Murphy and Pahlavan, 1979; Jones et al., 1981; Graybill et al., 1982; Moser et al., 1982; Domer et al., 1983; Hall et al., 1984). These preparations fall into three broad groups. The first type consists of antigens derived from disrupted *C. neoformans* cells such as cytoplasmic antigens (Salvin and Smith, 1961; Bennett et al., 1965; Muchmore et al., 1969; Graybill and Taylor, 1978; Hay and Reiss, 1978; Jones et al., 1981; Graybill et al., 1982; Domer et al., 1983), solubilized proteins from membrane-mitochondria fractions (Domer et al., 1983), or microsomal fractions (Jones et al., 1981). The second type of cryptococcal antigen is prepared by extraction of whole cryptococcal cells with reagents like urea (Atkinson and Bennett, 1968; Bennett, 1981) or alkali (Hall et al., 1984). The third type of preparation is the culture filtrate antigen (Murphy et al., 1974; Dykstra and Friedman, 1978; Hay and Reiss, 1978; Graybill and Taylor, 1978; Murphy and Pahlavan, 1979; Cauley and Murphy, 1987; Fidel and Murphy, 1988; Murphy and Cox, 1988, Murphy et al., 1988, Fidel and Murphy, in press). Only a few of the antigen preparations that have been used in studies of anticryptococcal CMI responses have been assessed for specificity for *C. neoformans*. Of the three that have been tested, two are relatively specific. These are the crude supernatant from disrupted cells which was originally prepared by Salvin and Smith (1961) and the culture filtrate antigen, CneF, as prepared by Murphy and coinvestigators (1974, 1979). As mentioned earlier, the third one, which is the urea extract antigen, was shown to be cross-reactive in animals and humans infected with other fungi (Atkinson and Bennett, 1968).

The anticryptococcal CMI response recognizes a mannoprotein constituent of *C. neoformans* (Murphy et al., 1988). This was demonstrated by immunizing mice with either CneF in complete Freunds adjuvant or whole killed *C. neoformans* cells, then footpad challenging to elicit a DTH response with various cryptococcal antigen fractions which included the primary serotype antigen GXM, galactoxylomannan (GalXM), and mannoprotein (MP). It was found that the MP fraction from two different isolates of *C. neoformans* elicited the strongest DTH responses; whereas extremely reduced to negative responses were observed after footpad challenging with GXM or GalXM. These data clearly demonstrate that the MP fraction contains the component(s) responsible for the specificity in the anticryptococcal DTH response (Murphy et al., 1988).

Suppression

Nonspecific Suppression

The findings discussed earlier strongly suggest that immunosuppression of the anticryptococcal CMI response occurs during the disease. One cannot completely attribute the observed suppression to inherent immunodeficiencies of the host, which were present before infection, because immunologically normal mice were shown to display suppressed CMI responses after infection with *C. neoformans* (Lim et al., 1980; Robinson et al., 1982). Several reports have confirmed that a *C. neoformans* infection in immunologically competent animals can induce suppressor cells which depress immune responses. For example, when C57Bl/6 mice were injected with either viable avirulent (Blackstock and Hall, 1982; Robinson et al., 1983) or virulent (Robinson et al., 1982; Blackstock and Hall, 1984) *C. neoformans* cells, immunosuppression was observed. Injection of the avirulent isolate induced a transient phase of suppression which depressed in vitro responses to T- and B-cell mitogens and in vivo humoral and cell-mediated immune responses to sheep red blood cells (SRBC) (Blackstock and Hall, 1982). In mice infected with the virulent isolate of *C. neoformans*, a stage of weak CMI reactivity, as measured by LT, to cryptococcal antigen was observed between 6 and 9 days after infection, and then that responsive phase was followed by an unresponsive phase (Robinson et al., 1982). The suppression in this infection model appeared to be mediated by both nylon wool- or plastic-adherent and -nonadherent splenic cells (Robinson et al., 1982). The suppression was not antigen-specific since the infected mice had a depressed primary humoral immune response and reduced DTH reactivity to SRBC as compared to similar responses in uninfected mice (Blackstock and Hall, 1984).

Another group of investigators have reported observing immunosuppression in rats infected with *C. neoformans* (Masih et al., 1986; Sotomayor et al., 1987). The suppression was regarded as nonspecific, because rats infected for 7 days with cryptococci could not mount as strong of a DTH response to human serum albumin (HSA) as could uninfected rats (Masih et al., 1986). Spleen cells from the *C. neoformans*-infected, HSA-immunized rats when adoptively transferred to naive rats suppressed the recipient's ability to respond to an immunization with HSA (Masih et al., 1986; Sotomayor et al., 1987). The cells which were capable of transferring the suppression of the anti-HSA DTH response were shown to be nylon wool-nonadherent cells which could bind HSA (Sotomayor et al., 1987). Although these data were interpreted to indicate that nonspecific afferent suppressor cells were induced during the course of cryptococcosis (Masih et al., 1986; Sotomayor et al., 1987), there exists the possibility that the cryptococcal infection caused a shift in the responsiveness so that the immunizing dose of HSA became a tolerizing dose and resulted in the induction of HSA-specific suppressor cells. The characteristics reported for these suppressor cells are quite similar to other antigen-specific, first-order T-suppressor (Ts1) cells (Dorf and Benacerraf, 1984). Further studies are needed to resolve this question.

To date, nonspecific suppression of immune responses has been demonstrated in *C. neoformans*-infected animals and not in animals which have been given cryptococcal antigen(s), so it may be possible that the nonspecific immune hyporesponsiveness demonstrated during an infection with cryptococci may be due to generalized adverse effects of the infection. In an animal stressed with disease, the proper environment for normal induction of an immune response may not be obtainable.

C. neoformans-Specific Suppression

Antigen-specific immunosuppression of the anticryptococcal CMI response may be a mechanism of considerable importance in cryptococcosis. In fact, it potentially could be the determining factor in whether or not the host recovers. A series of studies by two different groups of investigators have provided solid evidence that *C. neoformans*-specific suppressor cells can be induced in normal mice with cryptococcal antigen(s) at concentrations that can be found in patients with systemic cryptococcosis (Murphy and Moorhead, 1982; Morgan et al., 1983; Murphy et al., 1983; Murphy and Mosley, 1985; Mosley et al., 1986; Murphy, 1986a; Blackstock et al., 1987; Khakpour and Murphy, 1987; Fidel and Murphy, 1988, Murphy and Cox, 1988; Fidel and Murphy, in press).

Murphy and coinvestigators (1982, 1988) have shown that sera from *C. neoformans*-infected mice which have high titers for cryptococcal antigen induce *C. neoformans*-specific T-suppressor cells that depress the anticryptococcal CMI response. Similar suppressor cells can be induced in normal mice by an intravenous injection of a culture filtrate antigen of *C. neoformans*, i.e., CneF (Murphy and Moorhead, 1982; Murphy et al., 1983; Murphy and Mosley, 1985; Mosley et al., 1986; Murphy, 1986a; Khakpour and Murphy, 1987; Fidel and Murphy, 1988). This group of workers (Murphy and Moorhead, 1982; Murphy et al., 1983; Murphy and Mosley, 1985; Mosley et al., 1986; Murphy, 1986a; Khakpour and Murphy, 1987; Fidel and Murphy, 1988; Murphy and Cox, 1988) has demonstrated that suppression of the anticryptococcal CMI response is mediated by a complex series of suppressor cells and factors. Schemes for the induction of these various *C. neoformans*-specific suppressor cells and factors and their characteristics are depicted in Figures 3 and 4. Briefly, when mice are injected intravenously with cryptococcal antigen [20 μg-1 mg of CneF or serum from infected animals that have high titers of cryptococcal antigen (latex agglutination titer of 10^{-4} or higher)], first-order T-suppressor (Ts1) cells are induced in lymph nodes (Fig. 4A) (Murphy and Moorhead, 1982; Murphy et al., 1983). Ts1 cells mediate their effects through a soluble factor called TsF1. The Ts1 cells or TsF1 perform two functions: (1) they induce second-order T-suppressor (Ts2) cells in the spleens (Fig. 4B) (Murphy et al., 1983; Murphy and Mosley, 1985), and (2) they suppress the induction of the T lymphocytes responsible for the DTH response (T_{DH} cells) (Fig. 5) (Murphy, 1986). The Ts2 cells also mediate their effect through a soluble factor, TsF2 (Murphy and Mosley, 1985). TsF2 functions in concert with a third order T-suppressor (Ts3) cell to suppress the expression of T_{DH} cells (Fig. 5) (Khakpour and Murphy, 1987). As mentioned earlier, the Ts3 cell is induced during immunization with cryptococcal cells or antigen and arises concomitantly with T_{DH} cells and with a newly defined DTH amplifier cell population (Fig. 3). In the

Suppressor Cell Cascade

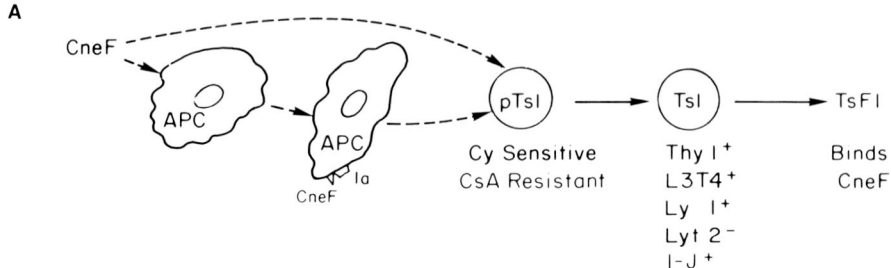

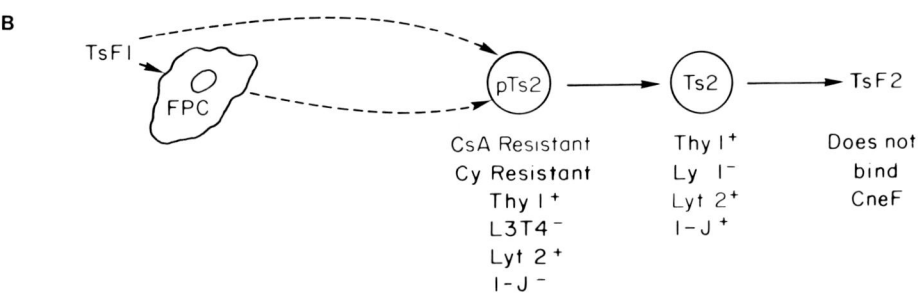

Figure 4 (A) Illustration of the induction pathway for the first-order suppressor T (Ts1) cells and soluble factor (TsF1) for cryptococcal antigen. (B) The proposed induction pathway for second-order suppressor T (Ts2) cells in the *C. neoformans*-specific suppressor cell cascade. Characteristics of each cell are listed below the cell. Dotted lines represent hypothetical aspects of the pathways; solid lines indicate aspects of the pathways for which there is experimental evidence.

immune animal, the Ts3 cell is nonfunctional until TsF2 is produced to activate it to suppress the efferent or expression limb of the DTH response (Fig. 5) (Khakpour and Murphy, 1987). The Ts1 and Ts2 cells of the cryptococcal suppressor cell cascade are specific for cryptococcal antigen inasmuch as they will not suppress DTH responses to *Mycobacterium tuberculosis* or 2,4-dinitrofluorobenzene (Murphy and Moorhead, 1982; Murphy and Mosley, 1985). For several of the cells in the pathways illustrated in Figures 3 and 4, the precursor cells have been assessed for sensitivity to cyclophosphamide and cyclosporin A, and those characteristics are indicated where appropriate. The phenotypes of these cells and antigen-binding capabilities of the suppressor factors are also provided in Figures 3 and 4. Figure 5 summarizes the anticryptococcal CMI response pathway and shows where the suppressor cells or factors block the response.

Immunoregulation in Cryptococcosis

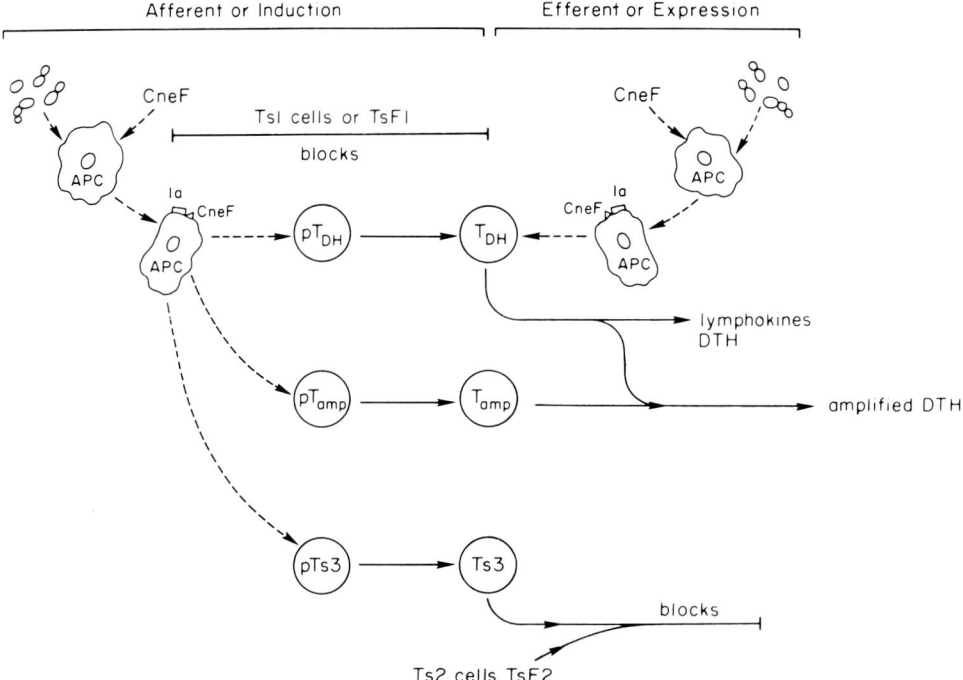

Figure 5 An illustration showing where Ts1 cells or TsF1 and Ts2-Ts3 cells or factors block the induction and expression phases of the delayed-type hypersensitivity response to cryptococcal antigen. Dotted lines represent hypothetical aspects; solid lines are proven aspects.

Murphy (1986b) has presented data showing that administration of cryptococcal antigen intravenously to mice prior to and during infection with *C. neoformans* adversely affects clearance of the cryptococci from tissues. In contrast, administration of cryptococcal antigen did not alter the clearance of *Listeria monocytogenes* or the anti-*Listeria* DTH response that develops during infection (Murphy, 1986b). These results are in accord with the specificity of suppression of the DTH response after i.v. administration of cryptococcal antigen (Murphy and Moorhead, 1982).

Using C57Bl/6 mice as a model, Blackstock, Hall, and coinvestigators (Morgan et al., 1983; Blackstock et al., 1987) also have presented data showing that antigen-specific T-suppressor cells can be induced either during an infection with *C. neoformans* or by injecting mice intraperitoneally with 25-100 µg of cryptococcal antigen. Suppressor cell induction was achieved with either a urea extract antigen or culture filtrate-type antigen of cryptococci. The functional assay, used by this group of investigators to demonstrate suppressor T cells, was different than that used by Murphy and coworkers. Rather than assessing the effects of suppressor cells on the DTH response or clearance of cryptococci from tissues, Blackstock and coinvestigators demonstrated that when the suppressor cells were restimulated in vitro with soluble cryptococcal antigen, a suppressive lymphokine was produced (Morgan et al., 1983; Blackstock et al., 1987). This suppressive lymphokine affects macrophages by inhibiting their ability to phagocytize particulate material, such as *C. neoformans* or *Saccharomyces cerevisiae* cells (Morgan et al., 1983). The suppressive lymphokine is heat-stable and trypsin-sensitive (Morgan et al., 1983). The induction of suppressor lymphocytes which produce a suppressive factor upon restimulation with antigen was not unique to cryptococcal antigens, because Blackstock et al. (1987) demonstrated that tolerogenic doses of levan and human gamma globulin also induce suppressor cells, which can be triggered to produce a similar suppressive lymphokine when stimulated with the homologous antigen.

It is not altogether clear if the suppressor cells described by this latter group of investigators (Morgan et al., 1983; Blackstock et al., 1987) fit into the antigen-specific suppressor cell circuit described by Murphy and coworkers (Figs. 3 and 4) or whether they represent another distinctly different suppressor cell pathway. There is some evidence that suggests that the suppressor cells described by the two different groups are different. Murphy and Cox (1988) have established that cryptococcal antigen will induce in BALB/c mice suppressor cells that suppress the anticryptococcal DTH response. In contrast, Morgan et al. (1983) reported that this mouse strain does not produce the suppressor cells that release the phagocytosis-inhibiting factor and that BALB/c macrophages were not responsive to the suppressive lymphokine produced by suppressor cells from (BALB/c × C57Bl/6)F1 hybrid mice (Morgan et al., 1983). Clearly, this is not sufficient evidence to say that the suppressor cells identified by the two different groups of workers represent different suppressor cell systems. Many more comparisons are required before it will be possible to speculate on the relationship of the various antigen-specific suppressor cells induced by cryptococcal antigen.

CONCLUSIONS

By considering information acquired from investigations on anticryptococcal CMI and its regulation in both humans and animals, the following sequence of immunological events in cryptococcosis could be proposed to occur in humans with apparently normal immune function. Early in infection with a low dose of cryptococci, the natural cellular defenses remove the majority of the organisms. As a part of the clearance process, cryptococcal antigen stimulates induction of the anticryptococcal antibody and CMI response. Once cell-mediated immunity is established, then the resistance mechanisms associated with the CMI response begin to attack the organisms and with the aid of the antibody clearance mechanisms, hopefully completely eliminate any remaining cryptococcal cells. Since the majority of individuals who become infected with *C. neoformans* clear the organism during the pulmonary phase of the disease and do not progress into extrapulmonary disease, it is reasonable to predict that the majority, if not all, of the cryptococci are eliminated during the pulmonary stage of disease. However, if the cryptococci begin to multiply faster than the host defenses in the lungs can eliminate them, the organisms may disseminate to other tissues. Once the organisms establish and multiply in extrapulmonary sites, especially in the central nervous system, antigen concentrations may rise to levels that induce suppression of both the humoral and CMI responses. When the anticryptococcal immune responses become suppressed, the host's abilities to clear the organisms are reduced; the cryptococcal antigen levels continue to rise; and the disease progresses to a more severe stage. If antifungal drugs are not given to break this cycle, then the host will succumb to the disease. This scenario would account for the fact that patients with systemic cryptococcosis who appear to be otherwise immunologically normal frequently have reduced levels of anticryptococcal humoral and CMI reactivities, i.e., antibodies, skin tests, LT, and IL-2 and MIF production, as compared to the reactivities of normal, skin test positive control subjects.

In immune-compromised hosts, the above scenario most likely is abbreviated, and the induction of suppression may be accelerated. If the host's natural defensive mechanisms in the lung are not able to eliminate the organisms completely and if the CMI response is defective, then the cryptococcal cells will proliferate, disseminate, and produce levels of antigen that will induce suppression of the remaining protective immune response. Thus, unless drugs alone are effective in eliminating the organism, the patient will be overcome by the infection. A sequence similar to this latter one is most likely occurring in AIDS patients who have cryptococcosis.

ACKNOWLEDGMENTS

I am grateful to Dr. Rebecca A. Cox and Michelle Hidore for reading the manuscript and providing helpful critiques. The work discussed from my laboratory was supported by National Institutes of Health Grant AI-15716.

REFERENCES

Abrahams, I., and Gilleran, T. G. (1960). Studies on actively acquired resistance to experimental cryptococcosis in mice. *J. Immunol. 85*:629-635.

Abrahams, I. (1966). Further studies on acquired resistance to murine cryptococcosis: enhancing effect of *Bordetella pertussis*. *J. Immunol. 96*: 525-529.

Adamson, D. M., and Cozad, G. C. (1969). Effect of antilymphocyte serum on animals experimentally infected with *Histoplasma capsulatum* or *Cryptococcus neoformans*. *J. Bacteriol. 100*: 1271-1276.

Atkinson, A. J., and Bennett, J. E. (1968). Experience with a new skin test antigen prepared from *Cryptococcus neoformans*. *Am. Rev. Resp. Dis. 97*: 637-643.

Bennett, J. E., Hasenclever, H. F., and Baum, G. L. (1965). Evaluation of a skin test for cryptococcosis. *Am. Rev. Resp. Dis. 91*: 616.

Bennett, J. E., and Hasenclever, H. F. (1965). *Cryptococcus neoformans* polysaccharide: studies of serologic properties and role in infection. *J. Immunol. 94*: 916-920.

Bennett, J. E. (1981). Cryptococcal skin test antigen: preparation variables and characterization. *Infect. Immun. 32*: 373-380.

Bennett, J. E., and Bailey, J. W. (1971). Control for rheumatoid factor in the latex test for cryptococcosis. *Am. J. Clin. Pathol. 56*: 360-365.

Bindschadler, D. D., and Bennett, J. E. (1968). Serology of human cryptococcosis. *Ann. Intern. Med. 69*: 45-52.

Blackstock, R., and Hall, N. K. (1981). Immunosuppression by avirulent, pseudohyphal forms of *Cryptococcus neoformans*. *Mycopathologia 80*: 95-99.

Blackstock, R., and Hall, N. K. (1984). Nonspecific immune suppression by *Cryptococcus neoformans* infection. *Mycopathologia 86*: 35-43.

Blackstock, R., McCormack, J. M., and Hall, N. K. (1987). Induction of a macrophage suppressive lymphokine by soluble cryptococcal antigens and its association with models of immunologic tolerance. *Infect. Immun. 55*: 233-239.

Bloomfield, N., Gordon, M. A., and Elmendorf, D. R. (1963). Detection of *Cryptococcus neoformans* antigen in body fluids by latex particle agglutination. *Proc. Soc. Exp. Biol. Med. 114*: 64-67.

Breen, J. F., Lee, I. C., Vogel, F. R., and Friedman, H. (1982). Cryptococcal capsular polysaccharide-induced modulation of murine immune responses. *Infect. Immun. 36*: 47-51.

Campbell, G. D. (1966). Primary pulmonary cryptococcosis. *Am. Rev. Resp. Dis. 94*: 236-243.

Cauley, L. K., and Murphy, J. W. (1979). Response of congenitally athymic (nude) and phenotypically normal mice to a *Cryptococcus neoformans* infection. *Infect. Immun. 23*: 644-651.

Cohen, J. (1982). The pathogenesis of cryptococcosis. *J. Infection. 5*: 109-116.

Cozad, G. C., Richey (Murphy), J. W., and Larsh, H. W. (1963). *Bact. Proc.*, p. 81.

Diamond, R. D., and Bennett, J. E. (1973). Disseminated cryptococcosis in man: decreased lymphocyte transformation in response to *Cryptococcus neoformans*. *J. Infect. Dis. 127*: 694-697.

Diamond, R. D., and Bennett, J. E, (1974). Prognostic factors in cryptococcal meningitis. A study of 111 cases. *Ann. Intern. Med. 80*: 176-180.

Diamond, R. D. (1977). Effects of stimulation and suppression of cell-mediated immunity on experimental cryptococcosis. *Infect. Immun. 17*: 187-194.

Domer, J. E., Lyon, F. L., and Murphy, J. W. (1983). Cellular immunity in a cutaneous model of cryptococcosis. *Infect. Immun. 40*: 1052-1059.

Dorf, M. E., and Benacerraf, B. (1984). Suppressor cells and immunoregulation. *Ann. Rev. Immunol. 2*: 127-158.

Dromer, F., Salamero, J., Contrepois, A., Carbon, C., and Yeni, P. (1987). Production, characterization, and antibody specificity of a mouse monoclonal antibody reactive with *Cryptococcus neoformans* capsular polysaccharide. *Infec. Immun. 55*: 742-748.

Dromer, F., Charreire, J., Contrepois, A., Carbon, C., and Yeni, P. (1987). Protection of mice against experimental cryptococcosis by anti-*Cryptococcus neoformans* monoclonal antibody. *Infect. Immun. 55*: 749-752.

Duperval, R., Hermans, P. E., Brewer, N. S., and Roberts, G. D. (1977). Cryptococcosis, with emphasis on the significance of isolation of *Cryptococcus neoformans* from the respiratory tract. *Chest 72*: 13-19.

Dykstra, M. A., and Friedman, L. (1978). Pathogenesis, lethality, and immunizing effect of experimental cutaneous cryptococcosis. *Infect. Immun. 20*: 446-455.

Eckert, T. F., and Kozel, T. R. (1987). Production and characterization of monoclonal antibodies specific for *Cryptococcus neoformans* capsular polysaccharide. *Infect. Immun. 55*: 1895-1899.

Emmons, C. W., Binford, C. H., Utz, J. P., and Kwon-Chung, K. J. (1977). In *Medical Mycology*, third edition. Lea and Febiger, Philadelphia, PA.

Eng, R., Chmel, H., Corrado, M., and Smith, S. M. (1983). The course of cryptococcal capsular polysaccharide elimination kinetics. *Infection 11*: 132-136.

Feigin, D. S. (1983). Pulmonary cryptococcosis: radiologic-pathologic correlates of its three forms. *Am. J. Roentgen. 141*: 1263-1272.

Fidel, P. L., Jr., and Murphy, J. W. (1988). Characterization of an in vitro-stimulated *Cryptococcus neoformans*-specific second-order suppressor T cell and its precursor. *Infect. Immun. 56*: 1267-1272.

Fidel, P. L., Jr., and Murphy, J. W. (1989). Effects of cyclosporin A on the cells responsible for the anticryptococcal cell-mediated immune response and its regulation. Infect. Immun. (in press, April 1989).

Fromtling, R. A., Blackstock, R., and Bulmer, G. S. (1979a). Immunization and passive transfer of immunity in murine cryptococcosis, p. 122-125. In *Proceedings of the VI Congress of ISHAM*. Edited by E. S. Kuttin and G. L. Baum. International Congress Series No. 480. Excerpta Medica, New York.

Fromtling, R. A., Blackstock, R., Hall, N. K., and Bulmer, G. S. (1979b). Immunization of mice with an avirulent pseudohyphal form of *Cryptococcus neoformans*. *Mycopathologia 68*: 179-181.

Fung, P. Y. S., and Murphy, J. W. (1982). In vitro interactions of immune lymphocytes and *Cryptococcus neoformans*. *Infect. Immun. 36*: 1128-1138.

Gadebusch, H. H. (1958). Active immunization against *Cryptococcus neoformans*. *J. Infect. Dis. 102*:219-226.
Goodman, J. S., Kaufman, L., and Koenig, M. G. (1971). Diagnosis of cryptococcal meningitis. Value of immunologic detection of cryptococcal antigen. *N. Engl. J. Med. 285*: 434-436.
Gordon, M. A. (1975). The epidemiology of cryptococcosis, p. 142-151. In *The Epidemiology of Human Mycotic Diseases*. Edited by Y. Al-Doory. Thomas, Springfield, Illinois, pp. 142-151.
Gordon, M. A., and Vedder, D. K. (1966). Serology test in diagnosis and prognosis of cryptococcosis. *J. Am. Med. Assoc. 197*: 131-137.
Goren, M. B., and Middlebrook, G. M. (1967). Protein conjugates of polysaccharide from *Cryptococcus neoformans*. *J. Immunol. 98*: 901-913.
Goren, M. B. (1967). Experimental murine cryptococcosis: effect of hyperimmunization to capsular polysaccharide. *J. Immunol. 98*: 914-922.
Graybill, J. R., and Drutz, D. J. (1978). Host defense in cryptococcosis in the nude mouse. *Cell. Immun. 40*: 263-274.
Graybill, J. R., Mitchell, L., and Drutz, D. J. (1979). Host defense in cryptococcosis. III. Protection of nude mice by thymus transplantation. *J. Infect. Dis. 140*: 546-552.
Graybill, J. R., and Alford, R. H. (1974). Cell-mediated immunity in cryptococcosis. *Cell. Immunol. 14*: 12-21.
Graybill, J. R., and Taylor, R. L. (1978). Host defense in cryptococcosis. I. An in vivo model for evaluating immune response. *Int. Arch. Allergy Appl. Immunol. 57*: 101-113.
Graybill, J. R., Straus, D. C., Nealon, T. J., Hague, M., and Paque, R. E. (1982). Immunogenic fractions of *Cryptococcus neoformans*. *Mycopathologia 78*: 31-39.
Graybill, J. R., Hague, M., and Drutz, D. J. (1981). Passive immunization in murine cryptococcosis. *Sabouraudia 19*: 237-244.
Hall, N. K., Maluf, K. C., and Blackstock, R. (1984). Functional testing and chemical composition of cryptococcal extracts. *Sabouraudia: J. Med. Vet. Mycol. 22*: 439-442.
Hatcher, C. R., Sehdeva, J., Waters, W. C. III, Schulze, V., Logan, D. W., Jr., Symbas, P., and Abbott, O. A. (1971). Primary pulmonary cryptococcosis. *J. Thorac. Cardiovasc. Surg. 61*: 39-49.
Hay, R. J., and Reiss, E. (1978). Delayed-type hypersensitivity responses in infected mice elicited by cytoplasmic fractions of *Cryptococcus neoformans*. *Infect. Immun. 22*: 72-79.
Henderson, D. K., Kan, V. L., and Bennett, J. E. (1986). Tolerance to cryptococcal polysaccharide in cured cryptococcosis patients: failure of antibody secretion in vitro. *Clin. Exp. Immunol. 65*: 639-646.
Henderson, D. K., Bennett, J. E., and Huber, M. A. (1982). Long-lasting, specific immunologic unresponsiveness associated with cryptococcal meningitis. *J. Clin. Invest. 69*: 1185-1190.
Jones, A. E., Reiss, E., and Spira, T. J. (1981). A microsomal fraction of *Cryp-*

tococcus neoformans induces lymphocyte blastogenesis in infected guinea pigs. *Mycopathologia 75*: 129-138.

Kauffman, C. A., Bergman, A. G., Severance, P. J., and McClatchey, K. D. (1981). Detection of cryptococcal antigen. Comparison of two latex agglutination tests. *Am. J. Clin. Pathol. 75*: 106-109.

Kerkering, T. M., Duma, R. J., and Shadomy, S. (1981). The evolution of pulmonary cryptococcosis. Clinical implications from a study of 41 patients with and without compromising host factors. *Ann. Intern. Med. 94*: 611-616.

Khakpour, F. R., and Murphy, J. W. (1987). Characterization of a third-order T suppressor cell (Ts3) induced by cryptococcal antigen(s). *Infect. Immun. 55*: 1657-1662.

Kovacs, J. A., Kovacs, A. A., Polis, M., Wright, W. C., Gill, V. J., Tuazon, C. U., Gelmann, E. P., Lane, H. C., Longfield, R., Overturf, G., Macher, A. M., Fauci, A. S., Parillo, J. E., Bennett, J. E., and Masur, H. (1985). Cryptococcosis in the acquired immunodeficiency syndrome. *Ann. Intern. Med. 103*: 533-538.

Kozel, T. R., and Cazin, J., Jr. (1972). Immune response to *Cryptococcus neoformans* soluble polysaccharide. I. Serological assay for antigen and antibody. *Infect. Immun. 5*: 35-41.

Kozel, T. R., Gulley, W. F., and Cazin, J., Jr. (1977). Immune response to *Cryptococcus neoformans* soluble polysaccharide: immunological unresponsiveness. *Infect. Immun. 18*: 701-707.

Kozel, T. R., and Cazin, J., Jr. (1974). Induction of humoral antibody response to soluble polysaccharide of *Cryptococcus neoformans*. *Mycopathol. Mycol. Appl. 54*: 21-30.

Lewis, J. L., and Rabinovich, S. (1972). The wide spectrum of cryptococcal infections. *Am. J. Med. 53*: 315-322.

Lim, T. S., and Murphy, J. W. (1980). Transfer of immunity to cryptococosis by T-enriched splenic lymphocytes from *Cryptococcus neoformans*-sensitized mice. *Infect. Immun. 30*: 5-11.

Lim, T. S., Murphy, J. W., and Cauley, L. K. (1980). Host-etiological agent interactions in intranasally and intraperitoneally induced cryptococcosis in mice. *Infect. Immun. 29*: 633-641.

Littman, M. L., and Zimmerman, L. E. (1956). *Cryptococcosis (Torulosis)*. Grune and Stratton, New York.

Masih, D. T., Rubinstein, H. R., Sotomayor, C. E., Ferro, M. E., and Riera, C. M. (1986). Non-specific immunosuppression in experimental cryptococosis in rats. *Mycopathologia 94*: 79-84.

Miller, G. P. G., and Puck, J. (1984). In vitro human lymphocyte responses to *Cryptococcus neoformans*. Evidence for primary and secondary responses in normals and infected subjects. *J. Immunol. 133*: 166-172.

Miller, G. P. G., and Lewis, D. E. (1987). In vitro effect of cyclosporine on Interleukin-2 receptor expression stimulated by *Cryptococcus neoformans*. *J. Infect. Dis. 155*: 799-802.

Monga, D. P., Kumar, R., Mohapatra, L. N., and Malaviza, A. N. (1979). Experimental cryptococcosis in normal and B cell deficient mice. *Infect. Immun. 26*: 1-3.

Morgan, M. A., Blackstock, R., Bulmer, G. S., and Hall, N. K. (1983). Modification of macrophage phagocytosis in murine cryptococcosis. *Infect. Immun. 40*: 493-500.

Moser, S. A., Lyon, F. L., Domer, J. E., and Williams, J. E. (1982). Immunization of mice by intracutaneous inoculation with viable virulent *Cryptococcus neoformans*: immunological and histopathological parameters. *Infect. Immun. 35*: 685-696.

Mosley, R. L., Murphy, J. W., and Cox, R. A. (1986). Immunoadsorption of *Cryptococcus*-specific suppressor T-cell factors. *Infect. Immun. 51*:844-850.

Muchmore, H. G., Fleton, F. G., Salvin, S. B., and Rhoades, E. R. (1969). Delayed hypersensitivity to cryptococcin in man. *Sabouraudia 6*:285-288.

Murphy, J. W. (1986a). Effects of first-order *Cryptococcus*-specific T suppressor cells on the induction of cells responsible for delayed-type hypersensitivity (T_{DH} cells). *Infect. Immun. 50*:50-57.

Murphy, J. W. (1986b). Clearance of *Cryptococcus neoformans* from immunologically suppressed mice. *6th International Congress of Immunology*, Abstract 5.16.8, p. 611.

Murphy, J. W., and Cozad, G. C. (1972). Immunological unresponsiveness induced by cryptococcal capsular polysaccharide assayed by the hemolytic plaque technique. *Infect. Immun. 5*:896-901.

Murphy, J. W., and Cox, R. A. (1988). Induction of antigen-specific suppression by circulating *Cryptococcus neoformans* antigen. *Clin. Exp. Immunol.*

Murphy, J. W., and Mosley, R. L. (1985). Regulation of cell-mediated immunity in cryptococcosis. III. Characterization of second-order T suppressor cells (Ts2). *J. Immunol. 134*: 577-584.

Murphy, J. W., Mosley, R. L., and Moorhead, J. W. (1983). Regulation of cell-mediated immunity in cryptococcosis. II. Characterization of first-order T suppressor cells (Ts1) and induction of second-order suppressor cells. *J. Immunol. 130*: 2876-2881.

Murphy, J. W., and Moorhead, J. W. (1982). Regulation of cell-mediated immunity in cryptococcosis. I. Induction of specific afferent T suppressor cells by cryptococcal antigen. *J. Immunol. 128*: 276-283.

Murphy, J. W., Mosley, R. L., Cherniak, R., Reyes, G., Kozel, T. R., and Reiss, E. (1988). Serological, electrophoretic, and biological properties of *Cryptococcus neoformans* antigens. *Infect. Immun. 56*: 424-431.

Murphy, J. W., and Pahlavan, N. (1979). Cryptococcal culture filtrate antigen for detection of delayed-type hypersensitivity in cryptococcosis. *Infect. Immun. 25*: 284-292.

Murphy, J. W., Gregory, J. A., and Larsh, H. W. (1974). Skin testing of guinea pigs and footpad testing of mice with a new antigen for detecting delayed hypersensitivity to *Cryptococcus neoformans*. *Infect. Immun. 9*: 404-409.

Neill, J. M., Suggs, J. Y., and McCauley, D. W. (1951). Serologically reactive material in spinal fluid, blood, and urine from a human case of cryptococcosis (Torulosis). *Proc. Soc. Exp. Biol. Med. 77*: 775-778.

Perfect, J. R., Durack, D. T., and Gallis, H. A. (1983). Cryptococcemia. *Medicine (Baltimore) 62*: 98-109.

Perfect, J. R., Lang, S. D., and Durack, D. T. (1981). Influence of agglutinating antibody in experimental cryptococcal meningitis. *Br. J. Exp. Pathol 62*: 595-599.

Randhawa, H. S., and Paliwal, D. K. (1977). Occurrence and significance of *Cryptococcus neoformans* in the oropharynx and in the skin of healthy human population. *J. Clin. Microbiol. 6*: 325-327.

Reiss, E., Cherniak, R., Eby, R., and Kaufman, L. (1984). Enzyme immunoassay detection of IgM to galactoxylomannan of *Cryptococcus neoformans. Diag. Immun. 2*: 109-115.

Rippon, J. W. (1982). *Medical Mycology*, second edition. Saunders, Philadelphia.

Robinson, B. E., Hall, N. K., Bulmer, G. S., and Blackstock, R. (1982). Suppression of responses to cryptococcal antigen in murine cryptococcosis. *Mycopathologia 80*: 157-163.

Salvin, S. B., and Smith, R. F. (1961). An antigen for detection of hypersensitivity to *Cryptococcus neoformans. Proc. Soc. Exp. Biol. Med. 108*: 498-501.

Schimpff, S. C., and Bennett, J. E. (1975). Abnormalities in cell-meidated immunity in patients with *Cryptococcus neoformans* infection. *J. Allergy Clin. Immunol. 55*: 430-441.

Scott, E. N., Muchmore, H. G., and Felton, F. G. (1981). Enzyme-linked immunosorbent assays in murine cryptococcosis. *Sabouraudia 19*: 257-265.

Sotomayor, C. E., Rubinstein, H. R., Riera, C. M., and Masih, D. T. (1987). Immunosuppression in experimental cryptococcosis in rats. Induction of afferent T suppressor cells to a nonrelated antigen. *J. Med. Vet. Mycol. 25*: 67-75.

Staib, F. (1981). The perfect state of *Cryptococcus neoformans, Filobasidiella neoformans*, on pigeon manure filtrate agar. *Zentralbl Bakteriol. (Orig. A) 248*: 575-578.

Walter, J. E., and Jones, R. D. (1968). Serodiagnosis of clinical cryptococcosis. *Am. Rev. Respir. Dis. 97*: 275-282.

Wold, L. E., Roberts, G. D., Brewere, N. S., Zeller, D. J., Fineman, L. D., Paschall, C. L., and Rosenlog, R. C. (1980). Massive antigenemia during disseminated cryptococcosis. *Mayo Clin. Proc. 55*: 313-315.

Young, E. J., Hirsh, D. D., Fornstein, V., and Williams, T. W. (1980). Pleural effusions due to *Cryptococcus neoformans* a review of the literature and report of two cases with cryptococcal antigen determinations. *Am. Rev. Resp. Dis. 121*: 743-747.

Zimmer, B. L., Hempel, H. O., and Goodman, N. L. (1981). Technique for the purification of the hyphae of *Filobasidiella neoformans. Mycopathologia 73*: 171-176.

14
Genetic Regulation of Pathogenesis and Host Responses in Fungal Infection

ROBERT B. ASHMAN
Clinical Immunology Research Unit, Princess Margaret Hospital, Perth, Western Australia

JOHN M. PAPADIMITRIOU
University of Western Australia, Perth, Western Australia

PREAMBLE

Fungi are among the most common of the opportunistic pathogens, and fungal infections represent a real and growing concern to the physician, particularly in the hospital environment. Of the mycoses, the yeast *Candida albicans* is probably unique in the diversity and severity of the infections for which it is responsible. These range from nosocomial systemic infections, which are increasing markedly in frequency (Bodey and Fainstein, 1985), to vaginal infections that affect as many as 30% of normal women (Anyon, Desmond, and Eastcott, 1971). Nosocomial infections are generally seen in debilitated or immunosuppressed patients (Odds, 1979, Bodey and Fainstein, 1985), and most clinical studies have shown that infection is associated with a deficient or compromised immune system (Smith, 1985). In contrast, there are diverse extrinsic factors, such as diabetes, antibiotic therapy, use of estrogenic antiovulants, and pregnancy, that are known to predispose women to vaginal candidiasis (Odds, 1979). However, severe recurrent infections also occur in women in whom these variables have been eliminated or definitively excluded (Sobel, 1984).

GENETICS OF HUMAN INFECTION

In the absence of other factors, susceptibility or resistance is ultimately determined by the genetic constitution of the host; however, very little definitive information is available about genetic associations in *Candida* infection. There are of course many reports of familial chronic mucocutaneous candidiasis (CMC) (reviewed by Kirkpatrick, 1984), but no overall pattern of genetic susceptibility has yet emerged from these studies. This probably reflects the fact that any genetic abnormality that affects the cell-mediated immune system will lead to an outbreak of the disease. There are no published reports of associations with particular alleles within the human major histocompatibility complex (MHC), although Kirkpatrick (1984) has referred to an increased frequency of HLA B7 in his patients with CMC.

The actual incidence of severe recurrent vaginal candidiasis is not known, but one out of every seven women can be expected to experience a clinically significant episode during her reproductive life (Jacobson, 1986), and of those treated for the disease, between 10 and 25% suffer a relapse within 30 days (Odds, 1982; Merkus et al., 1985). Thus, the overall frequency of chronic vaginal candidiasis in the female population can be estimated to be between 1 and 3%. This is a remarkably high incidence of an infectious disease in developed countries with a high standard of living; but despite this, the condition is widely, albeit incorrectly, regarded as being of minor significance. When all the ramifications of the disease are taken into account, the socioeconomic costs are immense (Odds, 1979), not to mention the physical distress and psychological trauma suffered by women with intractable recurrent vaginal candidiasis (Hurley, 1975). In fact, it is generally acknowledged that the medical community has largely failed to come to grips with the disease at either the clinical or research level (Sobel, 1985). Genetic factors in particular have, to date, not been investigated in any great detail.

MOUSE MODELS OF CANDIDA INFECTION

Reliable models of the human disease have been established in mice, rats, rabbits, and guinea pigs (Rogers and Balish, 1980; Louria, 1985), but in spite of the diversity of animal models utilized, only rarely have these studies adequately analyzed the genetic constitution of the host as a factor determining susceptibility to infection. Because of the great variety of inbred and congenic resistant strains now available, the mouse is obviously the animal of choice for a systematic investigation of genetic influences on experimental candidiasis. However, until recently, the pathogenesis of systemic candidiasis in mice had been described only in outbred animals; therefore, it is reasonable to ask how accurately the infection in inbred strains reflects the human disease. CBA/H mice are known to be highly susceptible to lethal intravenous infection with *C. albicans*, and histopathological assessment of the acute disease in this mouse strain confirmed that there was extensive colonization by the organism (Papadimitriou and Ashman, 1986). The organs most affected were the brain, kidney, and heart. All of these tissues have previously been reported to be sites of infection (Rogers and Balish, 1980; Louria, 1985); however, as the number of organisms used for challenge was reduced, the brain lesions remained almost constant in severity, whereas colonization and tissue damage in the kidney and other organs became milder and less obvious. This pattern of infection appears accurately to reflect the human disease as it occurs in endemic fungal areas (Parker et al., 1976) and in the altered host (Parker et al., 1976). Furthermore, it is clear from the high incidence of oral and vaginal candidiasis in patients with the acquired immune deficiency syndrome (AIDS) (Klein et al., 1984; Rhoads et al., 1987) that depletion of T cells in the systemic circulation also predisposes to the mucosal forms of the disease. Thus, it would seem to be valid to study experimental systemic candidiasis in the mouse with some confidence that the results will be relevant to the human disease.

EXPERIMENTAL SYSTEMIC CANDIDIASIS

Genetic regulation of host responses to *C. albicans* infection has been found to be quite complex. In a comprehensive survey of inbred strains, Hector et al. (1982) were able to demonstrate a hierarchy among strains in susceptibility to infection, and concluded that the hemolytic complement (Hc) gene was a major variable involved in susceptibility or resistance. Because BALB/c and DBA-2 mice both carry H-2^d alleles in the major histocompatibility complex (MHC) and the former was resistant, whereas the latter was susceptible, they further concluded that MHC genes had no effect on host responses to infection. This study, however, has certain flaws. First of all, conclusions about susceptibility to acute systemic infection were based upon cumulative mortality at 28 days postchallenge, and colonization in the brain and kidney at that single time point. Second, the genetic analysis did not use congenic resistant mice, and therefore results from "MHC-identical" mice could be influenced by the effects of "non-MHC," or "background," genes. More recently, Marquis et al. (1986) have evaluated the resistance to systemic infection of a number of inbred, congenic resistant, and mutant strains. In common with other reports (Bistoni et al., 1982; Ashman and Papadimitriou, 1987), they found that C57Bl/6 mice were more resistant to lethal intravenous infection than mice of other strains but were unable to demonstrate any MHC associations. In this series of experiments an attempt was made to correlate survival studies with tissue counts, so that fungal colonization in the kidneys of infected mice was assayed not at a discrete time point, but only after death of the animal. Since

Table 1 Comparative Mortality of Various Inbred and Congenic Resistant Mouse Strains after Intravenous Infection with *Candida* Blastospores[a]

Strain	H2 haplotype	Log_{10} inoculum		
		6.5	6.0	5.5
BALB/C	d	5/5[b] (100%)	16/22 (73%)	0.22 (0%)
BALB/c-H-2[b]	b	–	9/14 (64%)	3/13 (23%)
BALB/c-H-2[k]	k	–	9/14 (64%)	2/12 (17%)
B10.D2	d	17/19 (89%)	0/6 (0%)	–
C57B1/6J	b	27/57 (47%)	8/50 (16%)	–
B10.BR	k	13/17 (76%)	6/18 (33%)	–
CBA/H	k	–	5/5 (100%)	1/10 (10%)

[a]*Candida albicans* was grown in suspension culture in Sabouraud's broth, washed four times in phosphate-buffered saline, and adjusted to the appropriate concentration for i.v. injection in 0.2 ml. The animals were observed daily for 28 days.
[b]No dead/total injected.

the direct cause of death after lethal *C. albicans* infection is unknown and may be different for "resistant" and "susceptible" strains, colony counts at the time of death may not accurately reflect differences between strains during the development and resolution of an acute infection.

Our own work has evaluated acute systemic infection as assessed by the following variables: mortality; tissue damage; fungal colonization and clearance; and lymph node responses after local subcutaneous infection. Using specific pathogen free (SPF) mice that have been verified microbiologically to be free of *C. albicans*, we have established that mortality is a very poor and insensitive criterion of susceptibility or resistance. It was certainly possible to rank the strains in order of susceptibility, but the most resistant strain (C57B1/6J) differed by only one log from the most susceptible (CBA/H) (Table 1). No convincing MHC associations were demonstrable. Furthermore, the mean survival times were virtually identical, most animals dying within 5 days of infection, with few or no deaths thereafter. This is in marked contrast to other reports, which have recorded deaths occurring as long as 50 days postinfection (Bistoni et al., 1982; Marquis et al., 1986). The contradictions between these different patterns of mortality cannot be resolved at present. However, isolates of *C. albicans* are certainly heterogeneous in their virulence for the mouse, and the presence of endogenous viruses in animal colonies may influence the course of experimental candidiasis.

Although CBA/H mice were somewhat more susceptible to lethal intravenous infection than either BALB/c or C57Bl/6J mice, a completely different picture

Genetic Regulation of Pathogenesis and Host Response

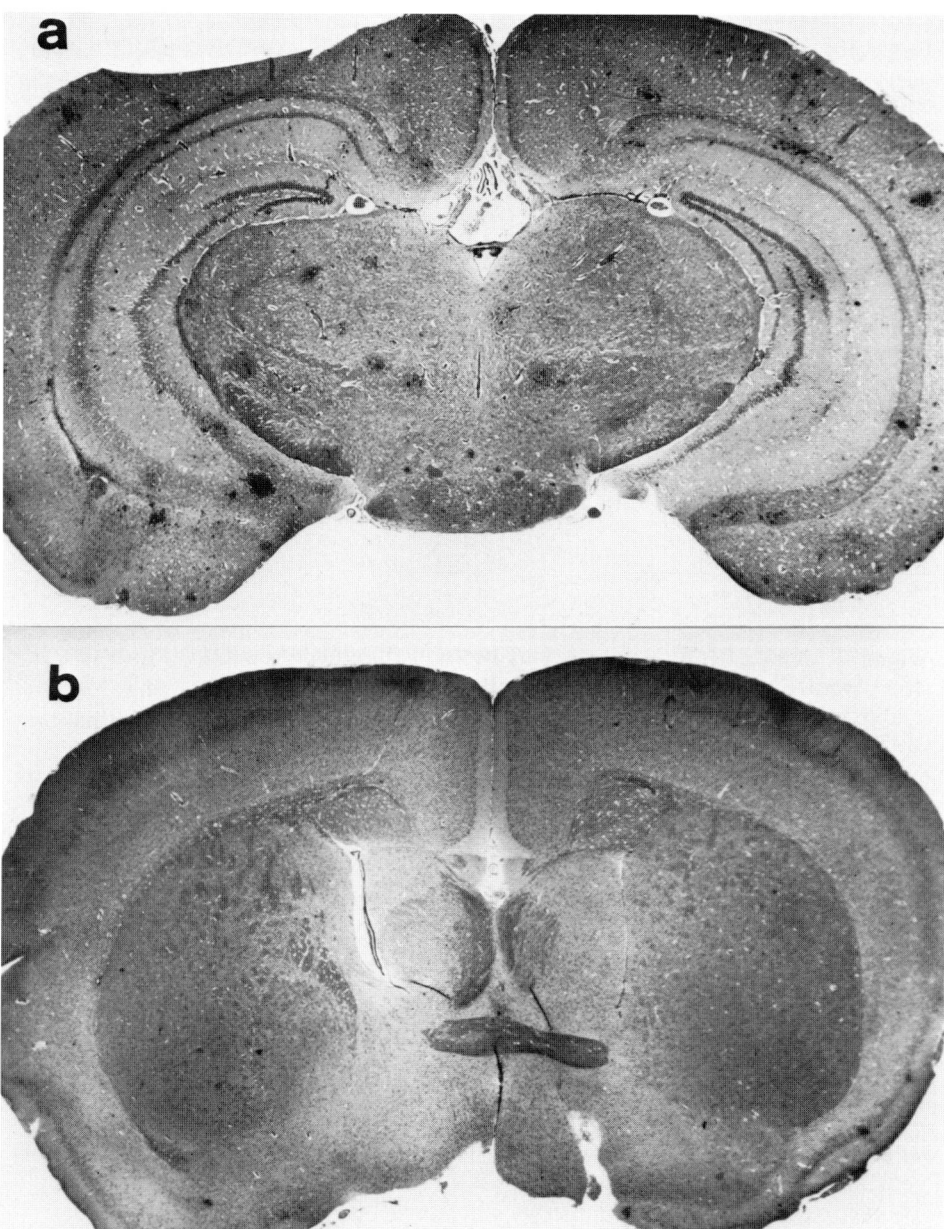

Figure 1 Transverse sections through the brain of a CBA/H (a) and a BALB/c (b) mouse 5 days after intravenous injection of 3×10^5 *Candida albicans* blastospores. There are many large abscesses present in the brain of the CBA/H mouse; however, the lesions in the BALB/c mouse are fewer and less severe. H&E ($\times 17$).

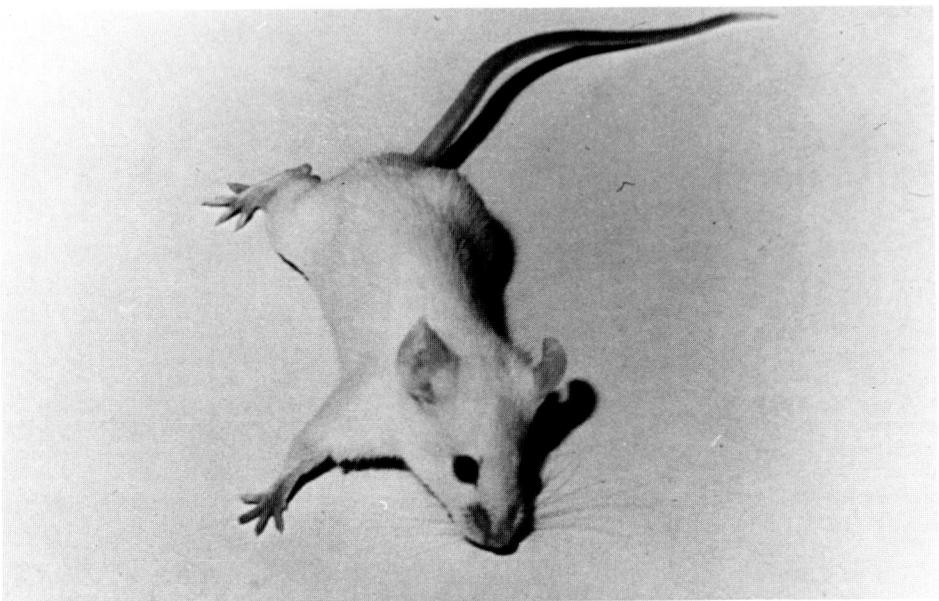

Figure 2 A BALB/c mouse displaying postural abnormalities after recovery from systemic *C. albicans* infection. Note particularly the attitude of the head and the splayed gait. This syndrome has been observed in only about 10% of BALB/c and C57Bl/6 mice, but occurs in all CBA/H mice.

emerged when tissue damage as assessed by histopathology was used as a criterion of susceptibility. The course of sublethal infection was carefully monitored in the following strains: CBA/H; C57Bl/6; BALB/c; BALB/c-H-2^k; DBA-2; and A/J. Somewhat surprisingly, only two patterns of infection were observed. Lesions in CBA/H mice were significantly more severe than in any of the other strains examined; however, there were no detectable differences between the rest. The most obvious pathological abnormalities were seen in the brain (Fig. 1), and it is noteworthy that, in both CBA/H and BALB/c mice, the cerebral abscesses remained almost constant in severity as the infecting inoculum was reduced, whereas tissue damage in other organs such as heart and kidney was either reduced or became inapparent. A singular feature of the sublethal infection observed uniformly in CBA/H mice, but only rarely in other strains, was a behavioral abnormality characterized by deafness, hyperactivity, titubation, and torticollis (Fig. 2). Otherwise, the mice ate and drank normally, displayed no signs of distress, and no deaths were recorded during a 3–7 month period of observation. This syndrome was finally found to be caused by a localized osteomyelitis of the petrous temporal (Fig. 3). The histopathology of the

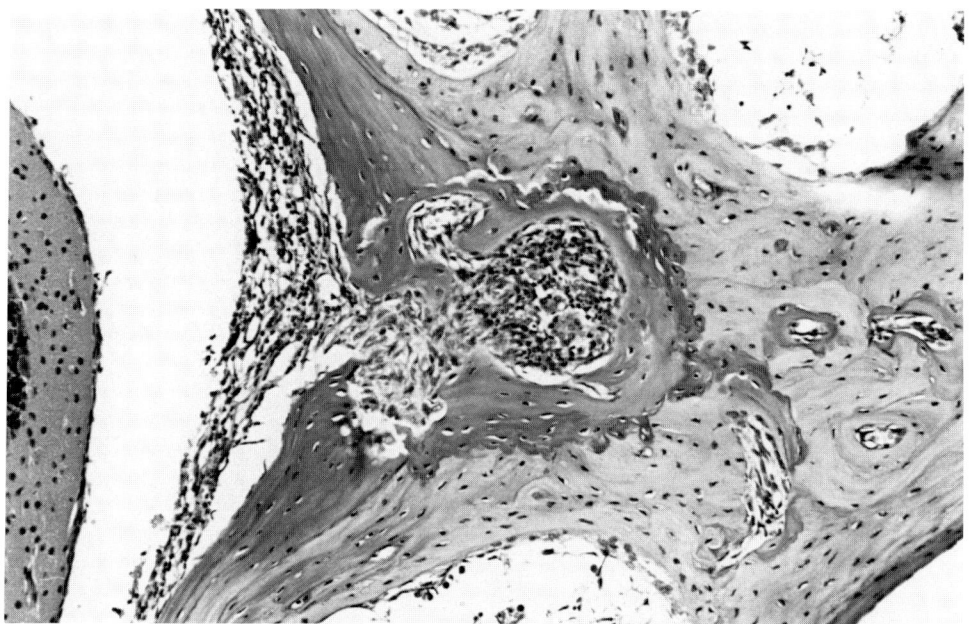

Figure 3 Petrous temporal bone of a CBA/H mouse 12 days after inoculation with 3×10^5 *C. albicans* blastospores. There is a pachymeningitis and osteomyelitis. In addition, the semicircular canals are dilated and infiltrated with leukocytes. H&E. (×160)

acute infection in BALB/c-H-2^k mice, which share MHC alleles with CBA/H, could not be distinguished from that in BALB/c mice, showing that the gene for susceptibility was located outside the MHC. Finally, comparison of the acute infection in [CBA/H × BALB/c] F1 hybrid mice with that in the parental strains showed that the genes regulating susceptibility and resistance were codominant.

GENETICALLY DETERMINED FACTORS PREDISPOSING TO SUSCEPTIBILITY

The clear difference in susceptibility between CBA/H and BALB/c mice thus provides a useful model system in which to examine intrinsic factors that may contribute or predispose to *Candida* infection. One of these is the ability of the organism to penetrate or lodge in various tissues and organs. This was assessed by the injection of ^{51}Cr-labeled *Candida* blastospores, and results representative

Table 2 Organ Distribution of Radiolabeled *Candida albicans* Blastospores[a]

Organs	CBA/H		C57Bl/6		BALB/c		BALB/c-H-2[k]	
				No. *Candida*/organ ($\times 10^5$)				
Brain	0.66[b]	(0.34)[c]	0.34	(0.18)	0.16	(.07)	.22	(.11)
Lungs	9.40	(18.06)	15.53	(13.13)	12.71	(16.26)	11.67	(20.06)
Heart	1.34	(1.44)	0.68	(0.64)	1.05	(1.04)	1.93	(1.92)
Spleen	1.78	(2.97)	1.01	(1.24)	1.11	(1.26)	1.72	(2.46)
Liver	31.63	(3.07)	24.14	(2.46)	23.96	(2.23)	24.06	(1.79)
Kidney	2.88	(2.52)	3.68	(3.27)	3.09	(1.99)	4.03	(2.05)

[a]10^{7} ^{51}Cr-labeled *Candida albicans* blastospores were injected intravenously and the recipient mice killed 2 h later. The results shown are the geometric means of values from 20 mice per strain. The data are reproduced from Ashman and Papadimitriou (1987).
[b]CBA/H mice were compared only with C57Bl/6, while the latter, and the BALB/c-H-2[k] mice, were compared only with BALB/c. The figures in boxes are significantly different from those with which they were compared, at at least the 5% level of significance.
[c]Number of *Candida*/mg tissue ($\times 10^3$).

Table 3 Number of Colony-Forming Units Present in the Brain of Normal and Immune BALB/c and CBA/H Mice Six Days after Intravenous Infection with *C. albicans* Blastospores[a]

	Mouse strains	
	BALB/c	CBA/H
Normal	4.00 ± 0.21 (10,046)[b]	4.95 ± 0.14 (88,512)
Immune	3.61 ± 0.16 (4,093)	2.76 ± 0.66 (571)

[a]Mice were immunized and challenged with 10^5 blastospores. At least 60 days were allowed to elapse between primary and secondary infections.
[b]Mean ± SD of the $\log_{10}$ colony counts in the brains of infected mice. At least 5 mice were used in each group. The geometric means are shown in brackets. The infection in immune mice of both strains is significantly less than in normal animals ($p < 0.05$).

of a number of experiments are shown in Table 2. CBA/H mice clearly acquired a greater fungal burden in the brain, heart, lungs, and spleen than mice of the resistant strains. However, this susceptibility can demonstrably be attributed to the influence of genes located outside the MHC, because apart from two minor differences, BALB/c-H-2^k mice display a pattern of colonization that is very close to that of BALB/c mice. The susceptibility of CBA/H mice has been confirmed by quantitation of the acute infection in the brain. Although there was considerable variation between experiments in the absolute numbers of fungal units recovered from the brain, CBA/H mice reproducibly showed counts approximately eightfold higher than did BALB/c (Table 3). So far, it would appear that the susceptibility of the CBA/H mice is associated with a higher level of deposition in the tissues and less "innate" resistance to the organism, permitting greater replication and invasion at the sites of infection.

MHC-REGULATION OF HOST RESPONSES

Unfortunately, the analysis has proved to be more complicated than this rather simplistic model would suggest. Host responses to primary infection have also been assessed by the inflammatory response in the popliteal lymph node after footpad inoculation of viable *C. albicans*. There were clear differences between mouse strains (Ashman and Papadimitriou, 1987) that could be attributed to the influence of genes within the major histocompatibility complex (MHC) (Fig. 4), and analysis of responses in [BALB/c × BALB/c-H-2^k] F1 hybrid mice has shown that low responsiveness is dominant (Fig. 5). However, "non-MHC" genes appear to be able to modulate this MHC-linked response, in that the pattern of responsiveness attributable to MHC genes was different when expressed on the

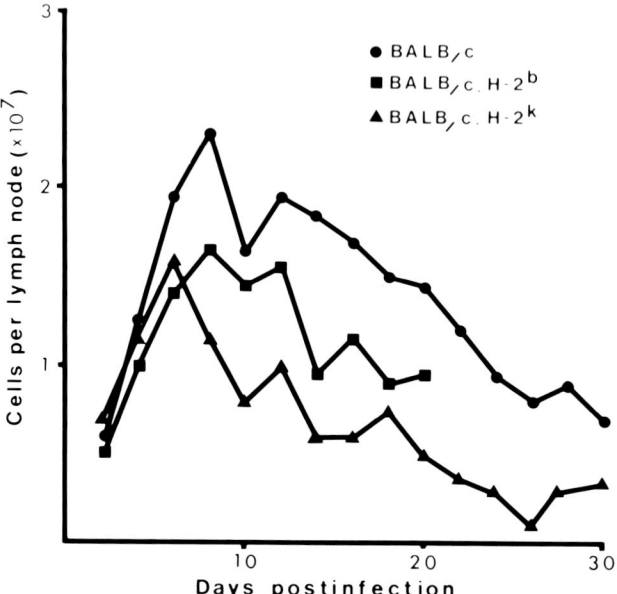

Figure 4 Inflammatory responses in the popliteal lymph nodes of congenic mice bred on the BALB background, after footpad injection of 10^7 *Candida* blastospores. Each point represents the geometric mean of the number of leukocytes per lymph node, averaged over a minimum of five animals.

C57Bl/10 (B10) background (Table 4). Nevertheless, because B10.A (2R) mice showed significantly lower responses than the B10.A strain, this particular system has enabled us to precisely map reactivity, at least in part, to a Class 1 MHC gene. This was somewhat unexpected, because proliferative responses to *Candida* antigens have been shown to require Class II-compatible antigen presenting cells (Nose et al., 1981); and most previous workers have tended to emphasize the role of delayed-type hypersensitivity (DTH) in the development of immunity to *Candida* infection (reviewed by Louria, 1985). Although DTH responses are in many cases regulated by Class II MHC genes, it has been shown very clearly that susceptibility to lymphocytic choriomeningitis virus, as well as production of cytotoxic T cells and the footpad swelling response (a classic measure of DTH), are all regulated by a Class 1 gene (H-2D) (Zinkernagel et al., 1985a, 1985b). Furthermore, in murine cytomegalovirus infection, the colony-stimulating factor, interleukin 3, is produced by Class 1-restricted T cells with the phenotype of cytotoxic cells (L3T4$^-$, Lyt 2$^+$) (Sinickas et al., 1985). Thus, T-cell-mediated responses in *Candida* infection may well follow similar pathways.

Genetic Regulation of Pathogenesis and Host Response

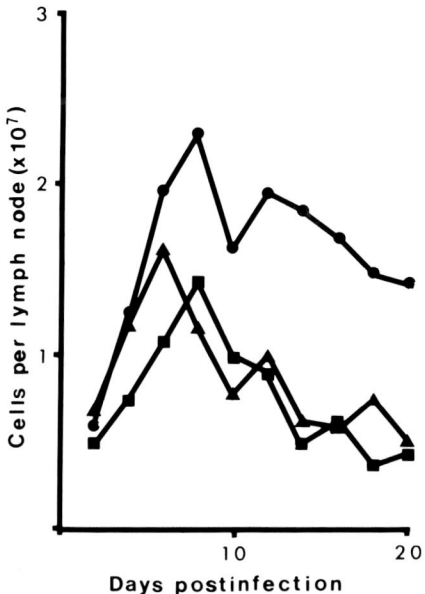

Figure 5 Comparison of lymph node responses in BALB/c (●), BALB/c-H-2^k (▲), and [BALB/c × BALB/c-H-2^k] F1 hybrid mice (■), after footpad injection of 1 × 10^7 *Candida* blastospores. The responses of BALB/c-H-2^k and F1 hybrid mice are virtually identical, showing that low responsiveness is dominant. Each point represents the geometric mean of the number of leukocytes per lymph node, averaged over a minimum of five animals.

Clinical research into candidiasis has almost invariably shown associations with defective cell-mediated immune responses; however, until recently, these observations have not been consistently reproduced in animal models such as adult-thymectomized, irradiated, bone marrow reconstituted (ATXBM) mice (Giger et al., 1978) or nude mice (Cutler, 1976; Rogers et al., 1976) that display deficient cell-mediated immune responses. In contrast, after treatment with a monoclonal Thy 1.2-specific antibody, the clearance of *Candida* cells from the spleen of BALB/c mice was significantly impaired (Table 5). Somewhat surprisingly, the tissue lesions in the antibody-treated mice were not distinguishable from those in control animals. This result clearly reinforces the dichotomy in this disease model between the role of the immune response in the clearance and elimination of the infectious organism, and "susceptibility" as measured by tissue damage. Obviously, the cell-mediated immune response does play a role in resistance to *Candida* infection, but the precise mechanisms by which it contributes to resolution of the infection are yet to be elucidated.

Table 4 H-2 Haplotypes and Responsiveness to *C. albicans* Infection of Various Congenic Resistant Mouse Strains Bred on the C57Bl/10 Background

Strain	H-2 haplotype	Allele								Mean area[a]
		K	Aα	Aβ	Eβ	J	Eα	S	D	
B10.D2	d	d	d	d	d	d	d	d	d	263.6
B10.A (2R)	h2	k	k	k	k	k	k	d	b	213.2
B10.A (4R)	h4	k	k	k	k	b	b	b	b	209.2
B10.A	a	k	k	k	k	k	k	d	d	303.3
B10.BR	k	k	k	k	k	k	k	k	k	248.6
B10.A (3R)	i3	b	b	b	b	b	k	d	d	234.2
B10.A (5R)	i5	b	b	b	b	k	k	d	d	259.7
C57Bl/6J	b	b	b	b	b	b	b	b	b	271.4

[a]The results represent the means of 5 to 10 areas, calculated by numerical integration of curves similar to those shown in Figure 4. The figures in boxes are significantly less ($p < 0.05$) than the response of B10.A mice. The data are reproduced from Ashman (1988).

Table 5 Effect of A Thy-1.2-Specific Antiserum on the Clearance of *Candida albicans* from the Spleen of BALB/c Mice[a]

Days postinfection	Treatment	
	Antibody	Medium
3	3.25 ± 0.08[b] (1767)	2.97 ± 0.15 (922)
10	2.70 ± 0.08 (497)	2.41 ± 0.22 (259)

[a]A Thy-1.2-specific monoclonal antibody was administered daily to the mice during the course of the experiment. The data are reproduced from Ashman (1987).
[b]Mean ± standard deviation of $\log_{10}$ colony counts from four animals. The geometric mean is shown in parentheses. On both days, the antibody-treated mice had significantly greater numbers of *Candida* cells in the spleen ($p < 0.05$).

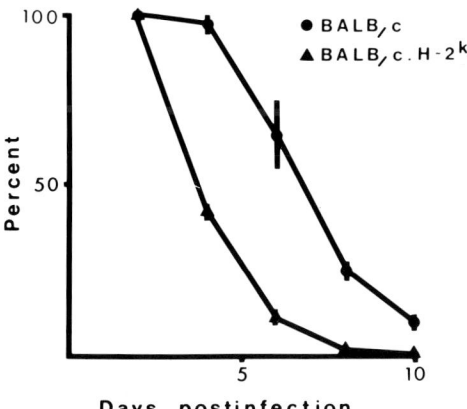

Figure 6 Clearance of *C. albicans* from the spleens of BALB/c and BALB/c-H-2^k mice infected with 10^5 blastospores i.v. The results are expressed as the percentage of organisms present in the spleen on day 2. The vertical bars represent the standard error of values derived from four animals.

There remains the unanswered question of the physiological basis for the observed difference in susceptibility and resistance to sublethal infection. Increased deposition of *C. albicans* blastospores in the tissues and the rapid growth of the organisms therein partly explain the experimental results. However, our data also show that susceptible mice cleared the organisms from the spleen more rapidly than the more resistant strains (Ashman and Papadimitriou, 1987), and this response is known to be regulated by MHC genes (Fig. 6). This poses something of a dilemma. Elicited granulocytes from CBA/H mice are no more effective at killing *Candida* than granulocytes from BALB/c mice, and the activity of spleen adherent cells, though low, is also comparable in the two strains (Ashman and Papadimitriou, unpublished data). It could therefore be inferred that the superior clearance of *Candida* from the spleen of CBA/H mice is due to augmentation of cellular candidacidal activity by factors secreted by T cells. Yet MHC-linked inflammatory responses are lower in CBA mice than in BALB/c (Ashman and Papadimitriou, 1987). As both T-suppressor cells (Stobo et al., 1976), as well as auto-antibodies (Mathur et a., 1980; Zouali et al., 1984), have been demonstrated in human candidiasis, we considered the possibility that susceptibility might be associated with a down-regulation of the immune response. The most likely causative factor seemed to be cross-reactivity between antigens of *Candida* and those expressed by the susceptible mouse strain. In order to test this hypothesis, spleen cells obtained from susceptible and resistant mice that had recovered from a primary infection were pulsed in vitro with a

Table 6 Strain-Specific Reactivity of Activated Spleen Cells from *Candida*-Immune Mice[a]

Recipient	Donor spleen cells	
	CBA/H (right footpad)	BALB/c (left footpad)
CBA	217.4 ± 7.0[b]	190.6 ± 5.9
BALB/c	194.5 ± 7.3	192.5 ± 7.9

[a]Spleen cells from previously infected mice were activated by in vitro culture for 5 days with a crude *Candida* antigen. CBA/H and BALB/c cells were then injected into the right and left footpads respectively of normal CBA/H and BALB/c recipients, in the absence of any further challenge with *Candida* antigen, and the response assessed by comparison of footpad thickness 24 h later. The results are expressed as the mean ± SD in mm × 10^{-2}.
[b]The response of CBA/H cells in syngeneic recipients was significantly greater than that of BALB/c cells in CBA/H mice ($p < 0.05$).

crude *Candida* antigen. These sensitized splenocytes were then injected into the footpad in the absence of any further exposure to *Candida* antigen to test their reactivity against antigens of the host. Activated cells from CBA/H mice displayed specific reactivity against CBA/H antigens as measured by the footpad swelling reaction, whereas those from BALB/c mice did not respond specifically to BALB/c antigens (Table 6). The activity of the CBA/H cells could be abrogated by treatment with a T-cell-specific antiserum, and the cells were further characterized as being of the $L3T4^-$, $Lyt2^+$ phenotype. However, they were not cytotoxic. The reactivity was directed against antigens expressed by CBA/H, but not by BALB/c-H-2^k mice, but it has not yet been possible to determine whether the response is MHC-restricted. The association between the potential to develop autoimmune responses and susceptibility to *C. albicans* infection throws a new light on the problem of recurrent vaginal candidiasis and allows a different interpretation of some of the more obscure associations, such as the influence of diet and the exacerbation of symptoms by desensitization procedures (Siegel, 1986).

THE GENERATION AND EXPRESSION OF IMMUNITY TO *C. Albicans*

There is considerable controversy, and more than a little misunderstanding, about immunity to *C. albicans*. First of all, it should be emphasized that the mechanisms of resistance in primary and secondary infection may not necessarily be the same. Classically, this is illustrated by the fact that recovery from primary viral infection is mediated almost exclusively by T cells, whereas protection against subsequent infections is conferred by antiviral antibodies. The point may seem self-evident, but many reports of the induction of "immunity" to *Candida* have not clearly distinguished between the two. For example, an attempt to protect an infected animal by passive transfer of spleen cells from an "immunized" donor is answering a different question than challenge of an animal that has recovered from a primary infection. Other variables that must also be taken into account are the strain in which these experiments are carried out, and the assay system used. As stated above, mortality does not accurately reflect the human disease, nor does it discriminate well between strains. Nonetheless, many, if not most, investigators have used it to measure the success of their "immunization" protocols. Overall, it seems that primary infection in most mouse strains results in the development of a degree of protection against subsequent reinfection (Rogers and Balish, 1980; Hector et al., 1982), but there are numerous contradictions and apparent anomalies in the published literature. At the one extreme, Hurtrel, Lagrange, and Michel (1980, 1981) failed to find any association between delayed-type hypersensitivity and protection against systemic candidiasis, and concluded that "non-immune" effector mechanisms, such as granulocytes, were the major factors involved in resistance to this disease. However opposite conclusions were presented in other studies. For example, immunization of CBA mice by subcutaneous injection of live organisms confers a degree of protection against lethal intravenous challenge, although without at least two immunizations the acquired immunity was variable and transient (Giger et al., 1978). A singular observation in this experimental model was that the lesions induced by subcutaneous infection in immunized mice were larger and contained a greater number of viable fungal units than those in unimmunized animals. In our experiments, systemic infection in CBA/H mice induced strong protective responses that markedly ameliorated the severity of lesions in the brain (Fig. 7) and dramatically reduced the levels of colonization (Table 3). On the other hand, resistant BALB/c mice showed no detectable alteration in the severity of tissue lesions after immunization, although there was a small but significant reduction in the number of fungal units in infected tissues, such as the brain (Table 3).

The basis for the difference in responses has not yet been identified. The problems associated with interpretation of passive transfer experiments without reference to the susceptibility or potential responsiveness of different strains has been exemplified by the experiments of Pearsall et al. (1978), who failed to demonstrate protection after passive transfer of syngeneic immune spleen cells in C57Bl/c and C3H mice, and consequently disregarded a significant response in

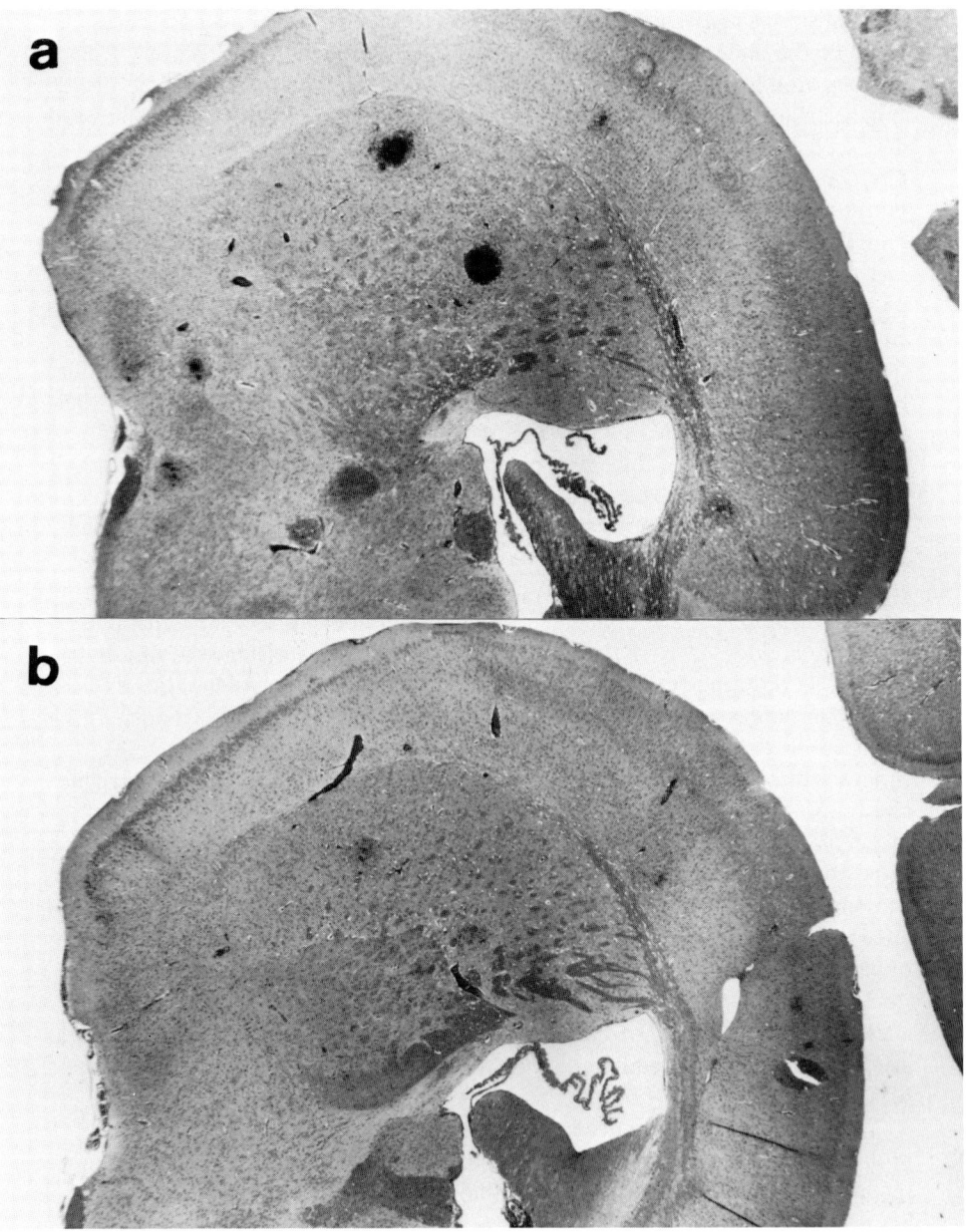

Figure 7 (a) Brain from CBA/H mouse taken 5 days after infection with 3×10^5 *C. albicans* blastospores. Several large abscesses can be seen. (b) Brain from CBA/H mouse immunized with 1×10^5 *C. albicans* blastospores, and sampled 5 days after reinfection with 3×10^5 blastospores. The abscesses are fewer and less severe. H&E. (×21)

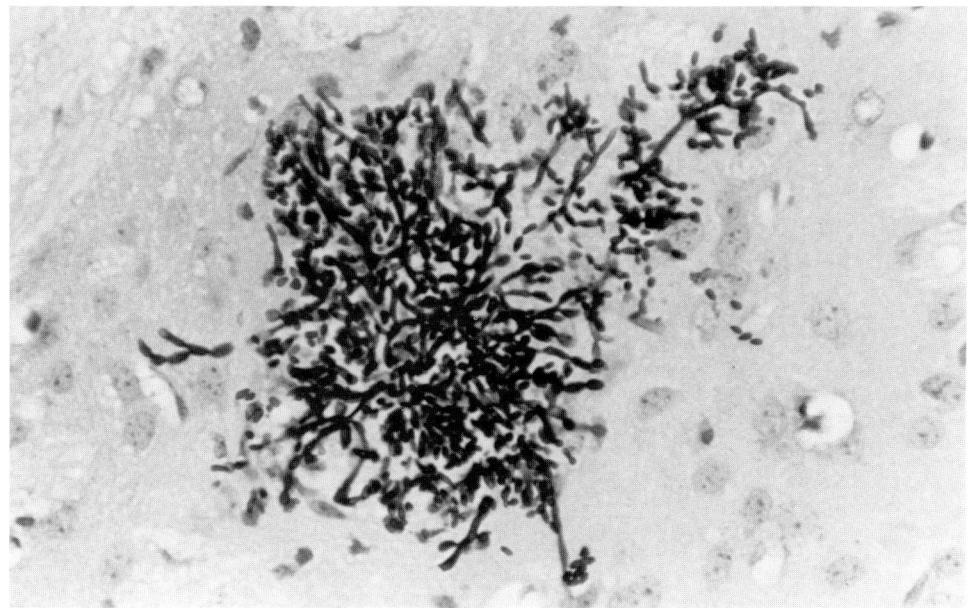

Figure 8 Brain from a normal CBA/H mouse, taken 5 days after infection with 3×10^5 blastospores. A large aggregate of fungal yeasts and pseudohyphae is present, but note the absence of a leukocytic exudate. PAS. ($\times$ 470)

the AKR strain. A separate investigation (Miyake et al., 1977) confirmed that intramuscular immunization of AKR mice with viable *Candida* blastospores resulted in reduced mortality, lower colony counts in the kidney and liver, and a reduction in the severity of the pyelonephritis in the kidney after secondary challenge. As in the study by Pearsall et al. (1978), the protection could be transferred by immune spleen cells. Although these passive transfer experiments confirm the importance of the cell-mediated immune response in recovery from primary infection, they do not necessarily mean that enhanced resistance to reinfection is solely a T-cell-dependent function. Antibody may enhance the function of immune lymphocytes (Kagaya et al., 1981), facilitate antibody-dependent cell-mediated cytotoxicity against fungal organisms (Diamond, 1974), or act directly to protect against infection. This has been demonstrated in both humans (Hiatt and Martin, 1946) and experimental animals (Mourad and Friedman, 1968; Al-Dorry, 1970; Pearsall et al., 1978). A possible role for antibody in inducing inflammatory responses at the site of *Candida* infection is also suggested by our own histopathological observations.

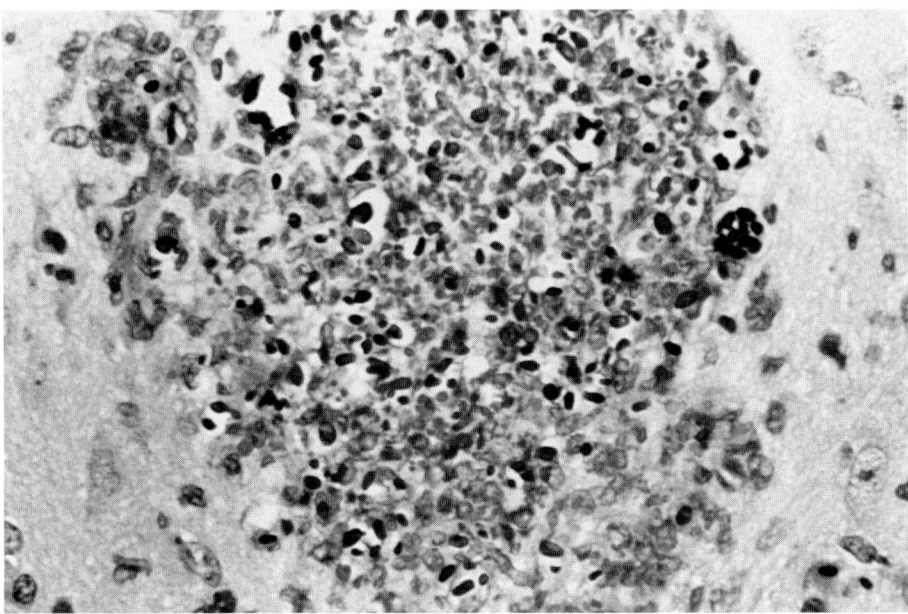

Figure 9 Brain from CBA/H mouse immunized with 1×10^5 blastospores, and sampled 3 days after challenge with 3×10^5 blastospores 60 days later. Note the dense leukocytic infiltrate. PAS. (×640)

In normal mice, yeasts can proliferate in the brain for up to five days without any evidence of an inflammatory response (Fig. 8). There appears to be no difference between susceptible and resistant mice in this regard, and this suggests that recruitment of granulocytes to the site of tissue invasion is not a major factor determining the different patterns of host responses in these two strains. This conclusion is consistent with other reports that susceptibility to fungal infections is related neither to the actual numbers of circulating granulocytes (Hector et al., 1982), nor to their effector functions (Hidore and Murphy, 1986). Nevertheless, these results are not entirely in agreement with clinical data (Smith, 1985), and the conflict has yet to be resolved. In contrast to the histological picture in normal animals, a leukocytic exudate rapidly accumulated around the fungal organisms in the brains of immunized mice (Fig. 9). Again, there was no appreciable difference between susceptible and resistant strains in the time of appearance of the inflammatory infiltrate, although as noted above, the severity of the lesions in CBA/H mice was very much reduced. These observations indicate that antibody may play a role in the increased resistance to challenge displayed by susceptible mice, and in this context it is pertinent to note that CBA mice produce higher titers of antibody after initial infection than do the more resistant BALB/c mice (Hector et al., 1982). However, the point has been made previously (Rogers and Balish, 1980) that human infections commonly occur in the presence of high titers of agglutinins and precipitins against *Candida* antigens, so it would be doing less than justice to a very confusing mass of data to attempt at present to make any definitive statements on "immunity" to *C. albicans*. Nevertheless, it seems clear from our histopathological data that susceptible mice do develop protective responses, and that specific antibody may be responsible, at least in part, for this acquired resistance.

CANDIDIASIS IN MUTANT AND IMMUNODEFICIENT MICE

Strains of inbred or congenic resistant mice often display defects in immune, phagocyte, or complement systems. These range from virtually complete elimination of the immune system, as in the case of the nude mouse, to minor and specific defects in phagocytosis. These mice, therefore, represent an invaluable tool for the analysis of effector mechanisms in infectious diseases, although results may need to be interpreted with caution, particularly in animals in which there are multiple defects. Nude mice are a good illustration. It was anticipated that, because of a generalized defect in cell-mediated immune responses that renders them unusually susceptible to a wide range of infectious agents (Kindred, 1981), they would also be a suitable model in which to demonstrate T-cell involvement in recovery from *Candida* infection. In fact, they were found to be more resistant than normal animals (Cutler, 1976; Rogers et al., 1976). Comparable results have been reported using *Listeria monocytogenes* and some other infectious agents, and this resistance has been attributed to the presence of activated macrophages (Cheers and Waller, 1975; Zinkernagel and Blanden, 1975). Nonetheless, it would be premature, in the absence of other evidence, to draw conclusions about the role of macrophages in defense against candidiasis in T-cell-sufficient animals.

The beige mutation, like the nude, has been found to affect a number of effector mechanisms, so interpretation of patterns of susceptibility is again quite difficult. Edin et al. (1974) reported that beige mice were significantly more susceptible to *C. albicans* infection than either C57Bl/6N or heterozygous mice; however, since the majority of their mice, in both experimental and control groups, died within 96 h, their aggressive challenge may not accurately reflect the influence of this mutation on sublethal infection. A more systematic approach (Marquis et al., 1986) has confirmed that *bg/bg* mice have a substantially shorter survival time than controls after lethal intravenous challenge (19 days compared to 55.7 days), but the levels of colonization in the kidney were not so dramatically different. This latter is more consistent with our own data, which showed a slight but definite increase in the severity of lesions in the homozygous mice. In terms of mechanisms of resistance to fungal infections, it is interesting that Hidore and Murphy (1986) have reported that granulocytes of beige mice are more effective in killing *Cryptococcus neoformans* than those from heterozygous animals, and we have confirmed this in our own experiments with *Candida*. These results clearly suggest that granulocytes by themselves, even in an "activated" state, cannot confer resistance to either *C. neoformans* or *C. albicans*, but analysis of the defect in beige mice suggests that it is too generalized to be able confidently to ascribe susceptibility to a deficiency in natural killer cells.

Defects in macrophage function have been described in a number of the common mouse strains, such as C3H/HeJ, BALB/c, and A/J (Vogel et al., 1981); however, the first two strains have not to date been found to show any impairment in their responses to *Candida* infection. The case of the A/J mouse is more difficult to interpret because it is also deficient in the fifth component of complement (Hc^0). It is generally acknowledged that these mice are extremely susceptible to lethal intravenous infection with both *C. albicans* (Hector et al., 1982) and *C. neoformans* (Rhodes et al., 1980); however, when the infecting inoculum was reduced, *Candida* lesions in the brain were found to be no more severe than in BALB/c mice treated similarly (Ashman and Papadimitriou, unpublished data). These observations tend to support our contention that mortality does not adequately reflect the ability of an animal to clear the organism from infected tissues and to mount effective inflammatory and immune responses against it. In addition, when the role of the Hc gene was evaluated in the congenic resistant B10.D2 "old" and "new" mice, the Hc^0 mice were slightly, but not markedly, more susceptible than those carrying the Hc^1 gene (Lyon et al., 1986). Of course, given the complexity of the interactions already demonstrated, it is not unlikely that the "resistance" gene present in B10 mice overrides the effect of the Hc gene in these strains, and it (the Hc gene) may well have a more dramatic effect when expressed on a susceptible background. Unfortunately, appropriate strain combinations in which to test this postulate are not yet available.

RECAPITULATION

The concept that the course and outcome of an infection is determined by the genetic constitution of the host is hardly new (reviewed by Skamene et al., 1980). However, investigation of the genetic factors which influence susceptibility to experimental murine candidiasis has provided some different insights into the respective roles of MHC and non-MHC genes, and the ways in which they may exert their effects. It is clear from work in such diverse infectious disease models as leishmania (Blackwell, 1982) and listeriosis (Skamene et al., 1979) that gross susceptibility to the disease is usually determined by genes mapping outside the MHC, and this also appears to be true for experimental candidiasis. Thus, although CBA/H mice are the most susceptible to systemic *Candida* infection, their MHC-linked inflammatory/immune responses are virtually identical to those in "resistant" BALB/c-H-2^k mice (Ashman and Papadimitriou, 1987). It is, therefore, possible to construct a paradigm relating these elements as follows: (1) the severity of the initial infection is determined by "innate" resistance factors, governed by "non-MHC" genes; (2) once infection has become established, *Candida*-specific T-cell responses are elicited. These augment the function of various effector cells, presumably by the production of soluble mediators such as interferon-gamma (Djeu et al., 1986) or colony-stimulating factors (Lopez et al., 1983; Vadas et al., 1983). However, cross-reactivity between antigens of *Candida* and those of certain hosts may lead to the development of autoimmune responses, which may compromise the generation of fully efficient immune responses against the organism; (3) the acquired immune responses occurring as a result of the infection are expressed within the context of innate resistance genes of the various mouse strains and are more readily seen in susceptible animals.

The notion that the potential to develop autoimmune responses may, at least partially, determine susceptibility to an infectious disease has not previously been considered. Murine candidiasis thus provides a unique experimental model in which the various factors that contribute to susceptibility and resistance can be analyzed and dissected at both genetic and immunological levels.

CONCLUSION

Failure to recognize the complexity of the genetic influences on both primary and secondary *Candida* infection has undoubtedly contributed to the current confusion in the field; it is clear that a further confounding factor has been the failure to discriminate between mechanisms of resistance in primary and secondary infections. The results presented here should facilitate the development of appropriate experimental models, and given the current prevalence of candidiasis in AIDS patients and in other immunologically compromised patients, it can be anticipated that rapid advances in the understanding of this common but most problematic disease will be made.

ACKNOWLEDGMENTS

The research presented in this paper has been supported by the National Health and Medical Research Council of Australia, the Channel 7 Telethon Trust, and the King Edward Memorial Hospital Research Foundation. We thank Mrs. M. Wesolowski and Miss K. Churack for technical assistance. This is Publication No. 278 from the Clinical Immunology Research Unit of the Children's Medical Research Foundation, Princess Margaret Hospital for Children.

REFERENCES

Al-Dorry, Y. (1970). An immune factor in baboon anti-*Candida* serum. *Sabouraudia* 8:31-47.

Anyon, C. P., Desmond, F. B., and Eastcott, D. F. (1971). A study of candida in one thousand and seven women. *N.Z. Med. J. 73*:9-13.

Ashman, R. B. (1987). Mouse candidiasis. II. Host responses are T-cell dependent and regulated by genes in the major histocompatibility complex. *Immunogenetics 25*:200-203.

Ashman, R. B. (1987). Murine candidiasis. III. Host inflammatory responses are regulated in part by Class 1 MHC genes. *J. Immunogenetics 19*:317-321.

Ashman, R. B., and Papadimitriou, J. M. (1987). Murine candidiasis. Pathogenesis and host responses in genetically distinct inbred mice. *Immunol. Cell. Biol.* 65:163-171.

Bistoni, F., Marconi, P., Frati, L., Bonmassor, E., and Garaci, E. (1982). Increase of mouse resistance to *Candida albicans* infection by thymosin α_1. *Infect. Immun. 36*:609-614.

Blackwell, J. M. (1982). Genetic control of recovery from visceral leishmaniasis. *Trans. R. Soc. Trop. Med. Hyg.* 76:147-151.

Bodey, G. P., and Fainstein, V. (1985). Systemic candidiasis. In *Candidiasis*. Edited by G. P. Bodey and V. Fainstein. New York, Raven Press, pp. 135-168.

Cheers, C., and Waller, R. (1975). Activated macrophages in congenitally athymic "nude" mice and in lethally irradiated mice. *J. Immunol. 115*:844-847.

Cutler, J. E. (1976). Acute systemic candidiasis in normal and congenitally thymic deficient "nude" mice. *J. Reticuloendothelial Soc. 19*:121-126.

Diamond, R. D. (1974). Antibody-dependent killing of *Cryptococcus neoformans* by human peripheral blood mononuclear cells. *Nature 247*:148-150.

Djeu, J. Y., Blanchard, D. K., Halkias, D., and Friedman, H. (1986). Growth inhibition of *Candida albicans* by human polymorphonuclear neutrophils: activation by interferon-gamma and tumor necrosis factor. *J. Immunol.* 137: 2980-2984.

Elin, R. J., Edelin, J. B., and Wolff, S. M. (1974). Infection and immunoglobulin concentrations in Chediak-Higashi mice. *Infect. Immun. 10*:88-91.

Giger, D. K., Domer, J. E., and McQuitty, J. T., Jr. (1978). Experimental murine candidiasis: pathological and immune responses to cutaneous inoculation with *Candida albicans. Infect. Immun. 19*:499-509.

Giger, D. K., Domer, J. E., Moser, S. A., and McQuitty, J. T., Jr. (1978). Experimental murine candidiasis: pathological and immune responses in T-lymphocyte-depleted mice. *Infect. Immun. 21*:729-737.

Hector, R. F., Domer, J. E., and Carrow, E. W. (1982). Immune responses to *Candida albicans* in genetically distinct mice. *Infect. Immun. 38*:1020-1028.

Hiatt, H. S., and Martin, D. S. (1946). Recovery from pulmonary moniliasis following serum therapy. *J. Am. Med. Assoc. 130*:205-206.

Hidore, M. R., and Murphy, J. W. (1986). Natural cellular resistance of beige mice against *Cryptococcus neoformans*. *J. Immunol. 137*:3624-3631.

Hurley, R. (1975). Inveterate vaginal thrush. *The Practitioner 215*:753-756.

Hurtrel, B., Lagrange, P. H., and Michel, J. C. (1980). Systemic candidiasis in mice. II. Main role of polymorphonuclear leucocytes in resistance to infection. *Ann. Immunol.* (Paris) *131C*:105-118.

Hurtrel, B., Lagrange, P. H., and Michel, J. (1981). Absence of correlation between delayed type hypersensitivity and protection in experimental systemic candidiasis in immunised mice. *Infect. Immun. 31*:95-101.

Jacobson, J. B. (1986). Current thinking on the pathogenesis of vulvovaginal candidiasis. *J. Reprod. Med. 31*(Suppl):645-646.

Kagaya, R., Shinoda, J., and Fukazawa, Y. (1981). Murine defense mechanisms against *Candida albicans* infection. *Microbiol. Immunol. 25*:647-654.

Kindred, B. (1981). Deficient and sufficient immune systems in the nude mouse. In *Immunologic Defects in Laboratory Animals*. Edited by Gershwin, M. E. and B. Merchant. Plenum Press, New York and London, pp. 215-265.

Kirkpatrick, C. H. (1984). Host factors in defense against fungal infections. *Amer. J. Med. (Suppl 4D)*:1-12.

Klein, R. S., Harris, C. A., Small, C. B., Moll, B., Lesser, M., and Friedland, G. H. (1984). Oral candidiasis in high-risk patients as the initial manifestation of the acquired immunodeficiency syndrome. *N. Engl. J. Med. 311*:354-358.

Lopez, A. F., Nicola, N. A., Burgess, A. W., Metchalf, D., Battye, F. L., Sewell, W. A., and Vadas, M. (1983). Activation of granulocyte cytotoxic function by purified mouse colony-stimulating factors. *J. Immunol. 131*:2983-2988.

Louria, D. B. (1985). Candida infections in experimental animals. In *Candidiasis*. Edited by G. P. Bodey, and V. Fainstein. Raven Press, New York, pp. 29-51.

Lyon, F. L., Hector, R. F., and Domer, J. E. (1986). Innate and acquired immune responses against *Candida albicans* in congenic B10.D2 mice with deficiency of the C5 complement component. *J. Med. Vet. Mycol. 24*:359-367.

Marquis, G., Montplaisir, S., Pelletier, M., Mousseau, S., and Auger, P. (1986). Strain-dependent differences in susceptibility of mice to experimental candidosis. *J. Inf. Dis. 154*:906-908.

Mathur, S., Melchers, J. T. III, Ades, E. W., Williamson, H. O., and Fudenberg, H. H. (1980). Anti-ovarian and anti-lymphocyte antibodies in patients with chronic vaginal candidiasis. *J. Reprod. Immunol. 2*:247-262.

Merkus, J. M. W. M., Bisschop, M. P. J. M., and Stolte, L. A. M. (1985). The proper nature of vaginal candidosis and the problem of recurrence. *Obstet. Gynecol. Survey 40*:493-504.

Miyake, T., Takeya, K., Nomota, K., and Muraoka, J. (1977). Cellular elements in the resistance to *Candida* infection in mice. I. Contribution of T lymphocytes and phagocytes at various stages of infection. *Microbiol. Immunol. 21*: 703-725.

Moser, S. A., and Domer, J. E. (1980). Effects of cyclophosphamide on murine candidiasis. *Infect. Immun. 27*:376-386.

Mourad, S., and Friedman, L. (1986). Passive immunisation of mice against *Candida albicans. Sabouraudia 6*:103-106.

Nose, Y., Komori, K., Inouye, H., Nomura, K., Yamamura, M., and Tsuji, K. (1981). Role of macrophages in T lymphocyte response to Candida allergen in man with special reference to HLA-D and DR. *Clin. Exp. Immunol. 45*: 152-157.

Odds, F. C. (1979). *Candida and Candidosis*. University Park Press, Baltimore.

Odds, R. C. (1982). Genital candidosis. *Clin. Exp. Dermatol. 7*:345-354.

Papadimitriou, J. M., and Ashman, R. B. (1986). The pathogenesis of acute systemic *Candida albicans* infection in a susceptible, inbred, mouse strain. *J. Pathol. 150*:257-265.

Parker, J. C. Jr., McCloskey, J. J., and Knauer, K. A. (1976). Pathobiologic features of human candidiasis. A common deep mycosis of the brain, heart and kidney in the altered host. *Am. J. Clin. Pathol. 65*:991-1000.

Parker, J. C. Jr., McCloskey, J. J., Solanki, K. V., and Goodman, N. L. (1976). Candidiasis: the most common cerebral mycosis in an endemic fungal area. *Surg. Neurol. 6*:123-128.

Pearsall, N., Adams, B., and Bunni, R. (1978). Immunologic responses to *Candida albicans*. III. Effects of passive transfer of lymphoid cells or serum on murine candidiasis. *J. Immunol. 120*:1176-1180.

Rhoads, J. L., Wright, D. C., Redfield, R. R., and Burke, D. S. (1987). Chronic vaginal candidiasis in women with human immunodeficiency virus infection. *JAMA 257*:3105-3107.

Rhodes, J. C., Wicker, L. S., and Urba, W. J. (1980). Genetic control of susceptibility to *Cryptococcus neoformans* in mice. *Infect. Immun. 29*:494-499.

Rogers, T. J., and Balish, E. (1980). Immunity to *Candida albicans. Microbiol. Rev. 44*:660-682.

Rogers, R. J., Balish, E., and Manning, D. D. (1976). The role of thymus-dependent cell-mediated immunity in resistance to experimental disseminated candidiasis. *J. Reticuloendothelial Soc. 20*:291-298.

Siegel, J. (1986). Clinical allergies and vulvo-vaginitis. *J. Reprod. Med. 31* (Suppl):647.

Sinickas, V. G., Ashman, R. B., Hodgkin, P. D., and Blanden, R. V. (1985). The cytotoxic response to murine cytomegalovirus. III. Lymphokine release and cytotoxicity are dependent upon phenotypically similar immune cell populations. *J. Gen. Virol. 66*:2551-2561.

Skamene, E., Kongshavn, P. A. L., and Landy, M. (1980). *Genetic Control of Natural Resistance to Infection and Malignancy.* Academic Press, New York.

Skamene, E., Kongshavn, P. A. L., and Sachs, D. H. (1979). Resistance to *Listeria monocytogenes* in mice is genetically controlled by genes which are not linked to the H-2 complex. *J. Inf. Dis. 139*:228-231.

Smith, C. B. (1985). Candidiasis: pathogenesis, host resistance, and predisposing factors. In *Candidiasis.* Edited by G. P. Bodey, and V. Fainstein. Raven Press, New York, pp. 53-70.

Sobel, J. D. (1984). Vulvovaginal candidiasis—what we do and do not know. *Ann. Intern. Med. 101*:390-392.

Sobel, J. D. (1985). Epidemiology and pathogenesis of recurrent vulvovaginal candidiasis. *Am. J. Obstet. Gynecol. 152*:924-935.

Stobo, J., Hull, S., VanScoy, R., and Hermans, P. (1976). Suppressor thymus-derived lymphocytes in fungal infection. *J. Clin. Invest. 57*:319-328.

Vadas, M. A., Nicola, N. A., and Metcalf, D. (1983). Activation of antibody-dependent cell-mediated cytotoxicity of human neutrophils and eosinophils by separate colony stimulating factors. *J. Immunol. 130*:795-799.

Vogel, S. M., Weinblatt, A. C., and Rosenstrich, D. L. (1981). Inherent macrophage defects in mice. In *Immunologic Defects in Laboratory Animals.* Edited by M. E. Gershwin and B. Merchant. Plenum Press, New York and London, pp. 327-357.

Zinkernagel, R. M., and Blanden, R. V. (1975). Macrophage activation in mice lacking thymus-derived (T) cells. *Experientia 31*:591.

Zinkernagel, R. M., Leist, T., Hengartner, H., and Althage, A. (1985a). Susceptibility to lymphocytic choriomeningitis virus isolates correlates directly with early and high cytotoxic T cell activity, as well as with footpad swelling reaction, and all three are regulated by H-2D. *J. Exp. Med. 162*:2125.

Zinkernagel, R. M., Pfau, C. J., Hengartner, H., and Althage, A. (1985b). Susceptibility to murine lymphocytic choriomeningitis maps to Class 1 MHC genes—a model for MHC/disease associations. *Nature* (Lond) *316*:814.

Zouali, M., Drouhet, E., and Eyquem, A. (1983). Evaluation of auto-antibodies in chronic mucocutaneous candidiasis without endocrinopathy. *Mycopathologica 84*:87-93.

15
Humoral Immune Response to *Candida albicans*

JEFFREY M. JONES
University of Wisconsin School of Medicine and William S. Middleton Memorial VA Hospital, Madison, Wisconsin

INTRODUCTION

The term candidiasis is applied to infections produced by yeasts of the genus *Candida*. *Candida albicans* and *C tropicalis* are the species most frequently causing human candidiasis; however, *C. stellatoidea, C. krusei, C. pseudotropicalis, C. guilliermondii, C. parapsilosis*, and *C. lusitaniae* are also medically important (Jones, 1984; Hadfield et al., 1987). Because *C. albicans* is clearly the most important human pathogen, I will be dealing here only with infections produced by this organism and immune responses against it; therefore, I will simply refer to *C. albicans* as *Candida* throughout this discussion.

Some authors consider *Candida* to be an obligatory animal saprophyte (Odds, 1979). Transfer of the organism from maternal mucous membranes to those of the newborn infant occurs during birth or soon thereafter, and colonization of mouth, intestines, and skin of intertriginous zones is completed soon after birth. Thus, in humans it is the breakdown of normal barriers to infection or overgrowth of the endogenous yeast flora which results in candidiasis.

Candida can reproduce as oval budding structures, called blastospores, measuring about 2.5 X 5 μm. In the environment of host tissues however, the organism produces cylindrical structures called germ tubes (Fig. 1). A germ tube forms no constriction at its point of origin. Elongation of the germ tube may continue with septae forming behind its extending apical end and blastospores budding from the resultant cylindrical septate hypha. *Candida* can also undergo a budding process in which elongated elements are formed that remain attached

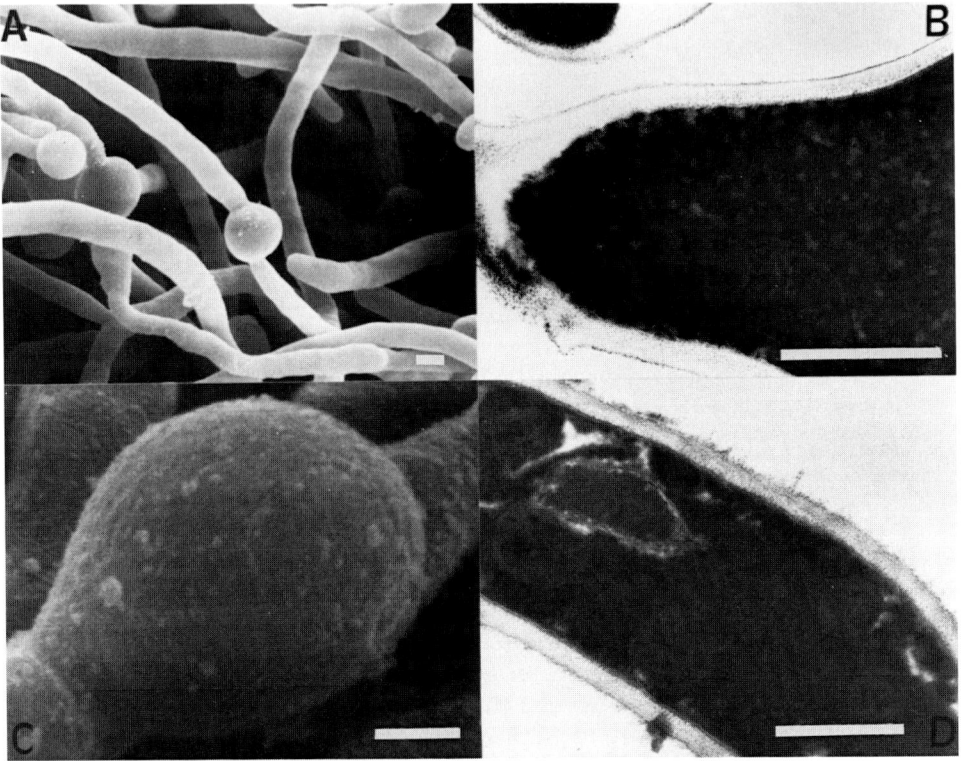

Figure 1 Morphology of mycelial phase *Candida albicans* demonstrated by scanning (A and C) and transmission (B and D) electron microscopy. Original magnification of A and C were 3200 X and 13,000 X, respectively. Original magnification of B and D was 56,000 X. Organisms shown in C and D had been treated with dithiothreitol (DTT). DTT treatment leads to loss of material from the outer electron dense layer of the cell wall, leaving it with roughened appearance.

to each other to produce pseudohyphae. *Candida* growing only as budding blastospores are said to be in the yeast phase, and when growth of germ tubes occurs, it is said to be growing in the mycelial phase. Depending upon conditions in culture or the tissues, organisms growing in the mycelial phase may also have budding blastoconidia and pseudohyphae.

Candida commonly produces infections of the skin and mucous membranes of humans which are self-limited or readily treated with topical antifungal drugs. These superficial infections include oral thrush, dermatitis of intertriginous areas and the feet, and vaginitis. Chronic mucocutaneous candidiasis, characterized by chronic infections of the oral mucosa and esophagus and by cutaneous granulomas, has been described commonly in patients with inherited defects in cellular immunity and in patients with acquired immunodeficiency syndrome (AIDS) (Edwards, 1978; Gottlieb, 1981). However, bloodborne dissemination to one or more distant sites including the kidney, liver, brain, retina, heart, skeletal muscle, abdominal viscera, bone, and skin is exceedingly rare in these patients with defective delayed hypersensitivity. Bloodborne dissemination happens commonly only in two situations, occurring singly or together. The first is a major disruption in the normal epithelial barriers to infection. Introduction of *Candida* via an intravenous catheter placed for hyperalimentation is an example of this problem (Edwards, 1974). The second is severe neutropenia, leading to an inability of the host to contain invasion that occurs when only minor breaches of the epithelial barriers occur. In patients with hematologic malignancies made neutropenic by chemotherapy, deeply invasive candidiasis has been observed in up to 20% and disseminated infections in up to 10% of neutropenic episodes (Winston et al., 1979; Meckstroth et al., 1981; Pizzo et al., 1982; Greenfield et al., 1983a). The tremendous importance of the neutrophil in preventing potentially lethal candidiasis leads one to consider the importance of the humoral immune system in host defense against candidiasis. Antibodies and complement are the two major components of humoral immunity responsible for recognizing antigens of invading microbes. There is a concensus that the neutrophil, and to a lesser extent tissue monocytes and macrophages, are effector cells for the humoral immune system. As the following discussion will show, specific antibodies against antigens of *Candida* are ubiquitous in human sera. Specific antibodies and complement are key in assuring chemotaxis of neutrophils, their attachment to the organism, and phagocytosis of the organism. Their cooperation with neutrophils, and to a lesser extent with tissue monocytes and macrophages, is essential in preventing deeply invasive candidiasis and bloodborne dissemination.

In the following sections I will discuss antigens of *Candida*, antibody responses to *Candida* antigens in humans, and interactions of human complement components with *Candida*. Finally, I will define the role of the humoral response in host defense in humans as can be ascertained by observations of human disease, in vitro experiments with human neutrophils, and pertinent laboratory animal models.

ANTIGENS OF *Candida albicans*

Cytoplasmic Antigens

Historically, antigen extracts from *Candida* were first prepared by breaking up the organism mechanically with a French press, Braun homogenizer, or sonicator.

The solubilized material obtained after the cell walls and membrane organelles were removed by centrifugation were assumed to be derived from the cytoplasm, and the extracts were referred to as cytoplasmic extracts or somatic extracts (Stallybras, 1964; Chew and Theus, 1967; Preisler et al., 1969; Taschdjian et al., 1969a-c; Axelsen, 1973; Evans et al., 1973; Syverson et al., 1975; Guinet and Gabriel, 1980). Subsequently, more careful analysis of these extracts showed that they almost always contained appreciable amounts of cell wall mannan (CWM) and therefore could not be viewed as containing only cytoplasmic antigens (Jones, 1980a). Probably most polysaccharide components in cytoplasmic extracts are derived from cell walls or represent precursors of cell wall structures; however, proteins present in high concentrations in cytoplasmic extracts are likely to be derived exclusively from the cytoplasm. Crude cytoplasmic extracts have been depleted of CWM by passage over concanavalin A (Jones, 1980a; Greenfield and Jones, 1981; Strockbine et al., 1984).

The large array of antigens present in cytoplasmic extracts was first demonstrated by crossed immunoelectrophoresis using sera from rabbits hyperimmunized with extracts. Axelsen (1973) demonstrated up to 78 precipitating antigens in cytoplasmic extracts using this technique. Manning and Mitchell (1980) grew organisms in [^{35}S] sulfate to internally label proteins and subjected extracts obtained to autoradiographic analysis using two-dimensional O'Farrell gel electrophoresis. Depending upon the extract studied, up to 168 proteins could beidentified by this technique. Although Evans et al. (1973) and Syverson et al. (1975) were able to demonstrate by crossed immunoelectrophoresis a limited number of antigens in mycelial phase cytoplasmic extracts that were not present in yeast phase extracts, Manning and Mitchell (1980) were unable to demonstrate such differences using the O'Farrell technique. Further studies are needed to resolve these divergent observations.

More recently, polyacrylamide gel electrophoresis (PAGE) or sodium dodecyl sulfate-PAGE (SDS-PAGE) have been used to analyze cytoplasmic extracts (Jones, 1980a; Greenfield and Jones, 1981; Strockbine et al., 1984). At least 30 protein staining bands can be recognized by PAGE techniques. Although the number of bands recognizable is smaller, SDS-PAGE offers several advantages to the other techniques for analysis. SDS-PAGE will reveal antigens present in extracts in appreciable amounts. Gels can be stained to reveal either protein or carbohydrate content of bands. In this way, proteins, glycoproteins, and polysaccharides can be distinguished. The molecular weight of individual components can be identified. Finally, Western blots of gels can be prepared and developed with a variety of ligands including lectins and antibodies. For example, the components in the extract capable of reacting with antibodies present in serum from an infected patient can be identified by their molecular weights.

Antigens of the Cell Wall

The cell wall is of paramount importance when considering host defenses against candidiasis. The surface of the cell wall and events occurring there are important in adherence to and invasion of host tissues, and there is growing evidence that immune responses are directed largely against cell wall antigens.

Considerable progress has been made in defining the structure and antigenic composition of the cell wall in the past 20 years. In transmission electron microscopy, up to nine layers can be identified in the cell wall of either blastospores or germ tubes. These can be conveniently grouped together as electron dense outer layers, electron transparent middle layers, and electron dense inner layers (Poulain et al., 1985a). By correlating observations made with electron microscopy after various extraction procedures with analyses of extracts, it has been possible to determine the location of various antigenic species within the cell wall (Figs. 1 and 2).

When intact organisms are treated with dithiothreitol, a complex array of proteins, glycoproteins, and polysaccharides is extracted (Smail and Jones, 1984; Ponton and Jones, 1986a), there is loss of material from the electron dense outer layers (Cassone et al., 1973), and the surface of the organism appears roughened (Figs. 1-3). Thus, there is a latticework of antigens in the outer layers, which is stabilized by disulfide bonds within or between proteins. The polysaccharides are mannans comprised of α-linked mannose units that also contain phosphate moieties (Okubo et al., 1979; Reiss, 1986). The extracted CWM is comprised of molecules of varying size; however, most CWM molecules are quite large, migrating in locations corresponding to protein markers > 100 kilodaltons (Kd) in SDS-PAGE gels (Fig. 3). CWM nearly devoid of protein content can be obtained by suitable techniques (Summers et al., 1964), and the areas of SDS-PAGE gels containing CWM do not stain with protein stains such as Coomassie blue (Fig. 3). Since CWM is shed from the cell wall (Jones, 1980b), a significant portion of CWM molecules may be held passively within the lattice of the outer cell wall; however, in the intact cell wall, many molecules must be attached to peptides or larger proteins. Multiple lines of evidence indicate that β-glucans are distributed throughout the cell wall, but the middle electron transparent layers are comprised almost entirely of β-glucans and chitin (Cassone et al., 1978; Tronchin et al., 1981). Poulain et al. (1985b) also noted that extraction of lipids from intact cells using methanol-chloroform led to loss of electron dense material from the outer layers of cell wall and visualization of fibrils in the intermediate layers of the wall corresponding to β-glucans and chitin. Thus, it is likely that the middle and outer cell wall layers of *Candida* contain lipids that are important to cell wall structure. The inner electron dense layers of the cell wall are rich in proteins and oligosaccharides (Poulain et al., 1985a). There are numerous enzyme systems associated with the cell membrane, including chitin

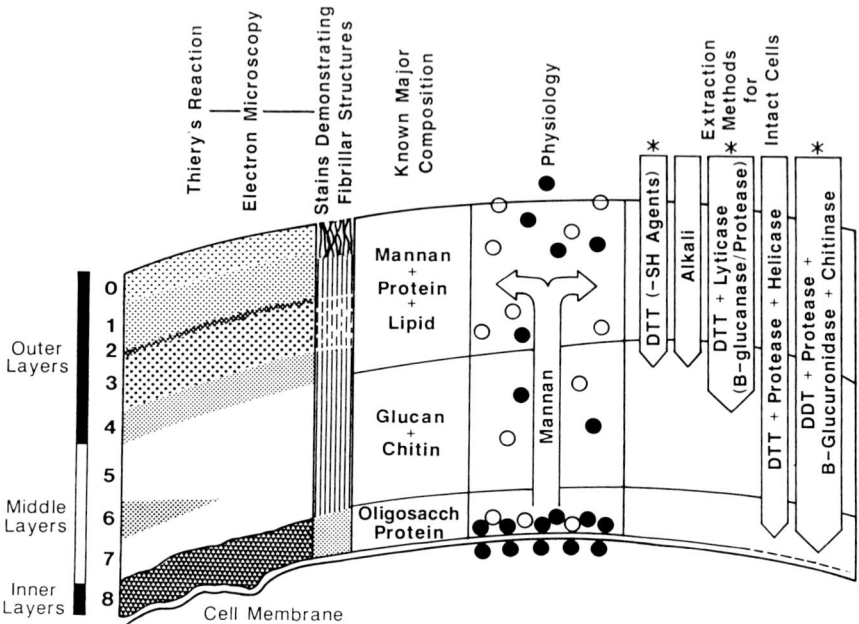

Figure 2 Representation of cell wall of *Candida albicans*. Layers visualized in electron microscopy are correlated with known information on the composition and physiology of the cell wall. Some molecules are held covalently or noncovalently within the cell wall matrix (○), whereas others (●) pass into the environment surrounding the cell. Arrows surrounding different extraction methods (right side of figure) overlie the areas of cell wall partially or completely removed by the given extraction procedure.

synthetase, mannan synthetase, and disulfide reductase (Marriott, 1975; Braun and Calderone, 1978; Arnold, 1981). Assembly of cell wall precursors must occur in this area of the cell wall. There is evidence that CWM passes through well-defined channels from inner layers to outer layers of the cell wall (Vogt et al., 1980), and presumably other structural components and enzymes pass through similar channels.

Nine enzymes activities were detected in DTT extracts that were screened for 19 common microbial enzymes (Ponton and Jones, 1986a). In addition, it is known that many strains of *C. albicans* secrete an acid protease (Ruchel, 1981). It is unknown if this enzyme is also held in high concentrations at the cell wall surface. Three phospholipases are secreted into culture media by *Candida* (Banno et al., 1985). Phospholipase activity can be detected in the cell wall histo-

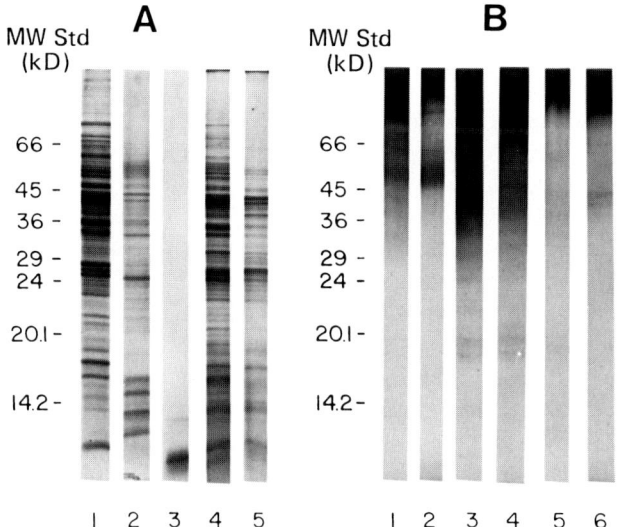

Figure 3 Analysis of DTT extracts from yeast (Y) and mycelial (M) phase *C. albicans* using a 12.5% SDS-PAGE gel. (A) Gel tracks stained for protein with Coomassie blue: 1, DTT extract of M cells; 2, DTT extract of Y cells; 3, DTT-protease extract of Y cells; 4, DTT-lyticase extract of M cells; and 5, DTT-lyticase extract of Y cells. (B) Gel tracks stained for polysaccharides with PAS: 1, DTT extract of M cells; 2, DTT extract of Y cells; 3, DTT-protease extract of M cells; 4, DTT-protease extract of Y cells; 5, DTT-lyticase extract of M cells; and 6, DTT-lyticase extract of Y cells. Figure reproduced with permission from Ponton and Jones (1986a).

chemically and is most concentrated in nascent buds and at the tips of germ tubes (Pugh and Cawson, 1975). Presumably, phospholipase action is involved in growth and restructuring of the cell walls. However, it has also been shown that in chick chorioallantoic membranes infected with *Candida*, dissolution of cell membranes of the chick cells occurs at locations corresponding to highest phospholipase activity (Pugh and Cawson, 1977). Many investigators feel that acid protease and phospholipase are important in invasion of *Candida* through epithelial cells and into deeper host tissues (Barrett-Bee et al., 1985; Kwon-Chung et al., 1985).

ANTIBODY RESPONSES TO *Candida albicans*

Antibody Responses to Cytoplasmic Antigens

Because of the increasing number of immunocompromised patients being encountered, interest in serodiagnosis of severe candidiasis intensified about 1960. It was felt that immune responses against cytoplasmic antigens would be elicited in two ways. First, some organisms invading tissues could undergo autolysis with release of the antigens. Alternatively, attack by host phagocytes could lead to disruption of cell walls of the organism with release of antigens. Many investigators hypothesized that appreciable serum levels of antibodies against cytoplasmic antigens would not be present in patients unless a severe infection was present. They sought to detect precipitating antibodies in human sera against antigens present in cytoplasmic extracts using agar gel immunodiffusion (Stallybras, 1964; Chew and Theus, 1967; Preisler et al., 1969; Taschdjian, 1969), counter immunoelectrophoresis (Dee and Rytel, 1975; Kozinn et al., 1978; Marier and Andriole, 1978), or crossed immunoelectrophoresis (Axelsen et al., 1975; Glew et al., 1978). These early studies are difficult to evaluate because, as mentioned earlier, cytoplasmic extracts typically contain CWM, and we now know the incidence of precipitins against CWM in sera from infected patients is higher than that for precipitins against cytoplasmic antigens. Nevertheless, using crossed immunoelectrophoresis, Axelsen et al. (1975) showed that in concentrates of normal human sera an average of 2.3 serum precipitins were detectable. Thus, it was likely that many normal human sera contain antibodies against one or more cytoplasmic antigens of *C. albicans*. In patients who had candidemia, chronic mucocutaneous candidiasis, or deep-seated candidiasis, they detected up to 20 precipitin arcs. More recently, it has been possible to use SDS-PAGE and Western blot techniques to identify presence of antibodies against protein antigens in cytoplasmic extracts. Matthews et al. (1984) probed Western blots of SDS-PAGE gels that had been used to resolve a cytoplasmic extract. A total of 26 bands were recognized by 201 sera taken serially from 45 patients with systemic candidiasis. Although some patients had only IgM or IgG antibodies against particular bands, in most instances where antibodies against a particular antigen were detected, patients made IgM and then IgG antibody against the antigen. Several laboratory groups have presented data indicating that there are a limited number of cytoplasmic antigens, having a molecular weight between 47 Kd and 59.7 Kd, which are immunodominant (Jones, 1980a,b; Greenfield et al., 1981, 1983; Karwowska et al., 1984; Strockbine et al., 1984; Au-Young et al., 1985). That is, in the course of deep-seated candidiasis, it is much more likely that patients will produce antibodies against one of these antigens than to other cytoplasmic antigens. Normal human sera appear to rarely have antibodies against these cytoplasmic components.

Pepys et al. (1968) performed skin tests with CWM and a dialyzed culture filtrate obtained by growing *Candida* in a medium containing dialyzable nutrients. They concluded that both preparations gave type I (immediate) and type III (Arthus) cutaneous reactions in most subjects. The filtrate also produced type IV (delayed hypersensitivity) reactions. Longbottom et al. (1976) removed CWM from a cytoplasmic extract by passing the extract over a concanavalin A sepharose column. They studied 12 patients by skin prick tests and found that 2 had positive tests to only CWM, 4 to only the purified cytoplasmic extract, and 4 to both antigens. Sera from the subjects were injected subcutaneously into monkeys and the latter were then challenged with intravenous injections of antigens (Prauznitz-Kustner tests). They determined that immediate reactions to CWM were mediated by heat-stable, short-term antibody, presumably of the IgG class. By contrast the reactions to the purified cytoplasmic extract were mediated by heat-labile, long-term sensitizing antibody, presumably IgE antibody.

In summary, it appears likely that soon after birth, self-limited infections of the skin and mucous membranes stimulate production of low levels of antibodies against cytoplasmic antigens of *C. albicans*. It is likely that these antibody levels are maintained through constant exposure to the fungus throughout life. The antibodies may very well participate in rapidly mediating reactions to the fungus when invasion of skin or mucosae occurs. Production of high levels of antibodies against certain immunodominant cytoplasmic antigens seems to require a severe infection as a stimulus. It is not known at this time if such responses add appreciably to host defenses.

Antibody Responses to Cell Wall Antigens

Agglutination of *Candida* blastospores by sera was commonly employed in early studies to demonstrate antibodies in human sera or sera from immunized animals. Hasenclever and Mitchell (1961) studied agglutination patterns of multiple *Candida* strains using rabbit antisera and showed that organisms could be divided into two serogroups, which they designated A and B. Serogroups A and B shared antigenic determinants, but serogroup A organisms also bore surface determinants not found on the surface of serogroup B cells. Because of this, most studies of antibodies against cell wall antigens have been performed using either serogroup A cells or extracts of serogroup A cells.

Using an alcohol precipitate of boiled blastospores as a source of antigen, Akiba et al. (1957) were the first to demonstrate precipitins against cell wall polysaccharides of *Candida* occurred in a small percentage of sera from normal human subjects. Stallybrass (1964) detected precipitins in serum from an infected patient using a formamide extract of cell walls. Summers et al. (1964) described isolation of CWM from *Candida* using the method of Peat et al. (1961), and this became the widely accepted method for obtaining purified CWM for construction of serologic tests. They also showed that the determinants responsible for serogroups were present in CWM.

In the past two decades, numerous publications have described application of agglutination tests and precipitin tests to the diagnosis of severe candidiasis (Preisler et al., 1969; Andersen and Stenderup, 1974; Harding et al., 1976; Felice et al., 1977). Although investigators were more concerned with showing changing titers of serum agglutinins or evolution of positive serum precipitin tests in the course of severe candidiasis, the studies were perhaps even more valuable in suggesting that significant levels of antibodies against CWM must occur in normal, healthy humans. For example, Chew and Theus (1967) detected precipitins in over 50% of serum concentrates from normal human subjects using Preer tubes. With the advent of radioimmunoassays (RIAs) and enzyme-linked immunosorbent assays (ELISAs), it became possible to quantitate levels of anti-CWM antibodies in human sera with these very sensitive methods. Using an RIA, my laboratory (Jones, 1980a) showed that anti-CWM antibody could be detected in virtually any normal human adult, antibody levels were normally distributed in a randomly selected adult population, and the 8–10% of normal subjects with highest anti-CWM levels measured by RIA gave positive precipitin tests using a standard plate agar gel diffusion assay. Using a normal human serum pool, it was shown that most antibody was of the IgG class. Using an ELISA, Greenfield et al. (1983a,b) detected IgG anti-CWM in sera of 100 randomly selected hospitalized patients and again found antibody levels were normally distributed. IgM anti-CWM was detected in only 11 of these patients. In neutropenic patients followed with serial determinations of anti-CWM levels, increasing levels of anti-CWM IgG occurred in most patients who developed severe candidiasis; however, a clear bimodal distribution in levels that could unequivocally separate uninfected from infected patients was not observed.

In many instances, antibodies against cell wall antigens have been demonstrated in human sera and secretions by techniques that do not allow one to determine whether or not antibodies detected were directed against CWM or another antigen. However, when immunoglobulin classes other than IgG were shown to bind to cell walls or antibodies binding to germ tubes and not to blastospores were detected, it is possible that antibodies against structures other than CWM were being detected. Cobb and Parratt (1978) used an RIA in which heat-killed *Candida* blastospores were used as the antigen target to study sera from blood donors and concluded that, in addition to IgG antibodies, 98% had IgM and 68% IgA antibodies at concentrations above the limit of detection by their assay. Using indirect immunofluorescent antibody (IFA) staining. Mathur et al. (1977a,b) demonstrated IgG, IgM, and secretory IgA in sera from women with a history of documented *Candida* vaginitis. When total IgE levels in sera and vaginal secretions from normal women were measured before and after absorption with blastospores, they found no IgE antibodies against *Candida* in any of the sera or secretions studied. However, using the same technique, they found that sera from patients convalescing from systemic candidiasis or recurrent vaginal candidiasis contained IgE anti-*Candida* antibodies. Vaginal washings from the latter patients also contained IgE anti-*Candida* antibodies. Using the IFA technique, Epstein et al. (1981) showed that saliva of most normal subjects who carry *Candida* on their oral mucous membranes have low titers of IgA and IgG anti-*Candida* antibodies. Patients with acute (thrush) or chronic (denture stomatitis) candidiasis have somewhat higher titers of these IgA and IgG antibodies. When treated with nystatin to eliminate *Candida*, titers in carriers fell to undetectable levels within 2 weeks, whereas titers of treated infected patients remained elevated.

Ho et al. (1976) used mycelial phase *Candida* to titer sera from normal subjects and patients with severe candidiasis. They found that normal sera contained antibodies that bound to both the blastospore and germ tube of targets but the titer of antiblastospore antibodies was higher than antigerm tube antibodies. During infection the titer of antibodies to both structures rose but the titer of anti-germ tube antibodies exceeded the antiblastospore titer. This suggested that certain antigens were expressed preferentially on the surface of germ tubes and blastospores. Recently, data has been reported from two laboratory groups (Sundston and Kenny, 1984, 1985; Smail and Jones, 1984; Ponton and Jones, 1986a,b; Sundstrom et al., 1987) identifying an antigen expressed on the surface of the germ tube but not on the surface of the blastospore that stimulates antibody production in humans. The antigen is a complex polysaccharide-protein antigen that can be demonstrated in dithiothreitol (DTT) extracts and DTT-lyticase extracts of mycelial phase *Candida* using SDS-PAGE and Western blot techniques. It is not present in similar extracts of yeast phase organisms. The antigen is quite large, having an apparent molecular weight greater than 200 Kd by SDS-PAGE. It is not presently known what the prevalence of antibodies against this antigen is in normal human sera; however, studies of a small panel of sera from patients with severe candidiasis showed that antibody to the antigen occurred late in the course of the infection (Ponton and Jones, 1986b). In addition to identifying an antigen unique to the germ tube surface that can stimulate antibody production in humans, SDS-PAGE and Western blot techniques have demonstrated that humans produce antibodies to a number of antigens in the outer cell wall layers of *Candida* that are clearly different from CWM. This was accomplished by absorbing sera from normal subjects or infected patients with CWM or blastospores and probing Western blots of gels that had been loaded with DTT extracts of yeast phase and mycelial phase organisms. Results of a representative experiment are shown in Figure 4.

In summary, antibodies against cell wall antigens of *C. albicans* are ubiquitous in human sera. Anti-CWM antibody can be considered a major component of the

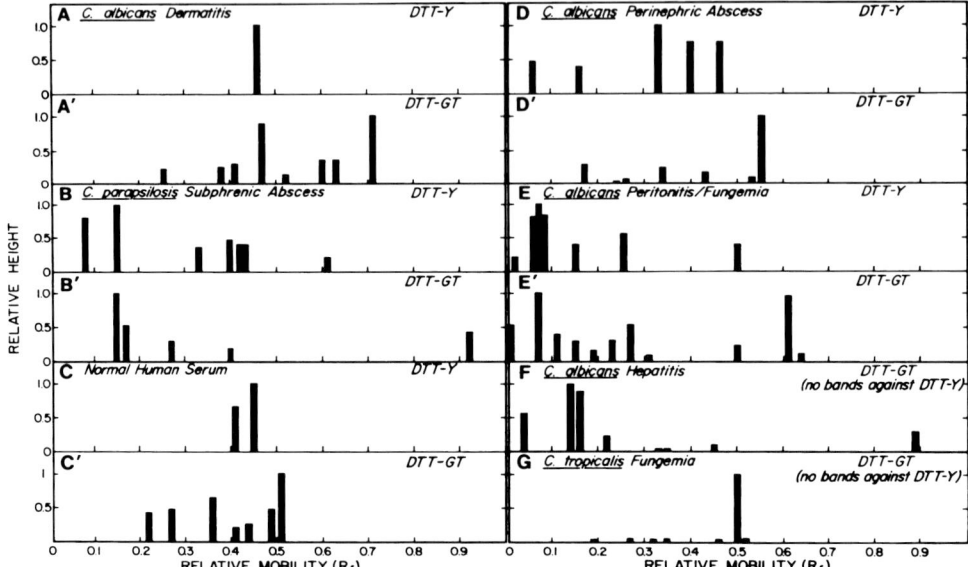

Figure 4 Densitometer readings of Western blots stained with normal serum or sera from infected humans. All sera were absorbed to neutralize anti-CWM antibodies prior to use in staining the blots. Relative peak heights were determined by standard techniques. R_f values establish the position of the band responsible for the peak along the electrophoretic track. Each panel (e.g., A–A′, B–B′) corresponds to a different serum, was used to stain blots of tracks of 12.5% SDS-gels (see Fig. 3) loaded with extracts of mycelial phase (DTT-GT) or yeast phase (DTT-Y) cells. Multiple antigens of varying molecular weights were recognized. Figure reproduced with permission from Ponton and Jones (1986a).

immunoglobulins present in normal human sera. Although data is not available on anti-*Candida* antibody responses in infants, it is likely that anti-CWM IgG is transmitted transplacentally to the infant, that a primary anti-CWM antibody response occurs in early infancy, and that anti-CWM antibody levels are maintained throughout life by colonization and/or self-limited mucocutaneous infections by the fungus. Antibodies against cell wall components other than CWM are present in normal human sera and responses against a complex array of cell wall antigens occurs during infections. It is likely that very low levels of IgA antibodies are present in saliva, vaginal secretions, and intestinal secretions of normal humans, and invasion rapidly stimulates production of higher antibody concentrations in these secretions. IgE antibody production appears to occur during significant infections, but it is unknown whether or not anti-cell wall

IgE antibodies are present in the skin or mucous membranes of normal humans bound to effector cells such as mast cells.

MATERIALS RELEASED FROM *CANDIDA* CAPABLE OF TRIGGERING HOST RESPONSES

Materials released in appreciable quantities from *Candida* could be important in triggering inflammatory reactions in the vicinity of the organism. Inflammatory responses could be triggered if the component combined with specific antibodies and the resultant complexes fixed complement or if the component could directly activate complement. Of the enzymes released by the organism, only antibody responses against acid protease have been investigated. Macdonald and Odds (1980) detected precipitating antibodies against acid protease in sera of 75% of a population of patients with severe candidiasis but not in sera of normal subjects. A more sensitive assay might have detected antibodies in the normals. Several investigators have demonstrated CWM in sera of up to 75% of patients with severe candidiasis (Meckstroth et al., 1981; Bailey et al., 1985; Kahn and Jones, 1986). Thus, it is very likely that CWM is present in high concentrations in tissues being invaded by *Candida*. Since antibodies against CWM are ubiquitous and CWM can activate the alternate pathway of complement (Ray et al., 1979), free CWM in tissues could be quite important in triggering host responses. It is likely that other cell wall antigens are released by *Candida* into infected tissues that could participate in triggering inflammatory responses. Although under investigation, at the present time it is not known whether or not any of the immunodominant cytoplasmic antigens are released in high concentrations into infected tissue.

COMPLEMENT ACTIVATION MEDIATED BY ANTIBODIES AND THE ORGANISM

Complement activation is the crucial link between the detection limb of the humoral response and the effector limb, namely, neutrophils and other phagocytes (Gordon and Hostetter, 1986). Fragments of C3 and C5 produce degranulation of mast cells with release of histamine and other mediators. This results in capillary permeability and vasodilatation, which facilitates flow of serum components and phagocytes into tissue. Fragments of C3, C5, and the C5,6,7 complex are chemotactic for neutrophils. Unstimulated neutrophils have 10–100 CR_1 complement receptors on their surface and the number of receptors increases with neutrophil activation. CR_1 binds to C3b and to a lesser extent with iC3b fragments of C3. Another receptor (CR_3) present on neutrophils recognizes iC3b. Engagement of CR_1 or CR_3 leads to activation of oxidative metabolism in neutrophils and phagocytosis. Monocyte surfaces have CR_1 and CR_3, and macrophages have these and CR_4 (recognizing iC3b and $C3_{dg}$) as well. Engagement of these receptors clearly stimulates metabolism of these cells.

Obviously, anti-*Candida* antibodies could attach to cell wall antigens or antigens released from *Candida* and activate the complement cascade via the "classic" pathway of complement. However, there is now strong evidence that mannan, either on the cell wall or shed from the cell wall, can efficiently activate complement via the "alternative" pathway (Thong et al., 1978; Ray et al., 1979). Each blastospore can bind $2.5-3 \times 10^5$ C3b (Kozel et al., 1987). Because there is little or no decay from C3b fixed to the organism to iC3b, maximal C3b binding occurs rapidly. Binding of C3b is completed within 5 minutes of exposure of blastospores to 20% normal serum in vitro. Although C3b binds poorly to germ tubes, iC3b and C3d bind extensively (Edwards et al., 1986). The in vitro rate of phagocytosis by neutrophils triggered by specific antibodies is more rapid than that observed for bacteria. Fixation of C3b amplifies this opsonization approximately threefold (Scribner and Fahney, 1976).

SEQUENCE OF HOST RESPONSES TRIGGERED BY THE HUMORAL IMMUNE SYSTEM DURING *CANDIDA* INFECTION

Depending upon the circumstances the humoral response can trigger host responses to *Candida* at numerous sites throughout the body. The reader can use Figure 5 to help visualize steps in superficial infection of skin and mucous membranes. Events in invasion of deeper tissues and bloodborne invasion are represented in Figure 6.

Candida can adhere to epithelial cells of the skin and mucous membranes through specific receptors (Figure 5A). Occlusion and a moist environment are important in predisposing skin to *Candida* invasion. Acid protease acts as a keratinase (Hattori et al., 1984) enabling the organism to invade cornified epithelium (Figure 5C). As noted before, phospholipase may facilitate penetration of blastoconidia and germ tubes through epithelial cells. Mucous membranes are quite resistant to infection unless they are colonized with large numbers of organisms and/or are defective in their ability to repair trivial mechanical damage. Denture stomatitis and esophageal candidiasis in patients who have received cytotoxic drugs exemplify these problems. Any physical defect in the mucosal epithelial barrier is likely to contain a fibrin-platelet meshwork to which *Candida* adhere avidly (Figure 5B; Rotrosen et al., 1986). *Candida* can rapidly breach the epithelial barrier from such sites. Normally, humoral responses to *Candida* trigger host responses even before the epithelial barriers are breached (Figure 5D). Intraepithelial pustules packed with neutrophils occur

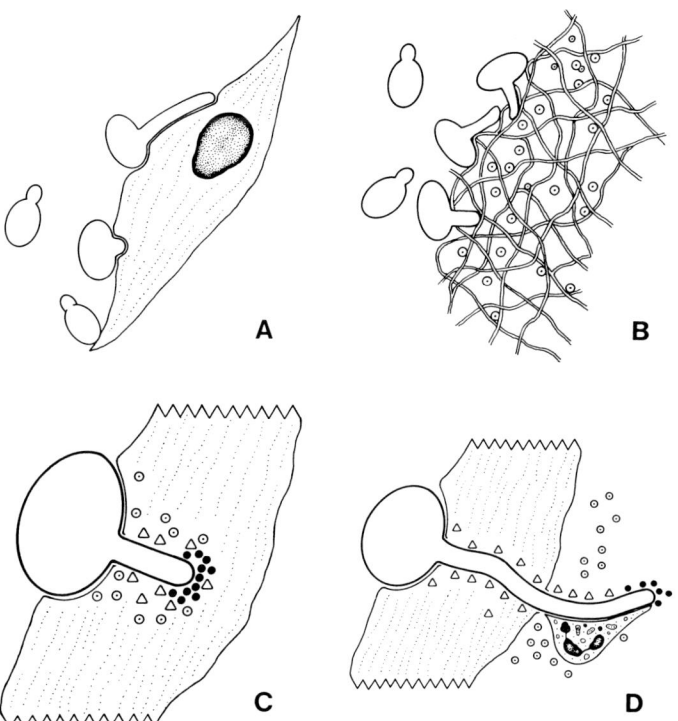

Figure 5 Representation of early steps in invasion of an epithelial membrane by *Candida albicans*. (A) Organisms attach to epithelial cells via receptors. (B) Microscopic or gross defects in the epithelium have fibrin at their bases to which *Candida* can adhere. (C) Blastospores remain attached to epithelial cells and germ tubes penetrate through cell membranes and cytoplasm. Cell wall materials (⊙) are shed from the organism and diffuse through epithelial membrane. Acid protease (△) and phospholipase (●) enhance penetration of the germ tube. Growth occurs at the germ tube tip. (D) Even before penetration through multiple layers of epithelial cells is achieved, soluble materials that have diffused from the organism can interact with antibodies and complement triggering an outpouring of neutrophils. These attach to and begin to attach organisms.

promptly after experimental infection of human skin beneath an occlusive dressing (Maibach and Klingman, 1962). Intraepithelial pustules coalesce and upper epithelial layers are sloughed. One is left with a reddened, moist surface containing proteins, phagocytes, and residual *Candida*. Most of the infecting *Candida* are lost with the sloughed cells and the number of *Candida* that must be killed by phagocytic cells is thereby decreased. Rodent models have been used to demon-

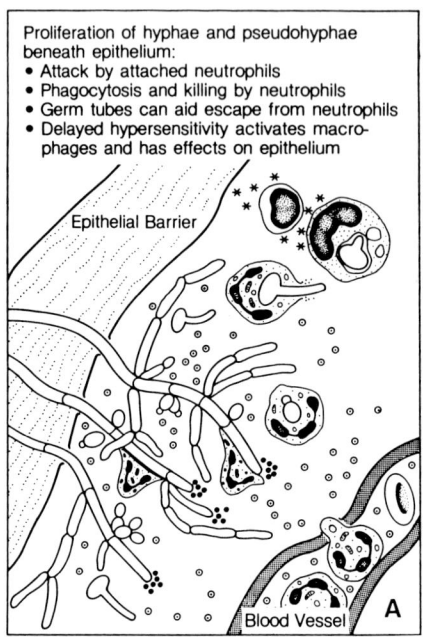 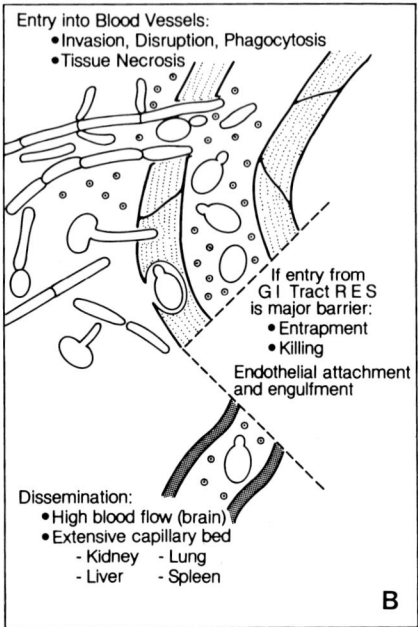

Figure 6 Representation of events in deeply invasive candidiasis. (A) Organisms themselves and soluble components bind antibodies and complement, triggering outpouring of neutrophils. This halts progress of the infection. Some organisms escape killing by neutrophils. Specific T lymphocytes produce lymphokines (*) that activate tissue macrophages and influence directly or indirectly epithelial proliferation. (B) In the neutropenic host, organisms can invade blood vessels and bloodborne dissemination can occur.

strate the mechanisms responsible for this outflow of neutrophils (Van Cutsem and Thienpont, 1971; Ray and Wuepper, 1978). When skin of neonatal mice are treated with staphylococcal epidermolysin, *Candida* applied to treated sites can readily penetrate into the intraepithelial clefts thus created. Twenty-four hours later, intraepithelial pustules have formed. In mice genetically deficient for C5 or depleted of complement by cobra venom, this outpouring of neutrophils does not occur. Thus, it appears likely that CWM and other materials shed from *Candida* can diffuse through the infected epithelium and activate the complement cascade, thereby producing chemotaxis of neutrophils.

Once *Candida* have penetrated through the epithelial barrier, an even more intense host response is triggered (Figure 6A). Vascular permeability induced by activated complement makes available both antibodies and complement from nearby capillaries. IgE-induced degranulation of mast cells and basophils may further increase vascular permeability as well as chemotaxis of neutrophils and monocytes. Although Diamond et al. (1978) have found that neutrophils can attach to and attack germ tubes of *Candida* in the absence of complement and antibodies, antibodies and complement undoubtedly amplify this neutrophilic attack in the host. Neutrophils contain a variety of materials important in killing *Candida*. It appears that the most important ones are contained within the large, azurophilic, nonspecific granules, sometimes referred to as "toxic granules," rather than in the specific neutrophilic granules. The following materials found in nonspecific granules would be toxic to *Candida*: α-mannosidase, β-glucuronidase, N-acetyl-β-glucosominidase, myeloperoxidase, lysozyme, chymotrypsinlike cationic proteins, and defensins (Diamond et al., 1978,1980; Ganz et al., 1985). The latter are newly described peptides, comprised of 20-30 amino acids, which are rich in arginine and contain 6 cysteine residues (Selsted et al., 1985a). Defensins are remarkably toxic to *Candida* in vitro, being capable of rapid killing of logarithmic numbers of organisms at concentrations 10-100 times less than those presumed to be present in the granules (Selsted et al., 1985b). Neutrophils that attach to germ tubes degranulate and undergo a respiratory burst; the free oxygen radicals generated can attack the organism even when phagocytosis does not occur (Diamond et al., 1980). Since neutrophils that degranulate on attachment to germ tubes and pseudohyphae release β-glucuronidase and lysozyme (Diamond et al., 1978), it is likely that they also release defensins. As noted earlier, specific antibodies and C3 complement components facilitate opsonization of *Candida* blastospores and killing in secondary lysosomes.

There are limitations to the ability of neutrophils to kill *Candida*. Because they can be phagocytized, blastospores are probably killed more efficiently than larger mycelial phase organisms. Also, in vitro studies have indicated that about 40% of vacuoles containing internalized blastospores remain "unsealed" (Cech and Lehrer, 1984). Unsealed vacuoles maintain functional communication with the cell's exterior, cannot achieve low pH levels, and can leak their contents. Seventy percent of blastospores in sealed vacuoles are killed in one hour, whereas only 14% of those in unsealed vacuoles are killed. Finally, complete binding of defensins does not occur at 0°C, indicating that metabolic activity of the fungus may be needed for this to occur (Lehrer et al., 1985). Thus, one can speculate that slowly growing organisms are poorly killed by defensins.

The relative importance of the humoral immune response and cellular (T-lymphocyte-mediated) immune responses in enabling the host to recover from infections of epithelial barriers and tissue invasion immediately beneath these barriers has been a subject of debate. In my view, T-cell-mediated immune responses, which lead to activation of monocytes, macrophages, and epithelial cell proliferation have evolved beautifully to eliminate organisms not killed by neutrophils at sites of infection. Although I will not deal with the intricacies of cellular immune response, I would like to briefly describe how I think humoral and T-lymphocyte-mediated responses interact in enabling the normal human to recover completely from mucocutaneous infections. As I have pointed out in the immediately preceding paragraphs, the impressive outpouring of neutrophils, which is humorally mediated, effectively terminates progress of the infection. Later in the course of infection, few *Candida* remain and the inflammatory infiltrate is comprised of monocytes and lymphocytes. Although complement can stimulate chemotaxis and phagocytosis by monocytes, it is widely accepted that lymphokines from specifically activated T lymphocytes are of major importance in chemotaxis and activation of monocytes. Mononuclear phagocytes may be ideally suited to kill the small numbers of organisms escaping the neutrophil attack. Also, lymphokines can trigger proliferation of epithelial cells; access of organisms in the superficial epithelial layers to deeper structures is thus thwarted and organisms are shed in epithelial scales.

I have represented the coordinated action of neutrophils and the cellular immune response in Figure 6A. As noted in the introduction, patients with severe neutropenia cannot terminate invasion rapidly enough and are at risk for bloodborne dissemination. Chronic mucocutaneous candidiasis results in two situations. In the first, defects in the T-lymphocyte immune system make it impossible for the host to eliminate the few organisms escaping neutrophil killing. In the second situation, there are adequate numbers of neutrophils, but there is a deficit in neutrophil chemotaxis (Van Scoy et al., 1975; Dahl et al., 1976) or in the fifth complement component (Drew, 1973; Rosenfeld et al., 1976). Here, the number of organisms escaping neutrophil killing is larger than normal and T-cell-mediated mechanisms have difficulty eliminating this burden of organisms.

Although animal models mirror poorly the events observed in human mucocutaneous infections, they do support the idea that humorally mediated neutrophil killing is the major defense limiting tissue invasion, and T-lymphocyte-mediated responses are involved in a "mopping up" operation. In the mouse thigh model of Pearsall et al. (1978), experimentally produced subcutaneous infection of the thigh is terminated by neutrophils within days. By 3 weeks very few organisms remain and a mononuclear inflammatory infiltrate persists. In immunized animals, neutrophils acted more rapidly in terminating the infection. Finally, antibodies, but not lymphocytes, from hyperimmunized animals could be used to passively transfer the capacity to rapidly terminate the subcutaneous infection. In the guinea pig model of Sohnle et al. (1976), production of superficial cutaneous infections by one technique triggers the alternative complement pathway and intraepithelial pustules containing neutrophils are seen, which are very similar to those seen in early stages of cutaneous candidiasis in humans.

When cutaneous infection is induced by another technique, pustules do not develop but a scaling of the skin occurs. This response is mediated by immune T-lymphocytes. Histologically, small numbers of *Candida* are seen in the epithelial scales thus produced and there is a mononuclear infiltrate of the dermis. The infection is terminated when organisms are shed in the scales. The latter is reminiscent of histologic features seen in cutaneous granulomas commonly seen in chronic cutaneous candidiasis. Here, there is a mononuclear infiltration and intense hyperkeratosis of overlying skin. Presumably, in these patient lymphokine production from T-lymphocytes is deficient or the capacity of monocytes to respond to lymphokines is inadequate so infection is not eliminated but lymphokine-induced proliferation of epithelial cells continues (Edwards et al., 1978).

Figure 6B depicts events that occur when *Candida* successfully invades blood vessels. Target organs for dissemination are mainly those with high blood flows and the main defense against dissemination is macrophages of the reticuloendothelial system. Observations in humans bearing on the role of humoral or cellular immune mechanisms in activation of these tissue macrophages are not available and one must rely on animal models for insight into the role of the immune system in blunting bloodborne dissemination. Isolated, perfused rabbit liver avidly cleared *Candida* suspended in heated normal rabbit serum and was nearly complete when fresh normal rabbit serum was used as the perfusate (Baine et al., 1974). This makes sense since hepatic macrophages (Kuppfer cells) should bear receptors for complement. There is strong evidence that antibody- and complement-mediated hepatic phagocytosis is a very important defense in neutropenic leukemics. Leukemics who are given cytotoxic drugs develop ulcerative lesions of the gastrointestinal tract that readily become infected with *Candida*. Their severe neutropenia allows organisms to enter the portal venous system. There is evidence that this is the major portal of entry for *Candida* in these patients and that dissemination occurs when the capacity of the liver to remove organisms is exceeded (Jones, 1981).

Observations made using murine models have led to conflicting results regarding the role of humoral and cellular immune responses in controlling bloodborne dissemination. These models have examined chiefly the survival of animals challenged with intravenous injections of *Candida*. Morelli and Rosenberg (1971) found that complement-positive CF-1 mice, infected with the organism, survived significantly longer than complement-deficient mice of the same strain, Kagaya et al. (1981) found that mice immunized with *Candida* demonstrated a high incidence of T-lymphocyte-mediated and a low incidence of humoral immune responses. Mice with positive delayed hypersensitivity reactions against a *Candida* extract showed protection against intravenous challenge compared to those with negative responses. Giving IgG from rabbits hyperimmunized with *Candida* to mice showing delayed hypersensitivity further enhanced protection but did not protect control mice against subsequent challenge. They concluded that cell-mediated immunity played a more important role than humoral immunity in defense against bloodborne dissemination. By contrast, Giger et al. (1978) found no difference in survival of immunized T-lymphocyte-depleted mice compared to normal mice challenged intravenously. Clearly, one must be careful in interpreting these murine experiments. It is evident that mice are not comparable to humans in their ability to produce anti-*Candida* antibodies. It is likely that once *Candida* reach the bloodstream in humans, the innate ability of macrophages to ingest and destroy the organism is of chief importance and that immune responses play a minor role in enhancing their activity.

FUTURE DIRECTIONS FOR RESEARCH

We need to have a better understanding of the structure and function of cell wall antigens. Knowing how they participate in adherence has obvious importance in understanding pathogenesis of candidal infections. We need to further dissect the complexities of antibody responses against these antigens and determine if molecules other than CWM can activate complement directly. As a practical matter it would be important to know if antibodies against protease or phospholipase can blunt tissue invasion in the absence of neutrophils. One can speculate that eventually it may be possible to synthesize large amounts of defensins or other naturally occurring molecules from phagocytes. If this were possible, packaging them in synthetic lipid bilayers armed with specific anti-*Candida* antibodies could enable their use as specific antimicrobial agents in compromised hosts.

REFERENCES

Akiba, T., Iwata, K., and Inouye, S. (1957). Studies on the serologic diagnosis of the deep-seated candidiasis. *Japan. J. Microb.* 1:11–18.

Andersen, P. L., and Stenderup, A. (1974). *Candida albicans* antibodies in candidiasis. *Scand. J. Infect. Dis.* 6:69–73.

Arnold, W. N. (1981). Enzymes. In *Yeast Cell Envelopes: Biochemistry, Biophysics, and Ultrastructure*, vol. 2. Edited by W. N. Arnold. CRC Press, Boca Ratan, Fl., pp. 1–46.

Au-Young, J. K., Troy, F. A., and Goldstein, E. (1985). Serologic analysis of antigen-specific reactivity in patients with systemic candidiasis. *Diagn. Microbiol. Infect. Dis.* 3:419–432.

Axelsen, N. H. (1973). Quantitative immunoelectrophoretic methods as tools for a polyvalent approach to standardization in the immunochemistry of *Candida albicans. Infect. Immun.* 7:949–960.

Axelsen, N. H., Buckley, H. R., Drouhet, E., Budtz-Jorgensen, E., Hatel, T., and Andersen, P. L. (1975). Crossed immunoelectrophoretic analysis of precipitins to *Candida albicans* in deep *Candida* infection. *Scand. J. Immunol. 4 (Suppl. 2)*:216–230.

Bailey, J. W., Sada, E., Brass, C., and Bennett, J. E. (1985). Diagnosis of systemic candidiasis by latex agglutination for serum antigen. *J. Clin. Microbiol. 21*:749–752.

Baine, W. B., Koenig, M. G., and Goodman, J. S. (1974). Clearance of *Candida albicans* from the bloodstream of rabbits. *Infect. Immun. 10*:1420–1425.

Banno, Y., Yamada, T., and Nozawa, Y. (1985). Secreted phospholipases of the dimorphic fungus, *Candida albicans*; separation of three enzymes and some biological properties. *Sabouraudia 23*:47–54.

Barret-Bee, K., Hayes, Y., Wilson, R. G., Ryley, J. F. (1985). A comparison of phospholipase activity, cellular adherence and pathogenicity of yeasts. *J. Gen. Microbiol. 131*:1217–1221.

Braun, P. C., and Calderone, R. A. (1978). Chitin synthesis in *Candida albicans*: comparison of yeast and hyphal forms. *J. Bacteriol. 135*:1472–1477.

Cassone, A., Simonetti, N., and Strippoli, V. (1973). Ultrastructural changes in the wall during germ-tube formation from blastospores of *Candida albicans*. *J. Gen. Microbiol. 77*:417–426.

Cassone, A., Mattia, E., and Boldrini, L. (1978). Agglutination of blastospores of *Candida albicans* by concanavalin A and its relationship with the distribution of mannan polymers and the ultrastructure of the cell wall. *J. Gen. Microbiol. 105*:263–273.

Cech, P., and Lehrer, R. I. (1984). Heterogeneity of human neutrophil phagolysosomes: functional consequences for candidacidal activity. *Blood 64*: 147–151.

Chew, W. H., and Theus, T. L. (1967). *Candida* precipitins. *J. Immunol. 98*: 220–224.

Cobb, S. J., and Parratt, D. (1978). Determination of antibody levels to *Candida albicans* in healthy and hospitalized adults using a radioimmunoassay. *J. Clin. Pathol. 31*:1161–1166.

Dee, T. H., and Rytel, M. W. (1975). Clinical application of counterimmunoelectrophoresis in detection of *Candida* serum precipitins. *J. Lab. Clin. Med. 85*: 161–166.

Diamond, R. D., and Krzesicki, R. (1978). Mechanisms of attachment of neutrophils to *Candida albicans* pseudohyphae in the absence of serum, and of subsequent damage to pseudohyphae by microbicidal processes of neutrophils in vitro. *J. Clin. Invest. 61*:360–369.

Diamond, R. D., Clark, R. A., and Haudenshild, C. C. (1980). Damage to *Candida albicans* hyphae and pseudohyphae by the myeloperoxidase system and oxidative products of neutrophil metabolism in vitro. *J. Clin. Invest. 66*: 908–917.

Drew, J. H. (1973). Chronic mucocutaneous candidiasis with abnormal function of serum complement. *Med. J. Aust. 2*:77–79.

Edwards, J. E., Jr., Foos, R. Y., Montgomerie, J. Z., and Guze, L. B. (1974). Ocular manifestations of *Candida* septicemia: review of seventy-six cases of hematogenous *Candida* endophthalmitis. *Medicine (Baltimore) 53*:47–75.

Edwards, J. E., Lehrer, R. I., Stiem, E. R., Fischer, T. J., and Young, L. S. (1978). Severe candidal infections: clinical perspective, immune defense mechanisms, and current concepts of therapy. *Ann. Intern. Med. 89*:91-106.

Edwards, J. E., Gaither, T. A., O'Shea, J. J., Rotrosen, D., Lawley, T. J., Wright, S. A., Frank, M. M., and Green, I. (1986). Expression of specific binding sites on *Candida* with functional and antigenic characteristics of human complement receptors. *J. Immunol. 137*:3577-3583.

Epstein, J. B., Pearsall, N. N., and Truelove, E. L. (1981). Oral candidiasis: effects of antifungal therapy upon clinical signs and symptoms, salivary antibody, and mucosal adherence of *Candida albicans*. *Oral Surg. 51*:32-36.

Evans, E. G. V., Richardson, M. D., Odds, F. C., and Holland, K. T. (1973). Relevance of antigenicity of *Candida albicans* growth phases to diagnosis of systemic candidiasis. *Br. Med. J. 4*:86-87.

Felice, G., Yu, B., and Armstrong, D. (1977). Immunodiffusion and agglutination tests for *Candida* in patients with neoplastic disease: inconsistent correlation of results with invasive infections. *J. Infect. Dis. 135*:349-357.

Ganz, T., Selsted, M. E., Szklarek, D., Harwig, S. L., Daher, K., Bainton, D. F., and Lehrer, R. I. (1985). Defensins: natural peptide antibiotics of human neutrophils. *J. Clin. Invest. 76*:1427-1435.

Giger, D. K., Domer, J. E., Moser, S. A., and McQuitty, J. T., Jr. (1978). Experimental murine candidiasis: pathological and immune responses in T-lymphocyte-depleted mice. *Infect. Immun. 21*:729-737.

Glew, R. H., Buckley, H. R., Rosen, H. M., Moellering, R. C., and Fischer, J. E. (1978). Serologic tests in the diagnosis of systemic candidiasis. Enhanced diagnostic accuracy with crossed immunoelectrophoresis. *Am. J. Med. 64*: 586-590.

Gordon, D. L., and Hostetter, M. K. (1986). Complement and host defence against microorganisms. *Pathology 18*:365-375.

Gottlieb, M. S., Schroff, R., Schanker, H. M., Weisman, J. D., Fan, P. T., Wolf, R. A., and Saxon, A. (1981). *Pneumocystis carinii* pneumonia and mucosal candidiasis in previously healthy homosexual men: evidence of a new acquired cellular immunodeficiency. *N. Engl. J. Med. 305*:1425-1431.

Greenfield, R. A., and Jones, J. M. (1981). Purification and characterization of a major cytoplasmic antigen of *Candida albicans*. *Infect. Immun. 34*:469-477.

Greenfield, R. A., Bussey, M. J., Stephens, J. S., and Jones, J. M. (1983a). Serial enzyme-linked immunosorbent assays for antibody to *Candida* antigens during chemotherapy for acute leukemia. *J. Infect. Dis. 148*:275-283.

Greenfield, R. A., Stephens, J. L., Bussey, M. J., and Jones, J. M. (1983b). Quantitation of antibody to *Candida* mannan by enzyme-linked immunosorbent assay. *J. Lab. Clin. Med. 101*:758-771.

Guinet, R. M. F., and Gabriel, S. M. (1980). *Candida albicans* group A-specific soluble antigens demonstrated by quantitative immunoelectrophoresis. *Infect. Immun. 29*:853-858.

Hadfield, T. L., Smith, M. B., Winn, R. E., Rinaldi, M. G., and Guerra, C. (1987). Mycoses caused by *Candida lusitaniae*. *Rev. Infect. Dis. 9*:1006-1012.

Harding, S. A., Sandford, G. R., and Merz, W. G. (1976). Three serologic tests for candidiasis. Diagnostic value in distinguishing deep or disseminated infection from superficial infection or colonization. *Am. J. Clin. Pathol. 65*:1001-1009.

Hasenclever, H. F., and Mitchell, W. O. (1961). Antigenic studies of *Candida*. I. Observation of two antigenic groups in *Candida albicans*. *J. Bacteriol. 82*: 570-753.

Hattori, M., Yoshiura, K., Negi, M., and Ogawa, H. (1984). Keratinolytic proteinase produced by *Candida albicans*. *Sabouraudia 22*:175-183.

Ho, Y. M., Teoh-Chan, C. H., Yue, P. C. K., and Huang, C. T. (1976). Indirect immunofluorescence assay for antibody to germ tube of *Candida albicans*—a new diagnostic test. *Clin. Pathol. 29*:1007-1010.

Jones, J. M. (1980a). Quantitation of antibody against cell wall mannan and a major cytoplasmic antigen of *Candida* in rabbits, mice, and humans. *Infect. Immun. 30*:78-89.

Jones, J. M. (1980b). Kinetics of antibody responses to cell wall mannan and a major cytoplasmic antigen of *Candida albicans* in rabbits and humans. *J. Lab. Clin. Med. 96*:845-860.

Jones, J. M. (1981). Granulomatous hepatitis due to *Candida albicans* in patients with acute leukemia. *Ann. Intern. Med. 94*:475-477.

Jones, J. M. (1984). Candidiasis. In *Tropical and Geographical Medicine*. Edited by K. S. Warren and A. A. F. Mahmoud. McGraw-Hill, New York, pp. 903-911.

Kagaya, K., Shinoda, T., and Fukazawa, Y. (1981). Murine defense mechanisms against *Candida albicans* infection. I. Collaboration of cell-mediated and humoral immunities in protection against systemic *C. albicans* infection. *Microbiol. Immunol. 25*:647-654.

Kahn, F. W., and Jones, J. M. (1986). Latex agglutination tests for detection of *Candida* antigens in sera of patients with invasive candidiasis. *J. Infect. Dis. 153*:579-585.

Karwowski, W., de Saint Basile, G., and Lisowska-Grospierre, B. (1984). Antibody levels to *Candida albicans* carbohydrate and major cytoplasmic antigens isolated from standard and patient strains. *Ann. Immunol. (Inst. Pasteur) 135D*:145-159.

Kozel, T. R., Brown, R. R., and Pfrommer, G. S. T. (1987). Activation and binding of C3 by *Candida albicans*. *Infect. Immun. 55*:1890-1894.

Kozinn, P. J., Taschdjian, C. L., Goldber, P. K., Protzmann, W. P., MacKenzie, D. W. R., Remington, J. S., Anderson, S., and Seelig, M. S. (1978). Efficiency of serologic tests in the diagnosis of systemic candidiasis. *Am. J. Clin. Pathol. 70*:893-898.

Kwon-Chung, K. J., Lehman, D., Good, C., and Magee, P. T. (1985). Genetic evidence for role of extracellular proteinase in virulence of *Candida albicans*. *Infect. Immun. 49*:571-575.

Lehrer, R. I., Szklarek, D., Ganz, T., and Selsted, M. E. (1985). Correlation of binding of rabbit granulocyte peptides to *Candida albicans* with candidacidal activity. *Infect. Immun. 49*:207-211.

Longbottom, J. L., Brighton, W. D., Edge, G., and Pepys, J. (1976). Antibodies mediating type I skin test reactions to polysaccharide and protein antigens of *Candida albicans*. *Clin. Allergy 6*:41–49.

Macdonald, F., and Odds, F. C. (1980). Purified *Candida albicans* proteinase in the serological diagnosis of systemic candidiasis. *JAMA 243*:2409–2411.

Mailback, H. I., and Kligman, A. M. (1962). The biology of experimental human cutaneous moniliasis (*Candida albicans*). *Arch. Dermatol. 85*:233–254.

Manning, M., and Mitchell, T. G. (1980). Analysis of cytoplasmic antigens of the yeast and mycelial phases of *Candida albicans* by two-dimensional electrophoresis. *Infect. Immun. 30*:484–495.

Marrier, R., and Andriole, V. T. (1978). Usefulness of serial determinations in diagnosis of candidiasis as measured by discontinuous counter immunoelectrophoresis using HS antigen. *J. Clin. Microbiol. 8*:15–22.

Marriott, M. S. (1975). Enzymic activity of purified plasma membranes from yeast and mycelial forms of *Candida albicans*. *J. Gen. Microbiol. 89*:345–352.

Mathur, S., Goust, J. M., Horger, E. O., and Fudenberg, H. H. (1977a). Immunoglobulin E anti-*Candida* antibodies and candidiasis. *Infect. Immun. 18*:257–259.

Mathur, S., Virella, G., Koistinen, J., Horger, E. O., Mahvi, T. A., and Fudenberg, H. H. (1977b). Humoral immunity in vaginal candidiasis. *Infect. Immun. 15*:287–294.

Matthews, R. C., Burnie, J. P., and Tabaqchali, S. (1984). Immunoblot analysis of the serological response in systemic candidosis. *Lancet 2*:1415–1418.

Meckstroth, K. L., Reiss, E., Keller, J. W., and Kaufman, L. (1981). Detection of antibodies and antigenemia in leukemic patients with candidiasis by enzyme-linked immunosorbent assay. *J. Infect. Dis. 144*:24–32.

Morelli, R., and Rosenberg, L. T. (1971). Role of complement during experimental *Candida* infection in mice. *Infect. Immun. 3*:521–523.

Odds, F. C. (1979). *Candida and Candidosis*. University Park Press, Baltimore, 382 pp.

Okubo, Y., Honma, Y., and Suzuki, S. (1979). Relationship between phosphate content and serological activities of the mannans of *Candida albicans* strains NIH A-207, NIH B-792, and J-1012. *J. Bacteriol. 137*:677–680.

Pearsall, N., Adams, B. L., and Bunni, R. (1978). Immunologic responses to *Candida albicans*. III. Effects of passive transfer of lymphoid cells or serum on murine candidiasis. *J. Immunol. 120*:1176–1180.

Peat, S., Turvey, J. R., and Doyle, D. J. (1961). Polysaccharides of baker's yeast. Part IV. Mannan. *J. Chem. Soc. 1*:29–34.

Pepys, J., Faux, J. A., Longbottom, J. L., McCarhy, D. S., and Hargreave, F. E. (1968). *Candida albicans* precipitins in respiratory disease in man. *J. Allergy 41*:305–318.

Pizzo, P. A., Robichaud, K. J., Gill, F. A., and Witebsky, F. G. (1982). Empiric antibiotic and antifungal therapy for cancer patients with prolonged fever and granulocytopenia. *Am. J. Med. 72*:101–111.

Ponton, J., and Jones, J. M. (1986a). Analysis of cell wall extract of *Candida albicans* by sodium dodecyl sulfate-polyacrylamide gel electrophoresis and Western blot techniques. *Infect. Immun. 53*:565–572.

Ponton, J., and Jones, J. M. (1986b). Identification of two germ-tube-specific cell wall antigens of *Candida albicans*. *Infect. Immun.* 54:864–868.
Poulain, D., Hopwood, V., and Vernes, A. (1985a). Antigenic variability of *Candida albicans*. *CRC Crit. Rev. Microbiol.* 12:223–269.
Poulain, D., Tronchin, G., and Vernes, A. (1985b). Application d'une méthode d'extraction des lipides aux parois des blastospores de *Candida albicans*. *Mycopathologia* 92:141–147.
Preisler, H. D., Hasenclever, H. F., Levitan, A. A., and Henderson, E. S. (1969). Serologic diagnosis of disseminated candidiasis in patients with acute leukemia. *Ann. Intern. Med.* 70:19–30.
Pugh, D., and Cawson, R. A. (1975). The cytochemical localization of phospholipase A and lysosphospholipase in *Candida albicans*. *Sabouraudia* 13:110–115.
Pugh, D., and Cawson, R. A. (1977). The cytochemical localization of phospholipase in *Candida albicans* infecting the chick chorio-allantoic membrane. *Sabouraudia* 15:29–35.
Ray, T. L., and Wuepper, K. D. (1978). Experimental cutaneous candidiasis in rodents. II. Role of the stratum corneum barrier and serum complement as a mediator of a protective inflammatory response. *Arch. Dermatol.* 114:539–543.
Ray, T. L., Hanson, A., Ray, L. F., and Wuepper, K. O. (1979). Purification of a mannan from *Candida albicans* which activates serum complement. *J. Infect. Dis.* 73:269–274.
Reiss, E. (1986). *Molecular Immunology of Mycotic and Actinomycotic Infections*. Elsevier, New York, pp. 191–239.
Rosenfeld, S. I., Kelly, M. E., and Leddy, J. P. (1976). Hereditary deficiency of the fifth component of complement in man. I. Clinical, immunochemical, and family studies. *J. Clin. Invest.* 57:1626–1634.
Rotrosen, D., Calderone, R. A., and Edwards, J. E. (1986). Adherence of *Candida* species to host tissues and plastic surfaces. *Rev. Infect. Dis.* 8:73–85.
Rüchel, R. (1981). Properties of a purified proteinase from the yeast *Candida albicans*. *Biochem. Biophys. Acta* 659:99–113.
Scribner, D. J., and Fahrney, D. (1976). Neutrophil receptors for IgG and complement: their roles in the attachment and ingestion phases of phagocytosis. *J. Immunol.* 116:892–897.
Selsted, M. E., Harwig, S. S. L., Ganz, T., Schilling, J. W., and Lehrer, R. I. (1985a). Primary structures of three human neutrophil defensins. *J. Clin. Invest.* 76:1436–1439.
Selsted, M. E., Szklarek, D., Ganz, T., Lehrer, R. I. (1985b). Activity of rabbit leukocyte peptides against *Candida albicans*. *Infect. Immun.* 49:202–206.
Smail, E. H., and Jones, J. M. (1984). Demonstration and solubilization of antigens expressed primarily on the surfaces of *Candida albicans* germ tubes. *Infect. Immun.* 45:74–81.
Sohnle, P. G., Frank, M. M., and Kirkpatrick, C. H. (1976). Mechanisms involved in elimination of organisms from experimental cutaneous *Candida albicans* infections in guinea pigs. *J. Immunol.* 117:523–530.

Stallybras, F. C. (1964). *Candida* precipitins. *J. Path. Bact. 87*:89-97.
Strockbine, N. A., Largen, M. L., Zweibel, S. M., and Buckley, H. R. (1984). Identification and molecular weight characterization of antigens from *Candida albicans* that are recognized by human sera. *Infect. Immun. 43*:715-721.
Summers, D. F., Grollman, A. P., and Hasenclever, H. F. (1964). Polysaccharide antigens of *Candida* cell wall. *J. Immunol. 92*:491-499.
Sundstrom, P. M., and Kenny, G. E. (1984). Characterization of antigens specific to the surface of germ tubes of *Candida albicans* by immunofluorescence. *Infect. Immun. 43*:850-855.
Sundstrom, P. M., and Kenny, G. E. (1985). Enzymatic release of germ tube-specific antigens from cell walls of *Candida albicans*. *Infect.Immun. 49*:609-614.
Sundstrom, P. M., Nichols, E. J., and Kenny, G. E. (1987). Antigenic differences between mannoproteins of germ tubes and blastospores of *Candida albicans*. *Infect. Immun. 55*:616-620.
Syverson, R. E., Buckley, H. R., and Campbell, C. C. (1975). Cytoplasmic antigens unique to the mycelial or yeast phase of *Candida albicans*. *Infect. Immun. 12*:1184-1188.
Taschdjian, C. L., Dobkin, G. B., Caroline, L., and Kozinn, P. J. (1969a). Immune studies relating to candidiasis. II. Experimental and preliminary clinical studies on antibody formation in systemic candiasis. *Sabouraudia 3*:129-139.
Taschdjian, C. L., Kozinn, P. J., and Caroline, L. (1969b). Immune studies in candidiasis III. Precipitating antibodies in systemic candidiasis. *Sabouraudia 3*:312-320.
Taschdjian, C. L., Kozinn, P. J., Cuesta, M. B., Caroline, L., and Kantrowitz, A. B. (1969c). Post mortem studies of systemic candidiasis I. Diagnostic validity of precipitin reaction and probable origin of sensitization to cytoplasmic candidal antigens. *Sabouraudia 7*:110-117.
Thong, Y. H., and Ferrante, A. (1978). Alternative pathway of complement activation by *Candida albicans*. *Aust. N. Z. J. Med. 8*:620-622.
Tronchin, G., Poulain, D., Herbaut, J., and Biguet, J. (1981). Localization of chitin in the cell wall of *Candida albicans* by means of wheat germ agglutinin. Fluorescent and ultrastructural studies. *Eur. J. Cell. Biol. 26*:121-128.
Van Cutsem, J., and Thinpoint, D. (1971). Experimental cutaneous *Candida albicans* infections in guinea pigs. *Sabouraudia 9*:17-20.
Van Scoy, R. E., Hill, H. R., Ritts, R. E., and Quie, P. G. (1975). Familial neutrophil chemotaxis defect, recurrent bacterial infections, mucocutaneous candidiasis, and hyperimmunoglobulinemia E. *Ann. Intern. Med. 82*:766-771.
Vogt, A., Batsford, S., Kuttin, E., Muller, J., and Takamiya, H. (1980). Immunoelectron microscopic localization of polysaccharide antigens in *Candida albicans*. In *Human and Animal Mycology: Proceedings of the Seventh Congress of the International Society of Human and Animal Mycology* (Jerusalem, Israel, March 11-16, 1979). E. S. Kuttin and G. L. Baum (Eds.). Excerpta Medica, Amsterdam, pp. 134-137.
Winston, D. J., Gale, R. P., Meyer, D. V., and Young, L. S. (1979). Infectious complications of human bone marrow transplantation. *Medicine 58*:1-31.

16
Humoral Response in Aspergillosis

ALAIN J-M. VERNES
bioMérieux, Marcy l'Etoile, France

INTRODUCTION

Since the first clinical report by Virchow (1856), pathology due to the *Aspergillus* species is still a great concern in spite of improvements in in vitro techniques for laboratory diagnosis of the disease and the advent of new antifungal drugs.

Traditional mycology, x-ray radiology, and immunology have contributed in turn to a better definition and more accurate diagnosis of the various clinical forms of aspergillosis.

The last two decades have been characterized, in the industrial countries, by a regular increase in the standard of living and consequently a higher level of hygiene, which has led to a decrease in the registered numbers of diseases such as tuberculosis and pneumoconiosis, which represent the major underlying conditions and risk factors for aspergilloma.

At the same time, progress in the treatment of cancer and leukemia and improvements in graft surgery have led to a greater incidence of aspergillosis in immunocompromised patients. Fisher et al. (1981) reported a series of 91 patients with an established diagnosis of invasive aspergillosis from July 1971 through December 1976 recorded in a single center. At present, invasive aspergillosis is considered the second most frequent fungal infection in immunocompromised hosts and is a great concern for clinicians and biologists owing to the difficulty of early diagnosis, which markedly affects the prognosis of the disease (Frazer et al., 1979). The incidence of invasive pulmonary aspergillosis varies, but it develops in up to 20% of patients admitted for chemotherapy of acute leukemia at some centers with overall mortality rates ranging from 67 to 100% as reported by several groups.

The mortality rate of invasive aspergillosis is largely dependent on an early diagnosis. This has been clearly demonstrated by Gerson et al. (1985), who proposed a discriminant scorecard allowing a more rapid clinical identification of patients which could lead to improved patient survival through earlier therapeutic intervention. Because of antigenic cross-reactions between the different *Aspergillii*, it is important to consider the different species known to cause the disease.

From the work conducted by Pepys et al. (1959a), Gernez-Rieux et al. (1964), Young et al. (1972), and Nalesnik et al. (1980) it is obvious that *A. fumigatus* is the most frequent etiologic agent, involved in at least two thirds of the recorded cases.

A. flavus is often encountered, and other species such as *A. niger, A. nidulans, A. terreus*, or *A. glaucus* are seen from time to time. These observations are not surprising since many opportunistic microorganisms are able to develop in immunocompromised patients. This is, of course, only an example, as we know that the distribution of the species can be affected by environmental conditions which vary from place to place.

ASPERGILLUS ANTIGENS

Aspergillii contain a very complex mosaic of protein, glycoprotein, and polysaccharide antigens. As many as 60 different antigenic components have been demonstrated in extracts from single isolates of *A. fumigatus*, and precipitating antibodies to more than 30 of these components have been found in sera from patients suffering from aspergilloma (Fig. 1).

The use of analytical techniques such as the Ouchterlony's gel precipitation (1948) and the immunoelectrophoresis designed by Grabar and William (1953) has greatly contributed to a better characterization of soluble *Aspergillus* antigens.

The first reports on precipitating antibodies to soluble extracts of *A. fumigatus* were done by Pepys et al. (1959b) and Biguet et al. (1962). These two groups have thoroughly studied the antigenic structure of *Aspergillus* and the immunology of aspergillosis. Using the immunoelectrophoresis technique, Biguet et al. (1965) showed that *Aspergillus* species have a complex antigenic structure. Rabbit hyperimmune sera raised against water-soluble extracts from different species of *Aspergillus* revealed from 16 to 26 antigenic components for each of the analyzed species.

Biguet et al. (1969) demonstrated that, although different strains of a same species may have minor quantitative or qualitative antigenic differences, the

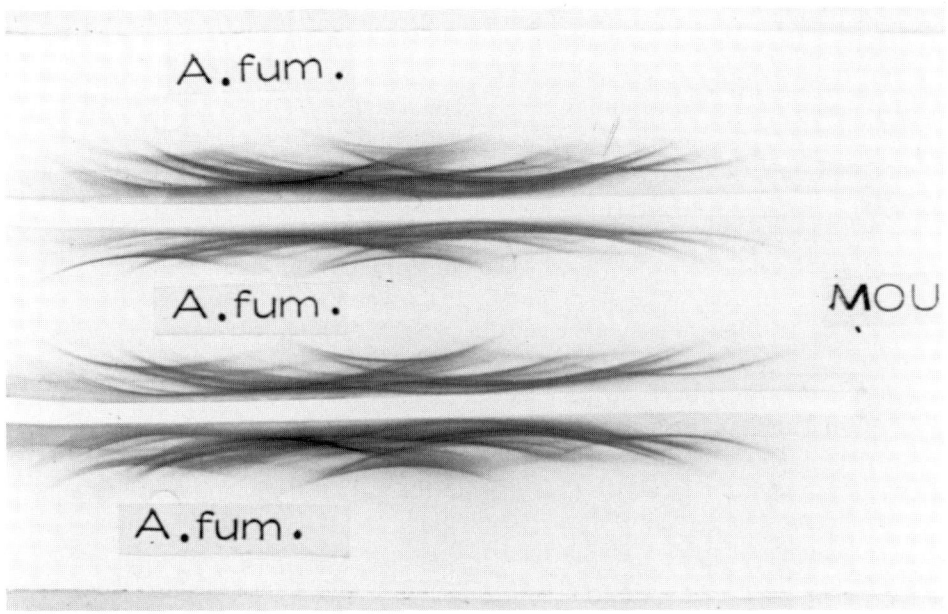

Figure 1 Immunoelectrophoregram of *Aspergillus fumigatus* antigens revealed by the serum of a patient suffering from fungus ball.

immunoelectrophoresis pattern of a given species is constant enough to be characteristic. A recent work of Longbottom and Austwick (1986) confirmed this point. These authors, using powerful analytical techniques such as crossed immunoelectrophoresis and radioimmunoelectrophoresis to study antigens and allergens of *A. fumigatus*, showed that differences in antigenicity and allergenicity were essentially quantitative rather than qualitative in nature. Their results indicated that both types of activity are concentrated in the protein fraction obtained by salt precipitation and that little activity is present in the predominantly polysaccharide supernatant.

The study of Biguet et al. (1969) conducted on 16 species belonging to 6 different groups of the *Aspergillus* classification has confirmed the heterogeneity of *Aspergillii*. This is particularly true when the comparison is made on species belonging to different groups of the classification. In the case of such species as *A. fumigatus, A. restrictus, A flavus, A. nidulans, A. terreus*, and *A. niger*, which are involved in human aspergillosis, common antigenic components are very minor, representing no more than 10-15% of the total antigens.

It is remarkable that all the immunological studies on *Aspergillii* reported by different research groups are in agreement with the botanical classification of

Raper and Fennell (1965). *A. fumigatus* was also compared by Biguet et al. (1965) with numerous species of fungi, including *Penicillium, Cladosporium, Mucor*, dermatophytes, *Candida*, and *Geotrichum*. As expected, these authors found very weak cross-reactions. Longbottom and Pepys (1964) described the presence of C substance in *A. fumigatus*. This point is of great practical importance on account of the presence in the sera of patients suffering from various diseases of the C-reactive protein, which reacts with C substance.

More knowledge of the *Aspergillus* antigens was obtained thanks to the techniques of Uriel (1963), allowing the detection of enzymatic activities within the precipitated antigen-antibody complexes. Tran Van Ky et al. (1966, 1970) described 19 different enzymatic activities on the *A. fumigatus* immunoelectrophoregram, including various oxydoreductases and hydrolases. Similar results were reported later by Tran Van Ky et al. (1971) for *A. flavus*.

Among the various enzymatic activities detected in *A. fumigatus* antigens, special attention was drawn to two of them, the catalase and chymotrypsin activities. As shown by Biguet et al. (1967), there are two chymotrypsic activities, the most important of them, the chymotrypsin II, being highly immunogenic and highly specific. This enzymatic activity corresponds to the C arc described by Biguet et al. (1962) and has been purified by Bout et al. (1973).

The other important activity of *A. fumigatus* is the catalase which has been shown by Tran Van Ky et al. (1968) to correspond to the J fraction. The catalasic activity of *A. fumigatus* shares partial identity with *A. flavus* and *A. niger* and total identity with *A. flavus*. Catalasic activities of *Aspergillus* species have been recently studied by Schønheyder et al. (1985). Their results were consistent with those previously reported by Tran Van Ky et al. (1968); moreover, these authors described a second catalasic activity specific for *A. flavus* that had not been reported before.

HUMORAL RESPONSE

Immunoglobulins Involved

All the classes of immunoglobulins are involved in aspergillosis. The most frequent is the IgG class, but the other classes participate differently in the humoral response according to the various clinical forms of the disease. Using a highly sensitive biotin-avidin immunosorbent assay, Brummund et al. (1987) showed that allergic bronchopulmonary aspergillosis (ABPA) is a disease characterized by a polyclonal antibody response to *Aspergillus* antigens and not only a response in the IgE and IgG antibody classes.

As shown by Slavin et al. (1970) and Stockes et al. (1981) each class of immunoglobulins is increased during ABPA, particularly cytophylic IgG_2 (Assem and Turner-Warwick, 1976). Part of these specific IgG are able to fix the C1q fraction of the complement (Schønheyder and Andersen, 1983a). Kauffman et al. (1986), using EIA and RIA techniques, found that IgG, IgA, and IgM antibodies titers to *A. fumigatus* were elevated in both sera from patient with ABPA and aspergilloma. Specific IgA to *A. fumigatus* are demonstrated in the various forms of the disease, but their diagnostic value is weak according to Patterson et al. (1980) and Schønheyder and Andersen (1984). Specific IgM are present in the various clinical forms of aspergillosis (Schønheyder et al., 1982a), mainly at the beginning of the disease. IgD have been shown by Luster et al. (1976) to participate in the total increase of immunoglobulins. Few antibodies are detected in the healthy population (Bardana, 1974). Only IgG are present in neonates. These antibodies disappear quickly during the first months of life to reappear during childhood, in association with IgM and IgA, and eventually to reach titers observed in adults (Schønheyder and Andersen, 1982b).

Because of their usefulness in the diagnosis of the allergic forms of aspergillosis, IgE have been extensively studied by several groups. An increased level of total or specific IgE is usually observed. As shown by Patterson et al. (1977), part of the elevation of IgE is nonspecific. These nonspecific IgE decrease quickly under corticosteroid therapy (Ricketti et al., 1984), and their kinetic is dependent upon the relapses of the disease (Rosenberg et al., 1977; Imbeau et al., 1982). Specific IgE can be quantitated using a radioimmunosorbent assay as shown by Dessaint et al. (1976).

Serological Diagnosis

Agglutination Techniques

The direct agglutination of conidia by patient's sera was used by Nicaud (1929). Latex agglutination was proposed later by Hipp et al. (1970) and Amos (1978). The specificity of latex agglutination is low, partly because of the use of parietal or cell wall galactomannans for the sensitization of latex particles. This technique is rarely used for the diagnosis of human aspergillosis but has been shown useful in veterinary practice by Corbel and Day (1978).

Passive hemagglutination has been used more often. Choosing a metabolic antigen for the sensitization of red blood cells, Senet and Brisset (1973) obtained results well correlated with immunoelectrophoresis. The test is sensitive and specific when the threshold of specificity is set at the 1:80 dilution. Purified chymotrypsin II (Senet et al., 1978a) and purified catalase (Senet et al., 1978b) have been used for the sensitization of the red blood cells without major improvements of the results. Other groups (Ikemoto and Shibata, 1973; Tonder and Rodsaether, 1974; Rodsaether and Tonder, 1975) confirmed the value of passive hemagglutination. Although the sensitivity of the technique is poor in invasive aspergillosis, passive hemagglutination can be in some cases the only positive test as demonstrated by Gold et al. (1980). Similar results have been reported for invasive aspergillosis by Morin et al. (1984).

Complement Fixation Test (CFT)

This was one of the first techniques described for the immunological diagnosis of aspergillosis by Matsumoto (1929) and then by Boe et al. (1939). Seeliger and Suhler (1974) described several cross-reactions with other deep mycoses. Walter and Jones (1968) and Weill et al. (1969) found the CFT more sensitive than the precipitation tests. Nevertheless, Negroni et al. (1972) and Gerber and Jones (1973) found a strong correlation of the CFT with precipitating antibodies. Biguet et al. (1970) in a comparative study on 212 patients found the CFT of less value than the detection of precipitating antibodies. Because of its weak advantages, the CFT is of little use.

Precipitation Techniques

Except for cases of invasive aspergillosis, the detection of precipitating antibodies to *Aspergillus*, perhaps associated with the revelation of the chymotrypsic activity, remains the most reliable technique. Immunodiffusion (ID) was used for the first time by Pepys et al. (1959b) and immunoelectrophoresis (IE) was described by Biguet et al. (1962).

ID was extensively studied by many authors (Stallybrass, 1963; Bronnestam and Hallberg, 1965; Cisek, 1967; Coleman and Kaufman (1972). As shown by Kaufman (1973), the ID test, when used with reference sera, is 100% specific. Sera from patients can be concentrated or not. Because of the weak antigenic similarity between the different pathogenic species of *Aspergillus*, it is recommended to test several antigens at the same time, the most often used being *A. fumigatus, A. flavus*, and *A. niger*. Under such conditions, specific precipitins can be found in 90% of aspergilloma and 70% of the cases of ABPA. The precipitating antibodies are found much less frequently in patients with invasive aspergillosis (Kaufman, 1980).

Immunoelectrophoresis, although fastidious, has some advantages over the ID test. The specificity is based on the qualitative demonstration of the C (chymotrypsic) and J (catalasic) precipitin lines. As shown by Biguet et al. (1970), the growth of the fungus in the patient's organism is revealed by the synthesis of antibodies reacting with the antigenic component bearing the chymotrypsic II activity. This is a means to separate such cases from those where the pathology is only due to sensitization to *Aspergillus* antigens. As most of the patients suffering from aspergilloma exhibit more than five precipitating systems (see Fig. 1), IE is more accurate than ID. Quantitative immunoelectrophoretic techniques such as crossed immunoelectrophoresis or crossed radio immunoelectrophoresis have been mainly used for standardization of antigens (Odds et al., 1983). Their use for detection of specific antibodies was shown to be sensitive (Drouhet et al., 1973; Malo et al., 1981); nevertheless, these sophisticated techniques are ill suited for routine diagnostic application.

On the contrary, counterimmunoelectrophoresis (CIE) appears as a practical improvement of precipitation techniques. First used by Gordon et al. (1971), CIE was confirmed as a quick and sensitive technique (Drouhet et al., 1973; Galussio et al., 1973; Dee, 1975; Mackenzie and Philpot, 1975; Bessieres et al., 1975; Phanuphak et al., 1975; Torres et al., 1976; Smith, 1977). The replacement of agar by cellulose acetate strips as suggested by Gentilini et al. (1972) and the possibility of revealing the chymotrypsic activity (Monjour et al., 1975) gave more simplicity and specificity to the technique. Pinon et al. (1977) with the enzyme-linked immunoelectrodiffusion assay and Lange et al. (1981) with the diffusion in gel enzyme-linked immunosorbent assay brought improvements of the technique allowing both an increased sensitivity and a class specific antibody characterization.

Immunofluorescence Assay (IFA)

Ambroise-Thomas et al. (1969) described an IFA in which antibodies to *Aspergillus* were detected on frozen sections of kidneys from rabbits infected with *A. fumigatus*. This technique was used with modifications of the antigen by other authors (Hocquet et al., 1972; Drouhet et al., 1972; Romburg et al., 1973; Franck and Dunan, 1977), who found a good correlation of their results with those of ID and IE. On the contrary, Negroni et al. (1977), evaluating the IFA, reported very disappointing results. In fact, results vary according to the nature of the antigens used in the test. The use of germ tubes as an antigen, as suggested by Gordon et al. (1974) and Warnock and Hann (1981), seems to give the best specificity. Nevertheless, the threshold of specificity needs to be carefully established according to the presence of specific antibodies in the healthy population, as shown by Warnock (1974) and Schønheyder and Andersen (1982b).

Radioimmunoassay (RIA)

RIA is mainly used for the detection of specific IgE, but has also been applied to the detection of IgG (Patterson et al., 1978; Dewair and Baur, 1984). Patterson et al. (1983) described an RIA index which allows a distinction between asthma and ABPA. The sensitivity of RIA is greater than the sensitivity of immunoprecipitation, and Marier et al. (1970) successfully used RIA for the diagnosis of invasive aspergillosis. Despite its great sensitivity, RIA is being progressively replaced for routine purposes by enzyme immunoassays, which are cheaper and eliminate the requirement to manipulate hazardous radioactive material.

Enzyme-Linked Immunosorbent Assay (ELISA)

Hommel et al. (1976) first introduced ELISA for the diagnosis of human aspergillosis. Ambroise-Thomas et al. (1978), Mantyjarvi et al. (1980), Greenberger et al. (1982), Richardson et al. (1982), Van Der Heide et al. (1984), and Shale and Faux (1985) described several modifications of ELISA which concern mainly the nature of antigens and the protocol for sensitization of the microplates. The results of ELISA depend largely on the antigens used in the reaction. Different procedures of antigen fractionation have been proposed, such as ammonium sulfate precipitation (Sepuvelda et al., 1979; Holmberg et al., 1980; Mishra et al., 1983), concanavalin A (Wilson and Hearn, 1983), and gel filtration (Schønheyder and Andersen, 1983b; Khan et al., 1984). The specificity of antibody detection increases with the degree of purification of antigens from *A. fumigatus*, but this results in failure to detect antibodies directed against other pathogenic *Aspergillus* species (Richardson and Warnock, 1983). Most of the authors agree that ELISA is more sensitive than immunoprecipitation and IFA. ELISA represents an improvement of *Aspergillus* serology, particularly for the detection of low levels of antibodies present in the sera of patients with invasive aspergillosis (Holmberg et al., 1980; Mishra et al., 1983).

Circulating Antigens

Despite their qualities, antibody detection methods fail to identify all the patients with *Aspergillus* infections. Eight to ten percent of cases of aspergillosis are not detected by serological means and require other immunological investigations. This is particularly true for immunocompromised patients undergoing an invasive aspergillosis. In such patients, even the most sensitive serological techniques are not reliable enough, and the percentage of false negative results can be over fifty percent in some series. Antibodies to galactomannans of *A. fumigatus* raised in rabbits or other laboratory animals have been evaluated for their ability to detect a few nanograms of *Aspergillus* polysaccharide antigens in the blood, urine, or bronchoalveolar fluids from patients and experimentally infected animals.

Reiss et al. (1979) with counterimmunoelectrophoresis and Shaffer et al. (1979) with radioimmunoassay were able to detect circulating galactomannan in all the patients studied. Nevertheless, the numbers of patients studied in these publications were small, and more recent data obtained in larger series of patients do not reach a 100% sensitivity. In a recent paper Talbot et al. (1987) found antigenemia in 16 out of 22 patients suffering from disseminated or invasive aspergillosis, a result very similar to those obtained earlier by Andrews et al. (1982) in a series of 9 patients.

EIA might offer in the future an interesting alternative for detection of circulating galactomannan, as shown recently by Dupont et al. (1987), who demonstrated that galactomannan can be detected either in the sera or in the urine of the patients.

CONCLUSION

The majority of the various clinical forms of aspergillosis are characterized by a strong humoral immune response in which the different antibody subclasses are present at significant levels. A large range of sensitive and specific techniques is available for the detection of these antibodies and allows an accurate serological diagnosis for most of the patients. The major problem is at present represented by the invasive forms of aspergillosis most often characterized by a very weak and even absent humoral antibody response, which invalidates the techniques devised for antibody detection. This disadvantage is increasingly counterbalanced by the detection of circulating antigens thanks to new technological improvements that detect nanograms of *Aspergillus* galactomannans in the body fluids.

ACKNOWLEDGMENTS

This work is dedicated to Professor Jean Biguet. I thank Mrs. S. Gamon for reviewing the manuscript and Miss S. Pestre for her excellent secretarial assistance.

REFERENCES

Ambroise-Thomas, P., Kien-Truong, T., Despeignes, J., Battesti, J., Gillot, B., and Goullier, A. (1969). Diagnostic sérologique de l'aspergillose par immunofluorescence sur coupes de reins de lapins inoculés avec *Aspergillus fumigatus*. Résultats preéliminaires. *Bull. Assoc. Dipl. Microbiol. Nancy 115*:27-31.

Ambroise-Thomas, P., Desgeorges, P. T., and Monget, D. (1978). Diagnostic immunoenzymologique (ELISA) des maladies parasitaires par une microméthode modifiée. 2. Résultats pour la toxoplasmose, l'amibiase, la trichinose, l'hydatidose et l'aspergillose. *Bull. W.H.O. 56*:797-804.

Amos, W. M. G. (1978). The extraction of fungal antigens and their use in serological tests as an aid to the diagnosis of bronchial disorders. *J. Med. Lab. Technol. 27*:18-32.

Andrews, C. P., and Weiner, M. H. (1982). *Aspergillus* antigen detection in bronchoalveolar lavage fluid from patients with invasive aspergillosis and aspergillomas. *Am. J. Med. 73*:372-380.

Assem, E. S. K., and Turner-Warnick, M. (1976). Cytophilic antibodies in bronchopulmonary aspergillosis, aspergilloma and cytogenic pulmonary eosinophilia. *Clin. Exp. Immunol. 26*:67-77.

Bardana, E. J., Jr. (1974). Measurement of humoral antibodies to *Aspergilli*. *Ann. N.Y. Acad. Sci. 221*:64-75.

Bessieres-Cathala, M. H., Recco-Sarracante, P., and Seguela, J. P. (1975). Intérêt de l'électrosynérèse dans le diagnostic de l'aspergillose. Valeurs et limites. *Bull. Soc. Fr. Mycol. Med. 4*:11-13.

Biguet, J., Tran Van Ky, P., Capron, A., and Fruit, J. (1962). Analyse immunochimique des fractions antigèniques solubles d'*Aspergillus fumigatus*. Ordre d'apparition des anticorps expérimentaux du lapin, comparaison de ces derniers avec des anticorps naturels humains. *C.R. Acad. Sci.*, Ser. D Paris *254*:3768-3770.

Biguet, J., Tran Van Ky, P., Andrieu, S., and Fruit, J. (1965). Analyse immunoélectrophorétique des antigènes fongiques et systématique des champignons. Répercussions pratiques sur le diagnostic des mycoses. *Mycopathol. Mycol. Appl. 26*:241-256.

Biguet, J., Tran Van Ky, P., Fruit, J., and Andrieu, S. (1967). Identification de l'activité chymostrypsique d'une fraction très spécifique d'*Aspergillus fumigatus*. Répercussions sur le diagnostic immunologique de l'aspergillose. *Rev. Immunol. 31*:317-328.

Biguet, J., Fruit, J., Andrieu, S., and Tran Van Ky, P. (1969). Structure antigènique et systématique des espèces du genre *Aspergillus*. *Bull. Soc. Fr. Mycol. Méd.* LXXXV:273-284.

Biguet, J., Fruit, J., Vernes, A., and Capron, A. (1970). La réaction du complément et l'immunoélectrophoèse appliquées au diagnostic immunologique de l'aspergillose pulmonaire. *Rev. Immunol. 34*:193-204.

Boe, J., Hartmann, O., and Thjotta, T. (1939). A serological study of *Aspergillus fumigatus*. *Acta Pathol. Microbiol. Scand. 16*:178-186.

Bout, D., Fruit, J., and Capron, A. (1973). Application de la chromatographie d'affinité à l'isolement des fractions antigèniques d'*Aspergillus fumigatus* supportant une activité chymotrypsique. *C.R. Acad. Sci.*, Série D *276*:2341-2345.

Bronnestsam, R., and Hallberg, T. (1965). Precipitins against an antigen extract of *Aspergillus fumigatus* in patients with aspergillosis or other pulmonary disease. *Acta Med. Scand. 177*:385-392.

Brummond, W., Resnick, A., Fink, J. N., and Kurup, V. P. (1987). *Aspergillus fumigatus*-specific antibodies in allergic bronchopulmonary aspergillosis and aspergilloma: evidence for a polyclonal antibody response. *J. Clin. Microbiol. 25*:5-9.

Cisek, J. (1967). Diagnostic value of the precipitin test with the *Aspergillus fumigatus* in cases of aspergilloma of the lung. *Pol. Med. J. 6*:1222-1230.

Coleman, M. R., and Kaufman, L. (1972). Use of the immunodiffusion test in the serodiagnosis of aspergillosis. *Appl. Microbiol. 23*:301-308.

Corbel, M. J., and Day, C. A. (1978). Examination of the immunoglobulin classes involved in the serological response of pregnant sheep to *Aspergillus fumigatus*. *Sabouraudia 16*:23-33.

Dee, T. H. (1975). Detection of *Aspergillus fumigatus* serum precipitins by counterimmunoelectrophoresis. *J. Clin. Microbiol. 2*:482-485.

Dessaint, J. P., Bout, D., Fruit, J., and Capron, A. (1976). Serum concentration of specific IgE antibodies against *Aspergillus fumigatus* and identification of the fungal allergen. *Clin. Immunol. Immunopath. 5*:314-319.

Dewair, M., and Baur, X. (1984). RAST for measurement of IgG antibodies to *A. fumigatus* in sera of patients with different lung diseases. *J. Immunol. Meth. 75*:117-128.

Drouhet, E., Camey, L., and Segretain, G. (1972). Valeur de l'immunoprécipitation et l'immunofluorescence indirecte dans les aspergilloses bronchopulmonaires. *Ann. Inst. Pasteur* Paris *123*:379-395.

Drouhet, E., Tabet-Derraz, O., Sanchez-Souza, A., and Viviani, M. A. (1973). Application de l'électrosynérèse et de l'immunoélectrophorèse bidimentionnelle au diagnostic des aspergilloses et à la standarisation des antigènes aspergillaires. *Bull. Soc. Fr. Mycol. Méd. 2*:7-10.

Dupont, B., Huber, M., Kim, S. J., and Bennett, J. E. (1987). Galactomannan antigenemia and antigenuria in aspergillosis: studies in patients and experimentally infected rabbits. *J. Inf. Dis. 155*:1-11.

Fisher, B. D., Armstrong, D., Yu, B., and Gold, J. W. M. (1981). Invasive aspergillosis. Progress in early diagnosis and treatment. *Am. J. Med. 71*:571-577.

Franck, J., and Dunan, S. (1977). Etude comparative de plusieurs types d'antigènes pour la réaction d'immunofluorescence indirecte dans l'aspergillose. *Med. Mal. Inf., 7*:73-76.

Frazer, D. W., Ward, J. I., Ajello, L., and Plikaytis, B. D. (1979). Aspergillosis and other systemic mycoses: the growing problem, *JAMA 242*:1631-1635.

Galussio, J. C., Fridman, J. L., and Negroni, R. (1973). Rapid diagnosis of pulmonary mycoses by counterimmunoelectrophoresis. *Mycophatol. Mycol. Appl. 51*:143-146.

Gentilini, M. L., Pinon, J. M., and Niel, G. (1972). Immunoélectrodiffusion sur membrane d'acétate de cellulose. Application en parasitologie, à propos de résultats préliminaires sur 600 essais. *Nouv. Presse Méd. 1*:1499-1501.

Gerber, J. D., and Jones, R. D. (1973). Immunologic significance of aspergillin antigens of six species of *Aspergillus* in the serodiagnosis of aspergillosis. *Am. Rev. Resp. Dis. 108*:1124-1129.

Gernez-Rieux, C., Biguet, J., Capron, A., Voisin, C., and Andrieu, S. (1964). Etude de la flore mycologique des bronches par examen de sécrétions bronchiques prélevées sous bronchoscopie chez 1120 malades de 1956 à 1964. *Rev. Tubercul. 28*:439-444.

Gerson, S. L., Talbot, G. H., Hurwitz, S., Lusk, E. J., Strom, B. L., and Cassileth, P. A. (1985). Discriminant scorecard for diagnosis of invasive pulmonary aspergillosis in patients with acute leukemia. *Am. J. Med. 79*:57-64.

Gold, J. W., Fischer, B., Yu, B., Chein, N., and Armstrong, D. (1980). Diagnosis of invasive aspergillosis by passive haemagglutination assay of antibody. *J. Infect. Dis. 142*:87-94.

Gordon, M. A., Almy, R. E., Greene, C. H., and Fenton, J. W. (1971). Diagnostic mycoserology by immunoelectroosmophoresis. *Am. J. Clin. Path. 56*:471-474.

Gordon, M. A., Lara, E. W., and Kane, J. (1974). Modified indirect fluorescent antibody test for aspergillosis. *J. Clin. Microbiol.* 6:161–165.

Grabar, P., and William, C. A. (1953). Méthode permettant l'étude conjuguée des propriétés électrophorétiques et immunochimiques d'un mélange de protéines. Application au sérum. *Biochem. Biophys. Acta. 10*:193–194.

Greenberger, P. A., and Patterson, R. (1982). Application of enzyme linked immunosorbent assay (ELISA) in diagnosis of allergic bronchopulmonary aspergillosis. *J. Lab. Clin. Med. 99*:288–293.

Hipp, S. S., Berns, P. S., Tompkins, V., and Buckley, H. R. (1970). Latex slide agglutination test for *Aspergillus* antibodies. *Sabouraudia* 8:237–241.

Hocquet, P., Senet, J. M., Aubert, S., and Brisset, C. (1972). Aspergillose et immunofluorescence indirecte, essais d'interprétation des résultats. *Bull. Soc. Fr. Mycol. Med. 1*:63–66.

Holmberg, K., Berdichewsky, M., and Young, L. S. (1980). Serologic immunodiagnosis of invasive aspergillosis. *J. Infect. Dis., 141*:656–664.

Hommel, M., Kien-Truong, T., and Bidwell, D. E. (1976). Techniques immunoenzymatiques (ELISA) appliquées au diagnostic sérologique des candidoses et des aspergilloses humaines. *Nouv. Presse Med.* 5:2789–2791.

Ikemoto, H., and Shibata, S. (1973). Indirect hemagglutination in pulmonary aspergilloma diagnosis. *Sabouraudia 11*:167–170.

Imbeau, S. A., Nichols, D., Flaherty, D., Dickie, H., and Reed, C. (1982). Relationship between prednisone therapy, disease activity and the total serum IgE level in allergic bronchopulmonary aspergillosis. *J. All. Clin. Immunol. 62*: 91–95.

Khan, Z.U., Richardson, M. D., and Warnock, D. W. (1984). Evaluation of a rapid ELISA for IgG antibodies to *Aspergillus fumigatus* in the serological diagnosis of allergic aspergillosis. *Int. Arch. All. Appl. Immunol. 73*:205–211.

Kaufman, L. (1973). Value of immunodiffusion test in the diagnosis of systemic mycotic diseases. *Ann. Clin. Lab. Sci. 3*:141–146.

Kaufman, L. (1980). Serodiagnosis of fungal diseases. In *Manual of Clinical Immunology*, 2nd Edition. Edited by N. R. Rose and H. Friedman. American Society of Microbiology, Washington, D.C., pp. 553–572.

Kauffman, H. F., Van Der Heide, S., Beaumont, F., Block, H., and De Vries, K. (1986). Class-specific antibody determination against *Aspergillus fumigatus* by means of the enzyme-linked immunosorbent assay. III. Comparative study: IgG, IgA, IgM ELISA titers, precipitating antibodies and IgE binding after fractionation of the antigen. *Int. Arch. Allergy Appl. Immun. 80*:300–306.

Lange, S., Nygren, H., Brorson, J. E., Holmberg, I., and Larsson, P. (1981). Diffusion in gel enzyme-linked immunosorbent assay (DIG-ELISA) for detection of class specific antibodies to *Aspergillus fumigatus* and *Candida albicans*. *Acta Pathol. Microbiol. Scand.*, Sect. C 89:387–389.

Longbottom, J. L., and Pepys, J. (1964). Pulmonary aspergillosis: diagnostic and immunological significance of antigens and C-substance in *Aspergillus fumigatus*. *J. Pathol. Bacteriol. 88*:141–151.

Longbottom, J. L., and Austwick, P. C. K. (1986). Antigens and allergens of *Aspergillus fumigatus*. I. Characterization by quantitative immunoelectrophoretic techniques. *J. Allergy Clin. Immunol. 78*:9-17.

Luster, M. I., Leslie, G. A., and Bardana, E. J. (1976). Structure and biological functions of human IgD. VI. Serum IgD in patients with allergic bronchopulmonary aspergillosis. *Int. Arch. All. Appl. Immunol. 50*:212-219.

Mackenzie, D. W., and Philpot, C. M. (1975). Counterimmunoelectrophoresis as a routine mycoserological procedure. *Mycopathologia 57*:1-7.

Malo, J. L., Paquin, P., and Longbottom, J. L. (1981). Prevalence of precipitating antibodies to different extracts of *Aspergillus fumigatus* in a North American asthmatic population. *Clin. Allergy 11*:33-341.

Mantyjarvi, R. A., Jouslathi, P., and Katila, M. L. (1980). Antibodies to *Aspergillus fumigatus* in farmer's lung patients measured by enzyme-linked immunosorbent assay (ELISA). *Clin. Allergy 10*:187-194.

Marier, R., Smith, W., Jansen, M., and Andriole, V. T. (1979). A solid phase radioimmunoassay for the measurement of antibody to *Aspergillus* in invasive aspergillosis. *J. Infect. Dis. 140*:771-779.

Matsumoto, T. (1929). The investigation of *Aspergilli* by serological methods. *Trans. Brit. Mycol. Soc. 14*:69.

Meyer, R. D., Young, L. S., Armstrong, D., and Yu, B. (1973). Aspergillosis complicating neoplastic disease. *Am. J. Med. 54*:6-15.

Mishra, S. K., Falkenberg, S., and Masihi, N. (1983). Efficacy of enzyme-linked immunosorbent assay in serodiagnosis of aspergillosis. *J. Clin. Microbiol. 17*: 708-710.

Monjour, L., Datry, J., Briend, A., and Gentilini, M. (1975). Diagnostic de l' aspergillose en électrosynérèse sur membrane d'acétate de cellulose par une méthode de caractérisation enzymatique des systèmes précipitants. *Bull. Soc. Fr. Mycol. Med. 4*:231-234.

Morin, O., Miegeville, M., Bouillard, C., Germaud, P., and Bouris, J. H. (1984). Aspergillose pulmonaire invasive chez des malades à haut risque. Difficultés du diagnostic. *Soc. Fr. Mycol. Med.*, Nov. 23.

Nalesnik, M. A., Myerowitz, R. L., Jenkins, R., Lenkey, J., and Herbert, D. (1980). Significance of *Aspergillus* species isolated from respiratory secretions in the diagnosis of invasive pulmonary aspergillosis. *J. Clin. Microbiol. 11*:370-376.

Negroni, R. D., Robles, A. M., and Galussio, J. C. (1972). Estudio comparativo de las reacciones serologicas cuantitativas con un antigenic metabolico de *Aspergillus fumigatus*. *Mycopatholgia 48*:275-287.

Negroni, R., De Flores, C., and Robles, A. M. (1977). Estudio sobre el valor diagnostica de la immunofluorescencia indirecta en la aspergillosis pulmonar. *Sabouraudia 15*:195-200.

Nicaud, P. (1929). Etudes des réactions humorales dans l'aspergillose. *Paris Médical 71*:531-534.

Odds, F. C., Ryan, M. D., and Sneath, P. H. A. (1983). Standardization of antigens from *Aspergillus fumigatus*. *J. Biol. Standard 11*:157-162.

Ouchterlony, O. (1948). Antigen-antibody reactions in gels. *Ark. Kemi. Sverige* 26:1.
Patterson, R., Rosenberg, M., and Roberts, M. (1977). Evidence that *Aspergillus fumigatus* growing in the airway of man can be a potent stimulus of specific and nonspecific IgE formation. *Am. J. Med.* 63:257-262.
Patterson, R., Wang, J. L. F., Roberts, M., and Zeiss, C. R. (1978). Comparison of radioimmunoassay techniques in the detection of IgE and IgG antibody activity against *Aspergillus fumigatus* antigens. *J. Immunol.* 120:66-71.
Patterson, R., Roberts, M., Ghory, A. C., and Greenberger, P. A. (1980). IgA antibody against *Aspergillus fumigatus* antigen in patients with allergic bronchopulmonary aspergillosis. *Clin. Exp. Immunol.* 42:395-398.
Patterson, R., Greenberger, P. A., Ricketti, A. J., and Roberts, M. (1983). A radioimmunoassay index for allergic bronchopulmonary aspergillosis. *Ann. Intern. Med.* 99:18-22.
Pepys, J., Riddel, R. W., Citron, K. M., Clayton, Y. M., and Short, E. I. (1959a). Clinical and immunological significance of *Aspergillus fumigatus* in the sputum. *Am. Rev. Resp. Dis.* 80:167-180.
Pepys, J., Riddell, R. W., and Clayton, Y. M. (1959b). Human precipitins against common pathogenic and nonpathogenic fungi. *Nature* 184:1328-1329.
Phanuphak, P., Salvaggio, J., Fink, J., and Kohler, P. (1975). Incidence of serum precipitins against organic dust antigens in different populations by counter-immunoelectrophoresis. *Chest* 68:753-758.
Pinon, J. M., Gorse, J. P., and Dropsy, G. (1977). Exploration de l'immunité humorale dans l'aspergillose: intérêt de l'ELIEDA. *Mycopathologia* 60:115-120.
Raper, K. B., and Fennell, D. I. (1965). *The genus Aspergillus.* The Williams and Wilkins Co., Baltimore.
Reiss, E., and Lehmann, P. F. (1979). Galactomannan antigenemia in invasive aspergillosis. *Infect. Immun.* 25:357-365.
Richardson, M. D., Stubbins, J. M., and Warnock, D. W. (1982). Rapid enzyme-linked immunosorbent assay for *Aspergillus fumigatus* antibodies. *J. Clin. Pathol.* 35:1134-1137.
Richardson, M. D., and Warnock, D. W. (1983). Enzyme-linked immunosorbent assay and its application to the serological diagnosis of fungal infections. *Sabouraudia* 21:1-14.
Ricketti, A. J., Greenberger, P. A., and Patterson, R. (1984). Serum IgE as an important aid in management of allergic bronchopulmonary aspergillosis. *J. All. Clin. Immunol.* 74:68-71.
Rodsaether, M., and Tonder, O. (1975). Antibodies to *Aspergillus fumigatus*. Characterization of a hemagglutinogen. *Acta Pathol. Microbiol. Scand.*, Sect. C 83:423-428.
Romburg, H., Vernede, A., and Kremer, M. (1973). Note préliminaire sur l'utilisation en immunofluorescence des coupes de culture d'*Aspergillus fumigatus. Bull. Soc. Fr. Mycol. Med.* 2:15-16.
Rosenberg, M., Patterson, R., and Roberts, M. (1977). Immunologic responses to therapy in allergic bronchopulmonary aspergillosis: serum IgE value as an indicator and predictor of disease activity. *J. Pediatrics* 91:914-917.

Schønheyder, H., Andersen, P., andStenderup, A. (1982a). Serum antibodies to *Aspergillus fumigatus* in patients with pulmonary aspergillosis detected by immunofluorescence. *Acta Pathol. Microbiol. Scand.*, Sect. B *90*:273-279.

Schønheyder, H., and Andersen, P. (1982b). An indirect immunofluorescence study of antibodies to *Aspergillus fumigatus* in sera from children and adults without aspergillosis. *Sabouraudia 20*:41-50.

Schønheyder, H., and Andersen, P. (1983a). Complement fixing antibodies to *Aspergillus fumigatus* in patients with pulmonary aspergillosis. *Acta Pathol. Microbiol. Immunol. Scand.*, Sect. B *91*:1-7.

Schønheyder, H., and Andersen, P. (1983b). Determination of antibodies to partially purified *Aspergillus* antigens by an enzyme-linked immunosorbent assay. *Intern. Arch. All. Appl. Immunol. 70*:108-111.

Schønheyder, H., and Andersen, P. (1984). Serum IgA antibodies to *Aspergillus fumigatus* determined by enzyme-linked immunosorbent assay. *Acta Pathol. Microbiol. Immunol. Scand.*, Sect. C *92*:279-282.

Schønheyder, H., Storgaard, L., and Andersen, P. (1985). Variation of a 470,000 daltons antigen complex and catalase antigen in clinical isolates of *Aspergillus fumigatus. J. Med. Vet. Mycology 23*:339-348.

Seeliger, H. P. R., and Suhler, H. (1974). Serologie der Aspergillose. *Zbl. Bakt. Hyg. Abt.*, Orig. A *229*:524-553.

Senet, J. M., and Brisset, C. (1973). The diagnosis of aspergillosis by passive hemagglutination. *Biomedicine 19*:365-368.

Senet, J. M., Girault, A., Robert, R., and Girault, M. (1978a). Le diagnostic de l'aspergillose par hémagglutination indirecte. I. Utilisation d'une fraction chymotrypsique purifiée d'*Aspergillus fumigatus. Bull. Soc. Fr. Mycol. Med. 7*:225-228.

Senet, J. M., Girault, A., Robert, R., and Girault, M. (1978b). Le diagnostic de l'aspergillose par hémagglutination indirecte. I. Utilisation d'une fraction catalasique purifiée d'*Aspergillus fumigatus. Bull. Soc. Fr. Mycol. Med. 7*: 229-232.

Sepuvelda, R., Longbottom, J. L., and Pepys, J. (1979). Enzyme-linked immunosorbent assay for IgG and IgE antibodies to protein and polysaccharide antigens of *Aspergillus fumigatus. Clin. Allergy 9*:359-371.

Shaffer, P. J., Kobayashi, G., and Medoff, G. (1979). Demonstration of antigenemia in patients with invasive aspergillosis by solid phase (protein A-rich *Staphylococcus aureus*) radioimmunoassay. *Am. J. Med. 67*:627-630.

Shale, D. J., and Faux, J. A. (1985). The evaluation of quantitative enzyme-linked immunosorbent assay for anti-*Aspergillus fumigatus* IgG. *J. Immunol. Methods 77*:197-205.

Slavin, R. G., Millon, L., and Cherry, J. (1970). Allergic bronchopulmonary aspergillosis. Characterization of antibodies and results of treatment. *J. Allergy 46*:150-155.

Smith, J. M. B. (1977). Counterimmunoelectrophoresis and opportunistic fungal infections. *Mycopathologia 60*:99-104.

Stallybrass, F. C. (1963). The precipitin test in human systemic aspergillosis. *Mycopathol. Mycol. Appl. 21*:272-278.

Stokes, T. C., Truton, C. W. G., and Turner-Warwick, M. (1981). A study of immunoglobulin G subclasses in patients with allergic bronchopulmonary aspergillosis. *Clin. Allergy 11*:209-216.

Talbot, G. H., Weiner, M. H., Gerson, S. L., Provencher, M., and Hurwitz, S. (1987). Serodiagnosis of invasive aspergillosis in patients with hematologic malignancy: validation of the *Aspergillus fumigatus* antigen radioimmunoassay. *J. Inf. Dis.155*:12-27.

Tonder, O., and Rodsaether, M. (1974). Indirect hemagglutination for demonstration of antibodies to *Aspergillus fumigatus*. *Acta Pathol. Scand.*, Sect. B *82*:871-878.

Torres, J. M., Guisantes, J. A., Da Luz, S., Lopez-Lemez, M., and Yarzabal, L. (1976). Aspergillosis respiratoria humana. Diagnostico para medio de la electrosineresis associada a la immunodiffusion secundaria. *Medicina Clinica 66*:155-158.

Tran Van Ky, P., Uriel, J., and Rose, F. (1966). Caractérisation des types d' activités enzymatiques dans les extraits antigèniques d'*Aspergillus fumigatus* après électrophorèse et immunoélectrophorèse en agarose. *Ann. Inst. Pasteur.*, Paris *111*:161-170.

Tran Van Ky, P., Biguet, J., and Vaucelle, T. (1968). Etude d'une fraction antigènique d'*Aspergillus fumigatus*, support d'une activité catalasique. Conséquences sur le diagnostic immunologique de l'aspergillose. *Rev. Immunol. 32*: 37-52.

Tran Van Ky, P., Vaucelle, T., and Biguet, J. (1970). Etude comparée de la structure antigénique par analyse immunoélectrophorétique des extraits antigèniques des champignons pathogènes du genre *Aspergillus* (*A. fumigatus, A. flavus, A. terreus, A. nidulans*). *Rev. Immunol. Therap. Antimicrob. 34*: 357-374.

Tran Van Ky, P., Biguet, J., Vaucelle, T., and Fruit, J. (1971). Analyse immuno-électrophorétique et caractérisation des activités enzymatiques des extraits antigèniques d'*Aspergillus flavus*. Répercussion sur le diagnostic différentiel des aspergilloses humaines. *Sabouraudia 9*:210-220.

Uriel, J. (1963). Color reactions for the identification of antigen-antibody precipitation in gel diffusion media. *N.Y. Acad. Sci. 103*:936-940.

Van Der Heide, S., Kauffman, H. F., and De Vries, K. (1984). Measurement of specific IgE and IgG antibodies against *Aspergillus fumigatus* antigen in patient sera by use of enzyme immunoassays. Influence of different procedures of antigen immobilization. *J. All. Clin. Immunol. 73*:813-818.

Virchow, R. (1856). Beiträge zur lehre von den beim menschen vorkommenden pflanzlichen Parasiten. *Arch. Path. Anat. 9*:557.

Wilson, E. V., and Hearn, V. M. (1983). Comparison of partially purified mycelial and culture filtrate antigens of *Aspergillus fumigatus* by enzyme-linked immunosorbent assay. *Sabouraudia 21*:195-203.

Walter, J. E., and Jones, R. D. (1968). Serologic tests in diagnosis of aspergillosis. *Dis. Chest. 53*:729-735.

Warnock, D. W. (1974). Indirect immunofluorescence test for the detection of *Aspergillus fumigatus* antibodies. *J. Clin. Pathol. 27*:911-912.

Warnock, D. W., and Hann, E. M. (1981). Further evaluation of indirect immunofluorescence methods for detection of antibodies against *Aspergillus fumigatus*. *Sabouraudia 19*:49-54.

Weill, C. P., Pesle, G. D., and Mehl, P. (1969). Diagnostic des infections à *Aspergillus* par déviation du complément. *Ann. Biol. Clin. 27*:87-91.

Young, R. C., Jenning, A., and Bennett, J. E. (1972). Species identification of invasive aspergillosis in man. *Am. J. Clin. Pathol. 58*:554-557.

17
Serological Procedures to Detect Dermatophyte Antigens

LUCIANO POLONELLI
University of Parma, Parma, Italy

GIULIA MORACE
Università Cattolica del Sacro Cuore, Rome, Italy

INTRODUCTION

The dermatophytes are specialized fungi which share the property of invading the keratinized tissues of humans and animals. In his inspired monograph, Ajello (1977) clearly emphasizes how, quite often, studies on dermatophytes paved the way for fundamental breakthroughs in microbiology and, particularly, medical mycology.

Our understanding of dermatophytes is constantly evolving. The acquisition of new knowledge occurs so rapidly that it is sometimes useful to stop and evaluate the impact of new technologies. In dermatophytosis, previously accepted concepts have been replaced by new ones in the realms of chemotherapy, genetics, and molecular biology, although the most dramatic changes have undoubtedly occurred in the field of immunology (Bloch, 1908); de Lamater and Benham, 1938; Huppert, 1955, 1962; Götz, 1962a; Salvin, 1963; Lepper, 1969; Grappel et al., 1974; Grappel, 1981; Jones and Artis, 1981; Kaaman, 1985b; Reiss, 1986).

SEROLOGICAL PROCEDURES

Serological procedures for the detection of dermatophyte antigens can be used for a variety of purposes. The complex antigenic structure of numerous dermatophytes has been investigated with different chemical and physical techniques, and studies have been performed to characterize the immunological activity of these components (Table 1).

Table 1 Selected Studies on Dermatophyte Structure

Acetone powdered mycelia: 129[a], 134
Acetyltrimethyl ammonium bromide precipitation: 23,24
Allergen evaluation: 10,22,27,155,156,308,408
Amino acid composition: 21,22,81,116,139,284,375,415
Autoclavated mycelium: 99,120,124,127
Cell wall disruption: 8,216,255,280,329,380
Complement fixation: 9,120-122,127
Concanavalin A affinity chromatography: 368
Crossed immunoelectrophoresis: 64,366,367,372,419
DEAE column: 21,35,37,99,155,171,172,284,285,412-414
Disc electrophoresis: 346,347
Double immunodiffusion I: 8,9,11,349,408,410
Electron microscopy: 151,152,280,283
Enzyme electrophoresis: 190
Environmental factor influence: 201,357
Erythrocyte inhibition isoagglutination: 410
Ethyleneglycol extraction: 21-23,25,27,66,70,99,134,137,155,156, 188,197,270,289,410,419
Extracellular enzyme detection: 59,74,320,322
Extracellular polysaccharide detection: 252,253
Freezing and thawing of mycelium: 258-260
Gel filtration: 8,270,284,285,382,408,412,414,415
Hemagglutination: 312,313
Hot diluted alkali: 34,36,126,404
Immunofluorescence staining: 67,164,357
Immunoperoxidase technique: 150-153
Ion exchange chromatography: 9,382
Isoelectric focusing: 183,228
Keratinase production: 67,119,342,381,382,412-415
Line immunoelectrophoresis: 267
Medium composition influence: 101
Methanol precipitation: 159
Methylation: 99,123
Mycelia homogenate: 7,144,157,286

Table 1 (Continued)

Paper chromatography: 81,213,343
Pathogenesis role: 221,263,316,321,349
Phenol extraction: 77,171-173,279,280,283-285
Polysaccharide composition: 82
Proteolytic activity: 8,9,21,22,41,63,119,263,278,280,283,284, 287,320,322,381,382,400,412-414
Proton magnetic resonance: 164
Quantitative immunoelectrophoresis: 11,120,125,267,368
Saline extraction: 32,125,126
SDS PAGE: 382
Sonicated mycelium: 86
Water extraction: 6,34-37,40,377
X-ray diffraction: 39,283

---

[a]Refer to number in parentheses following individual reference at end of chapter.

Sensitive serological procedures have been explored both for the study of in vivo and in vitro immunological responses and for the identification of clinical isolates. The study of delayed and immediate types of reactivity to dermatophyte antigens in man and animals has contributed remarkably to the fundamental understanding of the immunological responses to fungal antigens. The specificity and sensitivity of serological diagnosis in human patients and animals has been improved by such acquisitions. Moreover, the proper evaluation of the immunological activity of the different cultural components (cellular and metabolic) of the dermatophytes and the immunological response that they induce has made it possible to develop in animals and man a vaccination strategy (immunoprophylaxis and therapy), the application of which could exceed the specific area of dermatophytosis.

Although these approaches may have different specific aims, all of them require a profound knowledge of the complex antigenic nature of the dermatophytes.

Initial observations of the antigenic properties of dermatophyte extracts (trichophytin) were made before the fundamental work of Sabouraud (1910) on the taxonomy of dermatophytes based on morphological criteria. The two areas subsequently developed in parallel, and the antigenic studies of dermatophytes formed the basis for further acquisitions in the fields of chemical structure, host immunological response, in vivo antigen detection (Table 2), and acquired resistance to infections following vaccination.

Table 2 Selected Studies of In Vivo Identification

Immunoperoxidase technique:	150-152,[a]342
Indirect immunofluorescence:	67,143,154
Keratinase immunoelectron microscopy:	342

[a]Refer to number in parentheses following individual reference at end of chapter.

DERMATOPHYTE ANTIGENS DETECTION

Dermatophyte antigens are basically composed of glycopeptides, polysaccharides, and proteolytic enzymes. Their chemical structures have been investigated by means of different procedures, and their immunological activity has been evaluated primarily by skin testing in man and animals.

Study of the immunologic response in dermatophytosis has been going on for many years. Cell-mediated immunity is generally considered the major immunologic response elicited in both human and animal dermatophytoses, although clinical manifestations associated with humoral responses may also occur. Intradermal injections of trichophytin in patients with chronic dermatophytosis frequently cause immediate-type reactions, and the presence of specific IgE has actually been demonstrated in the sera of some of these patients (Tables 3 and 4).

It is presently believed that the major immunological response to dermatophytic infections is the stimulation of specific T-lymphocytes, which is manifested in vivo as a delayed-type hypersensitivity. Several in vitro immunological procedures have been used to detect the presence of the dermatophyte antigens. The role of the immune response and interferon protection in experimental infections has also been investigated (Nikiforov, 1978; Medvedev and Chistyakova, 1981; Kotsitadze et al., 1984). Indirectly, the sensitization toward the antigens might also be evaluated by considering the degree of acquired resistance of experimental animals and volunteers immunized with different kinds of antigens (Table 5).

Detection of dermatophyte antigens may be also used in the serological diagnosis of dermatophytosis in man and animals (Table 6).

Table 3 Selected Studies on Immediate-Type Reaction

Skin test:	8,[a]9,17,18,21,23,27,35-37,53,79,93,94,99,118,145,155, 177,180,185,188,192,238,241,243,244,248,272,276,284, 285,300,352,360,361,366,368,370

[a]Refer to number in parentheses following individual reference at end of chapter.

Table 4 Selected Studies on Delayed-Type Reaction

Allergic reactions:	10,[a]19,20,92,95,114,118,170,181,209,300,308-310,323,339,359,384,408
Autoimmune reactions:	55,293
Cell-mediated immunity:	12,112,186,353,363,366-368,397
Complement activation:	374
Dermatophytid eruptions:	15,43,69,84,111,114,118,148,149,166,169,174,175,198,230-232,268,291,325,358,360,361,389,405,406,409
Hemagglutination:	399
HLA system:	4,52,365,371
Leukocyte chemotaxis:	17,73,74,147,376
Leukocyte inhibition migration test:	13,71,73,87,93,96,98,112,141,142,147,158,241,396,397,399
Lymphocyte nuclear DNA:	92
Lymphocyte reactivity:	8,9,19,53,87,92,93,96-98,115,136,137,193,195,328,353,355,368,370,393
Lymphocyte transformation:	8,57,58,137,144,146,353,365,368,373
Macrophage stimulation test:	303
Skin test:	8-10,13,17,21-23,25,27,38,44,53,67,70,77,80,87,90,94,104,110,118,125,128,129,132,137,141,144,157,159,167,171,172,176,177,186,189,191-194,196,197,199,207,217-220,227,229,234-237,239,241,258,261,270,273,274,276,284,285,287,292,296,297,300,318,319,332,334,335,340,341,348,352,353,357,360,363,368,377,378,380,383,385,386,390,396,399,407,408
Skin test cross-reactivity:	16,121-123,192,411
T-cell function:	57,58,93,97,130,194,241,297,328

[a]Refer to number in parentheses following individual reference at end of chapter.

The serological procedures previously summarized have proven to be valid in an enormous number of studies with various aims. Interest in the separation and characterization of dermatophyte antigenic components has led to attempts at detecting group-specific and species-specific antigens. Immunological methods were then applied for the serological differentiation of species as well as of strains within the species. The development of these methods, with particular emphasis on the use of hybridoma technology for the production of monoclonal antibodies against dermatophyte antigens, is especially interesting (Table 7).

Table 5 Selected Studies on Vaccination

In Animals
  Acquired resistance: 2,[a]26,31,42,43,45,50,62,78,83,88,102,106,
            117,118,130,131,135,208,211,214,215,222,
            233,234,239,245,249-251,295,317,318,324,
            331,338,364,383,391,403
  Cross-immunity: 311,337
  Culture filtrates: 50,61,184,223
  Inactivated vaccines: 323,299
  Infected skin mixed with mycelium: 249-251
  Killed mycelium: 31,102,208,214,215,281,364,403
  Live attenuated cultures: 202
  Live vaccines: 47,399
  LTF-130: 1,113,299,327,354
  Temperature influence: 51
In Humans
  Culture filtrates: 28,242,262,273,359,389
  Disintegrated mycelium: 162
  Hydrolyzed mycelium: 105,182,242
  Killed powdered mycelium: 161,356,364
  Polysaccharide nucleic acid: 138
  Trichloroacetic mycelial extract: 247

[a]Refer to number in parentheses following individual reference at end of chapter.

The initial observations by Jadassohn et al. (1932) of group and species-specific antigens prompted many other attempts to differentiate dermatophyte species by serological methods. Early studies using mycelial extracts and rabbit antisera in precipitin tests showed that various cross-reactions could affect the potential of serology as a basis for species differentiation (Sharp, 1941; Kielstein, 1976; Sagara, 1968). Through the isolation of protein and carbohydrate antigens from culture filtrate and the use of a standardized synthetic medium for growing the cultures, it was possible to distinguish *Trichophyton rubrum* from *T. mentagrophytes* (Keeney and Ericksen, 1949) and to group *Trichophyton, Microsporum*, and *Epidermophyton* species (Tomomatsu, 1961a). Further studies using double immunodiffusion (Dyson and Landay, 1963; Rogers, 1968) and immunoelectrophoresis tests (Biguet et al., 1965; Andrieu et al., 1968; Christiansen and Svejgaard, 1976) confirmed the results of species differentiation by in vitro hair perforation tests and the closer relationship of *M. ferrugineum* to *Microsporum* than to the *Trichophyton* species. Fluorescent antibody techniques could be used only for the detection of dermatophytes in histological slides and their differentiation from other groups of fungi (Miura, 1963; Refai et al., 1976).

Table 6 Selected Studies on Serological Diagnosis

Agglutination test: 167,[a]168,298,319,387
Antibiotic relationships: 294
Block electrophoresis: 279
Charcoal agglutination: 118,125
Complement fixation: 14,46,54,60,68,93,103,107-110,124,125,131, 200,224,231,271,280,293,298,344,386,394
Concanavalin A affinity chromatography: 368
Counterimmunoelectrophoresis: 48,72,163,282
Crossed immunoelectrophoresis: 369,372
Cross-reactivity: 89,90,243,246,275,293,410,416,417
Direct immunofluorescence: 165,365,418
Double immunodiffusion: 14,41,65,68,125,127,167,231,262,271,296, 319,386,387,395,403
Enzyme-linked immunosorbent assay: 145,153,199,365
Gel filtration: 279,280
Hemagglutination: 319
Hemcoagulation indices: 225
Immune rosette formation: 18,93,97,240,241
Indirect immunofluorescence: 67,154,163,226,265,290,293,314,398
Lymphocyte transformation: 373
Lysozyme and complement activity: 93
Microcytotoxicity assay: 4
Oxygen uptake in vitro: 210
Passive agglutination: 91,93,210-212,280,319
Phagocytosis: 17.93,140,257,350
RAST: 18,276
Serum immunoglobulin contents: 2,3,17,18,29,30,93,125,136,144, 145,153,189,199,240,257,276,279, 302,351,368,370,372,401,409

[a]Refer to number in parentheses following individual reference at end of chapter.

Table 7 Selected Studies on Species Identification

Complement fixation: 126[a]
Crossed immunoelectrophoresis: 64,366,367
Crossed radioimmunoelectrophoresis: 372
Disc electrophoresis: 346,347
DNA hybridization: 75,76
Double immunodiffusion: 11,48,49,86,205,213,269,288,301,305-307, 326,333,336,349,362,379,392
Enzyme-linked immunosorbent assay: 134
Exoantigens: 205,306,336
Hemagglutination: 203,312
Immunoelectrophoresis: 11,32,33,362
Immunofluorescence: 85,164,264,266,365,379,392
Immunospot assay: 133,304
Isoelectric focusing: 183
Monoclonal antibodies: 133,269,304,305,307
Monoclonal antibody affinity chromatography: 304
Precipitin test: 134,207,345,384
Protein patterns: 228
Proton magnetic resonance: 164
Schultz-Dale technique: 100,177-179
Western immunoblotting: 133,134,304
Zymogram patterns: 190

[a]Refer to number in parentheses following individual reference at end of chapter.

The success of the exoantigen technique in identifying mycelial-form cultures of fungi pathogenic to man (Kaufman and Standard, 1978) provided a way for separating dermatophyte species by serological techniques. Despite rigid control of growth conditions, optimization of reagent concentration and the use of multiple serological methods, dermatophytes have proven to be closely related antigenically, and extensive cross-reactions have occurred among the species (Fischer, 1956; Reyes and Friedman, 1966; Scheckter et al., 1968; Takahashi, 1971; Turner and Kaplan, 1974; Drouhet et al., 1976; Refai et al., 1977; Böhme et al., 1978a; Philpot, 1978; Kaufman and Lopez, 1980). Partial improvement in the serological differentiation of a limited number of morphologically related dermatophyte species has been achieved by using adsorbed monospecific antisera

(Polonelli and Morace, 1985a). The separation of secreted dermatophyte proteins by analytical isoelectric focusing has shown that the information derived may be valid when applied for the taxonomic differentiation of *Microsporum* and *Trichophyton* species (Jeffries et al., 1984).

The antigenic constitution of dermatophytes has proven to be quite complex, and it is this complexity which has frequently hindered the development of immunodiagnostic tests with sufficient sensitivity and specificity. The advent of hybridoma technology might provide an invaluable tool for the purification and characterization of specific fungal antigens.

One of the most interesting uses of monoclonal antibodies in medical mycology is that of probes to separate and identify antigenic fractions from complex mixtures of fungal macromolecules.

Use of Monoclonal Antibodies

For the production of monoclonal antibodies against dermatophyte antigens, a decision must be made prior to selecting immunization procedures between the use of mycelium homogenate or soluble antigen as immunogen. The alternatives are spending time purifying the antigen and then obtaining the relevant hybridomas with relative ease or spending time searching among several hundred hybridomas for the one that produces the required antibody.

In the immunoidentification of dermatophytes by monoclonal antibodies, the use of a soluble antigen, and specifically a reference exoantigen, could be highly effective. This type of antigen has consistently proven to be valid in the serological differentiation of mycelial cultures of pathogenic fungi (Polonelli and Morace, 1984). It is of fundamental importance, moreover, that exoantigens are very easily and rapidly prepared from the dermatophyte cultures to be identified.

In a series of studies devoted to the serological analysis of a large number of dermatophyte antigens by monoclonal antibodies, Polonelli and Morace (Polonelli and Morace, 1985b; Morace et al., 1986; Polonelli et al., 1986) adopted a reference strain of *M. canis* for the production of the reference antigen. It was derived from a merthiolate-treated solution, obtained from numerous Sabouraud dextrose agar slant cultures of different ages, which had been concentrated by lyophilization. The protein and carbohydrate contents of the concentrated reference antigen were standardized to ensure reproducible results. Chemical and serological analysis had previously demonstrated that this antigen was comparable to the one obtained via an acetone-treated filtrate (Morace et al., 1986). Furthermore this antigen was serologically detectable in all of the cultures of the reference strain of *M. canis* tested at 5-day intervals from 5 to 45 days in age.

A standard protocol was used for immunizing the mice, fusing syngenic myeloma cells with spleen cells and screening the hybrid fluids by enzyme-linked immunosorbent assay (ELISA) (Polonelli and Morace, 1985b).

Hybridoma cultures secreting the desired antibody were propagated and cloned by double limiting dilution in the presence of mouse thymocytes. Clones that continue to produce the antibody were expanded. The monoclonality of the antibody produced by the various clones was ascertained by double immunodiffusion against rabbit anti-mouse immunoglobulin antisera. Ascites fluids were produced by intraperitoneal injection of hybridoma cells in syngenic mice with pristane treated. Immune ascites was purified by ammonium sulfate precipitation of the immunoglobulins following centrifugation and dialysis. The clones were frozen and maintained in liquid nitrogen (Polonelli and Morace, 1985b).

Monoclonal antibodies of the immunoglobulin G class were produced. When the antigen was electrophoretically separated in denaturing gels and immobilized on nitrocellulose strips (Western blotting technique), a diversity of monoclonal antibodies to fungal glycoproteins was detected, and the study of the fungal polypeptides reactive with the different monoclonal antibodies allowed investigation of several properties of the antigens. Antibody-rich ascitic fluids reacted in double immunodiffusion, allowing the detection of unexpected distribution of antigenic determinants in the dermatophyte isolates. The production of ascitic fluids was necessary in double immunodiffusion tests in order to obtain the sufficient amounts of monoclonal antibody to precipitate with the standard concentration of the reference antigen. Ascitic fluid produced from one type of hybridoma-producing monoclonal antibody reacted in double immunodiffusion with all of the heterologous and homologous dermatophyte antigens. The presence of an antigen with a determinant shared by all of the dermatophyte isolates tested is obviously suggestive of a very common function. Another type of monoclonal antibody did not react with the antigens obtained from three different isolates of *M. canis*, demonstrating the existence of different serotypes within the species. The exoantigens of the same three *M. canis* isolates that were negative with the latter type of monoclonal antibody regularly reacted with their homologous rabbit polyvalent antiserum. This result suggested the possibility that different serotypes may also exist within other species. And, in fact, a third type of monoclonal antibody produced against the reference antigen of *M. canis* allowed the detection of different serotypes within the species *T. rubrum* (Morace et al., 1986).

Species-specific monoclonal antibody should more readily differentiate between the serotypes. It is conceivable, in fact, that some isolates lack the antigenic determinant necessary for reactivity with a specific monoclonal antibody.

Antigen diversity in the dermatophyte *T. rubrum* has also been subsequently demonstrated by de Haan and Wikler (1986b). In order to investigate the antigenic composition in culture filtrates and in an ethyleneglycol extract of *T. rubrum*, they isolated different culture filtrates and one ethyleneglycol extract of acetone-dried material. Antisera against a mixture of the seven culture filtrates and against acetone-dried material were raised in rabbits. Antigen diversity was demonstrated by using precipitation test, ELISA, ELISA-inhibition, and immunoblotting techniques.

These data show that monoclonal antibodies are powerful reagents for serotyping dermatophyte isolates and provide clear demonstration of intratypic antigenic variation. The effectiveness of monoclonal antibodies as serotyping reagents would be considerably enhanced by the use of a panel of type-specific antibodies. It is expected that the number of subgroups identified will depend on the number of monoclonal antibodies used for analysis, and additional groupings are likely to emerge as more hybridomas are used.

Monoclonal antibodies should be readily adopted as serotyping reagents and may prove valid for epidemiological studies on dermatophyte infections. Subdivision of species by serotyping provides valuable epidemiological data, particularly when other systems are not available.

Polonelli et al. (1986) verified, moreover, the feasibility and usefulness of using monoclonal antibodies for the immunoidentification and antigenic characterization of dermatophytes by the Western blotting technique (Otcenasek et al., 1975; Burnette, 1980).

Western Blotting Technique

The potential of the Western blotting technique may be completely exploited by using a gel filtration calibration kit of substances of known molecular weight. The markers can be easily mixed together with the soluble antigens before the test. Each strip of nitrocellulose can be differentially stained either by Coomassie blue staining for proteins or immunoperoxidase for evidence of antigen-monoclonal antibody binding (Polonelli et al., 1986).

Consequently, the Western blotting technique permits the easy determination of the molecular weight of single antigenic determinant reacting with each monoclonal antibody. The specificity of a monoclonal antibody derives from its purity, but does not preclude cross-reactivity due to recognition of similar antigens or identical determinants located in different antigenic mixtures. This is particularly relevant in a complex situation, such as the expression of cell surface antigens in which a common antigenic determinant can be expressed on different molecules (a common segment of evolutionarily related proteins, a carbohydrate moiety, etc.). On the other hand, lack of reactivity with a monoclonal antibody does not prove absence of the antigenic determinant. The reactivity may be affected by changes in the environment of the antigen.

These properties were particularly evident for relatively nonspecific monoclonal antibodies. One *M. canis* monoclonal antibody recognized all of the *M. canis* exoantigens tested by reacting with two specific antigenic determinants of the same molecular weight. The detection of other reactive antigenic determinants in the different isolates within the species clearly shows the existence of numerous serotypes. These antigenic determinants were not simultaneously found in heterologous cross-reactive species.

The Western blotting behavior of another *M. canis* monoclonal antibody was similar, although it characterized single *M. canis* exoantigens either by species-specific or strain-specific antigenic determinants of different molecular weights.

A third *M. canis* monoclonal antibody detected only one species-specific antigenic determinant. All of the isolates of *M. gypseum* tested as exoantigens showed a common domain at different molecular weight.

Of all of the *M. canis* monoclonal antibodies produced, one proved to be particularly interesting and specific. It reacted with only a low molecular weight antigenic determinant in each *M. canis* exoantigen tested. The same reaction was also observed in the *M. distortum* isolate tested, thus confirming the identity of the two species as demonstrated by mating experiments (Matsumoto et al., 1983; Weitzman et al., 1986). Cross-reactions, although at different molecular weights, were observed only in the *M. ferrugineum, M. gallinae*, and *T. soudanense* exoantigens tested, which are considered to be closely related morphologically (Polonelli et al., 1986).

De Haan et al. (1986a) produced monoclonal antibodies against an antigen of a *T. rubrum* isolate and evaluated their specificity by determining different reactions with culture filtrates of other *T. rubrum* isolates by immunoblotting. Cross-reactions with culture filtrates of other dematophytes as well as of unrelated fungi were observed by immunospot assay and were not inhibited by α-methylmannoside.

These results clearly show that by using antigens electrophoretically separated in denaturing gels and then immobilized on nitrocellulose strips, it is possible to detect a greater diversity of monoclonal antibodies to dermatophyte glycoproteins than when techniques of immunoprecipitation of soluble nondenaturated dermatophyte antigens or ELISA are used. The primary advantage of the Western blotting technique is in the detection of nonprecipitating antibodies and of antibodies to poorly soluble antigens not available for reactions in other preparations. Studies of the dermatophyte proteins reactive with a panel of monoclonal antibodies indicated that the technique can be used to investigate several properties of the separated antigens.

Complex antigenic mixtures, on the other hand, may contain dominant immunogenic components. These may effectively decrease the variability of the response to a given antigen, in which case partial purification of the antigen in question may become essential. One general way to achieve such partial purification is by using monoclonal antibodies as immunoadsorbents to remove such impurities.

With only a few ml of soluble antigen, monoclonal antibody affinity chromatography could provide powerful reagents for study.

Polonelli et al. (1986) found that single monoclonal antibodies were directed against single peaks of the antigen mixture of the *M. canis* reference exoantigen used for immunization. After affinity chromatography with an immunoadsorbed *M. canis* monoclonal antibody, it was possible to separate the peak corresponding to the unadsorbed antigen from the adsorbed antigen. Reversed-phase high performance liquid chromatography of the unadsorbed antigen showed the disappearance of a single peak from the original antigenic pattern.

The immunospot assay carried out with the monoclonal antibody used in the affinity chromatography confirmed the loss of reactivity only for unadsorbed antigen.

Studies of dermatophyte antigens by monoclonal antibodies clearly demonstrated the tremendous potential of such reagents in the immunoidentification of dermatophytes. Their effectiveness may be enhanced by using them with a highly sensitive analytical procedure such as the Western blotting technique. The primary results will be largely conditioned on finding relatively specific monoclonal antibodies devoid of too many cross-reactions. The procedure of affinity chromatography however, may also overcome difficulties by immuno-adsorbing the nonspecific monoclonal antibody to obtain a purified antigen depleted of the common antigenic determinant. The production of new second-generation monoclonal antibodies may make it possible to achieve the desired specificity or to continue the purification of the antigen.

REFERENCES

Aavik, K. (1978). The production of TF-130 and LTF-130 vaccines at Khal'yal-askogo bioworks in Estonia. *Byulleten' Vsesoyuznogo Ordena Lenina Instituta Eksperimental'noi Veterinarii No. 32*: 22-23.

Ahmed, A. R. (1982). Immunology of human dermatophyte infections. *Arch. Dermatol. 118*: 521-525. (2)

Ahmed, A. R., and Ajello, J. (1983). Chronic tinea pedis: presence of high serum IgE levels, anti-ICS antibody, and ANA. *Clin. Exp. Dermatol. 8*: 414-420. (3)

Ahmed, A. R., Schreiber, P., Ajello, J., Tiwari, J. L., and Terasaki, P. I. (1985). A preliminary report on the role of some immunologic factors in persistence of chronic tinea pedis. *Clin. Exp. Dermatol. 10*: 45-50. (4)

Ajello, L. (1977). Milestones in the history of medical mycology: the dermatophytes. In *Recent Advances in Medical and Veterinary Mycology*. Edited by K. Iwata. University of Tokyo Press, pp. 3-11. (5)

Alfes, H., Bishop, C. T., and Blank, F. (1963). The water soluble polysaccharides of dermatophytes. II. A glucan from *Microsporum quinckeanum*. *Can. J. Chem. 41*:2621-2627. (6)

Allen, D. E., Snyderman, R., Meadows, L., and Pinnell, S. R. (1977). Generalized *Microsporum audouinii* infection and depressed cellular immunity associated with a missing plasma factor required for lymphocytes blastogenesis. *Am. J. Med. 63*: 991-1000. (7)

Al-Rammahi, A. K., Levy, J. E., and Kelly, B. (1978a). Isolation and partial characterization of immunologically reactive fractions from chitinase digested cell wall of *Trichophyton mentagrophytes*. *Mycopathologia 66*: 129-137. (8)

Al-Rammahi, A. K., Levy, J. E., and Kelly, B. (1978b). Characterization of immunologically active peptides from the cell wall of *Trichophyton mentagrophytes*. *Mycopathologia 66*: 139-146. (9)

Andersson, B. Å., Hellgren, L., and Vincent, J. (1976). Allergic delayed skin reactions from lipid fractions of trichophytin. *Sabouraudia 14*: 237-241. (10)

Andrieu, S., Biguet, J., and Laloux, B. (1968). Analyse immunoéléctrophorétique comparée des structures antigéniques de 17 especes de dermatophytes. *Mycopathol. Mycol. Appl. 34*: 161-185. (11)

Arnold, M. T., Grappel, S. F., Lerro, A. V., and Blank, F. (1976). Peptido polysaccharide antigens from *Trichophyton mentagrophytes* var. *granulosum*. *Infect. Immun. 14*: 376-382. (12)

Asahi, M., Ueda, S., Kwakazu, M., and Urabe, H. (1982). Purification and characterization of a new peptide antigen extracted from dermatophyte mycelia. *J. Invest. Dermatol. 78*: 38-43. (13)

Attapattu, M. C., and Clayton, Y. M. (1982). Humoral antibody in dermatophytosis—factors affecting the antibody response in *Trichophyton rubrum* infections. *Sabouraudia 20*: 273-279. (14)

Ayers, S., and Anderson, N. P. (1934). Inhibition of fungi in cultures by blood serum from patients with "phytid" eruptions. *Arch. Dermatol. 29*: 537-547. (15)

Baldo, B. A., Fletcher, T. C., and Pepys, J. (1977). Isolation of a peptido-polysaccharide from the dermatophyte *Epidermophyton floccosum* and a study of its reaction with human C-reactive protein and a mouse antiphosphorylcholine myeloma serum. *Immunology 32*: 831-842. (16)

Balogh, É., Fórizs, E., and Debreczenni, M. (1985). Investigations on the chemotactic and phagocytic activities of circulating monocytes in chronic dermatophytosis. *Mykosen 28*: 490-494. (17)

Balogh, É., Fórizs, E., Debreczenni, M., and Szabolcsy, M. (1981). Serum IgE level and T-cell count in chronic dermatophytosis. *Mykosen 24*: 84-89. (18)

Balogh, É., Mészáros, C., and Halmy, K. (1971). Die Anwendung des Lymphocytentransformationstestes bei der Untersuchung der mykotischen Sensibilization. *Mykosen 14*: 207-211. (19).

Bannikov, E. A., Bakhtilin, V. Ya., Gerasimova, N. M., and Kardakova, T. A. (1981). (On immunological features of paraallergic reactions developing in combined experimental mycosis and dermatitis from dinitrochlorobenzene.) *Vestn. Dermatol. Venerol. No. 1*: 53-57. (20)

Barker, S. A., Basarab, O., and Cruickshank, C. N. D. (1967). Galactomannan peptides of *Trichophyton mentagrophytes. Carbohyd. Res. 3*: 325-332. (21)

Barker, S. A., Cruickshank, C. N. D., and Holden, J. H. (1963). Structure of a galactomannan-peptide allergen from *Trichophyton mentagrophytes. Biochim. Biophys. Acta 74*: 239-246. (22)

Barker, S. A., Cruickshank, C. N. D., Morris, J. H., and Wood, S. R. (1962). The isolation of trichophytic glycopeptide and its structure in relation to the immediate and delayed reactions. *Immunology 5*: 627-632. (23)

Barker, S. A., Stacey, M., and Zweifel, G. (1957). The separation of neutral polysaccharides. *Chem. Ind.* (London) *330*. (24)

Barker, S. A., and Trotter, M. D. (1960). Isolation of purified trichophytin. *Nature* (London) *188*: 232-233. (25)

Barlow, A. J. E., and Chattaway, F. W. (1958). The parasitism of ringworm group of fungi. *Arch. Dermatol. 77*: 399-405. (26)

Basarab, O., How, M. J., and Cruickshank, C. N. D. (1968). Immunological relationships between glycopeptides of *Microsporum canis, Trichophyton rubrum, Trichophyton mentagrophytes* and other fungi. *Sabouraudia 6*: 119-126. (27)

Bazyka, A. P. (1966). (Changes in the reactivity of the organism and in the morphological structures of the affected skin in patients with mycoses of the feet under the effect of fungus immune preparations.) *Vestn. Dermatol. Venerol. 40*: 3-9. (28)

Bazyka, A. P., and Tret'yakova, T. F. (1980). (The serum immunoglobulin content in patients with dermatophytosis.) *Vestn. Dermatol. Venerol. No. 9*: 18-22. (29)

Belukha, U. K., Bidrat, M. S., Rynkova, T. V., Luk'yanova, A. S., and Abidova, Z. M. (1982). (Dynamics of the level of serum immunoglobulins in patients with mycoses of the feet under hot climatic conditions.) *Vestn. Dermatol. Venerol. No. 11*: 66-68. (30)

Beznos, T. I. (1966). (An experience of immunization of animals with antigens prepared from *Trichophyton cultures.) Vestn. Dermatol. Venerol. 40*:14-17. (31)

Biguet, J., Andrieu, S., and Laloux, B. (1965). Les affinités taxonomiques de *Microsporum ferrugineum* appréciées par l'analyse immunoélectrophorétique de sa structure antigénique. *Ann. Parasitol. Hum. Comp. 40*: 365-370. (32)

Biguet, J., Andrieu, S., and Tran Van Ky, P. (1961). Application des méthodes électrophorétiques et immunologiques à l'étude des fractions antigeniques des Dermatophytes. *C. R. Acad. Sci. 253*: 167-169. (33)

Bishop, C. T., Blank, F., and Hranisavljevic-Jakovljevic, M. (1962). The water-soluble polysaccharides of dermatophytes. I. A galactomannan from *Trichophyton granulosum. Can. J. Chem. 40*: 1816-1825. (34)

Bishop, C. T., Perry, M. B., and Blank, F. (1966a). The water-soluble polysaccharide of dermatophytes. V. Galactomannans II from *Trichophyton granulosum, Trichophyton interdigitale, Microsporum quinckeanum, Trichophyton rubrum*, and *Trichophyton schoenleinii. Can. J. Chem. 44*: 2291-2297. (35)

Bishop, C. T., Perry, M. B., Blank, F., and Cooper, F. P. (1965). The water-soluble polysaccharides of dermatophytes. IV. Galactomannans I from *Trichophyton granulosum, Trichophyton interdigitale, Microsprum quinckeanum, Trichophyton rubrum*, and *Trichophyton schoenleinii. Can. J. Chem. 43*: 30-39. (36)

Bishop, C. T., Perry, M. B., Hulyalkar, R. K., and Blank, F. (1966b). The water-soluble polysaccharide of dermatophytes. VI. Glucans from *Trichophyton granulosum, Trichophyton interdigitale, Microsporum quinckeanum, Trichophyton rubrum*, and *Trichophyton schoenleinii. Can. J. Chem. 44*: 2299-2303. (37)

Blamoutier, J. (1980). Scléro-kératite et sensibilization aux dermatophytes. *Rev. Fr. Allergol. Immunol. Clin. 20*: 97. (38)

Blank, F. (1953). The chemical composition of the cell walls of dermatophytes. *Biochim. Biophys. Acta 10*: 110-113. (39)

Blank, F., and Perry, M. B. (1964). The water-soluble polysaccharides of dermatophytes. III. A galactomannan from *Trichophyton interdigitale. Can. J. Chem. 42*: 2862-2871. (40)

Blank, H., Taplin, D., and Zaias, N. (1960). Cutaneous *Trichophyton mentagrophytes* infection in Vietnam. *Arch. Dermatol. 99*: 135-144. (41)

Bloch, B. (1908). Zur Lehre von den Dermatomykosen. *Arch. Dermatol Syph. 93*: 157-220. (42)

Bloch, B. (1928). Allgemeine und experimentelle Biologie der Dermatomykosen und die Trichophytide. In *Handbuch der Haut- und Geschlechtskrankheiten*. Edited by J. Jadassohn, vol. 11. J. Springer, Berlin, pp. 300-376, 564-606. (43)

Bloch, B., Labouchére, A., and Schaaf, F. (1925). Versuche einer chemischen Charakterisierung und Reindarstellung des Trichophytins (des aktiven, antigenen Prinzips pathogener Hautpilze). *Arch. Dermatol. Syph. 148*: 413-424. (44)

Bloch, B., and Massini, R. (1909). Studien über Immunität und Überempfindlichkeit bei Hyphomycetenerkrankungen. *Z. Hyg. Infektionskr. 63*: 68-69. (45)

Blumenthal, F., and Haupt von, A. (1922). Über Vorkommen von Antikörpen im Blutserum trichophytiekranker Menschen. *Dermatol. Z. 36*: 313-336. (46)

Böhm, K. H., and Floerke, I. (1984). Versuche zur immunisierung mit equinen Dermatophyten. *De. Vet. Med. Gesellschaft*: 44-54. (47)

Böhme, H., Barthelmes, H., Grundoff, R., Tausch, I., and Ziegler, H. (1978a). Vergleich verschiedener Antigenpräparate aus Dermatophyten in Immunodoppeldiffusion test. *Dermatol. Monatschr. 164*: 799-803. (48)

Böhme, H., Grundhoff, R., Tausch, I., and Ziegler, H. (1978b). Nachweis präzipitierender Antikörpen gegen Dermatophyten mit Hife der Gegenstromelektrophorese. *Dermatol. Monatschr. 164*: 416-422. (49)

Bonk, A. F., Friedman, L., and Derbes, V. J. (1962). Experimental dermatophytosis. *J. Invest. Dermatol. 39*: 281-286. (50)

Borisova, L. I. (1982). (Effect of temperature of desiccation on the quality of S—P—I vaccine.) *Byulleten' Vsesoyuznogo Nauchno-issledovatel' skogo Instituta Eksperimental' noi Veterinarii No. 48*: 88-89. (51)

Braathen, L. R., and Kaaman, T. (1983). Human epidermal Langerhans cells induce cellular immune response to trichophytin in dermatophytosis. *Br. J. Dermatol. 109*: 295-300. (52)

Brahmi, Z., Liautaud, B., and Marill, F. (1980). Depressed cell-mediated immunity in chronic dermatophytic infections. *Ann. Immunol. 131*: 143-153. (53)

Brocq-Rousseu, D., Urbain, A., and Barotte, J. (1927). Etude des teignes du cheval et de l'immunité dans les teignes expérimentales. *Ann. Inst. Pasteur Paris 41*: 513-551. (54)

Brusilovskaia, D. A. (1970). (Immunoallergic reactions of the organism in mycoses caused by *Trichophyton rubrum* and certain zoophilic fungi.) *Vestn. Dermatol. Venerol. 44*: 47-51. (55)

Burnette, W. N. (1980). "Western blotting": electrophoretic transfer of proteins from sodium dodecyl sulphate-polyacrylamide gels to unmodified nitrocellulose and radiographic detection with antibody and radioiodinated protein A. *Anal. Biochem. 112*: 195-203. (56)

Calderon, R. A., and Hay, R. J. (1984a). Cell-mediated immunity in experimental murine dermatophytosis. I. Temporal aspects of T-suppressor activity caused by *Trichophyton quinckeanum*. *Immunology 53*: 457-464. (57)

Calderon, R. A., and Hay, R. J. (1984b). Cell-mediated immunity in experimental murine dermatophytosis. II. Adoptive transfer of immunity to dermatophyte infection by lymphoid cells from donors with acute or chronic infections. *Immunology 53*: 465-472. (58)

Calvo, M. A., Bruguera, T., Cabañes, F. J., Calvo, R. M., Trape, J., and Abarca, L. (1985). Brief communication: extracellular enzymatic activities of dermatophytes. *Mycopathologia 92*: 19-22. (59)

Carol, W. L. L. (1918). De complementbindingsreactie bij trichophytie. *Ned. Tjdschr. Geneesk. 62*: 784-792 (60)

Catanei, A. (1945). Les effets de l'inoculation intrapéritonéale d'un dermatophyte au cobaye. Existence d'une prémunition d'origine mycosique. *Arch. Inst. Pasteur Algerie 23*: 21-44. (61)

Catanei, A. (1946). Nouvelles reserches expérimentales sur la résistance aux reinfections dans les teignes. Effets des réinoculations successives. *Arch. Inst. Pasteur Algerie 24*: 32-43. (62)

Chattaway, F. W., Ellis, D. A., and Barlow, A. J. E. (1963). Peptides of dermatophytes. *J. Invest. Dermatol. 41*: 31-37. (63)

Christiansen, Å. H., and Svejgaard, E. (1976). Studies of the antigenic structure of *Trichophyton rubrum, T. mentagrophytes, Microsporum canis* and *Epidermophyton floccosum* by crossed immunoelectrophoresis. *Acta Path. Microbiol. Scand. Sect. C. 84*: 337-341. (64)

Citron, J. (1905). Über das Verhalten der Favus und Trichophytonpilze im Organismus. *Z. Hyg. Infektionskr. 49*: 120-134. (65)

Codner, R. C., Cruickshank, C. N. D., Trotter, M. D., and Wood, S. R. (1961).

The production of trichophytin antigen in submerged culture of *T. mentagrophytes*. *Sabouraudia 1*: 116-122. (66)

Collins, J. P., Grappel, S. F., and Blank, F. (1973). Role of keratinases in dermatophytosis. II. Fluorescent antibody studies with keratinase II of *Trichophyton mentagrophytes*. *Dermatologica 146*: 95-100. (67)

Cox, W. A., and Moore, J. A. (1968). Experimental *Trichophyton verrucosum* infections in laboratory animals. *J. Comp. Pathol. 78*: 35-41. (68)

Cruickshank, C. N. D. (1966). Allergy. In *Modern Trends in Dermatology*, Edited by R. M. B. MacKenna, vol. 3. Butterworth, Washington, D.C., pp. 52-83. (69)

Cruickshank, C. N. D., Trotter, M. D., and Wood, S. R. (1960). Studies on trichophytin sensitivity. *J. Invest. Dermatol. 35*: 219-223. (70)

David, J. R., Al-Askari, S., Lawrence, H. S., and Thomas, L. (1964). Delayed hypersensitivity in vitro. The specificity of inhibition of cell migration by antigens. *J. Immunol. 93*: 265-273. (71)

Davies, R. R., Ganderton, M. A., and Savage, M. A. (1983). Human nail dust and precipitating antibodies to *Trichophyton rubrum* in chiropodists. *Clin. Allergy 13*: 309-315. (72)

Davies, R. R., and Zaini, F. (1984a). *Trichophyton rubrum* and the chemotaxis of polymorphonuclear leukocytes. *Sabouraudia: J. Med. Vet. Mycol. 22*: 65-71. (73)

Davies, R. R., and Zaini, F. (1984b). Enzymic activities of *Trichophyton rubrum* and the chemotaxis of polymorphonuclear leukocytes. *Sabouraudia: J. Med. Vet. Mycol. 22*: 235-241. (74)

Davison, F. D., and Mackenzie, D. W. R. (1984). DNA homology studies in the taxonomy of dermatophytes. *Sabouraudia: J. Med. Vet. Mycol. 22*: 117-123. (75)

Davison, F. D., Mackenzie, D. W. R., and Owen, R. (1980). Deoxyribonucleic acid base compositions of dermatophytes. *J. Gen, Microbiol. 118*: 465-470. (76)

Dejima, S. (1969). Studies on the immunologically-active polysaccharide-peptide complexes of a dermatophyte, *Trichophyton mentagrophytes*. *Acta Sch. Med. Gifu 16*: 711-742. (77)

Desai, S. C., Bhat, M. L. A., and Modi, P. J. (1963). Biology of *Trichophyton rubrum* infections. *Ind. J. Med. Res. 51*: 233-243. (78)

Desai, S. C., Khokani, A. C., and Modi, P. J. (1966). Evaluation of trichophytin reactivity in *T. rubrum* infections and its use as an exposure index. *Ind. J. Med. Res. 54*: 148-156. (79)

Desai, S. C., Modi, P. J., and Bhat, M. L. A. (1962). Dysproteinemia in chronic dermatomycosis due to *T. rubrum* with negative anergy to trichophytin. *J. Invest. Dermatol. 31*: 365-367. (80)

Dobromỹsløv, V. V., and Duganova, N. V. (1965a). (Amino acid composition of antigens from dermatophytes.) *Trudỹ Leningr. Khim-farm. Inst. 18*: 48-52. (81)

Dobromỹsløv, V. V., and Duganova, N. V. (1965b). (Comparative polysaccharide composition of antigens from dermatophytes.) *Trudy Leningr. Khimfarm . Inst. 18*: 53-56. (82)

Dokudovski, E. G. (1961). (Vaccination of animals as a method of combating ringworm.) *Veterinariya 38*: 53. (83)

Dostrowsky, A., Kallner, G., Raubitschek, F., and Sagher, F. (1955). Tinea capitis. An epidemiologic, therapeutic, and laboratory investigation of 6,390 cases. *J. Invest. Dermatol. 24*: 195-200. (84)

Drouhet, E., Papachristou-Moratti, A., and Minas, A. (1976). Essays d'identification des dermatophytes par l'immunofluorescence. *Bull. Soc. Fr. Mycol. Med. 5*: 133-136. (85)

Dyson, J. E., Jr., and Landay, M. E. (1963). Differentiation of *Trichophyton rubrum* from *Trichophyton mentagrophytes*. *Mycopathol. Mycol. Appl. 20*: 81-97. (86)

Eleuterio, M. K., Grappel, S. F., Caustic, C. A., and Blank, F. (1973). Role of keratinases in dermatophytosis. III. Demonstration of delayed hypersensitivity to keratinases by the capillary tube migration test. *Dermatologica 147*: 255-260. (87)

Epstein, S., and Grünmandel, S. (1930). Untersuchungen über die spontane Abheilung von oberflächlichen Trichopytien. *Arch. Dermatol. Syph. 161*: 395-428. (88)

Evolceanu, R., Alteras, I., and Nicolau, G. (1965). New data concerning the immunological properties of geophilic dermatophytes. *Mycopathol. Mycol. Appl. 24*: 333-336. (89)

Falk, C. (1948). Source of error in trichophytin test. Response to trichophytin of tuberculin positive subjects, with or without tuberculosis. *Acta Derm. Venereol. 28*: 342-351. (90)

Farid, A., Soliman, R., and Refai, M. (1976). Application of passive haemoagglutination test for detection of antibodies to *Trichophyton verrucosum* in immunized rabbits and naturally infected cattle. *Castellania 4*: 241-242. (91)

Fedotov, V. P. (1981). (Immunosuppressive state of T and B lymphocyte systems in patients with onychomycosis caused by *Trichophyton rubrum*.) *Vestn. Dermatol. Venerol. No. 3*: 30-35. (92)

Fedotov, V. P. (1982). (Justification of clinico-immunological groups in mycosis due to *Trichophyton rubrum* and a method for their determination.) *Vestn. Dermatol. Venerol. No. 10*: 15-20. (93)

Fedotov, V. P. (1984a). (Immunological mechanisms of skin reactions to specific antigens in patients with *Trichophyton rubrum* infection.) *Vestn. Dermatol. Venerol. No. 6*: 17-22. (94)

Fedotov, V. P., Nalaya, V. Ya., and Chetvertushkin, V. V. (1979). (Provocative-basophilic test in the evaluation of allergic reactions in patients with fungal infections.) *Lab-Delo No. 7*: 441-443. (95)

Fedotov, V. P., Syrtsov, V. K., and Sichevoĭ, V. P. (1980). (Estimation of the functional activity of lymphocytes in patients with mycotic infection from the data of autoradiography using ^{3}H-thymidine.). *Vestn. Dermatol. Venerol. No. 7*: 19-23. (96)

Fedotov, V. P., Tsepkova, E. A., and Dyudyun, A. D. (1984b). (Dynamics of lymphocyte and neutrophil rosette formation in patients with microbic eczema and mycosis of the feet.) *Vestn. Dermatol. Venerol. No. 8*: 16-19 (97).

Fiedotow, W. P. (1980). (The T-lymphocyte system in *T. rubrum* infections.) *Przeglad Dermatologiczny 67*: 587-591. (98)

Fielder, R. J., Grappel, S. F., Bishop, C. T., and Blank, F. (1972). The occurrence of pyruvic and ketal in a glucan from a fungus, *Microsporum quinckeanum*. *Can. J. Microbiol. 18*: 701-704. (99)

Fischer, M. E. (1956). Antigenanlytische und tierexperimentelle Untersuchungen zur Mykologie der Erreger der Interdigitalmykosen. *Arch. Klin. Exp. Dermatol. 203*: 270-310. (100)

Flaherty, D. K., and Burrel, R. (1970). Further environmental factors affecting the antigenicity of *Trichophyton rubrum*. *Mycopathol. Mycol. Appl. 42*: 165-175. (101)

Florian, E., Nemeseri, L., and Lovas, G. (1964). Active immunization of calves against ringworm. *Magy. Allatorv. Lapja 19*: 529-530. (102)

Földvári, F. (1935). Undersuchungen uber Komplementbindung bei Pilzkrankheiten. *Dermatol. Z. 70*: 260-265. (103)

Fomin, K. F., and Afans'eva, D. (1981). (Some features of immunological reactivity and its dynamics in patients with zooanthroponotic microsporiosis.) *Vestn. Dermatol. Venerol. No. 7*: 63-64. (104)

Fonseca da, O., and Leao de Area, A. E. (1931). Vaccinotherapia das epidermophyceas e das tinhas tonsurantes. Contribuicao para o estudo das clasovaccinas. *Rev. Med. Cir. Brazil 39*: 269-277. (105)

Friedman, L., and Derbes, V. J. (1960). The question of immunity in ringworm infections. *Ann. N.Y. Acad. Sci. 89*: 178-183. (106)

Fuké, F. (1941a). Studien über Polysaccharide aus Pathogenen Pilzen I. *Mitteilung. J. Biochem. 34*: 39-48. (107)

Fuké, F. (1941b). Studien über Polysaccharide aus Pathogenen Pilzen II. *Mitteilung. J. Biochem. 34*: 251-256. (108)

Fuké, F. (1941c). Studien über Polysaccharide aus Pathogenen Pilzen III. *Mitteilung. J. Biochem. 35*: 259-362. (109)

Fuké, F. (1942). Studien über Polysaccharide aus Pathogenen Pilzen IV. *Mitteilung. J. Biochem. 35*: 263-270. (110)

Gans, O., and Steigleder, G. K. (1957). *Histologie der Hautkrankheiten*, vol. 2, 2nd ed. Springer-Verlag, Berlin-Göttingen-Heidelberg. (111)

Garciá de Lomas, J., Rodriguez, F., Cavas, M. L., López, L., and Altuna, A. (1983). Immunology of dermatophytes. Experimental study. *Mycopathologia 82*: 29-32. (112)

Gorinov, Yu. M. (1978). (Production of the immunogenic preparation TF-130 against ringworm in cattle.) *Byulleten' Vsesoyuznogo Ordena Lenina Instituta Eksperimental' noi Veterinarii 32*: 20-22. (113)

Götz, H. (1962a). Die Trichophytinallergie. In *Handbuch der Haut- und Geschlechtskrankheiten Erganzungswerk*, J. Jadassohn (ed). Springer-Verlag, Berlin-Göttingen-Heidelberg, pp. 123-133. (114)

Götz, H., and Heitmann, H. J. (1967). Über "in-vitro" Reaktionen von Lymphocyten Gegenwart von Trichophytin. *Hauttarzt. 1*: 23-24. (115)

Götz, H., and Pascher, G. (1962b). Über die chemische Zusammensetzung des *Trichophyton mentagrophytes* var. *granulosum, T. schöenleinii* und *T. rubrum* im Vergleich zu Schimmelpilzen (*Penicillium, Cladosporium herbarum* und *Fusarium*). *Dermatologica 124*: 31-41. (116)

Grappel, S. F. (1981). Immunology of surface fungi: Dermatophytes. In *Immunology of Human Infection, Part I: Bacteria, mycoplasmae, Chlamydiae and fungi*, Edited by A. J. Nahmias, and R. J. O'Reilly. Plenum Medical Book Company, New York, pp. 495-524. (117)

Grappel, S. F., Bishop, C. T., and Blank, F. (1974). Immunology of dermatophytes and dermatophytosis. *Bacteriol. Rev. 38*: 222-250. (118)

Grappel, S. F., and Blank, F. (1972a). Role of keratinases in dermatophytosis. I. Immune responses of guinea pigs infected with *Trichophyton mentagrophytes* and guinea pigs immunized with keratinases. *Dermatologica 145*: 245-255. (119)

Grappel, S. F., Blank, F., and Bishop, C. T. (1967). Immunological studies on dermatophytes. I. Serological reactivities of neutral polysaccharides with rabbit antiserum to *Microsporum quinckeanum. J. Bacteriol. 93*: 1001-1008. (120)

Grappel, S. F., Blank, F., and Bishop, C. T. (1968a). Immunological studies on dermatophytes. II. Serological reactivities of mannans prepared from galactomannans I and II of *Microsporum quinckeanum, Trichophyton granulosum, Trichophyton interdigitale, Trichophyton rubrum*, and *Trichophyton schoenleinii. J. Bacteriol. 95*: 1238-1242. (121)

Grappel, S. F., Blank, F., and Bishop, C. T. (1968b). Immunological studies on dermatophytes. III. Further analyses of the reactivities of neutral polysaccharides with rabbit antisera to *Microsporum quinckeanum, Trichophyton schoenleinii, Trichophyton rubrum, Trichophyton interdigitale*, and *Trichophyton granulosum. J. Bacteriol. 96*: 70-75. (122)

Grappel, S. F., Blank, F., and Bishop, C. T. (1969). Immunological studies on dermatophytes. IV. Chemical structures and serological reactivities of polysaccharides from *Microsporum praecox, Trichophyton ferrugineum, Trichophyton sabouraudii*, and *Trichophyton tonsurans. J. Bacteriol. 97*: 23-26. (123)

Grappel, S. F., Blank, F., and Bishop, C. T. (1971a). Circulating antibodies in human favus. *Dermatologica 143*: 271-278. (124)

Grappel, S. F., Blank, F., and Bishop, C. T. (1972b). Circulating antibodies in dermatophytosis. *Dermatologica 144*: 1-11. (125)

Grappel, S. F., Buscavage, C. A., Blank, F., and Bishop, C. T. (1970). Comparative serological reactivities of twenty-seven polysaccharides from nine species of dermatophytes. *Sabouraudia 8*: 116-125. (126)

Grappel, S. F., Fethère, A., and Blank, F. (1971b). Effect of antibodies on growth and structure of *Trichophyton mentagrophytes. Sabouraudia 9*: 50-55. (127)

Green, F., III, Anderson, J. W., and Balish, E. (1980). Cutaneous basophil hypersensitivity after cutaneous *Trichophyton mentagrophytes* infections. *Infect. Immun. 29*: 758-767. (128)

Green, F., III, and Balish, E. (1979). Suppression of in vitro lymphocyte transformation during an experimental dermatophyte infection. *Infect. Immun. 26*: 554-572. (129)

Green, F., III, Weber, J. K., and Balish, E. (1983). The thymus dependency of acquired resistance to *Trichophyton mentagrophytes* dermatophytosis in rats. *J. Invest. Dermatol. 81*: 31-38. (130)

Greenbaum, S. S. (1924). Immunity in ringworm infections. I. Active acquired immunity; with a note on complement fixation tests in superficial ringworm infections. *Arch. Dermatol. 10*: 279-282. (131)

Greenberg, J. H., King, R. D., Krebs, S., and Field, R. (1976). A quantitative dermatophyte infection model in the guinea pig—a parallel to the quantitated human infection model. *J. Invest. Dermatol. 67*: 704-708. (132)

Haan de, P., Raay-Helmen van den, E., and Wikler, J. (1986a). Monoclonal antibodies against fungal (dermatophyte) antigens and their cross-reactions with related and nonrelated antigens. *First International Symposium on Fungal Antigens*, Paris, 17-19 November, Abstract. (133)

Haan de, P., and Wikler, J. (1986b). Antigen diversity in culture filtrates in ethyleneglycol extracts of a dermatophyte, *Trichophyton rubrum*. *First International Symposium on Fungal Antigens*, Paris, 17-19 November, Abstract. (134)

Hanawa, S. (1913). Histologische Untersuchungen über Trichophytie-Heilung und Allergie beim Meerschweinchen. *Dermatol. Wochenschr. 57*: 941–955. (135)

Hanifin, J. M., and Ray, L.F. (1973). Immunological studies in human *Trichophyton* infections. *Clin. Res. 21*: 477. (136)

Hanafin, J. M., Ray, L. F., and Lobitz, W. C., Jr. (1974). Immunological reactivity in dermatophytosis. *Br. J. Dermatol. 90*: 1-8. (137)

Harada, S. (1969). Vaccine therapy of mycosis. *Nippon Ika Daigaku. Zasshi. 36*: 464-465. (138)

Hare, P. J., and Hopkins, M. D. (1953). Amino-acid composition of eight dermatophyte fungi. *Lancet 265*: 1238. (139)

Hauck, H., Skǒrepová, M., Simon, M., Jr., and Djawari, D. (1985). Intracellular killing of *Trichophyton mentagrophytes* microconidia by normal human polymorphonuclear leukocytes. *Arch. Dermatol. Res. 278*: 77-78. (140)

Hay, R. J. (1979). Failure of treatment in chronic dermatophyte infections. *Postgrad. Med. J. 55*: 607-610. (141)

Hay, R. J., and Brostoff, J. (1977). Immune response in patients with chronic *Trichophyton rubrum* infections. *Clin. Exp. Dermatol. 2*: 373-380. (142)

Hay, R. J., and Saeed, E. N. (1981). The immunofluorescence staining of fungi in chronic dermatophyte infections. *Clin. Exp. Dermatol. 6*: 155-158. (143)

Hay, R. J., and Shennan, G. (1982). Chronic dermatophyte infections II. Anti-

body and cell-mediated immune responses. *Br. J. Dermatol. 106*: 191-198. (144)
Hay, R. J., and Shennan, G. (1984). Antibody responses in tinea imbricata: the role of immunoglobulin E. *Trans. Roy. Soc. Trop. Med. Hyg. 78*: 653-655. (145)
Helander, I. (1978a). The lymphocyte transformation test in dermatophytosis. *Mykosen 21*: 71-80. (146)
Helander, I. (1978b). Leukocyte migration agarose test in dermatophytosis. *Mykosen 21*: 167-175. (147)
Henrici, A. T. (1939). Experimental trichophytid in guinea pigs. *Proc. Soc. Exp. Biol. Med. 41*: 349-353. (148)
Henrici, A. T. (1940). Experimental trichophytid. In *Proc. 3rd Int. Cong. Microbiol.* Baltimore, Waverley Press, p. 567-569. (149)
Holden, C. A., Hay, R. J., and MacDonald, D. M. (1980). Localization, in situ, of antigens of cultured and invading *Trichophyton rubrum. Br. J. Dermatol. 103 (Suppl. 18)*: 23-24. (150)
Holden, C. A., Hay, R. J., and MacDonald, D. M. (1981a). The antigenicity of *Trichophyton rubrum*: in situ studies by an immunoperoxidase technique in light and electron microscopy. *Acta Derm. Venereol. 61*: 207-211. (151)
Holden, C. A., Hay, R. J., and MacDonald, D. M. (1981b). A method for identification of dermatophyte antigens in situ by an immunoperoxidase technique in light and electron microscopy. *Clin. Exp. Dermatol. 6*: 311-316. (152)
Honbo, S., Jones, H.E., and Artis, W. M. (1984). Chronic dermatophyte infection: evaluation of the Ig class-specific antibody response reactive with polysaccharide and peptide antigens derived from *Trichophyton mentagrophytes. J. Invest. Dermatol. 82*: 287-290. (153)
Hopper, R. L., Grappel, S. F., and Blank, F. (1975). Antibodies with affinity for epithelial tissue in chronic dermatophytosis. *Dermatologica 151*: 135-143. (154)
How, M. J., Withnall, M. T., and Cruickshank, C. N. D. (1972). Allergenic glucans from dermatophytes. Part I. Isolation, purification and biological properties. *Carbohyd. Res. 25*: 341-353. (155)
How, M. J., Withnall, M. T., and Somers, P. J. (1973). Allergenic glucans from dermatophytes. Part II. Enzymatic degradation. *Carbohyd. Res. 26*: 21-31. (156)
Hunjan, B. S., and Cronholm, L. S. (1979). An animal model for cell-mediated immune response to dermatophytes. *J. Allergy Clin. Immunol. 63*: 361-369. (157)
Hunjan, B. S., Silverman, I. H., Curlovich, K. M., Osuamkpe, C. O., Melvin, J. R., Jr., and Cronholm, L. S. (1981). A simplified in vitro assay of delayed hypersensitivity in diagnosis of dermatomycoses. *J. Allergy Clin. Immunol. 67*: 485-494. (158)
Hunziker, N., and Brun, R. (1980). Lack of delayed reaction in presence of cell-mediated immunity in trichophytin hypersensitivity. *Arch. Dermatol. 116*: 1266-1268. (159)

Huppert, M. (1955). Immunoserological studies of the dermatophytes. *Diss. Abstr. 15*: 1294. (160)

Huppert, M. (1962). Immunization against superficial fungous infection. In *Fungi and Fungous Diseases*, Edited by G. Dalldorf. Charles C. Thomas, Publisher, Springfield, Ill., pp. 239-253. (161)

Huppert, M., and Keeney, E. L. (1959). Immunization against superficial fungous infection. *J. Invest. Dermatol. 32*: 15-19. (162)

Hussin, Z., and Smith, J. M. B. (1980). An indirect fluorescent antibody test for the detection of humoral antibodies in dermatophytosis. *Proc. Univ. Otago Med. School 58*: 14-16. (163)

Ikutomi, M. (1979). (Antigenic analysis of *Trichophyton mentagrophytes* complex.) *Jap. J. Med. Mycol. 20*: 266-272. (164)

Imamura, S., Tanaka, M., and Watanabe, S. (1975). Use of immunofluorescence staining in kerion. *Arch. Dermatol. 111*: 906-909. (165)

Inaba, K. (1961). Experimental studies on hematogenous reinfection with *Trichophyton metagrophytes* var. *asteroides*. VI. Experimental trichophytid (generalized) produced by hematogenous infection with *Trichophyton mentagrophytes* var. *asteroides* as seen from comparative exanthematology. *Bull. Pharm. Res. Inst. Osaka 34*: 1-13. (166)

Ito, K. (1963). Immunologic aspects of superificial fungous diseases. Trichophytin; skin and serological reactions. In *Proc. XII Int. Cong. Dermatol.*, Edited by D. M. Pillsbury, and C. S. Livingwood. Excerpta Medica Foundation, New York, pp. 563-567. (167)

Ito, K., and Kashima, R. (1957). Prolonged observation of immunobiologic follow up in *Trichophyton purpureum* rabbits receiving repeated reinoculations and of immunobiologic relation between *T. purpureum* and *T. asteroides*. *Bull. Pharm. Res. Inst. Osaka 14*: 16-25. (168)

Ito, K., and Kuroda, K. (1962). Experimental study of mycotic epidermal pathology. I. Epidermal pathology of experimental dermatophytid (produced at the ears of rabbits by dorsal cutaneous reinfection with *Trichophyton*). *Bull. Pharm. Res. Inst. Osaka 36*: 7-19, (169)

Ito, K., and Nishitami, M. (1961). Artificial production of systemic basculitis due to repeated sensitization by fungus antigens. *Bull. Pharm. Res. Inst. Osaka 35*: 13-31. (170)

Ito, Y. (1965a). On the immunologically active substances of the dermatophytes. I. *J. Invest. Dermatol. 45*: 275-284. (171)

Ito, Y. (1965b). On the immunologically active substances of the dermatophytes. II. *J. Invest. Dermatol. 45*: 285-294. (172)

Ito, Y. (1967). Immunologically active substances of *Trichophyton*. *Jap. J. Med. Mycol. 8*: 278-283. (173)

Jadassohn, J. (1981). Über die Trichophytien. *Berliner Klin. Wochenschr. 21*: 489-494. (174)

Jadassohn, W. (1927). Beitrag zur Genese der Allergie bei Impfmykosen der Übertritt von Sporen aus dem cutanen Impfhert ins Blut mit Entwicklung von hematogenen Hautmetastasen. *Arch. Dermatol. Syph. 153*: 476–481. (175)

Jadassohn, W. (1962). Delayed hypersensitivity to fungal antigens. In *Proc. 4th Int. Cong. Allergol., N.Y. 1961*, E. A. Brown (ed). MacMillan, New York, pp. 220-226. (176)

Jadassohn, W., Schaaf, F., and Laetsch, W. (1935). Antigen-analytische Untersuchungen an Trichophytinen. *Arch. Dermatol. Syph. 171*: 461-468. (177)

Jadassohn, W., Schaaf, F., and Sulzberger, M. B. (1932). Der Schultz-Dalesche Versuche mit Trichophytin. *Klin. Wochenschr. 20*: 1-10. (178)

Jadassohn, W., Schaaf, F., and Wohler, G. (1937). Analyses of composite antigens by the Schultz-Dale technic: further experimental analyses of trichophytins. *J. Immunol. 32*: 203-227. (179)

Jadassohn, W., and Suter, M. (1951). A propos de la réaction urticarienne immédiate à la trichophytine. *Acta Allergol. 4*: 150-157. (180)

Jaksch, W. (1963). Die Diagnose der Dermatomykosen der Haustiere mittels der allergischen Dermoreaktion. *Int. Vet. Cong.* Hannover, pp. 1247-1250. (181)

Jausion, H., and Sohier, R. (1930). Les claso-vaccines. Vaccinothérapie des dermatomycoses et de leurs séquelles allergiques. *Press. Med. 30*: 621-625. (182)

Jeffries, C. D., Reiss, E., and Ajello, L. (1984). Analytical isoelectric focusing of secreted dermatophyte proteins applied to taxonomic differentiation of *Microsporum* and *Trichophyton* species (preliminary studies). *Sabouraudia: J. Med. Vet. Micol. 22*: 369-379. (183)

Jessner, M., and Hoffmann, H. (1926). Untersuchungen über subcutane *Trichophyton*-inokulationen. *Arch. Dermatol. Syph. 151*: 98-105. (184)

Jillson, O. F., and Huppert, M. (1949). The immediate weal and the 24-48 h tuberculin type edematous reactions to trichophytin. *J. Invest. Dermatol. 12*: 179-185. (185)

Jones, H. E. (1986). Cell-mediated immunity in the immunopathogenesis of dermatophytosis. *Acta Derm. Venereol. Suppl. 121*: 73-83. (186)

Jones, H. E., and Artis, W. M. (1981). Dermatophytosis. *Comprehensive Immunology 7*, Edited by R. A. Good, and S. A. Day. Plenum Publishing Corporation, New York, pp. 514-524. (187)

Jones, H. E., Reinhardt, J. H., and Rinaldi, M. G. (1973). A clinical mycological and immunological survey for dermatophytosis. *Arch. Dermatol. 108*: 61-65. (188)

Jones, H. E., Reinhardt, J. H., and Rinaldi, M. G. (1974). Acquired immunity to dermatophytes. *Arch. Dermatol. 109*: 840-848. (189)

Jones, M. G., and Noble, W. C. (1982). An electrophoretic study of enzymes as a tool in the taxonomy of the dermatophytes. *J. Gen. Microbiol. 128*: 1101-1107. (190)

Kaaman, T. (1978). The clinical significance of cutaneous reactions to trichophytin in dermatophytosis. *Acta Derm. Venereol. 58*: 139-143. (191)

Kaaman, T. (1981a). Cell-mediated reactivity in dermatophytosis: differences in skin responses to purified trichophytin in tinea pedis and tinea cruris. *Acta Derm. Venereol. 61*: 119-123. (192)

Kaaman, T. (1985a). Skin reactivity in atopic patients with dermatophytosis. *Mykosen 28*: 183-190. (193)

Kaaman, T. (1985b). Dermatophyte antigens and cell-mediated immunity in dermatophytosis. In *Current Topics in Medical Mycology*, Edited by M. R. McGinnis. Springer-Verlag, New York, pp. 117-134. (194)

Kaaman, T., Petrini, B., and Wasserman, J. (1979). In vivo and in vitro immune responses to trichophytin in dermatophytosis. *Acta Derm. Venereol. 59*: 222-233. (195)

Kaaman, T., Stedingk von, L. V., Stedingk von, M., and Wasserman, J. (1981b). ELISA-determined serological reactivity against purified trichophytin in dermatophytosis. *Acta Derm. Venereol. 61*: 313-317. (196)

Kaaman, T., Stedingk von, L. V., and Wasserman, J. (1976). An evaluation of delayed hypersensitivity in guinea pigs to various trichophytin preparations. *Acta Derm. Venereol. 56*: 283. (197)

Kaaman, T., and Torssander, J. (1984). Dermatophytid—a misdiagnosed entity? *Acta Derm. Venereol 63*: 404-408. (198)

Kaaman, T., and Wasserman, J. (1981c). Cell-mediated cross-reactivity in vivo and in vitro to purified dermatophyte antigen preparations in sensitized guinea pigs. *Acta Derm. Venereol. 61*: 213-217. (199)

Kallenberg, F. J. (1956). Zum Antikörpenachweis mit der Komplementbindungreaktion im Serum und Liquor von Pilzkranken und gesunden. *Inaug. Dissertation*, Köln. (200)

Kamalan, A., and Thambiah, A. S. (1980). Growth pattern and constituents of dermatophytes in varied substrates. *Mykosen 23*: 141-150. (201)

Kamyszek, F. (1980). (Prevention and treatment of ringworm in cattle with "Trichovac" vaccine.) *Medycyna Weterynaryjna 36*: 436-438. (202)

Kashkin, A. P., and Makovenko, L. V. (1978). (Use of indirect haemoagglutination reaction to detect antibodies to proteolytic enzymes on dermatophytes.) *Vestn. Dermatol. Venerol. No. 6*: 26-30. (203)

Katsitadze, A. G., and Bakhutashvili, V. I. (1984). (Interferon in experimental mycotic infection caused by *Trichophyton rubrum*.) *Vestn. Dermatol. Venerol. No. 9*: 29-32. (204)

Kaufman, L., and Lopez, R. B. (1980). Immunodiffusion studies of morphologically similar dermatophyte species. *Proc. 5th Int. Conf. Mycoses*, PAHO. Washington, D.C., pp. 159-173. (205)

Kaufman, L., and Standard, P. (1978). Immunoidentification of cultures of fungi pathogenic to man. *Curr. Microbiol. 1*: 135-140. (206)

Keeney, E. L., and Ericksen, N. (1949). The chemical isolation and biologic assay of extracellular antigenic fractions from pathogenic fungi. *J. Allergy 20*: 172-184. (207)

Keeney, E. L., and Huppert, M. (1959). Immunization against superficial fungous infection. *J. Invest. Dermatol. 32*: 7-13. (208)

Kielstein, P. (1967a). A study of spontaneous healing of cattle trichophytosis. In *Recent Advances of Human and Animal Mycology*, Edited by L. Chmel. Slovak Acad. Sci., Bratislava, pp. 85-89. (209)

Kielstein, P. (1967b). Zur Immunobiologie der Rindertrichophytie. *Mykosen 10*: 205. (210)
Kielstein, P. (1968). Immunologische Untersuchungen während der spontanen und experimentellen Rindertrichophytie. *Wiss. Z. Karl Marx Univ. Leipzig 17*: 177-178. (211)
Kielstein, P. (1971). Zur Prophylaxe und Bekämpfung der Rindertrichophytie im Jungtier-Grossbeständen. *Fortpfl. Aufz. Haustiere 7*: 46-56. (212)
Kielstein, P., and Erler, W. (1966). Papierchromatographische Untersuchungen serologisch reagierender Polysaccharide einiger Dermatophyten. *Mykosen 9*: 4-10. (213)
Kielstein, P., and Richter, W. (1970a). Versuche zur Immunoprophylaxe der Rindertrichophytie. *Arch. Exp. Vet. Med. 24*: 1205-1218. (214)
Kielstein, P., and Richter, W. (1970b). Zur Prophylaxe der Rindertrichophytie in Grossbeständen durch aktive Immunisierung (Vorläufige, Mitteilung). *Montash. Vet. Med. 25*: 334-337. (215)
Kitajima, Y., and Nozawa, Y. (1975). Isolation, ultrastructure and chemical composition of the outmost layer ("exo-layer") of the *Epidermophyton floccosum* cell wall. *Biochim. Biophys. Acta 394*: 558-568. (216)
Kitamura, J. (1962). An experimental study of the antigenicity of crude fractions (obtained from *Trichophyton mentagrophytes* var. *asteroides*, mechanically disintegrated). I. Histopathologic observation of the skin allergy evoked by erude fractions (obtained from *Trichophyton mentagrophytes* var. *asteroides*, mechanically disintegrated). *Bull. Pharm. Res. Inst. Osaka 40*: 1-19. (217)
Kitamura, J. (1963). Chemical analysis of partially "purified" fraction of a *Trichophyton* species (obtained from *Trichophyton mentagrophytes* var. *asteroides*, mechanically disintegrated). *Bull. Pharm. Res. Inst. Osaka 44*: 1-7. (218)
Kligman, A. M., and Ginsberg, D. (1950a). Immunity of the adult scalp to infection with *Microsporum audouinii. J. Invest. Dermatol. 14*: 345-356. (219)
Kligman, A. M., and Lamater de, E. D. (1950b). The immunology of the human mycoses. *Annu. Rev. Microbiol. 4*: 283-312. (220)
Kocik, T. (1981). (Pathogenesis, immunogenic properties and antigenic structure of domestic strains of *Trichophyton verrucosum*). *Pol. Arch. Weterynaryjna 23*: 17-30. (221)
Kogoj, F. (1926). Experimentelle Beiträge zur Lehre von den Dermatomykosen mit besonderer Berücksichtigung der Lokalisationsbestimmung hämatogener Infektionen. *Arch. Dermatol. Syph. 150*: 333-355. (222)
Kogoj, F. (1928). Hämatogene Dermatomykosen nach subcutaner und intramuskularer Infektion. *Arch. Dermatol. Syph. 154*: 463-475. (223)
Kolmer, J. A., and Strickler, A. (1915). Complement fixation in parasitic skin diseases. *JAMA 64*: 800-804. (224)
Korchma, N. L. (1982). (Haemocoagulation indices in patients with eczema and mycosis of the feet with allergic manifestations.) *Vestn. Dermatol. Venerol. No. 4*; 48-51. (225)

Kotrajaras, O. (1969). Pathogenesis of dermatophytosis in Bangkok, Thailand. *Dermatol. Int. 8*: 63-65. (226)

Krebs, S., Greenberg, J., and Jesrani, K. (1977). Temporal correlation of lymphocyte blastogenesis, skin test responses and erythema during dermatophyte infections. *Clin. Exp. Immunol. 27*: 526-530. (227)

Krempl-Lamprecht, L., Krempl, H., and Nohel, S. (1984). Enzymproteinmuster als Merkmal zur Beurteilung von Art und Verwandtschaft bei Dermatophyten. *Mykosen 27*: 279-283. (228)

Kull, E., Brun, R., and Jadassohn, W. (1951). A propos de trichophytines sèches. *Int. Arch. Allergy Appl. Immunol. 2*: 252-259. (229)

Kuroda, K. (1953). Serological studies of experimental trichophytosis. I. Serological reactions using a mechanically prepared antigen. *Ann. Tuberc. 4*: 15-19. (230)

Kuroda, K. (1958). Serological studies of experimental trichophytosis. IX. Complement fixation and precipitating reactions in patient with dermatomycosis using polysaccharide from *T. asteroides* mechanically disintegrated as antigen. *Bull. Pharm. Res. Inst. Osaka 15*: 5-8. (231)

Kuroda, K. (1959). Experimental studies on fungus granuloma. VI. Fungus granuloma, fungus "id" reaction experimentally produced. *Bull. Pharm. Res. Inst. Osaka 21*: 7-18. (232)

Kusonoki, F. (1913). Experimentelle und klinische Studien zur Lehre der Dermatomykosen (Infektion, Prophylaxe, Immunität). *Arch. Dermatol. Syph. 114*: 1-88. (233)

Lamater de, E. D. (1941). Experimental studies with the dermatophytes. III. Development and duration of immunity and hypersensitivity in guinea pigs. *J. Invest. Dermatol. 4*: 143-158. (234)

Lamater de, E. D. (1942). Experimental studies with the dermatophytes. IV. The influence of age upon the allergic response in experimental ringworm in the guinea pigs. *J. Invest. Dermatol. 5*: 423-429. (235)

Lamater de, E. D., and Benham, R. W. (1938). Experimental studies with the dermatophytes. II. Immunity and hypersensitivity produced in laboratory animals. *J. Invest. Dermatol. 1*: 469-488. (236)

Leao de Area, A. E. (1940). Fisiopatogenia das tinhas. *Acta Med. Rio de Janeiro 5*: 283-299. (237)

Lepper, A. W. D. (1969). Immunological aspects of dermatomycoses in animals and man. *Rev. Med. Vet. Mycol. 6*: 435-446. (238)

Lepper, A. W. D. (1972). Experimental bovine *Trichophyton verrucosum* infection. Preliminary clinical, immunological, and histological observations in primarily infected and reinoculated cattle. *Rev. Vet. Sci. 13*: 105-115. (239)

Lestschenko, W. M., and Fedotow, W. P. (1982). Defizit der Immunesysteme als eine der Ursachen der von Dermatophyten verursachten Onychomykose. *Mykosen 25*: 237-240. (240)

Lestschenko, W. M., and Fedotow, W. P. (1986). Die Entwicklungsmechanismen der cutanen Immunantwort bei der *Trichophyton rubrum* infektion. *Mykosen 29*: 329-336. (241)

Lewis, G. M., and Hopper, M. E. (1937). Ringworm of the scalp. IV. a) Comparative reactions to cutaneous tests with trichophytin in children with and without ringworm of the scalp; b) evaluation of therapy with stock vaccines in types of infection resistant to treatment. *Arch. Dermatol. 36*: 821-832. (242)

Lewis, G. M., and Hopper, M. E. (1948). *An introductory to medical mycology*, 3rd ed. The Year Book Publishers, Chicago. (243)

Lewis, G. M., Hopper, M. E., Wilson, J. W., and Plunkett, O. A. (1958). *An introductory to medical mycology*, 4th ed. The Year Book Publishers, Chicago. (244)

Lombardo, C. (1911). Ricerche sulla ipersensibilità ed immunità in alcune dermatomicosi. *G. Ital. Mal. Vener. Pelle 52*: 70-88. (245)

Longbottom, J. L., and Pepys, J. (1964). Pulmonary aspergillosis: diagnostic and immunological significance of antigens and C-substances in *Aspergillus fumigatus*. *J. Pathol. Bacteriol. 88*: 141-151. (246)

Longhin, S., and Olaru, V. (1970). The value of immunotherapy in dermatomycosis. In *Proc. 2nd Int. Symp. Med. Mycol.*, Edited by W. Sowinski. Poznam, Poland, pp. 181-183. (247)

Marcussen, P. V. (1937). Relationship of the urticarial to inflammatory reaction to trichophytin. *Arch. Dermatol. 36*: 494-514. (248)

Martenstein, H. (1921). Experimentelle Beiträge zur Frage der Uberempfindlichkeit des Meerschweinchens nach überstandener Trichophytie. *Arch. Dermatol. Syph. 131*: 180-192. (249)

Marteinstein, H. (1922). Über die *Achorion Quinckeanum*-Erkrankung der graviden Meerschweinchen und über das Verhalten ihrer Nachkommen zur Allergie. *Arch. Dermatol. Syph. 140*: 329-335. (250)

Martenstein, H. (1923). Weitere experimentelle Untersuchungen über die Allergie des Meerschweinchens nach der Impfung mit *Achorion Quinckeanum*. *Arch. Dermatol. Syph. 142*: 279-304. (251)

Martin, S. M. (1958). Production of extracellular polysaccharides by *Mucor racemosus*. *Can. J. Microbiol. 4*: 317-319. (252)

Martin, S. M., and Adams, G. A. (1956). A survey of fungal polysaccharides. *Can. J. Microbiol. 2*: 715-721. (253)

Matsumoto, T., Padhye, A. A., and Ajello, L. (1983). Successful mating of *Microsporum distortum* with *Nannizia otae*. *Trans. Br. Mycol. Soc. 81*: 645-650. (254)

McNall, E. G., Sternberg, T. H., Newcomer, V. D., and Sorensen, L. J. (1961). Chemical and immunological studies on dermatophyte cell wall polysaccharides. *J. Invest. Dermatol. 36*: 155-157. (255)

Medvedev, Yu. A., and Chistyakova, E. V. (1981). (Role of immune responsiveness of the organism in the formation of superficial and infiltrative zooanthroponotic trichophytosis.) *Vestn. Dermatol. Venerol. No. 11*: 39-41. (256)

Medvedev, Yu. A., Chistyakova, E. V., and Yarmukhamedova, A. G. (1984). (Immunological factors in superficial infiltrative and suppurative forms of zooanthroponotic trichophytosis.) *Vestn. Dermatol. Venerol. No. 6*: 62-64. (257)

Merkel, M. (1957). The chemical and antigenic structure of *Trichophyton gypseum. Bull. Acad. Pol. Sci. Cl. 2, 5*: 341-345. (258)

Merkel, M. (1958). Investigations of the chemical structure of the genus *Achorion*. I. Chemical structure of *Achorion schoenleinii. Bull. Acad. Pol. Sci. Cl. 2, 6*: 417-422. (259)

Merkel, M. (1959). Investigations of the chemical structure of the genus *Achorion*. II. Chemical structure of *Achorion quinckeanum. Bull. Acad. Pol. Sci. Cl. 2, 7*: 501-505. (260)

Meyer, J., Sartori, R., and Touillier, J. (1952). Sur la nature proteique des trichophytines et leur composition en amino-acides. *C. R. Acad. Sci. 234*: 2224-2226. (261)

Miller, H. E., Stewart, R. A., and Kimura, F. (1941). "Undernatured trichophytin." Preparation and clinical application. *Arch. Dermatol. 44*: 804-815. (262)

Minocha, Y., Pastricha, J. S., Mohapatra, L. N., and Kandhari, K. C. (1972). Proteolytic activity of dermatophytes and its role in the pathogenesis of skin lesions. *Sabouraudia 10*: 79-85. (263)

Miura, T. (1963). Studies on dermatophytes by means of the fluorescent antibody technique. *Jap. J. Dermatol. Ser. B 73*: 338-354. (264)

Miura, T. (1968). Clinical application of the fluorescent antibody technic to dermatophytosis. *Tohoku J. Exp. Med. 96*: 165-170. (265)

Miura, T., and Kasai, T. (1967). Immunofluorescent studies on *Trichophyton asteroides* compared with *Trichophyton interdigitale. Tohoku J. Exp. Med. 93*: 49-55. (266)

Montclos de, H., and Guinet, R. M. F. (1982). Antigenic structure of dermatophytes as demonstrated by quantitative immunoelectrophoresis. *Mykosen 25*: 705-710. (267)

Montgomery, H. (1967). *Dermopathology*, vol. I. Harper and Row, New York. (268)

Morace, G., Amalfitano, G., and Polonelli, L. (1986). Serotyping of fungal isolates by precipitating monoclonal antibodies. *Mycopathologia 94*: 53-57. (269)

Moser, S. A., and Pollack, J. D. (1978). Isolation of glycopeptides with skin test activity from dermatophytes. *Infect. Immun. 19*: 1031-1046. (270)

Nathan, E. (1920). Zur Kenntnis der Immunitätsvorgänge bei der Trichophytie des Menschen. *Dermatol. Wochenschr. 71*: 439-448. (271)

Neering, H., and Horst van der, J. C. (1980). Immunologische aspecten van huidinfecties door *Trichophyton rubrum. Nederlands Tijdschrift voor Geneeskunde 124*: 996-999. (272)

Neisser, A. (1902). Plato's Versuche über die Erstellung und Verwendung von "Trichophytin." *Arch. Dermatol. Syph. 60*: 63-76. (273)

Neves, H. (1962). Trichophytin reaction in natural and experimental dermatophytosis and tuberculosis. *Sabouraudia 1*: 197-202. (274)

Nielsen, P. G. (1983). The prevalence of dermatophyte infections in hereditary palmo-plantar keratoderma. *Acta Derm. Venereol. 63*: 439-441. (275)

Nielsen, P. G. (1984). Immunological aspects of dermatophyte infections in hereditary palmo-plantar keratoderma. *Acta Derm. Venereol. 64*: 296-301. (276)

Nikiforov, L. I. (1978). (Immunity to ringworm in furbearing animals.) *Byulleten' Vsesoyuznogo Ordena Lenina Instituta Eksperimental'noi Veterinarii No. 31*: 27-28. (277)

Noguchi, T., Banno, Y., Watanabe, T., Nozawa, Y., and Ito, Y. (1975). Carbohydrate composition of the isolated cell walls of dermatophytes. *Mycopathologia 55*: 71-76. (278)

Noguchi, T., Hattori, T., Shimonaka, H., and Ito, Y. (1971a). Immunochemical studies of *Trichophyton mentagrophytes*. Isolation of immunoglobulins and their immune responses. *Jap. J. Exp. Med. 41*: 401-410. (279)

Noguchi, T., Kitazima, Y., Nozawa, Y., and Ito, Y. (1971b). Isolation, composition and structure of cell walls of *Trichophyton mentagrophytes*. *Arch. Biochem. Biophys. 146*: 506-512. (280)

Noskov, A. I. (1959). (Fungal antigen for treatment of ringworm in cattle.) *Trudÿ Vses. Nauchno-issled. Inst. Vet. Sanit. Ektoparazit. 14*: 86-88. (281)

Novick, N. L., Tapia, L., and Bottone, E. J. (1987). Invasive *Trichophyton rubrum* infection in an immunocompromised host. *Am. J. Med, 82*: 321-325. (282)

Nozawa, Y., Kitajima, Y., and Ito, Y. (1973). Chemical and ultrastructural studies of isolated cell walls of *Epidermophyton floccosum*. Presence of chitin inferred from x-ray diffraction analysis and electron microscopy. *Biochim. Biophys. Acta 307*: 92-103. (283)

Nozawa, Y., Noguchi, T., Ito, Y., Sudo, N., and Watanabe, S. (1971). Immunochemical studies on *Trichophyton mentagrophytes*. *Sabouraudia 9*: 129-138. (284)

Nozawa, Y., Noguchi, T., Uesaka, H., and Ito, Y. (1970). Studies on the immunologically active substances of the dermatophytes. Enzymatic digestion of polysaccharide-peptide complexes isolated from *Trichophyton mentagrophytes* and their immunochemical properties. *Jap. J. Med. Mycol. 11*: 159-165. (285)

Nozawa, Y., Noguchi, T., Uesaka, H., Hattori, T.,and Ito, Y. (1968). Studies on immunologically active substances of *Trichophyton mentagrophytes*. *Jap. J. Med. Mycol. 9*: 258-266. (286)

O'Sullivan, J., and Mathison, G. E. (1971). The localization and secretion of a proteolytic enzyme complex by the dermatophytic fungus *Microsporum canis*. *J. Gen. Microbiol. 68*: 319-326. (287)

Otcenasek, M., Tomsikova, A., and Novackova, D. (1975). Die Möglichkeit der serologischen Unterscheidung von *Trichophyton rubrum* Varietäten. *Dermatol. Monatschr. 161*: 196-200. (288)

Ottaviano, P. J., Jones, H. E., Jaeger, J., King, R. D., and Bibel, D. (1974). Trichophytin extraction: biological comparison of trichophytin extracted from *Trichophyton mentagrophytes* grown in a complex medium and a defined medium. *Appl. Environ. Microbiol. 28*: 271-275. (289)

Papini, M., and Simonetti, S. (1985). Humoral immunofluorescent antibodies in subjects with dermatophytes. *Mykosen 28*: 419-429. (290)

Peck, S. M. (1950). Fungus antigens and their importance as sensitizers in the general population. *Ann. N.Y. Acad. Sci. 50*: 1362-1375. (291)

Peck, S. M., Glick, A., and Weissbard, E. (1941). Trichophytin. II. The apparent separation of the skin reactive factor from the therapeutic principle of trichophytin. *Arch. Dermatol. Syph. 44*: 816-836. (292)

Peck, S. M., Osserman, K. E., and Rule, A. H. (1972). Intercellular antibodies: presence in a *Trichophyton rubrum* infection. *J. Invest. Dermatol. 58*: 133-138. (293)

Peck, S. M., Siegal, S., and Bergamini, R. (1947). Immunologic relationships of the antibiotics and trichophytin. Clinical observations and animal experiments. *J. Invest. Dermatol. 9*: 165-185. (294)

Pecori, G. (1914). *Sulla immunità nella tricofizia.* Coop. Tip. Manuzia, Roma. (295)

Pepys, J. (1959). Allergic hypersensitivity to fungi. *Postgrad. Med. J. 35*: 436-440. (296)

Petrini, B., and Kaaman, T. (1981). T-lymphocyte subpopulations in patients with chronic dermatophytosis. *Int. Arch. Allergy Appl. Immunol. 66*: 105-109. (297)

Petrovic, S. V. (1978). (The complement fixation test on horses with ringworm.) *Byulleten, Vsesoyuznogo Ordena Lenina Instituta Eksperimental'noi Veterinarii No. 32*: 16-17. (298)

Petrovic, S. V., Golovina, N. P., Ivanova, L. G., and Polyakov, I. D. (1980). (LTF-130 vaccine produced in the USA with Soviet permission.) *Veterinariya, Moskow USSR No. 9*: 35-37. (299)

Petrovic, S. V., Polyakov, I. D., Runova, V. F., and Polyakova, T. M. (1982). (Allergic reactions in dermatomycoses of horses.) *Veterinariya, Moskow USSR No. 10*: 22-23. (300)

Philpot, C. M. (1978). Serological differences among the dermatophytes. *Sabouraudia 16*: 247-256. (301)

Pipkin, Ya. S., Fedotov, V. P., Logunov, V. P., Lesnitskii, A. I., and Mattvienko, D. A. (1980). (Serum immunoglobulins in patients with dermatomycosis pyoderma and microbial eczema.) *Vestn. Dermatol. Venerol. No. 1*: 42-46. (302)

Pohler, H. (1976). Der Makrophegen-Stimulations-Test (MST)—Ein neuer in vitro-Test zum Nachweiss der mykotischen Sensibilization. *Dermatol. Monatschr. 162*: 104-107. (303)

Polonelli, L., Castagnola, M., and Morace, G. (1986). Identification and serotyping of *Microsporum canis* isolates by monoclonal antibodies. *J. Clin. Microbiol. 23*: 609-615. (304)

Polonelli, L., and Morace, G. (1984). Rapid immunoidentification of pathogenic fungi. In *New Horizons in Microbiology*, Edited by A. Sanna, and G. Morace Elsevier Sciences Publishers, Amsterdam, pp. 203-219. (305)

Polonelli, L., and Morace, G. (1985a). Antigenic characterization of *Microsporum canis, M. distortum, M. equinum, M. ferrugineum* and *Trichophyton soudanense* cultures. *Mycopathologia 92*: 7-10. (306)

Polonelli, L., and Morace, G. (1985b). Serological analysis of dermatophyte isolates with monoclonal antibodies produced against *Microsporum canis*. *J. Clin. Microbiol. 21*: 138-139. (307)

Polyakov, I. D. (1981a). (Activity of allergens from dermatophytes.) *Veterinariya, Moskow USSR No. 9*: 37-39. (308)

Polyakov, I. D. (1981b). (Demonstration of delayed-type hypersensitivity to allergens from spores, mycelium and products of metabolism of *Trichophyton equinum.*) *Byulletin, Vsesoyuznogo Instituta Eksperimental'noi Veterinarii No. 38*: 58-60. (309)

Polyakov, I. D. (1981c). (Comparative estimation of allergens prepared from dermatophyte cultures.) *Byulleten' Vsesoyuznogo Institute Eksperimental'noi Veterinarii No. 38*: 61-63. (310)

Rasulev, Sh. T., and Turdiev, Sh. (1975). (Immunological relationship between *Trichophyton verrucosum* and *T. mentagrophytes*.) *Veterinariya, Moskow USSR No. 11*: 33-35. (311)

Refai, M., Shalaby, M. A., and Saber, M. S. (1977). Studies on the physical, chemical and serological properties of haemagglutinins of dermatophytes. *Mykosen 20*: 198-202. (312)

Refai, M., Soliman, R., Attia, S., and El-Far, F. (1985). Studies on haemoagglutinins of *Microsporum gypseum. Mykosen 28*: 85-89. (313)

Refai, M., Soliman, R., and Farid, A. (1976). Application of the fluorescent antibody technique (FAT) for detection of antibodies to *Trichophyton verrucosum* in immunized rabbits and naturally infected cattle. *Castellania 4*: 197-199. (314)

Reiss, E. (1986). Dermatophytes. In *Molecular Immunology of Mycotic and Actinomycotic Infections*. Edited by E. Reiss. Elsevier, New York, Amsterdam, London, pp. 157-175. (315)

Reiss, F. (1944). Successful inoculation of animals with *Trichophyton rubrum*: course of disease and immunologic and histologic features. *Arch. Dermatol. Syph. 49*: 242-246. (316)

Reiss, F., and Leonard, L. (1955). Experimental *Microsporum lanosum* infection in dogs, cats and rabbits. II. Studies on the course of the infection. *J. Invest. Dermatol. 24*: 589-594. (317)

Reiss, F., and Leonard, L. (1956). Failure of active immunization against *Trichophyton gypseum* infection in guinea pigs. *J. Invest. Dermatol. 26*: 449-452. (318)

Reyes, A. C., and Friedman, L. (1966). Concerning the specificity of dermatophyte-reacting antibody in human and experimental animal sera. *J. Invest. Dermatol. 47*: 27-34. (319)

Rippon, J. W. (1968a). Extracellular collagenase from *Trichophyton schoenleinii. J. Bacteriol. 95*: 43-46. (320)

Rippon, J. W., and Garber, E. D. (1969). Dermatophyte pathogenicity as a

function of mating type and associated enzymes. *J. Invest. Dermatol. 53*: 445-448. (321)
Rippon, J. W., and Varadi, V. P. (1968b). The elastases of pathogenic fungi and actinomycetes. *J. Invest. Dermatol. 50*: 54-58. (322)
Rivalier, E. (1929). Recherches expérimentales sur l'allergie et l'immunité trichophytiques. *Ann. Dermatol. Syphiligr. 10*: 618-640. (323)
Rivalier, E. (1960). Phénomènes d'immunité dans les dermatophyties. *Pathol. Biol. 8*: 307-317. (324)
Rogachefsky, H., Rosenthal, S. A., Litt, J. Z., and Baer, R. L. (1956). Trichophytid reaction after exposure to dermatophytes. *AMA Arch. Dermatol. 74*: 606-609. (325)
Rogers, A. L. (1968). A study of the relationships of selected dermatophytes using subcellular fractions as antigens in immunodiffusion techniques. *Dis. Abstr. 28 (12B)*: 4894. (326)
Rotermund, H. (1980). LTF-130—eine wirksame Vakzine gegen Rindertrichophytie. *Monatsh. Vet. Med. 35*: 334-335. (327)
Rubins, A. Ya. (1981). (The status of some indices of immunity and nonspecific reactions in patients with rubromycosis during therapy.) *Vestn. Dermatol. Venerol. No. 2*: 65-70. (328)
Rzucidlo, L, Stachow, A., Nowakovska, A., and Kubica, J. (1958). Chemical and biological properties of cell walls of *Candida krusei, Trichophyton gypseum* and *Penicillium notatum. Bull. Acad. Pol. Sci. Cl., II Ser. Sci. Biol. 6*: 15-20. (329)
Sabouraud, R. (1910). Maladies du cuir chevelu. III. Les maladies cryptogamiques. *Les teignes*, Masson et Cie, Paris. (330)
Saeves, J. (1915). Experimentelle Beiträge zur Dermatomykosenlehre. *Arch. Dermatol. Syph. 121*: 161-236. (331)
Saferstein, H. L., Strachn, A. A., Blank, F., and Bishop, C. T. (1968). Trichophytin activity and polysaccharides. *Dermatologica 136*: 151-154. (332)
Sagara, J. (1968). Immunological studies on dermatophytes: Antigenic relationships between *Trichophyton interdigitale* and *Trichophyton asteroides. Tohoku Med. J. 78*: 103-113. (333)
Salvin, S. B. (1963). Immunologic aspects of the mycoses. *Progr. Allergy 7*: 213-331. (334)
Sampei, H. (1950). (On the antigenicity of trichophytin.) *Jap. J. Dermatol. 60*: 617. (335)
Sanchez de, T. I., and Mackenzie, D. W. R. (1983). Exoantigens of dermatophytes. *Sabouraudia 21*: 159-161. (336)
Sanke, R. (1979). Untersuchungen über Infektionsablauf und Immunitätsverhältnisse bei der experimentellen Trichophytie von Kaninchen und Ratte. *Inaugural Dissertation*, Hannover, Federal German Republic; Tierärztliche Hochschule, p. 105. (337)
Sarkisov, A., Petrovich, Kh. S. V., Nikiforov, L. I., Yablochnik, L. M., and Korolov, V. P. (1971). (Immunization of cattle against ringworm.) *Veterinariya 47*: 54-56. (338)

Schetsiruli, L. T. (1971). Klinish-experimentelles Studium des spezifischen Reaktivitätszustandes des Organismus bei durch *Trichophyton* (*Epidermophyton*) *rubrum* hervorgerufenen Mykosen. *Mykosen 14*: 429-432. (339)
Seeliger, H. P. R. (1958). *Mykologische Serodiagnostik*. J. A. Barth, Leipzig. (340)
Seeliger, H. P. R. (1965). Standardization and assay of skin test antigens for mycotic diseases. *Progress in Immunobiological Standards*, vol. 2. S. Karger, Basel and New York, pp. 154-163. (341)
Sei, Y., and Takiuchi, I. (1982). (Studies on the effect of keratinase on the keratinized structure and the localization of keratinase.) *Jap. J. Med. Mycol. 23*: 308-313. (342)
Shah, V. K., and Knight, S. C. (1968). Chemical composition of hyphal walls of dermatophytes. *Arch. Biochem. Biophys. 127*: 229-234. (343)
Sharp, W. B. (1941). Extraction of antigen from molds. *J. Invest. Dermatol. 4*: 205-217. (344)
Sharp, W. B. (1945). Serologic relationship among the dermatophytes. *Texas Rep. Biol. Med. 3*: 159-169. (345)
Shechter, Y., Landau, J. W., Dabrowa, N., and Newcomer, V. D. (1966). Comparative disc electrophoretic studies of protein from dermatophytes. *Sabouraudia 5*: 144-149. (346)
Shechter, Y., Landau, J. W., Dabrowa, N., and Newcomer, V. D. (1968). Disc electrophoretic studies of intraspecific variability of proteins from dermatophytes. *Sabouraudia 6*: 133-137. (347)
Sherwin, W. K., Ross, T. H., Rosenthal, C. M., and Petrozzi, J. W. (1979). An immunosuppressive serum factor in widespread cutaneous dermatophytosis. *Arch. Dermatol. 115*: 600-604. (348)
Shukla, N. P., and Agarwal, G. P. (1983). Antigenic relationship between down and granular forms of *Trichophyton mentagrophytes* and *T. rubrum*. *Microbiol. Immunol. 27*: 311-314. (349)
Skŏrepovà, M., Hauch, H., Hornestein, O. P., Simon, M., Jr., Djawari, D., and Freymuller, G. (1985). Phagocytosis and intracellular killing of *Trichophyton mentagrophytes* microconidia by polymorphonuclear leukocytes in patients with tinea infections and controls. *Mykosen 28*: 485-489. (350)
Sohnle, P. G., Collins-Lech, C., and Huhta, K. E. (1983). Cross-specific antibodies in young and aged humans against organisms producing superficial fungal infections. *Br. J. Dermatol. 108*: 69-70. (351)
Sorensen, G. W., and Jones, H. E. (1976). Immediate and delayed hypersensitivity in chronic dermatophytosis. *Arch. Dermatol. 112*: 40-42. (352)
Stahl, D., and Svejgaard, E. (1982). Lymphocyte transformation in vitro in acute dermatophytosis: a follow-up study. *Acta Derm. Venereol. 62*: 289-293. (353)
Stankushev, H., Duparinova, L., Kostov, G., and Gradinarski, I. (1979). (Comparative immunological investigations and assessment of the effectiveness of the Russian "LTF-130" ringworm vaccine in cattle.) *Veterinarnomedisinski Nauki 16*: 67-73. (354)

Stepanova, Zh. V. (1980). (Lymphocyte blast transformation test in patients with chronic mycoses.) *Vestn. Dermatol. Venerol. No. 9*: 14-16. (355)
Strickler, A. (1915). The vaccine treatment of ringworm of the scalp. *JAMA 65*: 224-227. (356)
Stuka, A. J., and Burrell, R. (1967). Factors affecting the antigenicity of *Trichophyton rubrum. J. Bacteriol. 94*: 914-918. (357)
Sulzberger, M. B. (1928). The pathogenesis of trichophytids. The spontaneous passage of formed elements (spores) from the primary lesion into the circulating blood. *Arch. Dermatol. 18*: 891-901. (358)
Sulzberger, M. B. (1932). Experiments in passive transference of urticarial hypersensitiveness to fungous extracts. (A) Immunological grouping of Trichophytons and Monilia. (B) Refractoriness of certain individuals to sensitization by reagins. *J. Immunol. 23*: 73-79. (359)
Sulzberger, M. B. (1940). *Dermatologic Allergy*. Charles C. Thomas Publisher, Springfield, IL. (360)
Sulzberger, M. B., and Kerr, P. S. (1930). *Trichophyton* hypersensitiveness of urticarial type, with circulating antibodies and passive transference. *J. Allergy 2*: 11-16. (361)
Sundaram, B. M., Sirsi, M., and Rao, G. R. (1977). Antigenicity and cross-reactivity in *Epidermophyton floccosum. Curr. Sci. 46*: 867-868. (362)
Suter, E., and Ramseier, H. (1964). Cellular reactions in infection. *Adv. Immunol. 4*: 117-173. (363)
Sutter, E. (1917). Weitere Beiträge zur Lehre von der Immunität und Überempfindlichkeit bei Trichophytieerkrankungen. *Dermatol. Z. 24*: 65-99. (364)
Svejgaard, E. (1985a). Immunologic investigations of dermatophytes. *Seminars in Dermatology 4*: 201-221. (365)
Svejgaard, E. (1986a). Humoral antibody responses in immunopathogenesis of dermatophytosis. *Acta Derm. Venereol. Suppl. 121*: 85-89. (366)
Svejgaard, E. (1986b). Recent trends in the immunology of dermatophytosis. *Microbiol. Sci. 3*: 154-159. (367)
Svejgaard, E. (1986c). Immunologic properties of a fraction of *Trichophyton rubrum* with affinity to concanavalin A. *J. Med. Vet. Mycol. 24*: 271-280. (368)
Svejgaard, E., and Christiansen, Å. H. (1979). Precipitating antibodies in dermatophytosis demonstrated by crossed immunoelectrophoresis. *Acta Path. Microbiol. Scand. Sect. C 87*: 23-27. (369)
Svejgaard, E., Christiansen, Å. H., Stahl, D., and Thomsen, K. (1984). Clinical and immunological studies in chronic dermatophytosis caused by *Trichophyton rubrum. Acta Derm. Venereol. 64*: 493-500. (370)
Svejgaard, E., Jakobsen, B., and Svejgaard, A. (1983). HLA studies in chronic dermatophytosis caused by *Trichophyton rubrum. Acta Derm. Venereol. 63*: 254-255. (371)
Svejgaard, E., and Lowenstein, H. (1985b). *Trichophyton rubrum* specific IgE in serum in patients with chronic *T. rubrum* infection as demonstrated by crossed radioimmunoelectrophoresis. *Acta Derm. Venereol. Suppl. 120*: 72-75. (372)

Svejgaard, E., Thomsen, M., Morling, N., and Christiansen, Å. H. (1976). Lymphocyte transformation in vitro in dermatophytosis. *Acta Path. Microbiol. Scand. 84*: 511-519. (373)

Swan, J. W., Dahl, M. V., Coopo, P. A., and Hammerschmidt, D. E. (1983). Complement activation by *Trichophyton rubrum*. *J. Invest. Dermatol. 80*: 156-158. (374)

Swanson, R., and Stock, J. J. (1966). Biochemical alterations of dermatophytes during growth. *Appl. Microbiol. 14*: 438-444. (375)

Tagami, H., Natsume, N., Aoshima, T., Inoue, F., Suehisa, S., and Yamada, M. (1982). Analysis of transepidermal leukocyte chemotaxis in experimental dermatophytosis in guinea pigs. *Arch. Dermatol. Res. 273*: 205-217. (376)

Tagami, H., Watanabe, S., and Ofugi, S. (1973). Trichophytin contact sensitivity in guinea pigs with experimental dermatophytosis induced by a new inoculation method. *J. Invest. Dermatol. 61*: 237-241. (377)

Tagami, H., Watanabe, S., Ofugi, S., and Minami, K. (1977). Trichophytin contact sensitivity in patients with dermatophytosis. *Arch. Dermatol. 113*: 1409-1414. (378)

Takahashi, Y. (1971). Mycotic disease of the skin in northern Japan. *Jap. J. Dermatol. 81*: 187-213. (379)

Takeuchi, H. (1962). Antigenicity of relatively "purified" polysaccharide from *Trichophyton mentagrophytes* var. *asteroides* mechanically disintegrated. IV. Antigenicity for skin testing of relatively "purified" polysaccharide (from *Trichophyton mentagrophytes* var. *asteroides* mechanically disintegrated). *Bull. Pharm. Res. Inst. Osaka 39*: 1-4. (380)

Takiuchi, I., Higuchi, D., Sei, Y., and Koga, M. (1982). Isolation of an extracellular proteinase (keratinase) from *Microsporum canis*. *Sabouraudia 20*: 281-288. (381)

Takiuchi, I., Higuchi, D., Sei, Y., and Koga, M. (1983). Immunological studies of extracellular keratinase. *J. Dermatol. 10*: 327-330. (382)

Thardshimanjanz, A. (1910). Experimentelle Untersuchungen über Immunität und Allergie bei Hyphomycetenerkrankungen. *Inaugural Ph.D. dissertation*. Univ. Basel, Switzerland. (383)

Tomomatsu, S. (1961a). A serological study of experimental dermatomycoses using an antigen prepared from fungus mechanically disintegrated. III. Cross precipitation reactions among some fungi belonging to *Trichophyton, Epidermophyton* and *Microsporum* genus. *Bull. Pharm. Res. Inst. Osaka 34*: 14-23. (384)

Tomomatsu, S. (1961b). A serological study of experimental dermatomycoses using an antigen prepared from fungus mechanically disintegrated. IV. Skin allergy and humoral antibody production in experimental trichophytosis, using crude fractions (obtained from *Trichophyton mentagrophytes* var. *asteroides* mechanically disintegrated) as antigen. *Bull. Pharm. Res. Inst. Osaka 35*: 1-12. (385)

Tomomatsu, S. (1962). A serological study of experimental dermatomycoses using an antigen prepared from fungus mechanically disintegrated. V. Production of humoral antibody and skin hypersensitivity in test animals and man during griseofulvin therapy. *Bull. Pharm. Res. Inst. Osaka 36*: 20-35. (386)

Tomomatsu, S., and Inaba, K. (1956). Serological studies of experimental trichophytosis. IV. Precipitin production in infected rabbits. *Bull. Pharm. Res. Inst. Osaka 11*: 6-10. (387)

Towbin, H., Staehelin, T., and Gordon, J. (1979). Electrophoretic transfer of protein from polyacrylamide gels to nitrocellulose sheets procedure and some applications. *Proc. Natl. Acad. Sci. USA 76*: 4350-4354. (388)

Traub, E. F., and Tolmach, J. A. (1935). Dermatophytosis. Its treatment with trichophytin. *Arch. Dermatol. 32*: 413-421. (389)

Truffi, M. (1904). Richerche sulla tricofitina. *Clin. Med. Ital. 64*: 377-408. (390)

Tschernogubow, N., and Muskatblut, E. (1929). Klinische Beobachtungen über die Immunitat bei oberflachlichen Pilzerkrankungen der behaarten Kopfhaut. *Arch. Dermatol. Syph. 159*: 46-72. (391)

Turner, W. E., and Kaplan, W. (1974). The development and evaluation of immunodiffusion, immunofluorescence and physiological tests for the differentiation of atypical *Trichophyton rubrum* and *T. mentagrophytes* isolates. *Mycopathol. Mycol. Appl. 53*: 183-200. (392)

Utmelize, O. G. (1982). (Plasmocellular reaction during immunization of white mice with associated antigens from causal agents of trichophytosis in horses.) *Byulleten' Vsesoyuznogo Nauchno-Issledovatel'skog Instituta Eksperimental-noi Veterinarii 48*: 45-47. (393)

Verotti, G. (1916). Un caso di micosi microsporica del cuoio capelluto da *Microsporum lanosum*. *G. Ital. Mal. Ven. 57*: 84-92. (394)

Voldánová, K., Dvoissky, K., Bartak, P., Bilek, J., and Jilek, M. (1963). Serum antibodies in nodular granulomatose perifolliculitis of mycotic origin. In *Recent Advances of Human and Animal Mycology*. Edited by L. Chmel. Slovak Acad. Sci., Bratislava, pp. 309-312. (395)

Walters, B. A. J., Beardmore, G. L., and Halliday, W. J. (1976). Specific cell-mediated immunity in the laboratory diagnosis of dermatophytic infections. *Br. J. Dermatol. 94*: 55-61. (396)

Walters, B. A. J., Chick, J. E. D., and Halliday, W. J. (1974). Cell-mediated immunity and serum blocking factors in patients with chronic dermatophytic infections. *Int. Arch. Allergy 46*: 849-857. (397)

Walzer, R.A., and Einbinder, J. (1962). Immunofluorescent studies in the dermatophyte infection. *J. Invest. Dermatol. 39*: 165-168. (398)

Wawrzkiewicz, K., and Wawrzkiewicz, J. (1984). (Evaluation of the immunogenic properties of live and inactivated vaccines against ringworm in cattle.) *Medycyna Weterynaryjna 40*: 33-36. (399)

Weary, P. E., and Canby, C. M. (1969). Further observations on keratinolytic activity of *Trichophyton schoenleinii* and *Trichophyton rubrum*. *J. Invest. Dermatol. 53*: 58-63. (400)

Weary, P. E., and Guerrant, J. L. (1967). Chronic urticaria in association with dermatophytosis. *Arch. Dermatol. 95*: 400-401. (401)

Weitzman, I., McGinnis, M. R., Padhye, A. A., and Ajello, L. (1986). The genus

Arthroderma and its later synonym *Nannizia*. *Mycotaxon 25*: 505-518. (402)
Wharton, M. L., Reiss, F., and Wharton, D. R. A. (1950). Active immunization against *Trichophyton purpureum*. *J. Invest. Dermatol. 14*: 291-303. (403)
Whistler, R. L., and DeMiller, J. N. (1958). Alkaline degradation of polysaccharides. *Advan. Carbohyd. Chem. 13*: 289-329. (404)
Williams, C. M. (1926). Dermatophytid complicating dermatophytosis of the glabrous skin. *Arch. Dermatol. 13*: 661-669. (405)
Williams, C. M. (1927). The enlarging conception of dermatophytosis. *Arch. Dermatol. 15*: 451-469. (406)
Wilson, J. W., Plunkett, O. A., and Gregersen, A. (1954). Nodular granulomatous perifolliculitis of the legs caused by *Trichophyton rubrum*. *Arch. Dermatol. 69*: 258-277. (407)
Woloszyn, S., and Umiński, M. (1985). (Serological and allergenic activity of purified and crude trichophytin.) *Medycyna Weterynaryjna 41*: 394-398. (408)
Wood, S. R., and Cruikshank, C. N. D. (1962). The relation between trichophytin and fungal infection. *Br. J. Dermatol. 74*: 329-336. (409)
Young, E., and Roth, F. J. (1979). Immunological cross-reactivity between a glycoprotein isolated from *Trichophyton mentagrophytes*, and human isoantigen A. *J. Invest. Dermatol. 72*: 46-51. (410)
Yu, R. J., Bishop, C. T., Cooper, F. P., Hasenclever, H. F., and Blank, F. (1967). Structural studies and mannans from *Candida albicans* (serotypes A and B), *Candida stellatoidea*, and *Candida tropicalis*. *Can. J. Chem. 45*: 2205-2211. (411)
Yu, R. J., Harmon, S. R., and Blank, F. (1968). Isolation and purification of an extracellular keratinase of *Trichophyton mentagrophytes*. *J. Bacteriol. 96*: 1435-1436. (412)
Yu, R. J., Harmon, S. R., and Blank, F. (1969a). Hair digestion by a keratinase of *Trichophyton mentagrophytes*. *J. Invest. Dermatol. 53*: 166-171. (413)
Yu, R. J., Harmon, S. R., Grappel, S. F., and Blank, F. (1971). Two cell-bound keratinases of *Trichophyton mentagrophytes*. *J. Invest. Dermatol. 56*: 27-32. (414)
Yu, R. J., Harmon, S. R., Wachter, P. E., and Blank, F. (1969b). Amino acid composition and specificity of a keratinase of *Trichophyton mentagrophytes*. *Arch. Biochem. Biophys. 135*: 336-370. (415)
Zamiechowska-Miazcova, J. (1964a). Chemical investigations of fungi of the genus *Trichophyton*. *Bull. Acad. Pol. Sci. Cl. II. Ser. Sci. Biol. 72*: 67-72. (416)
Zamiechowska-Miazcova, J. (1964b). Immunological investigations of fungi of the genus *Trichophyton*. *Bull. Acad. Pol. Sci. Cl. II Ser. Sci. Biol. 72*: 73-76. (417)
Zaslow, L., and Derbes, V. J. (1969). The immunologic nature of kerion Celsi formation. *Dermatol. Int. 8*: 1-4. (418)
Ziegler, H., and Böhme, H. (1976). Zur Problematik der Präparation von definierten Pilzantigenen. *Dermatol. Monatschr. 162*: 87-92. (419)

18
Fungal Skin Test Antigens

PIERRE AUGER
University of Montreal, Montreal, Quebec, Canada

CUTANEOUS HYPERSENSITIVITY

Immune mechanisms are classified into four types (Coombs and Gell, 1975):

1. *Type I reactions or anaphylactic.* These reactions, also termed immediate hypersensitivity, are reagin-dependent. They may be seen in skin testing as pruritic erythematous wheals with serpiginous borders and blanched centers. Clinical example of this type of reaction is penicillin allergy.
2. *Type II reactions or cytotoxic.* The target for the cytotoxic reactions may be within the skin at the dermal-epidermal junction where deposition of antibodies occurs; this reaction is termed pemphigoid disease.
3. *Type III reactions or Arthus type.* The classical Arthus reaction results in a vasculitis. Following intradermal injection of the antigen into a sensitized animal, local swelling and erythema occur in 2 h, increase for 6 h, and disappear within 12 h. This phenomenon has been described with mumps and aspergillosis.
4. *Type IV reactions or delayed hypersensitivity.* The classical lesion of this type reaction is the delayed skin reaction.

The thymus gland, which generates T cells, is the principal organ responsible in man for delayed hypersensitivity or cell-mediated immunity. Its functions play a major role in defense and limitation of spread and growth of fungi (Mason and Kirkpatrick, 1983). The immune deficiency syndromes provide illustration of the clinical significance of the immune system. The main characteristics of T cells are:

1. They are responsible for delayed cutaneous hypersensitivity, for graft rejection, and for surveillance of tumor.
2. They have a longer life than B cells and, in contrast to those cells, have no immunoglobulins on their surface.
3. They produce mediators called lymphokines: lymphocyte transformation factor, macrophage activating or inhibitory factors, transfer factor, etc.

The functional activity of T cells is measured in vivo by the production of delayed cutaneous hypersensitivity after stimulation with a specific antigen and in vitro by various methods like thymidine incorporation and inhibition of leukocyte migration. Delayed-type hypersensitivity is the most common phenomenon examined with skin testing. This reaction requires 24 and sometimes 48 h to be demonstrable. The reaction is due to cell infiltration of the superficial dermis at the site of antigen inoculation. It has been demonstrated that delayed cutaneous hypersensitivity can be transferred to previously nonreactive individuals by injecting those persons with leukocytes from sensitized donors. The most common method to administer skin tests is to inject 0.1 ml of antigen in the epidermis to cause an immediate wheal of 10 mm in diameter. If it is to be used in a patient with an allergic manifestation of the disease, like erythema nodosum in the course of coccidioidomycosis, a dilution of the antigen should be employed. As with any biological product, epinephrine should be immediately available, as an acute hypersensitivity reaction may occur. This type of anaphylaxis was recently reported in one patient who had been skin tested with a group of delayed hypersensitivity antigens; in this case the adverse reaction was triggered by trichophytin (Geller and Geller, 1980). The reaction must be read 24-72 h after antigen inoculation. An area of induration of at least 5 mm (for most antigens) and present for longer than 24 h indicates a positive reaction. In the majority of patients, dermal inflammation at the site of inoculation is resolved in 4-7 days. In some instances where the delayed hypersensitivity is more extreme, erythema, vesiculation, ulceration, or even necrosis with pain and pruritis can persist for many weeks. This may even result in scarring. Dysfunction of T cells with or without impairment of B cells is often associated with fungal diseases: aspergillosis, candidiasis, cryptococcosis. Recently, one third of subjects with cryptococcosis are AIDS patients. However, the fact that some infections like blastomycosis, coccidioidomycosis, and histoplasmosis are more prevalent in immunocompetent populations than in patients with immunodeficiency states remains unexplained.

The time period required before a delayed hypersensitivity state can be first demonstrated varies between 10 days and more than 10 weeks: the shortest time occurs in histoplasmosis, while the longest period has been reported with leprosy. Delayed hypersensitivity usually persists for many years, sometimes for life, but it has a tendency to diminish with age.

A positive test, when properly conducted, establishes exposure to the specific antigen in the recent or distant past. If the subject is part of a group of persons in which it is unlikely that a skin test to that specific antigen would be positive, a positive test is almost diagnostic, whereas it is difficult to interpret the same result when found in people living in a geographic area where there is high frequency of skin test reactivity. A negative skin test is of little value except in epidemiologic surveys. The results of a skin test must always be analyzed in the epidemiologic context in which it occurs. Some problems have been mentioned with the use of skin tests:

1. The test must be done properly, i.e., truly intradermally, using nondisposable syringes and needles because blastomycin, coccidioidin, and histoplasmin are not inactivated by autoclaving.
2. The largest diameter of induration, not erythema, must be recorded at 24, 48, and 72-h intervals.
3. Cross-reactivity is encountered between closely related organisms: patients with blastomycosis or histoplasmosis cross-react with coccidioidin. Because fungi share common antigens, skin tests with several antigens such as coccidioidin and histoplasmin should be performed simultaneously. The results must be interpreted in accordance with the clinical status and other biological data. The magnitude of the reaction does not necessarily indicate the homologous reaction (Kaufman and Reiss, 1985).
4. Several processes or events can diminish or abolish skin test reactivity (Betts, 1985). Delayed hypersensitivity may be reduced in older patients, in uremic patients, in malnutrition (even marginal), in ascorbic acid deficiency, or in patients suffering from severe unrelated illness: diabetes mellitus, liver disease, malignancy, measles, mononucleosis; in patients who have been recently vaccinated with live viral vaccines (i.e., measles, mumps, polio, and rubella), in patients undergoing treatments that suppress immunity (corticosteroid agents, irradiation, etc.), in patients in whom skin tests are done earlier than 2 weeks after exposure and, finally, in patients with the disseminated form of a fungal infection.
5. Repeated skin tests may increase, in immunocompetent patients, the skin test response.
6. Some infections, like brucellosis, result in development of stronger response to other antigens.
7. Clinicians should be aware that batches of commercially available skin test reagents vary in potency, so that they are extremely cautious in the interpretation of information obtained with various lots of reagents.

Skin test conversions, which refers to the occurrence of reactivity in a person known to be previously nonreactive to a specific antigen, confirm that, between the time of the previous test and the current skin test, infection caused by a particular infecting agent has occurred. A special problem observed with skin testing is the fact that in many infections, where the serological response is the most useful method of diagnostic, the dose of antigen administered with the skin test procedure in previously infected patients may result in an anamnestic humoral response. Thus, serological changes which could have been quite helpful can no longer be used for diagnostic purpose. Skin tests are most useful for the detection of anergy and to assess the prognosis in diseases caused by dimorphic fungi. In the disseminated form of coccidioidomycosis, skin tests are usually negative and a reversion to a positive result indicates that the infection comes under control.

Finally, detection of hypersensitivity is used to assess the level of immunocompetence in various states of immunodeficiency. The simultaneous use of several skin test antigens, candidin, mumps, streptokinase-streptodornase, trichophytin, or old tuberculin can detect impairments in the ability to mount a delayed hypersensitivity response. A commercial preparation marketed by Rhone-Poulenc Pharma Inc. is available in North America under the trade name of "Multitest C.M.I." This is a disposable applicator consisting of eight test heads preloaded with seven skin test antigens and glycerine as negative control. Two of the seven antigens are fungal extracts—*Candida* antigen and trichophytin antigen—the five others being old tuberculin, tetanus toxoid, diphtheria toxoid, and *Streptococcus* and *Proteus* antigens. Twenty four to 48 h after administration of the test, reading is done by recording at each site the size of induration (Kniker et al., 1979). Clinical studies with this test have shown differences in the ability to mount a delayed hypersensitivity response between healthy persons and those with cancer. Moreover, periodic testing can be conducted to determine if a state of anergy persists or if skin test reactivity returns.

ASPERGILLOSIS

Antigens for skin testing are commercially available* (Emmons et al., 1977) and these tests have been studied to establish a diagnosis of aspergillosis in patients with various forms of the disease: allergic aspergillosis, aspergilloma, bronchopulmonary infection, etc. A positive skin test to *Aspergillus* antigen can be obtained in most immunocompetent patients. Longbottom and Pepys (1964) have correlated the various forms of aspergillosis with immediate and delayed-type hypersensitivity (Rippon, 1982). Immediate- and Arthus-type

---

*Hollister-Strier Lab.

skin test reactions are useful criteria for the diagnosis of allergic aspergillosis. Cross-reactivity has been shown to occur not only between various species of *Aspergillus* but also with *Cladosporium* sp. (Conant et al., 1971).

BLASTOMYCOSIS

There is no commercially available skin test for the diagnosis of blastomycosis (Emmons et al., 1977; Rippon, 1982). The use of blastomycin in patients suffering from active disease has produced inconclusive results in most studies. Blastomycin is prepared by growing the fungus (mycelial phase) in a synthetic medium and the preparation is used at a 1:1000 dilution (Conant et al., 1971). A yeast form reagent has also been used, but results with this skin test were not superior to those obtained with mycelial blastomycin.

CANDIDIASIS

Because *Candida albicans* is a commensal inhabitant of the mucous membranes the great majority of human adults have innate immunity to infection by this yeast unless predisposing factors are present (Odds, 1979). Thus, many normal individuals have both low levels of antibody to *Candida* and react to the skin test antigen. Patients with iron deficiency anemia and those who receive treatment with corticosteroids or other immunosuppressive drugs show a substantially depressed candidin response when skin is tested. The reactivity to candidin has also been observed to be somewhat reduced in hospitalized patients compared with normal subjects (Alfort, 1973).

An intact cell-mediated immune response is important in host defense against various forms of *Candida* infection, and multiple defects in cell-mediated immunity to *Candida* have been noted in subjects with chronic mucocutaneous candidiasis. A positive skin test is not sufficiently related to a present illness to be useful in the diagnosis of candidiasis and is mainly used for the detection of anergy. Shannon et al. (1966) were the first to demonstrate that a positive delayed cutaneous hypersensitivity response to *Candida* antigens was so common in normal persons that *Candida* skin tests could be used to detect general deficiencies in cell-mediated immunity. However, new methods to prepare candidin seem promising: soluble substances derived from disrupted cells and polysaccharides may be more useful for skin testing of patients with this yeast infection. Few data are yet available in individuals with systemic candidiasis like septicemia, endocarditis, etc. In patients with uveitis, it has been observed that reactivity to candidin is higher than in a control group (Bloch-Michel and Timsit, 1985). The authors suggest that certain cases of uveitis may result from an interaction between microorganisms, the immune system, and the uveal tract

(Wacker et al., 1979). Cell-mediated immune responses to *Candida* antigen have been studied extensively in individuals with chronic mucocutaneous candidiasis. Less than 10% of these individuals showed a positive delayed cutaneous hypersensitivity reaction to *Candida* antigens, and only about 40% responded to stimulation with *Candida* antigens in the lymphocyte transformation assay. About 40% of these patients could be sensitized to dinitrochlorobenzene, and a vast majority responded in vitro to nonspecific mitogen stimulation.

COCCIDIOIDOMYCOSIS

Delayed cutaneous hypersensitivity reaction to *Coccidioides immitis* antigens can be useful in epidemiologic studies and has been applied to delineate endemic areas in the world. There are two commercially available antigen preparations: coccidioidin and spherulin. Coccidioidin (method of Smith et al., 1956) is prepared by growing at 37°C, 10 strains of *C. immitis* (mycelial phase) for several weeks in a static glucose-asparagine broth culture. Merthiolate is added and mycelial elements are removed by filtration; 0.1 ml of 1:100 dilution of this preparation is injected intradermally. A positive delayed hypersensitivity reaction is manifested by an induration exceeding 5 mm or more in diameter at 24 to 48 h (maximum reactivity is usually seen at 36 h). A positive reaction to coccidioidin is detected in most cases of coccidioidomycosis 3 days to 3 weeks after the onset of symptoms and may persist for more than 20 years. A tine test using coccidioidin has also been recently developed and may facilitate large-scale epidemiologic surveys.

Reactivity to coccidioidin may decrease or even disappear when disseminated disease is present to eventually reappear following total or partial recovery. The skin test is therefore useful in monitoring the patient's response to antifungal therapy. Cross-reactivity to blastomycin, histoplasmin, paracoccidioidin, and haplosparangin (prepared from *Emmonsia parva*) has been encountered in some patients.

In patients with suspected allergic manifestations of coccidioidomycosis, such as erythema nodosum, erythema multiforme, or desert rheumatism, coccidioidin must be diluted 10 to 100 times more than usual concentration to avoid severe local or systemic reactions. These allergic reactions are mainly seen in adult white women. Skin testing may trigger the appearance of erythema nodosum or aggravate the syndrome. Repeated skin testing does not seem to induce a state of delayed hypersensitivity or a rise in antibody titer if the individual has never been exposed to the fungus. Spherulin has been prepared from spherules of *C. immitis*. Several studies suggest that this skin test is more sensitive in detecting sensitization to *C. immitis* than coccidioidin but, since experience with coccidioidin is greater, coccidioidin still remains the standard antigen preparation for skin testing. Moreover, in a report summarizing experience with coccidioidin and spherulin antigens in persons living in an endemic area (southeastern United States), Woodruff et al. (1984) suggested that spherulin may be less specific than coccidioidin. In another study, when both antigens were used for skin testing in patients with proved coccidioidomycosis, spherulin failed to offer a greater sensitivity in the diagnosis of coccidioidal infection over two dosages of coccidioidin (1:10 and 1:100 dilutions) (Gifford and Catanzaro, 1981). Finally, spherulin produced fewer positive reactions in a group of 54 diabetic subjects studied by Campbell and Smith (1982). It is currently used as a complementary diagnostic tool in patients presenting a suggestive clinical syndrome with a negative coccidioidin response.

CRYPTOCOCCOSIS

Skin testing is not used for the diagnosis of cryptococcosis because all available antigenic preparations of this fungus are unsatisfactory for this purpose. Serodiagnosis with the combined use of latex agglutination, tube agglutination, and indirect fluorescent antibody tests is an established procedure with good diagnostic and prognostic values, although cross-reactions are frequent (Chandler et al., 1980). An improved method of extraction of cryptococcin by urea has been described by Atkinson and Bennett (1968) and modified by Bennett (1981). With this antigen, a positive delayed cutaneous hypersensitivity response has been noted in persons with active cryptococcal disease but also in subjects infected with dimorphic fungi (blastomycosis, coccidioidomycosis, histoplasmosis). Skin test is still an imperfect research and diagnostic tool for this yeast infection.

DERMATOPHYTOSIS

The mechanisms responsible for protective immunity against the dermatophytes have not yet been clearly defined. Natural as well as partial and local resistance are known. A positive delayed hypersensitivity response to trichophytin has been described in infected humans and many species of animals. Being a mixture of dermatophyte extracts, trichophytin is common to nearly all strains of dermatophytes and therefore is not species-specific. Furthermore, saprobic fungi found in man's environment such as *Aspergillus* and *Penicillium* species produce trichophytinlike substances. Glycopeptides appear to be responsible for cutaneous reactivity to trichophytin. Intradermal inoculation of trichophytin elicit either type I (IgE-mediated) or type IV (cell-mediated) reactions. Patients with chronic infections caused by *Trichophyton rubrum* failed to mount delayed hypersensitivity responses to trichophytin but frequently manifested immediate wheal-type reactions to this antigen (Hanifein et al., 1974). Type I reactions were also commonly seen in atopic patients free of dermatophytic infection (Hunziker and Brun, 1980). These patients are known to be more susceptible to chronic dermatophytoses. In a recent study made by Kaaman (1985), atopic patients with chronic dermatophytosis had little reactivity not only to trichophytin but also to tuberculin. Patients with dermatophytids, which are sterile vesicles most commonly found along the rides of the fingers, are responsive to trichophytin. Passive transfer of delayed hypersensitivity to trichophytin has been achieved in guinea pigs by injecting cells of sensitized donors to naive animals. Accrued resistance to subsequent challenge has been demonstrated in laboratory animals recovering from dermatophytosis. This phenomenon is correlated with the establishment of delayed hypersensitivity, which occurs 15 to 20 days followng disappearance of the primary lesion.

Skin testing with trichophytin is presently of limited value for establishing the diagnosis or the prognosis of dermatophytosis in humans.

HISTOPLASMOSIS

Histoplasmin is prepared by growing in liquid culture the saprobic phase of the fungus for 2 to 4 months at 25°C. The mycelial elements are removed by filtration and the preparation obtained is injected at 1:100 or 1:1000 dilutions intradermally. The test is considered positive when at least 5 mm of induration is present after 48 h. Positive delayed cutaneous reactivity is established within 2 weeks after the onset of the infection and, in most persons, the state of hypersensitivity tends to wane 10 years after cessation of exposure. False negatives (acute fulminating histoplasmosis, older patients, etc.) and false positives (cross-reactions in patients infected with other dimorphic fungi such as *B. dermatitidis* and *C. immitis*) are so frequent with histoplasmin that this skin test has little diagnostic value. An additional drawback is the fact that skin testing with histoplasmin may produce in a histoplasmin-positive reactor a one- or twofold rise in the titer of antibodies. Therefore, blood samples should always be drawn prior to the administration of the skin test. A new commercially available reagent, HISTOLYN-CYL,* prepared by hypotonic lysis of yeast phase cells at 37°C (Owens and Scalarone, 1984), induces no detectable changes in complement-fixing or immunodiffusion tests. Histoplasmin has also been prepared from broth cultures of *Histoplasma capsulatum* var. *duboisii*. Results obtained in patients who have been skin tested with both types of histoplasmin (*capsulatum* and *duboisii*) are so far inconclusive. Tine test and jet injection have also

---

*Berkeley Biological.

been used for skin testing with histoplasmin. In summary, skin testing with histoplasmin is most useful as an epidemiologic tool indicating, if positive, a past or recent exposure.

PARACOCCIDIOIDOMYCOSIS

Microscopic demonstration of *Paracoccidioides brasiliensis* is relatively easy so that serologic or skin tests are not essential to establish the diagnosis. Antigens for skin testing have been prepared from yeast cultures (the filamentous phase of *P. brasiliensis*) and from tissues of infected guinea pigs.

Cross-reactions have been described mostly with histoplasmosis and blastomycosis but also with coccidioidomycosis. Like in other mycosis, patients with disseminated infection are usually anergic.

SPOROTRICHOSIS

There is no commercially available preparation of sporotrichin at the present time.

This antigen has been prepared from suspensions of heat-killed yeast cells and from filtrates. A polysaccharide derived from these preparations has been studied and appears to be more specific. Recently, Rodrigez-Vidas (1986) has demonstrated by skin testing of 54 patients suffering from sporotrichosis that the polysaccharides derived from *Sporothrix schenckii* and *Ceratocystis stenoceras* are closely related. Skin testing with sporotrichin is still considered an experimental tool (Emmons et al., 1977).

CONCLUDING REMARKS

Skin testing has also been used in some patients infected with one of the agents of chromoblastomycosis (*F. pedrosoi*) and in *Nocardia* infections (nocardin). It has been shown that cattle naturally infected by *N. asteroides* have a positive skin test to nocardin. There is no antigen for studying delayed-type cutaneous hypersensitivity in entomophthoromycosis, geotrichosis, lobomycosis, mucormycosis, mycetoma, rhinosporidiosis, or tinea versicolor.

REFERENCES

Alfort, R. H. (1973). Transformation of lymphocytes of normal and hospitalized adults by *Candida albicans* extract. *Proc. Soc. Exp. Biol. Med. 144*: 826-829.

Atkinson, A. J., and Bennett, J. E. (1968). Experience with a new skin test antigen prepared from *Cryptococcus neoformans*. *Am. Rev. Resp. Dis. 97*: 637-643.

Bennett, J. E. (1981). Cryptococcal skin test antigen: preparation variables and characterization. *Infect. Immun. 32*: 373-380.

Betts, R. F. (1985). Skin testing. In *Principles and Practice of Infectious Diseases*, second edition. Edited by Mandel, G. L., Douglas, G. R., Jr., and Bennett, J. E. Wiley Medical, New York, pp. 149-153.

Bloch-Michel, E., and Timsit, J. C. (1985). Uvietis with allergy to candidin. *Ophthalmologica* (Basel) *191*: 102-106.

Campbell, S. C., and Smith, J. P. (1982). An evaluation of reactivity to *Coccidioides immitis* skin tests in subjects with diabetes mellitus. *Mycopathologia 80*: 133-136.

Chandler, F. W., Kaplan, W., and Ajello, L. (1980). *A Color Atlas and Text of the Histopathology of Mycotic Diseases*. Year Book Medical Publishers, Chicago, pp. 54-58.

Conant, N. F., Smith, D. T., Baker, R. D., and Callaway, J. L. (1971). *Manual of Clinical Mycology*, third edition. W. B. Saunders Co., Philadelphia, pp. 117-123, 396-397.

Coombs, R. R. A., and Gell, P. G. H. (1975). Classification of allergic reactions responsible for clinical hypersensitivity and disease. In *Clinical Aspects of Immunology*. Edited by Gell, P. G. H., Coombs, R. R. A., and Lachmann, P. J. Blackwell Scientific Publications, Oxford, pp. 761-781.

Emmons, C. W., Binford, C. H., Utz, J. P., and Kwon-Chung, K. J. (1977). *Medical Mycology*, third edition. Lea-Febiger, Philadelphia, pp. 68-69, 298.

Geller, M., and Geller, M. (1980). Letters to the editors. *Ann. Allergy 45*: 381.

Gifford, J., and Catanzaro, A. (1981). A comparison of coccidioidin and spherulin skin testing in the diagnosis of coccidioidomycosis. *Am. Rev. Resp. Dis. 124*: 440-444.

Hanifin, J. M., Ray, L. F., and Lobitz, W. C., Jr. (1974). Immunological reactivity in dermatophytosis. *Br. J. Dermatol. 90*: 1-8.

Hunziker, N., and Brun, R. (1980). Lack of delayed reaction in presence of cell-mediated immunity in trichophytin hypersensitivity. *Arch. Dermatol. 116*: 1266-1268.

Kaaman, T. (1985). Skin reactivity in atopic patients with dermatophytosis. *Mykosen 28*: 183-190.

Kaufman, L., and Reiss, E. (1985). Serodiagnosis of fungal diseases. In *Manual of Clinical Microbiology*, fourth edition. Edited by Lennette, E. H., Balows, A., Hansler, W. J., Jr., and Shadomy, H. J. American Society for Microbiology, Washington, D.C., pp. 924-944.

Kniker, W. T., Anderson, C. T., and Roumiantzeff, M. (1979). The MULTI-TEST system: A standardized approach to evaluation of delayed hypersensitivity and cell mediated immunity. *Ann. Allergy 43*; 73-79.

Longbottom, J. L., and Pepys, J. (1964). Pulmonary aspergillosis. *J. Pathol. Bacteriol. 88*: 141-151.

Mason, U. G., and Kirkpatrick, C. H. (1983). The lymphoid system in mycotic and mycobacterial diseases. In *Advances in Host Defense Mechanisms*. Edited by Gallin, J. I., and Fauci, A. S. Raven Press, New York, pp. 143-177.

Odds, F. C. (1979). *Candida and Candidiasis*. University Park Press, Baltimore, pp. 209-227.

Owens, R. D., and Scalarone, G. M. (1984). Immunodiffusion studies on the antigens of *Histoplasma capsulatum*: comparison of mycelial histoplasmin with the yeast phase reagent Histolyn-CYL. *Sabouraudia: J. Med. Veter. Mycol. 22*: 487-491.

Rippon, J. W. (1982). *Medical Mycology. The Pathogenic Fungi and the Pathogenic Actinomycetes*, second edition. W. B. Saunders Co., Philadelphia, pp. 449-451, 587-488.

Rodriguez-Vindas, J. (1986). Intradermoreactions aux antigènes polyosidiques de *Sporothrix schenckii* et *Ceratocystis stenoceras* de malades sporotrichosiques et de sujets témoins. *Bull. Soc. Fr. Mycol. Med. 15*: 473-476.

Shannon, D. C., Johnson, G., Rosen, F. S., and Austen, K. F. (1966). *Candida albicans* antigen. *N. Engl. J. Med. 275*: 690-693.

Smith, C. E., Saito, M. T., Simons, S. S., and Berkeley, M. P. H. (1956). Pattern of 39,500 serologic tests in coccidioidomycosis. *JAMA 160*: 546-552.

Wacker, W. B., Rao, W. A. and Marak, G. E. (1979). In *Immunology and Immunopathology of the Eye*. Edited by Silverstein, A. M., and O'Connor, R. G. Masson Publ., New York, pp. 121-126.

Woodruff, W. W., Buckley, C. E., Gallis, H. A., Cohn, J. R., and Wheat, R. W. (1984). Reactivity to spherule-derived coccidioidin in the southeastern United States. *Infect. Immun. 43*: 860-869.

19
Circulating Antigens in Systemic Fungal Diseases

LOUIS de REPENTIGNY
University of Montreal and Ste-Justine Hospital, Montreal, Quebec, Canada

INTRODUCTION

The serological diagnosis of systemic fungal infection has received considerable attention because patients often present with few distinctive clinical signs and microbiological techniques frequently fail to isolate the causative organism (de Repentigny and Reiss, 1984; Anonymous, 1986; Hopwood and Warnock, 1986; Bennett, 1987). Recent efforts have been directed to the rapid diagnosis of the mycoses by the detection of circulating antigens and metabolites, and to the identification of fungal antigens which elicit a specific humoral response.

A number of studies have substantiated the presence of circulating antigen in invasive candidiasis, invasive aspergillosis, disseminative histoplasmosis and coccidioidomycosis. However, in contrast to the prolonged antigenemia observed in cryptococcosis (Prevost and Newell, 1978), *Candida* mannan and *Aspergillus* galactomannan antigens are rapidly removed from the circulation by the formation of immune complexes (Jones, 1980; Reiss et al., 1982; Burges et al., 1983) and mannose receptor-mediated endocytosis by Kupffer cells in the liver (Bennett et al., 1987). The resulting transient presence of antigen in serum may provide insufficient sensitivity for disease, especially if a single serum sample is tested at the time when fungal infection is clinically suspected. Improved detection of *Candida* and *Aspergillus* antigens may thus result mainly from repeated serum (Burnie and Williams, 1985; Kahn and Jones, 1986) or concentrated urine (Dupont et al., 1987) samplings rather than from the development of more sensitive immunoassays.

The usefulness of antibody detection in the diagnosis of fungal infection has been limited by the shared antigenicity of several genera and species of pathogenic fungi, and the inability to distinguish between superficial colonization and deep infection. Immunoblot analysis of the serological response is a useful tool for the identification of immunogenic fungal components which elicit a specific antibody response in invasive disease. This method has been successfully applied to the study of the immune response to several fungi including *Candida, Aspergillus*, and *Rhizopus*.

In this review, a survey will be made of recent developments in antigen detection in the diagnosis of systemic fungal infection.

ANTIGEN DETECTION IN INVASIVE CANDIDIASIS

One approach to early diagnosis of invasive candidiasis in profoundly immunosuppressed patients involves the detection of circulating antigens of the *Candida* species (de Repentigny and Reiss, 1984; Bennett, 1987). In candidiasis, methods have been devised for the detection of the cell wall polysaccharide antigen mannan in serum by monoclonal (Reiss et al., 1986) or polyclonal (de Repentigny et al., 1985; Fujita et al., 1986) enzyme immunoassay (EIA), radioimmunoassay (RIA) (Poor and Cutler, 1979, Weiner and Coats-Stephen, 1979b), latex agglutination (Bailey et al., 1985; Kahn and Jones, 1986; Greenfield et al., 1988), or coagglutination (Koshi et al., 1987); of a 48-kD cytoplasmic protein antigen by monoclonal EIA (Strockbine et al., 1984a, 1984b; Matthews et al., 1987) or dot immunoassay (Matthews and Burnie, 1988); and of a heat-labile glycoprotein antigen by latex agglutination (Gentry et al., 1983; Bailey et al., 1985; Burnie and Williams, 1985; Kahn and Jones, 1986; Price and Gentry, 1986). Commercial kits are under evaluation for the detection of the latter two antigens.

Mannan

A number of factors need to be considered in evaluating published clinical studies on the detection of serum mannan: (1) the study design may be either prospective, with sera drawn at predetermined intervals from a given number of patients at risk, or retrospective, using a limited number of frozen sera from patients known to have had the disease in the recent past. The number of sera available per patient is usually greater in a prospective study, and if antigenemia is transient this will tend to produce a falsely high estimate of sensitivity for disease, especially if the test is ultimately ordered on clinical suspicion of candidiasis. In a prospective study by Kahn and Jones (1986), the sensitivity of a latex agglutination test for serum mannan was much greater (78%) when several sera were obtained prospectively, as compared to sera collected only at the time candidiasis was clinically suspected (22%). These authors also showed that the presence of mannan in serum is often transient (Kahn and Jones, 1986), and optimal diagnostic sensitivity would require at least weekly serum samplings.

(2) The proximity of sera to the time of diagnosis or death can influence diagnostic sensitivity, especially if antigenemia is not sustained but transient. This time interval has varied greatly and often complicates the comparison of published studies. (3) The case definition of invasive candidiasis is a critical aspect which deserves close attention. Clearly, the most well-defined study population would be composed of patients at risk, with or without invasive candidiasis at autopsy. This case definition has several advantages. There is no doubt about the diagnosis, either in patients with invasive candidiasis or in control patients. The extent of dissemination can also be ascertained. However, it is very difficult to assemble a study population based on these criteria who also have a sufficient number of sera in the weeks preceding death. There may also be a bias toward selecting the more fulminant forms of the disease with greater candidal organ content and antigenemia. At the opposite extreme, some case definitions of invasive candidiasis have relied on repeated culture of *Candida* species as mucosal sites of patients at risk, and the diagnosis of disease in these studies is in grave doubt. A more realistic approach is to stratify patients according to the probability of disease (Kahn and Jones, 1986). (4) Patient populations studied have varied with regard to underlying disease, and the degree of immunosuppression of these populations may influence the production of antibodies to mannan and consequently the clearance of circulating antigen. (5) In most clinical studies the serotype of *Candida albicans* was not determined and the ability to detect serum mannan in invasive candidiasis caused by *Candida albicans* serotype B could not be ascertained (Reiss et al., 1986). (6) Immunoassay formats and antisera have also varied greatly, and these factors can have a profound influence on the outcome of clinical studies. (7) The dissociation of immune complexes increases the sensitivity of immunoassays for serum mannan, and thus the overall sensitivity of the test for disease. For example, Bailey et al. (1985) detected mannan by latex agglutination in 17 of 21 patients with disseminated candidiasis when sera were treated with protease and heat, while antigen was found in only 3 of the 21 patients when this step was omitted.

Despite these limitations in comparing published studies, the following conclusions can be drawn about the value of serum mannan detection for the diagnosis of invasive candidiasis (de Repentigny and Reiss, 1984). Available methods, including latex agglutination (Kahn and Jones, 1986; Bailey et al., 1985) and EIA (de Repentigny et al., 1985), have moderate sensitivity (50-70%) but high specificity ($>99\%$) for invasive candidiasis. For example, our retrospective evaluation of the double-antibody sandwich EIA in cancer patients with or without invasive candidiasis at autopsy showed a sensitivity of 65% and a specificity of 100% using 1 ng/ml as the cutoff value (de Repentigny et al., 1985). Further gains in sensitivity for detecting disease may be obtained from more frequent blood samplings (Kahn and Jones, 1986) because of the transient nature of antigenemia, or from detection of mannan in serial samples of urine. In aspergillosis, detectable levels of galactomannan were found more frequently in concentrated urine than in serum (Dupont et al., 1987). Available methods (EIA, latex agglutination) are sufficiently simple to be performed in most diagnostic laboratories, but the widespread use of assays for serum mannan awaits the development of a reliable commercial kit.

Cytoplasmic Proteins

Cytoplasmic protein antigens also circulate in the serum of patients with invasive candidiasis (Stevens et al., 1980; Araj et al., 1982; Matthews et al., 1987; Matthews and Burnie, 1988). For example, Araj et al. (1982) detected a *Candida albicans* cytoplasmic antigen that appears to be a heat-stable protein in sera of patients with cancer. Likewise, Stevens et al. (1980) described a radioimmunoassay for the detection of an antigen that may be a cytoplasmic protein and which does not cross-react with mannan. Finally, the major 48-kD cytoplasmic protein antigen (Strockbine et al., 1984a, 1984b), was isolated by affinity chromatography from the sera of patients with disseminated *Candida albicans* infections and appears to be a heat-stable breakdown product of several larger heat-labile components (Matthews et al., 1987). It offers promise as a specific reagent to detect antibody in invasive candidiasis (Strockbine et al., 1984a, 1984b; Matthews et al., 1987), and has been detected by a monoclonal enzyme immunoassay in sera of mice with experimental invasive candidiasis (Eng, M., Walsh, T., Maret, M., Lockatell, V., Johnson, D., Rosenstein, R., Buckley, H. Evaluation of a monoclonal ELISA for detection of invasive candidiasis in mice. Program and Abstracts of the Annual Meeting of the American Society for Microbiology, 1986;F31:402) and by dot immunoassay in 77% of neutropenic patients with invasive candidiasis (Matthews and Burnie, 1988). The latter method provided greater sensitivity for diagnosis of disease than a latex agglutination test for the detection of a partially heat-labile circulating antigen (Burnie, 1985; Matthews and Burnie, 1988). The value of these methods awaits further clinical trials.

Heat-Labile Glycoprotein

The tests described thus far rely on the prior selection of a specific antigen as the candidal component most likely to be detected in systemic disease. A different approach was used by Gentry et al. (1983) in which a circulating heat-labile antigen was detected using latex sensitized with serum from rabbits immunized with whole heat-killed *Candida albicans* blastoconidia. The circulating antigen was sensitive to heat (56°C for 30 min), pronase, 2-mercaptoethanol, and sodium periodate, which suggested that it might be a glycoprotein; its definite characterization as a component of the fungus was not reported. The sensitivity of the antigen to heat and the inability of sensitized latex to agglutinate mannan (Gentry et al., 1983) suggest that the circulating antigen is other than mannan.

The latex agglutination test developed by Gentry et al. (1983) has been commercialized as the Cand-Tec Candida Detection System. It has the advantages of being commercially available and of being easy and rapid (10 min) to perform in almost any diagnostic laboratory. However, kits do not contain control latex sensitized with normal rabbit serum, and thus false-positive results due to rheumatoid factor (Burnie and Williams, 1985) are not excluded. Furthermore, published clinical evaluations (Gentry et al., 1983; Burnie and Williams, 1985; Bailey et al., 1985; Kahn and Jones, 1986; Price and Gentry, 1986; Fung et al., 1986) of the Cand-Tec test suggest that a titer of 1:8 is usually diagnostic of invasive disease while a titer of 1:4 occurs more commonly in invasive disease but is also found in colonized patients. The test appears insensitive (19-71%) (Gentry et al., 1983; Burnie and Williams, 1985; Bailey et al., 1985; Kahn and Jones, 1986; Price and Gentry, 1986; Fung et al., 1986) using a titer of 1:8 which exludes most false-positive results, and this lack of sensitivity is especially frequent on single serum specimens (Burnie and Williams, 1985; Kahn and Jones, 1986). For example, Burnie and Williams (1985) found that in 10 patients who had invasive candidiasis and an antigen titer of 1:8 or greater, only 18 out of 108 sera (26%) were positive at this level. Likewise, Kahn and Jones (1986) found that the sensitivity of the Cand-Tec test was 48% in a panel of 355 sera sequentially collected from 75 leukemic patients, while antigenemia was detected in only 19% of patients with invasive candidiasis in a second panel of 364 sera collected from 150 patients with a variety of underlying diseases only at the time candidiasis was suspected. Despite the lack of sensitivity of the Cand-Tec procedure, it seems highly specific for invasive candidiasis at a titer of 1:8. Furthermore, the method detects antigenemia in patients infected with *Candida albicans* serotypes A and B (Burnie and Williams, 1985), *Candida tropicalis* (Gentry et al., 1983; Bailey et al., 1985; Kahn and Jones, 1986), and *Candida parapsilosis* (Gentry et al., 1983; Burnie and Williams, 1985). The Cand-Tec test was compared to a latex agglutination method for serum mannan in two studies (Bailey et al., 1985; Kahn and Jones, 1986), and the mannan test was found to have greater sensitivity and equivalent specificity for invasive candidiasis.

ANTIGEN DETECTION IN INVASIVE ASPERGILLOSIS

The unsatisfactory predictive value of antibody detection in invasive aspergillosis has prompted the development of immunoassays for *Aspergillus* antigen (de Repentigny and Reiss, 1984; Hopwood and Warnock, 1986; Weiner et al., 1986; Bennett, 1987). The advantages of this approach are that *Aspergillus* antigenemia might correlate with disease activity and that the assay would not be dependent on host humoral responses which could be attenuated or delayed (Weiner et al., 1986). Lehmann and Reiss (1978) and Reiss and Lehmann (1979) first reported the detection of galactomannan polysaccharide antigen by counterimmunoelectrophoresis in the serum and urine of patients and experimentally infected rabbits with invasive aspergillosis. Further refinement of immunoassays for *Aspergillus* antigenemia led to the development of increasingly sensitive and rapid RIAs (Shaffer et al., 1979a, 1979b; Weiner and Coats-Stephen, 1979a, Weiner, 1980, 1985, Andrews and Weiner, 1981, 1982; Weiner et al., 1983, 1986, Dupont et al., 1987, Talbot et al., 1987) and more recently EIAs (Sabetta et al., 1985, de Repentigny et al., 1987; Dupont et al., 1987, Wilson et al., 1987, Patterson et al., 1988).

Shaffer et al. (1979a,b) devised a RIA using protein A-rich *Staphylococcus aureus* as solid phase which detected an undefined polysaccharide antigen in a rabbit model as well as three patients with invasive aspergillosis. The minimum sensitivity of the prototype RIA was 500 ng/ml, later refined to 10-100 ng/ml. The extensive studies of Weiner et al. in experimentally infected rabbits (Weiner and Coats-Stephen, 1979a; Andrews and Weiner, 1981) and in humans (Weiner, 1980, 1985; Andrews and Weiner, 1982; Weiner et al., 1983, 1986; Talbot et al., 1987) evaluated a RIA which used a purified cell wall carbohydrate antigen. Circulating antigen was detected in 78% of 51 rabbits with invasive aspergillosis but was absent in serum of 76 normal rabbits or 25 rabbits with invasive candidiasis (Weiner and Coats-Stephen, 1979a). In clinical studies (Weiner, 1980; Weiner et al., 1983; Talbot et al., 1987), the RIA had a sensitivity of 70-80%, a specificity of 90%, a positive predictive value of 82%, and a negative predictive value of 85% (Talbot et al., 1987). In the most recent study, antigen was detected before invasive aspergillosis was suspected during 30% of admissions and prior to laboratory confirmation of disease in 46% (Talbot et al, 1987). Detection of polysaccharide antigen in bronchoalveolar lavage fluid (Andrews and Weiner, 1981, 1982) may provide added diagnostic sensitivity in patients with major pulmonary involvement and falsely negative sera. The major drawbacks of the RIA are the absence of detectable antigen in about 25% of patients and the use of radioactive material. In addition, antigen concentrations in patient sera ranged from 186 to 682 ng/ml in an initial study (Weiner, 1980), but were substantially lower in later reports (7-28 ng/ml). Reasons for these differences are unknown.

In contrast to the results obtained by Weiner et al., Dupont et al. (1987) detected purified galactomannan antigen by RIA or EIA in the serum of only 4 of 12 rabbits lethally infected with *Aspergillus fumigatus*. Serum antigen was detected in only two of 12 patients with invasive aspergillosis (Dupont et al., 1987). Results of assay for galactomannan in urine were far more encouraging. Urinary galactomannan was detectable in all 16 lethally infected rabbits and in urine from seven of 13 patients with invasive aspergillosis (Dupont et al., 1987). The molecular weight of galactomannan in rabbit urine, 18 kD, was consistent with clearance by glomerular filtration. EIA and RIA gave comparable results. The transient presence of antigen in serum (Wilson et al., 1987) was explained by the formation of immune complexes and by the rapid clearance by mannose receptor-mediated endocytosis by Kupffer cells in the liver (Bennett et al., 1987). *Aspergillus fumigatus* galactomannan contains a mannan core substituted with short immunodominant side chains of galactose (Reiss and Lehmann, 1979; Barreto-Bergter et al., 1981, Bennett et al., 1985). Twenty to 53% of mannosyl units were unsubstituted (Reiss and Lehmann, 1979; Barreto-Bergter et al., 1981; Bennett et al., 1985) and D-manno-pyranose terminal residues are thought to be recognized by the macrophage mannosyl receptor, resulting in a short circulating half-life for *Aspergillus fumigatus* galactomannan. In rabbits injected with radiolabeled galactomannan, the liver contained 35% of the injected dose within 1 hr, but excretion of the same percentage of injected dose in urine was delayed to 24 hr. Immunization of rabbits with live conidia increased the hepatic uptake of galactomannan (Bennett et al., 1987), probably by the formation of immune complexes, which circulate in patients with invasive aspergillosis (Wilson et al., 1987) and which must be dissociated prior to immunoassay for *Aspergillus* antigen. The dissociation of immune complexes can be achieved using dilutions of serum and heat (Dupont et al., 1987; Wilson et al., 1987). precipitation of protein with trichloroacetic acid and then dialysis of the supernatant (Shaffer, 1979a; Sabetta et al., 1985), treatment with citric acid, heat, pepsin, and a 10-fold dilution (Weiner, 1980), or briefly boiling in EDTA (de Repentigny et al., 1987).

Three other groups of investigators also reported on the detection of *Aspergillus* antigen by increasingly sensitive EIAs. Sabetta et al. (1985) devised a competitive EIA which detected an undefined carbohydrate in five of six immunosuppressed and infected rabbits, and 11 of 19 patients with invasive aspergillosis. The dissociation of immune complexes, achieved by a lengthy dialysis step, was later replaced with boiling in trichloroacetic acid (Patterson et al., 1988). de Repentigny et al. (1987), using purified galactomannan as sensitizing agent for detector plates, detected circulating antigen in experimentally infected and immunosuppressed rabbits which bound to concanavalin A and had an apparent molecular weight of 50-100 kD. This immunoassay could be performed in 4 hr and detected as little as 10 ng/ml of galactomannan. Finally, Wilson et al. (1987), also using a competitive EIA, surveyed the presence of circulating *Aspergillus* antigen in immunocompromised patients with proven or suspected invasive disease. The results, in agreement with those of Dupont et al. (1987), showed that tests on single serum specimens were often negative and that multiple specimens from the same patient greatly increased the frequency of detection. Repeated monitoring of sera from a single patient showed wide fluctuations in antigen concentration, which may be related in part to antifungal therapy (Patterson et al., 1988) and to the appearance of specific anti-*Aspergillus* antibody (Wilson et al., 1987).

The overall results indicate that antigen detection in the serum of patients with invasive aspergillosis is moderately sensitive but highly specific for disease. The relatively prolonged excretion of galactomannan antigen in urine compared to the transient presence of antigen in serum may provide greater sensitivity for invasive aspergillosis. Prospective studies for the detection of galactomannan antigen in serial samples of urine from immunocompromised patients at risk of invasive aspergillosis are clearly warranted. In addition, the complex antigenic structure of *Aspergillus* suggests that other components may also circulate in patients with invasive aspergillosis. Sustained antigenemia, useful clinically, would be predicted from antigens which are not rapidly cleared as immune complexes or by receptors for mannose.

ANTIGEN DETECTION IN CRYPTOCOCCOSIS

Since its inception (Bloomfield et al., 1963), the latex agglutination test for the detection of cryptococcal capsular polysaccharide antigen (Bhattacharjee et al., 1984) has proven highly sensitive and specific for the diagnosis of meningeal and disseminated forms of cryptococcosis (Prevost and Newell, 1978; Kaufman et al., 1981). The high sensitivity of the latex agglutination test for disease and its ease of performance make it one of the most reliable mycoserological tests. Occasional false-negative results caused by a prozone effect can be corrected by dilution of the specimen (Stamm and Polt, 1980). The appearance of capsule-deficient *Cryptococcus neoformans* in AIDS patients results in low antigen concentrations, which may eventually require the enhanced sensitivity of EIA (Scott et al., 1980, 1981). False-positive results have been rarely encountered in the cerobrospinal fluid of patients with septicemia due to DF-2 (Westerink et al., 1987) or of patients with malignancy (Hopfer et al., 1982), and in the serum of a patient with disseminated infection caused by *Trichosporon beigelii* (McManus and Jones, 1985). The pronase treatment modification of the latex agglutination test increased the sensitivity of antigen detection in some cerebrospinal fluid and most serum specimens, and is a useful and simple modification of the assay (Stockman and Roberts, 1983; Gray and Roberts, 1988).

ANTIGEN DETECTION IN HISTOPLASMOSIS

A novel approach to the serologic diagnosis of disseminated histoplasmosis involves the detection of low-molecular weight polysaccharide antigen in serum and urine by sandwich RIA (Wheat et al., 1986). In a retrospective evaluation, antigenuria was noted in 20 of 22 episodes of disseminated histoplasmosis that occurred in 16 patients, in 6 of 32 patients with self-limited infection, in 2 of 32 patients with cavitary histoplasmosis, and in 4 of 8 patients with a sarcoid-like illness caused by *Histoplasma capsulatum*. Antigen was also detected but less frequently in serum and appeared to decrease after initiation of antifungal therapy. Blinded prospective studies will be needed to evaluate the RIA's sensitivity and specificity, and to determine whether the antigen is shared among strains of *Histoplasma capsulatum* from different goegraphic regions. The dissociation of immune complexes may also improve the detection of antigen in serum.

ANTIGEN DETECTION IN COCCIDIOIDOMYCOSIS

As with other fungal diseases, specific immune complexes also circulate in coccidioidomycosis and have led to the development of immunoassays for antigen in serum. Yoshinoya et al. (1980) first reported the detection of immune complexes in 16 of 22 patients with chronic pulmonary or disseminated coccidioidomycosis, which were found in only 7 of 54 healthy controls. Coccidioidin antigen was revealed in the immune complexes by an antigen-specific RIA. Weiner (1983) then obtained an immunoreactive pool of 200-230 kD by gel filtration of coccidioidin, compatible with a polymeric glycoprotein, and devised an inhibition RIA assay for circulating antigen. The dissociation of immune complexes by an acid-heat extraction procedure produced a sensitivity of 24 ng/ml. Retrospective clinical evaluation of the RIA demonstrated the presence of antigen in the serum of 5 of 9 patients with active coccidioidomycosis but was absent in the serum of 106 control patients with other fungal diseases and normal donors. Three of four patients with coccidioidomycosis who did not have detectable antigen had received antifungal therapy. A single false-positive detection of antigen was found in a patient with histoplasmosis. Thus, the overall results suggested that specific glycoprotein antigen circulates in active, progressive forms of coccidioidomycosis. Antigenemia has also been reported in primary coccidioidomycosis (Galgiani et al., 1984). Galgiani et al. (1984) used spherulin and antiserum to spherulin to construct an inhibition EIA. In contrast to the two previous reports, circulating antigen was detected in primary infection. Twenty-one of 27 sera from 19 infected patients contained antigen within two months of onset of symptoms. IgM antibody against spherulin occurred in parallel with antigenemia, suggesting the formation of immune complexes. This same group of investigators later reported that acidification extraction improves detection (Wack et al., 1988), which is consistent with this view. The nature of the antigen which circulates in early, primary coccidioidomycosis is not known.

REFERENCES

Andrews, C. P., and Weiner, M. H. (1981). Immunodiagnosis of invasive pulmonary aspergillosis in rabbits: fungal antigen detected by radioimmunoassay in bronchoalveolar lavage fluid. *Am. Rev. Respir. Dis. 124*: 60-64.

Andrews, C. P., and Weiner, M. H. (1982). *Aspergillus* antigen detection in bronchoalveolar lavage fluid patients with invasive aspergillosis and aspergillomas. *Am. J. Med. 73*: 372-380.

Anonymous (1986). Serodiagnosis of *Candida* infections. *Lancet ii*: 1373-1374 [editorial].

Araj, G. F., Hopfer, R. L., Chesnut, S., Fainstein, V., and Bodey, G. P. (1982). Diagnostic value of the enzyme-linked immunosorbent assay for detection of *Candida albicans* cytoplasmic antigen in sera of cancer patients. *J. Clin. Microbiol. 16*: 46-52.

Bailey, J. W., Sada, E., Brass, C., and Bennett, J. E. (1985). Diagnosis of systemic candidiasis by latex agglutination for serum antigen. *J. Clin. Microbiol. 21*: 749-752.

Barreto-Bergter, E., Gorin, P. A. J., and Travassos, L. R. (1981). Cell constituents of mycelia and conidia of *Aspergillus fumigatus*. *Carbohydr. Res. 95*: 205-218.

Bennett, J. E. (1987). Rapid diagnosis of candidiasis and aspergillosis. *Rev. Infect. Dis. 9*: 398-402.

Bennett, J. E., Bhattacharjee, A. K., and Glaudemans, C. P. J. (1985). Galactofuranosyl groups are immunodominant in *Aspergillus fumigatus* galactomannan. *Mol. Immunol. 22*: 251-254.

Bennett, J. E., Friedman, M. M., and Dupont, B. (1987). Receptor-mediated clearance of *Aspergillus* galactomannan. *J. Infect. Dis. 155*: 1005-1010.

Bhattacharjee, A. K., Bennett, J. E., and Glaudemans, C. P. J. (1984). Capsular polysaccharides of *Cryptococcus neoformans*. *Rev. Infect. Dis. 6*: 619-624.

Bloomfield, N., Gordon, M. A., and Elmendorf, D. F. (1963). Detection of *Cryptococcus neoformans* antigen in body fluid by latex particle agglutination. *Proc. Soc. Exp. Biol. Med. 114*: 64-67.

Burges, G., Holley, H. P., and Virella, G. (1983). Circulating immune complexes in patients with *Candida albicans* infections. *Clin. Exp. Immunol. 53*: 165-174.

Burnie, J. (1985). A reverse passive latex agglutination test for the diagnosis of systemic candidiasis. *J. Immunol. Meth. 82*: 267-280.

Burnie, J. P., and Williams, J. D. (1985). Evaluation of the Ramco latex agglutination test in the early diagnosis of systemic candidiasis. *Eur. J. Clin. Microbiol. 4*: 98-101.

de Repentigny, L., Boushira, M., Ste-Marie, L., and Bosisio, G. (1987). Detection of galactomannan antigenemia by enzyme immunoassay in experimental invasive aspergillosis. *J. Clin. Microbiol. 25*: 863-867.
de Repentigny, L., Marr, L. D., Keller, J. W., Carter, A. W., Kuykendall, R. J., Kaufman, L., and Reiss, E. (1985). Comparison of enzyme immunoassay and gas-liquid chromatography for the rapid diagnosis of invasive candidiasis in cancer patients. *J. Clin. Microbiol. 21*: 972-979.
de Repentigny, L., and Reiss, E. (1984). Current trends in immunodiagnosis of candidiasis and aspergillosis. *Rev. Infect. Dis. 6*: 301-312.
Dupont, B., Huber, M.,Kim, S.J., and Bennett, J. E. (1987). Galactomannan antigenemia and antigenuria in aspergillosis: studies in patients and experimentally infected rabbits. *J. Infect. Dis. 155*: 1-11.
Fujita, S., Matsubara, F., and Matsuda, T. (1986). Enzyme linked immunosorbent assay measurement of fluctuations in antibody titer and antigenemia in cancer patients with and without candidiasis. *J. Clin. Microbiol. 23*: 568-575.
Fung, J. C., Donata, S. T., and Tilton, R. C. (1986). Candida detection system (Cand-Tec) to differentiate between *Candida albicans* colonization and disease. *J. Clin. Microbiol. 24*: 542-547.
Galgiani, J. N., Dugger, K. O., Ito, J. I., and Wieden, M. A. (1984). Antigenemia in primary coccidioidomycosis. *Am. J. Trop. Med. Hyg. 33*: 645-649.
Gentry, L. D., Wilkinson, I. D., Lea, A. S., and Price, M. F. (1983). Latex agglutination test for detection of *Candida* antigen in patients with disseminated disease. *Eur. J. Clin. Microbiol. 2*: 122-128.
Gray, L. D., and Roberts, G. D. (1988). Experience with the use of pronase to eliminate interference factors in the latex agglutination test for cryptococcal antigen. *J. Clin. Microbiol. 26*: 2450-2451.
Greenfield, R. A., Troutt, D. L., Rickard, R. C., and Altmiller, D. H. (1988). Comparison of antibody, antigen, and metabolite assays in rat models of systemic and gastrointestinal candidiasis. *J. Clin. Microbiol. 26*: 409-417.
Hopfer, R. L., Perry, E. V., and Fainstein, V. (1982). Diagnostic value of cryptococcal antigen in the cerebrospinal fluid of patients with malignant disease. *J. Infect. Dis. 145*: 915.
Hopwood, V., and Warnock, D. W. (1986). New developments in the diagnosis of opportunistic fungal infection. *Eur. J. Clin. Microbiol. 5*: 379-388.
Jones, J. M. (1980). Kinetics of antibody responses to cell wall mannan and a major cytoplasmic antigen of *Candida albicans* in rabbits and humans. *J. Lab. Clin. Med. 96*: 845-860.
Kahn, F. W., and Jones, J. M. (1986). Latex agglutination tests for detection of *Candida* antigens in sera of patients with invasive candidiasis. *J. Infect. Dis. 153*: 579-585.
Kauffman, C. A., Bergman, A. G., Severance, P. J., and McClatchey, K. D. (1981). Detection of cryptococcal antigen: comparison of two latex agglutination tests. *Am. J. Clin. Pathol. 75*: 106-109.
Koshi, G , Viswanathan, A., Chandy, M., and Jairaj, P. S. (1987). Development

of a coagglutination technic to detect *Candida* antigenemia. *Am. J. Clin. Pathol. 88*: 429-435.

Lehmann, P. F., and Reiss, E. (1978). Invasive aspergillosis: antiserum for circulating antigen produced after immunization with serum from infected rabbits. *Infect. Immun. 20*: 570-572.

Matthews, R., and Burnie, J. (1988). Diagnosis of systemic candidiasis by an enzyme-linked dot immunobinding assay for a circulating immunodominant 47-kilodalton antigen. *J. Clin. Microbiol 26*: 459-463.

Matthews, R. C., Burnie, J. P., and Tabaqchali, S. (1987 . Isolation of immunodominant antigens from sera of patients with systemic candidiasis and characterization of serological response to *Candida albicans. J. Clin. Microbiol. 25*: 230-237.

McManus, E. J., and Jones, J. M. (1985). Detection of a *Trichosporon beigelii* antigen cross-reactive with *Cryptococcus neoformans* capsular polysaccharide in serum from a patient with disseminated *Trichosporon* infection. *J. Clin. Microbiol. 21*: 681-685.

Patterson, T. F., Miniter, P., Ryan, R. L., and Andriole, V. T. (1988). Effect of immunosuppression and amphotericin B on *Aspergillus* antigenemia in an experimental model. *J. Infect. Dis. 158*: 415-422.

Poor, A. H., and Cutler, J. E. (1979). Partially purified antibodies used in a solid-phase radioimmunoassay for detecting candidal antigenemia. *J. Clin. Microbiol. 9*: 362-368.

Prevost, E., and Newell, R. (1978): Commercial cryptococccal latex kit: clinical evaluation in a medical center hospital. *J. Clin. Microbiol. 8*: 529-533.

Price, M. F. and Gentry, L. O. (1986). Incidence and significance of *Candida* antigen in low-risk and high-risk patient populations. *Eur. J. Clin. Microbiol. 5*: 416-419.

Reiss, E., de Repentigny, L., Kuykendall, R. J., Carter, A. W., Galindo, R., Auger, P., Bragg, S. L., and Kaufman, L. (1986). Monoclonal antibodies against *Candida tropicalis* mannan: antigen detection by enzyme immunoassay and immunofluorescence. *J. Clin. Microbiol. 24*: 796-802.

Reiss, E., Kuykendall, R. J., and Kaufman, L. (1986). Antigenemia in rabbits infected with *Candida albicans* serotype B: detection by enzyme immunoassay and preliminary characterization of the antigen. *J. Med. Vet. Mycol. 24*: 259-269.

Reiss, E., and Lehmann, P. F. (1979). Galactomannan antigenemia in invasive aspergillosis. *Infect. Immun. 25*:357-365.

Reiss, E., Stockman, L., Kuykendall, R. J., and Smith, S. J. (1982). Dissociation of mannan-serum complexes and detection of *Candida albicans* mannan by enzyme immunoassay variations. *Clin. Chem. 28*: 306-310.

Sabetta, J. R., Miniter, P., and Andriole, V. T. (1985). The diagnosis of invasive aspergillosis by an enzyme-linked immunosorbent assay for circulating antigen. *J. Infect. Dis. 152*: 946-953.

Scott, E. N., Muchmore, H. G., and Felton, F. G. (1980). Comparison of enzyme immunoassay and latex agglutination methods for detection of *Cryptococcus neoformans* antigen. *Am. J. Clin. Pathol. 73*: 790-794.

Scott, E. N., Muchmore, H. G., and Felton, F. G. (1981). Enzyme-linked immunosorbent assays in murine cryptococcosis. *Sabouraudia 19*: 257-265.

Shaffer, P. J., Kobayashi, G. S., and Medoff, G. (1979a). Demonstration of antigenemia in patients with invasive aspergillosis by solid phase (protein A-rich *Staphylococcus aureus*) radioimmunoassay. *Am. J. Med. 67*: 627-630.

Shaffer, P. J., Medoff, G., and Kobayashi, G. S. (1979b). Demonstration of antigenemia by radioimmunoassay in rabbits experimentally infected with *Aspergillus. J. Infect. Dis. 139*: 313-319.

Stamm, A. M., and Polt, S. J. (1980). False-negative cryptococcal antigen test. *J Am. Med. Assoc. 244*: 1359.

Stevens, P., Huang, S., Young, L. S., and Berdischewsky, M. (1980). Detection of *Candida* antigenemia in human invasive candidiasis by a new solid phase radioimmunoassay. *Infection 8*: S334-S338.

Stockman, L., and Roberts, G. D. (1983). Corrected version. Specificity of the latex test for cryptococcal antigen: a rapid, simple method for eliminating interference factors. *J. Clin. Microbiol. 17.* 000-000.

Strockbine, N. A., Largen, M. T., and Buckley, H. R. (1984a). Production and characterization of three monoclonal antibodies to *Candida albicans* proteins. *Infect. Immun. 43*: 1012-1018.

Strockbine, N. A., Largen, M. T., Zweibel, S. M., and Buckley, H. R. (1984b). Identification and molecular weight characterization of antigens from *Candida albicans* that are recognized by human sera. *Infect. Immun. 43*: 715-721.

Talbot, G. H., Weiner, M. H., Gerson, S. L., Provencher, M., and Hurwitz, S. (1987). Serodiagnosis of invasive aspergillosis in patients with hematologic malignancy: validation of the *Aspergillus fumigatus* antigen radioimmunoassay. *J. Infect. Dis. 155*: 12-27.

Wack, E. E., Dugger, K. O., and Galgiani, J. N. (1988). Enzyme-linked immunosorbent assay for antigens of *Coccidioides immitis*: human sera interference corrected by acidification-heat extraction. *J. Lab. Clin. Med. 111*: 560-565.

Weiner, M. H. (1980). Antigenemia detected by radioimmunoassay in systemic aspergillosis. *Ann. Intern. Med. 92*: 793-796.

Weiner, M. H. (1983). Antigenemia detected in human coccidioidomycosis. *J. Clin. Microbiol. 18*: 136-142.

Weiner, M. H. (1985). Immunodiagnosis of invasive aspergillosis and aspergilloma. In *Aspergillosis*. Edited by Y. Al-Doory, G. E. Wagner. Charles C. Thomas, Springfield, IL, pp. 147-155.

Weiner, M. H., and Coats-Stephen, M. (1979a). Immunodiagnosis of systemic aspergillosis. I. Antigenemia detected by radioimmunoassay in experimental infection. *J. Lab. Clin. Med. 93*: 111-119.

Weiner, M. H., and Coats-Stephen, M. (1979b). Immunodiagnosis of systemic candidiasis: mannan antigenemia detected by radioimmunoassay in experimental and human infection. *J. Infect. Dis. 140*: 989-993.

Weiner, M. H., Talbot, G. H., Gerson, S. L., Fetchick, R., Andrews, C., Peacock, J. E., Filice, G., Cohen, M., Provencher, M., and Cassileth, P. (1986). Detection of fungal antigen in body fluids for diagnosis of invasive aspergillosis. *Zentralbl. Bakteriol. Parasitenkd. Infektionskrankheiten Hyg. (A) 261*: 517-522.

Weiner, M. H., Talbot, G. H., Gerson, S. L., Filice, G., and Cassileth, P. A. (1983). Antigen detection in the diagnosis of invasive aspergillosis: utility in controlled, blinded trials. *Ann. Intern. Med. 99*: 777-782.

Westerink, M. A. L., Amsterdam, D., Petell, R. J., Stram, M. N., and Apicella, M. A. (1987). Septicemia due to DF-2: cause of a false-positive cryptococcal latex agglutination result. *Am. J. Med. 83*: 155-158.

Wheat, L. J., Kohler, R. B., and Tewari, R. P. (1986). Diagnosis of disseminated histoplasmosis by detection of *Histoplasma capsulatum* antigen in serum and urine specimens. *N Engl. J. Med. 314*: 83-88.

Wilson, E. V., Hearn, V. M., and Mackenzie, D. W. R. (1987). Evaluation of a test to detect circulating *Aspergillus fumigatus* antigen in a survey of immunocompromised patients with proven or suspected invasive disease. *J. Med. Vet. Mycol. 25*: 365-374.

Yoshinoya, S., Cox, R. A., and Pope, R. M. (1980). Circulating immune complexes in coccidioidomycosis. *J. Clin. Invest. 61*: 655-663.

Index

Acid phosphatase, 21-23
Acid proteinases, 23-24
Agglutination techniques for diagnosis of aspergillosis, 405-406
Alkali-extracted rhamnomannans of *Sporothrix schenckii*, 196-197
Anaphylactic hypersensitivity (Type I reaction), 459
Antibody responses to *Candida albicans*, 382-387
  to cell wall antigens, 383-387
  to cytoplasmic antigens, 382-383
Anticryptococcal CMI response, development of, 326-333
Antigenic structure of *Aspergillus* species, 87-111
  antigens with enzymic functions, 101-102
  antigens located at the cell surface or in the wall of *Aspergillus*, 89-90
  Aspergillus fumigatus antigens in vivo, 99-101
  galactomannans, 90-93

[Antigenic structure of *Aspergillus* species]
  glycoprotein antigens with concanavalin A-binding properties, 93-97
  glycoprotein antigens of undefined carbohydrate composition, 97-99
Antigenic structure of *Candida albicans*, 3-35
  cell wall architecture of the blastospore, 7-11
  cell wall enzymes, 21-25
    acid phosphatase, 21-23
    acid proteinases, 23-24
    β-D-glucanases, 24
    phospholipase A and lysophospholipase, 24
  disposition of wall components, 26-28
  dynamic variations in the organization of the cell wall, 7
  presence of a cell wall coat, 13-15

[Antigenic structure of *Candida albicans*]
  presence of receptors for fibrinogen and for complement components within the cell wall, 25-26
  ultrastructural changes in the cell wall in association with budding and germination, 11-13
  ultrastructural localization of cell wall components by histochemical and immunocytochemical methods, 15-21
Antigenic structure of *Coccidioides immitis*, 133-170
  immunoreactive antigens, 141-162
    antigenic comparisons of morphological forms of *C. immitis*, 161-162
    antigens reactive in cell-mediated immune assays, 155-160
    CF antigen, 152-154
    crude lysate antigens, 141-142
    heat-stable exoantigen, 154-155
    immunomodulatory antigens, 160-161
    TP antigen, 142-152
  morphogenesis and ultrastructure, 134-141
Antigenic structure of *Cryptococcus neoformans*, 63-86
  areas for future study, 81
  association between serotype and epidemiology of cryptococci, 79
  biological properties of cryptococcal polysaccharides, 65-67
  extracellular polysaccharides of *C. neoformans*, 64-65
  immunochemical relationship between cryptococcal polysaccharides, 72-76
  role of cryptococcal polysaccharides in diagnosis of cryptococcosis, 80-81

[Antigenic structure of *Cryptococcus neoformans*]
  serotypes of cryptococcal polysaccharide, 67
  structures of cryptococcal polysaccharides, 68-69, 70-72
  structures of immunodeterminants on cryptococcal polysaccharides, 76-79
Antigenic structure of *Paracoccidioides brasiliensis*, 171-192
  cell wall antigens, 172-173
  exocellular (metabolic) antigens, 180-185
  intracellular (somatic) antigens, 173-180
  origin and nature of *P. brasiliensis* antigens, 185-187
Antigenic structure of *Sporothrix schenckii*, 193-221
  antigenic polysaccharides, 200-208
  complement activation, 213-215
  hypersensitivity reactions, 211-212
  indirect detection of antigens, 208-211
  isolation of antigenic structures, 194-200
    from culture supernatants, 197-199
    from isolated cell walls, 200
    from whole cells, 194-197
Arthus type hypersensitivity (Type III reaction), 459
Aspergillosis, 245-246
  humoral responses in, 401-417
    Aspergillus antigens, 402-404
    immunoglobulins involved, 404-405
    serological diagnosis, 405-409
  skin tests for, 462-463
Aspergillus antigens, 402-404
Aspergillus conidia

[*Aspergillus conidia*]
 alveolar macrophage defense against, 249-250
 circulating monocyte defense against, 253
Aspergillus fumigatus antigens in vivo, 99-101
Aspergillus hyphase, alveolar macrophage and circulating monocyte defense against, 253-254
Aspergillus species, antigenic structure of, 87-111
 antigens with enzymic functions, 101-102
 antigens located at the cell surface or in the wall of *Aspergillus*, 89-90
 Aspergillus fumigatus antigens in vivo, 99-101
 galactomannans, 90-93
 glycoprotein antigens with concanavalin A-binding properties, 93-97
 glycoprotein antigens of undefined carbohydrate composition, 97-99

Blastomycosis:
 neutrophils and macrophages in resistance to, 274-277
 skin tests for, 463

Candida
 immunodominant epitopes of *Candida* species mannans, 50-58
 immunomodulation in response to, 193-317
  clinical studies suggesting immunosuppressive phenomena, 294-300
  immunomodulation in experimental systems, 300-308

[*Candida*]
 polymorphonuclear leukocyte defense against, 259-260
Candida albicans
 antigenic structure of, 3-35
 cell wall architecture of the blastospore, 7-11
 cell wall enzymes, 21-25
 disposition of wall components, 26-28
 dynamic variations in the organization of the cell wall, 7
 presence of a cell wall coat, 13-15
 presence of receptors for fibrinogen and for complement components within the cell wall, 25-26
 ultrastructural changes in the cell wall in association with budding and germination, 11-13
 ultrastructural localization of cell wall components by histochemical and immunocytochemical methods, 15-21
 complement receptors of, 228-234
 generation and expression of immunity to, 360-365
 humoral immune response to, 375-400
 antibody response to *C. albicans*, 382-387
 antignes of *C. albicans*, 377-382
 complement activation mediated by antibodies and the organism, 387-388
 future directions for research, 394
 materials released from *Candida* capable of triggering host responses, 387
 sequence of host responses triggered by the humoral immune system during *Candida* infection, 388-394

[*Candida albicans*]
serotypes of, 38-40
Candida albicans serotypes A and B strain, specificity of agglutinating monoclonal antibodies against, 40-41
Candida blastoconidia
alveolar macrophage defense against, 255-258
circulating monocytes defense against, 258
Candida hyphase, alveolar macrophage and circulating monocyte defense against, 238-259
Candida infection, role of complement in, 225-227
Candida species mannans, immunodominant epitopes of, 50-58
Candidiasis
pulmonary, 247
serum-dependent cellular immune defects in patients with, 296-297
skin tests for, 463-464
Cell-mediated immune assays, antigens reactive in, 155-160
Cell walls of *Candida albicans*
cell wall architecture of the blastospore, 7-11
disposition of wall components, 26-28
dynamic variations in the organization of, 7
enzymes of, 21-25
presence of a cell wall coat, 13-15
presence of receptors for fibrinogen and for complement components within, 25-26
ultrastructural changes, in association with budding or germination in, 11-13

[Cell walls of *Candida albicans*]
ultrastructural localization of cell wall components by histochemical and immunocytochemical methods, 15-21
CF antigen, 152-154
Circulating antigens in diagnosis of aspergillosis, 408-409
Coccidioides immitis, antigenic structure of, 133-170
immunoreactive antigens, 141-162
antigenic comparison of morphological forms of *C. immitis*, 161-162
antigens reactive in cell-mediated immune assays, 155-160
CF antigen, 152-154
crude lysate antigens, 141-142
heat-stable exoantigen, 154-155
immunomodulatory antigens, 160-161
TP antigen, 142-152
morphogenesis and ultrastructure, 134-141
Coccidioidomycosis
neutrophils and macrophages in resistance to, 277-279
skin tests for, 464-465
Complement fixation test (CFT) in diagnosis of aspergillosis, 406
Complement role in host resistance to systemic fungal infection, 225-242
complement and cutaneous candidiasis, 227-228
complement and disease progression, 237-238
complement-*Candida* interactions, 225-227
complement interactions with other pathogenic fungi, 234-237

Index

[Complement role in host resistance to systemic fungal infection]
complement receptors of *C. albicans*, 228-231
complement receptors of *C. albicans* for iC3b and C3d, 231-234
C. neoformans-complement interactions, 234
Crude lysate antigens, 141-142
Cryptococcal polysaccharides
areas for future study, 81
biological properties of, 65-67
immunochemical relationship between, 72-76
role in diagnosis of cryptococcosis, 80-81
serotypes of, 67
structures of, 68-69, 70-72
structures of immunodeterminants on, 76-79
Cryptococcosis
immunoregulation in, 319-345
regulation of the cell-mediated immune response, 326-338
regulation of the humoral immune response, 322-326
pulmonary, 247-248
role of cryptococcal polysaccharides in diagnosis of, 80-81
skin tests for, 465
Cryptococcus
alveolar macrophage defense against, 260-262
polymorphonuclear leukocyte and circulating monocyte defense against, 262
Cryptococcus neoformans
antigenic structure of, 63-86
areas for future study, 81
association between serotype and epidemiology of cryptococci, 79

[*Cryptococcus neoformans*]
biological properties of cryptococcal polysaccharides, 65-67
extracellular polysaccharides of *C. neoformans*, 64-65
immunochemical relationship between cryptococcal polysaccharides, 72-76
role of cryptococcal polysaccharides in diagnosis of cryptococcis, 80-81
serotypes of cryptococcal polysaccharide, 68-69, 70-72
structures of immunodeterminants on cryptococcal polysaccharides, 76-79
complement interactions with, 234
Cutaneous candidiasis, role of complement in, 227-228
Cutaneous hypersensitivity, 459-462
Cytoplasmic antigens of *Candida albicans*, 377-378
Cytotoxic hypersensitivity (Type II reaction), 459

Delayed hypersensitivity (Type IV reaction), 459
Dermatophyte antigens, 113-132
discussion, 130-131
isolation of, 115-116
monoclonal antibody reaction with commercial *Trichophyton rubrum* and *Trichophyton mentagrophytes* antigens, 127
monoclonal antibody reaction with nondermatophytes, 127-130
monoclonal antibody reaction with other dermatophytes—cross-reactivity, 127
preparation of monoclonal antibodies, 118-125

[Dermatophyte antigens]
  selection of antigens to be used for immunization of mice, 116-118
  serological procedures to detect, 419-457
    detection of dermatophyte antigens, 422-431
    serological procedures, 419-422
Dermatophytosis, skin tests for, 465-466

Enzyme-linked immunosorbent assay (ELISA) in diagnosis of aspergillosis, 408
Extracellular polysaccharides of *Cryptococcus neoformans*, 64-65

Fungal skin test antigens, 459-469
  aspergillosis, 462-463
  blastomycosis, 463
  candidiasis, 463-464
  coccidioidomycosis, 464-465
  cryptococcosis, 465
  cutaneous hypersensitivity, 459-462
  dermatophytosis, 465-466
  histoplasmosis, 466-467
  paracoccidioidomycosis, 467
  sporotrichosis, 467

Galactomannans (GM) of *Aspergillus* species, 90-93
Galactomannans of *Sporothrix schenckii*, 198-199, 203
Genetic regulation of pathogenesis and host responses, 347-371
  candidosis in mutant and immunodeficient mice, 365-366

[Genetic regulation of pathogenesis of host responses]
  experimental systemic candidiasis, 349-353
  generation and expression of immunity to *C. albicans*, 360-365
  genetically determined factors predisposing to susceptibility, 353-355
  genetics of human infection, 347-348
  MHC-regulation of host responses, 355-360
  mouse models of *Candida* infection, 348-349
  recapitulation, 366-367
β-D-Glucanases, 24
Glucuronoxylomannan (GXM), 64-67
  of *C. neoformans*, structures of, 68-69
Glycoprotein antigens
  with concanavalin A-binding proteins, 93-97
  of undefined carbohydrate composition, 97-99

Heat-stable exoantigen, 154-155
Histoplasmosis
  neutrophils and macrophages in resistance to, 281-283
  skin tests for, 466-467
Host resistance to systemic fungal infection, role of complement in, 225-242
  complement and cutaneous candidiasis, 227-228
  complement and disease progression, 237-238
  complement-*Candida* interactions, 225-227

Index

[Host resistance to systemic fungal infection, role of complement in]
  complement interactions with other human pathogenic fungi, 234-237
  complement receptors of *C. albicans*, 228-231
  complement receptors of *C. albicans* for iC3b and C3d, 231-234
  C. neoformans-complement interactions, 234
Humoral immune response to *Candida albicans*, 375-400
  antibody response to *C. albicans* 382-387
    antibody response to cell wall antigens, 383-387
    antibody responses to cytoplasmic antigens, 382-383
  antigens of *C. albicans*, 377-382
  complement activation mediated by antibodies and the organism, 387-388
  future directions for research, 394
  materials released from *Candida* capable of triggering host responses, 387
  sequence of host response triggered by the humoral immune system during *Candida* infection, 388-394
Humoral response in aspergillosis, 401-417
  Aspergillus antigens, 402-404
  immunoglobulins involved, 404-405
  serological diagnosis, 405-409
Hypersensitivity reactions of *Sporothrix schenckii*, 211-212

Immunofluorescence assay (IFA) in the diagnosis of aspergillosis, 407
Immunoglobulins involved in humoral response in aspergillosis, 404-405
Immunomodulation in response to *Candida*, 293-317
  clinical studies suggesting immunosuppressive phenomena, 294-300
    alterations in lymphocyte subsets, 299-300
    serum-independent phagocytic cell defects, 299
    serum/plasma factors, 295-299
  immunomodulation in experimental systems, 300-308
    antitumor effects, 307-308
    in vitro studies in animal models, 303-307
    in vitro studies with human cells, 301-303
Immunomodulatory antigens, 160-161
Immunoreactive antigens of *Coccidioides immitis*, 141-162
  antigenic comparisons of morphological forms of *C. immitis*, 161-162
  antigens reactive in cell-mediated immune assays, 155-160
  CF antigen, 152-154
  crude lysate antigens, 141-142
  heat-stable exoantigen, 154-155
  immunomodulatory antigens, 160-161
  TP antigen, 142-152
Immunoregulation in cryptococcosis, 319-345

[Immunoregulation in cryptococcosis]
  regulation of the cell-mediated immune response, 326-338
    development of the anticryptococcal CMI response, 326-333
    suppression, 333-338
  regulation of the humoral immune response, 322-326

Lysophospholipase, 24

Macrophages in host resistance, see Neutrophils and macrophages in host resistance
Major histocompatibility complex (MHC)-regulation of host responses, 333-360
Mannans in yeast, serological specificity of, 37-62
  immunodominant epitopes of *Candida* species mannans, 50-58
  immunodominant epitopes of *S. cerevisiae* mannans, 45-50
  methods used for the immunochemistry of yeast mannans, 41-45
  serotyping of yeasts, 38-41
Medically important yeasts, antigenic structures of, 38
Monoclonal antibodies used in detection of dermatophyte antigens, 427-429
Morphological forms of *ides immitis*, antigenic comparisons of, 161-162
Mucormycosis, 246-247

Neutrophils and macrophages in host resistance, 273-289
  blastomycosis, 274-277

[Neutrophils and macrophages in host resistance]
  coccidioidomycosis, 277-279
  histoplasmosis, 281-283
  paracoccidioidomycosis, 279-281

Opportunistic pulmonary fungal infections, 245-248
  aspergillosis, 245-246
  mucormycosis, 246-247
  pulmonary candidiasis, 247
  pulmonary cryptococcosis, 247-248

Paracoccidioides brasiliensis, antigenic structure of, 171-192
  cell wall antigens, 172-173
  exocellular (metabolic) antigens, 180-185
  intracellular (somatic) antigens, 173-180
  origin and nature of *P. brasiliensis* antigens, 185-187
Paracoccidioidomycosis
  neutrophils and macrophages in resistance to, 279-281
  skin tests for, 467
Peptidorhamnomannans of *Sporothrix schenckii*, 196
Phagocytic cells in pulmonary defenses, 248-262
  Aspergillus and *Rhizopus*, 249-255
  Candida, 255-260
  Cryptococcus, 260-262
Phospholipase A, 24
Polymorphonuclear leukocyte defense against *Candida*, 259-260
Polymorphonuclear neutrophils (PMN), 273-274
  role in host resistance to systemic fungal infections, 273-283

Index 493

Precipitation techniques in diagnosis of aspergillosis, 406-407
Pulmonary candidiasis, 247
Pulmonary cryptococcosis, 247-248
Pulmonary defense mechanisms, 243-271
  clinical aspects of opportunistic pulmonary fungal infections, 245-248
    aspergillosis, 245-246
    mucormycosis, 246-247
    pulmonary candidiasis, 247
    pulmonary cryptococcosis, 247-248
  phagocytic cells in pulmonary defenses, 248-262
    Aspergillus and *Rhizopus*, 249-255
    Candida, 255-260
    Cryptococcus, 260-262

Radioimmunoassay (RIA) in diagnosis of aspergillosis, 407
Rhamnomannans of *Sporothrix schenckii*, 201-202
Rhamnomannan-containing antigens of *Sporothrix schenckii*, 198
Rhizopus hyphae, alveolar macrophage and circulating monocyte defense against, 253-254
Rhizopus spores
  alveolar macrophage defense against, 251-252
  circulating monocyte defense against, 253

Saccharomyces cerevisiae mannans, immunodominant epitopes of, 45-50
Serological procedures to detect dermatophyte antigens, 419-457

[Serological procedures to detect dermatophyte antigens]
  detection of dermatophyte antigens, 422-431
  use of monoclonal antibodies, 427-429
  Western blotting techniques, 429-431
  serological procedures, 419-422
Serological specificity of mannans in yeast, 37-62
  immunodominant epitopes of *Candida* species mannans, 50-58
  immunodominant epitopes of *S. cerevisiae* mannans, 45-50
  methods used for the immunochemistry of yeast mannans, 41-45
  serotyping of yeasts, 38-41
Serum-dependent cellular immune defects in patients with candidiasis, 296-297
Sporothrix schenckii, antigenic structures of, 193-221
  antigenic polysaccharides, 200-208
  complement activation, 213-215
  hypersensitivity reactions, 211-212
  indirect detection of antigens, 208-211
  isolation of antigenic structures, 194-200
    from culture supernatants, 197-199
    from isolated cell walls, 200
    from whole cells, 194-197
Sporotrichosis, skin tests for, 467
Systemic fungal infection, role of complement in host resistance to, 225-242
  complement and cutaneous candidiasis, 227-228

[Systemic fungal infection, role of complement in host resistance to]
complement and disease progression, 237-238
complement-*Candida* interactions, 225-227
complement interactions with other human pathogenic fungi, 234-237
complement receptors of *C. albicans* for iC3b and C3d, 231-234
C. neoformans-complement interactions, 234

TP antigen, 142-152

Western blotting technique for detection of dermatophyte antigens, 429-431

Yeast mannans, serologic specificity of, 37-62
immunodominant epitopes of *Candida* species mannans, 50-58
immunodominant epitopes of *S. cerevisiae* mannans, 45-50
methods used for the immunochemistry of yeast mannans, 41-45
serotyping of yeasts, 38-41
Yeasts, medically important, antigenic structures of, 38